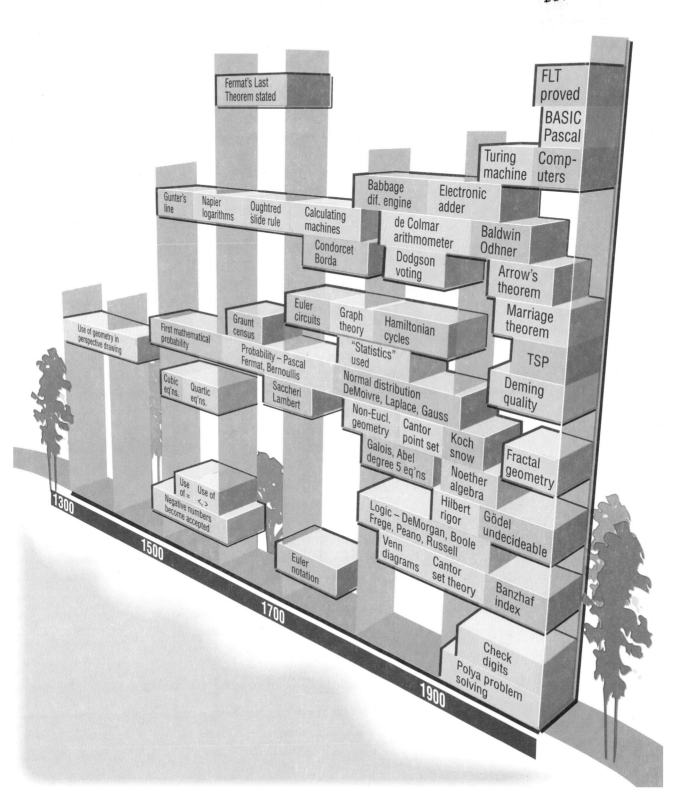

MATHEMATICS
ONE OF THE LIBERAL ARTS

MATHEMATICS
ONE OF THE LIBERAL ARTS

THOMAS J. MILES
Central Michigan University

DOUGLAS W. NANCE
Central Michigan University

BROOKS/COLE PUBLISHING COMPANY
I(T)P® An International Thomson Publishing Company

Pacific Grove • Albany • Belmont • Bonn • Boston • Cincinnati • Detroit • Johannesburg • London •
Madrid • Melbourne • Mexico City • New York • Paris • Singapore • Tokyo • Toronto • Washington

PRODUCTION CREDITS

Composition and Artwork
The Clarinda Company

Copyediting
Elliot Simon, Simon and Associates

Interior Design
Diane Beasley

Page Layout and Cover Design
DeNee Reiton Skipper

Photo Research
Kathy Ringrose

Cover Image
Restored historic skylight of the Rookery Building. Owner: Baldwin Development Company. Restoration architect: McClier Corporation. Photo by: Nick Merrick of Hedrich Blessing.

Etymologies
Reprinted from *The Words of Mathematics* by Steven Schwartzman with the permission of The Mathematical Association of America, Washington, DC, 1994.

Production, Prepress, Printing and Binding by West Publishing Company

Text and Photo Credits Follow Index

FOR MORE INFORMATION, CONTACT

BROOKS/COLE PUBLISHING COMPANY
511 Forest Lodge Road
Pacific Grove, CA 93950
USA

International Thomson Publishing Europe
Berkshire House 168-173
High Holborn
London WC1V 7AA
England

Thomas Nelson Australia
102 Dodds Street
South Melbourne, 3205
Victoria, Australia

Nelson Canada
1120 Birchmount Road
Scarborough, Ontario
Canada M1K 5G4

International Thomson Editores
Campos Eliseas 385, Piso 7
Col. Polanco
11560 México D. F. México

International Thomson Publishing GmbH
Königswinterer Strasse 418
53227 Bonn
Germany

International Thomson Publishing Asia
221 Henderson Road
#05-10 Henderson Building
Singapore 0315

International Thomson Publishing Japan
Hirakawacho Kyowa Building, 3F
2-2-1 Hirakawacho
Chiyoda-ku, Tokyo 102
Japan

British Library Cataloguing-in-Publication Data. A catalogue record for this book is available from the British Library.

Copyright © 1997 by Brooks/Cole Publishing Company
A Division of International Thomson Publishing Inc.
I(T)P   The ITP Logo is a trademark under license.

Printed in the United States of America

04   03   02   01   00   99   98   97   8   7   6   5   4   3   2   1   0

LIBRARY OF CONGRESS CATALOGING-IN-PUBLICATION DATA

Miles, Thomas J.
    Mathematics : one of the liberal arts / Thomas J. Miles, Douglas W. Nance.
       p.   cm.
    Includes bibliographical references and index.
    ISBN 0-314-09576-4 (hard : alk. paper)
     1. Mathematics.   I. Nance, Douglas W.   II. Title.
QA39.2.M548   1997
510—dc21

96-29496
CIP

To my family.
They stuck with me while
Mathematics: One of the Liberal Arts
was one of the laboral arts.
TJM

To Helen.
She understands, accepts, and
participates in the seemingly endless
process of textbook preparation.
DWN

BRIEF CONTENTS

CONTENTS

CHAPTER

SETS 65

CHAPTER

LOGIC 105

CHAPTER

ALGEBRA   153

CHAPTER

FUNDAMENTALS OF GEOMETRY 215

CHAPTER

TOPICS IN GEOMETRY 261

CHAPTER

CONSUMER MATHEMATICS 389

Prologue 389

CHAPTER

DISCRETE MATHEMATICS 423

Prologue 423

CHAPTER

MATHEMATICS AND COMPUTERS 473

Prologue 473

CHAPTER

MATHEMATICS AND THE OTHER LIBERAL ARTS 531

PREFACE

This text is written for students who are taking mathematics for liberal arts or general education purposes, including quantitative literacy and mathematics competency. A variety of courses, some described below, can be taught from its contents. We presume students have taken at least one year of high school algebra. A year of geometry is preferred, but not presumed.

OVERVIEW

We have included topics that many college mathematics teachers believe are important. We develop some algebra, geometry, and set theory topics that students may have seen before. We then extend these topics to material students probably have not seen such as population growth, non-euclidean geometry, fractal geometry, and power in weighted voting. In addition to basic concepts that review high school mathematics, the text presents mathematics as an integral part of the development of intellectual ideas and as a subject which is used to model real world situations in almost every discipline.

STATEMENT OF PHILOSOPHY

The present day liberal arts have their roots in the *quadrivium* of the ancient Greeks. It included arithmetic and geometry. Through 25 centuries a liberal arts background has been regarded as an essential component of a well-rounded education and mathematics has been regarded as a vital part of that liberal arts background. In order to emphasize this, several features have been included. They are:

- *Doing mathematics.* Students are encouraged to become active learners by formulating some concepts themselves, either cooperatively in groups or individually. Nearly every section has one or more "Your Formulation" activities to encourage this. In addition, numerous exercises are included for doing mathematics.
- *Historical development.* Topics are presented in the context of their historical development.
- *Cultural development.* How have societal changes prompted or hindered mathematical development? How have mathematical developments affected society?
- *Human ideas.* Mathematics is something developed by people. We describe those people and we describe mathematics as a creative human activity.

Written assignments. The formation and subsequent expression of ideas is fundamental to educated individuals. Learning is assisted by expressing what we know. Accordingly, we have included suggestions for written assignments throughout the text.

MAJOR THEMES

Problem solving is a major objective of students' mathematical experiences. The National Council of Teachers of Mathematics (NCTM) *Curriculum and Evaluation Standards for School Mathematics* and related publications put problem solving as a main objective of K–12 mathematics instruction. The report *Crossroads in Mathematics: Standards for Introductory College Mathematics Before Calculus,* published by the American Mathematical Association of Two-Year Colleges (AMATYC) in 1995, has as Standard I-1, "Students will engage in substantial mathematical problem solving." For students who have already developed problem solving skills, this book offers material to add to those skills. For those who have not worked on developing problem solving skills, they can begin here. The book begins with two sections on problem solving. Throughout the book students are presented with a number of new and unfamiliar situations in which they are asked to apply problem solving skills in order to arrive at a solution.

Mathematical modeling is synonymous with applications of mathematics. The authors have found that students respond much better to modeling-oriented mathematics than to just extending their algebra, geometry, trigonometry, or mathematical functions development. In the report *Quantitative Reasoning for College Graduates: A Complement to the Standards,* approved by the Committee on the Undergraduate Program (CUPM) of the Mathematical Association of America (MAA), conclusion 2 is that "Colleges and universities should expect every college graduate to be able to apply simple mathematical methods to the solution of real-world problems." AMATYC's Standard I-2 says, "Students will learn mathematics through modeling real-world situations." Accordingly, we have developed basic mathematical concepts and then used them to model various real world situations. Section 1.3 is devoted to mathematical modeling, but ideas of modeling are used throughout the book.

POSSIBLE COURSES

The material in this book and the above emphases can be organized in several different ways to fit several types of courses.

Mathematical modeling. Selections from Sections 1.3 (Mathematical Modeling), 2.5 (Check Digits), part of 3.2 on weighted voting, 3.6 (Applications of Sets), much of Chapter 5 (Algebra), Chapter 8 (Probability), Chapter 9 (Statistics), Chapter 10 (Consumer Mathematics), Chapter 11 (Discrete Mathematics), and Chapter 13 (Mathematics and the Other Liberal Arts).

Quantitative literacy. The MAA's *Quantitative Reasoning for College Graduates* makes clear that a required mathematics course or two is not sufficient for quantitative literacy, but the report lists five capabilities desired in a quantitatively literate college graduate (p. 10). These capabilities can be developed by material in Chapter 1 (Problem Solving and Mathematical Modeling), Section 3.6 (Applications of Sets), Chapter 5 (Algebra), Chapter 6 (Geometry), and Chapter 9 (Statistics). In addition, such a course should "immerse [students] in doing quantitative reasoning of a nonroutine nature." (p. 14) Such topics can be selected from Section 2.5 (Check Digits), Chapter 11 (Discrete Mathematics), and parts of Chapter 13 (Mathematics and the Other Liberal Arts).

- *Mathematical competency.* Chapter 1 (Problem Solving and Mathematical Modeling), Chapter 3 (Sets), Chapter 4 (Logic), Chapter 5 (Algebra), and Chapter 6 (Geometry).
- *Survey of mathematics.* Topics selected from any of the chapters.
- *Historical or cultural approach to mathematics.* Chapter 2 (Numbers and Numerals), Section 4.1 (Introduction [to Logic]), part of Section 5.1 ([Algebra as] Generalized Arithmetic), Section 5.4 (Further Developments in Algebra), Section 6.1 (Overview and History [of Geometry]), part of Section 7.1 (Geometry as an Axiomatic System), Section 7.2 (Non-Euclidean Geometry), part of Section 7.3 (Fractals), Section 8.1 (Historical Background [of Probability]), much of Chapter 12 (Mathematics and Computers), parts of Chapter 13 (Mathematics and the Other Liberal Arts).

FEATURES

The text has the following features designed to motivate and aid learning.

- *Prologue.* Each chapter begins with a prologue designed to pique the student's interest by relating some of the material in the chapter to a concept that the student might not have associated with the chapter material.
- *Etymology.* Each chapter contains the history of the key word or words in the chapter title. The authors are grateful to Steven Schwartzman and the MAA for permission to use lightly edited etymological references from Schwartzman's delightful book *The Words of Mathematics: An Etymological Dictionary of Mathematical Terms Used in English.*
- *Goals and Objectives.* Each section begins with a statement of up to four goals for the section. The instructor's manual contains a more detailed list of objectives. It can be modified to meet the objectives of your course. This modified list can serve as a study guide for students.
- *Your Formulation.* Most sections have one or more "Your Formulation" activities in which students are asked to take an active part in formulating concepts. These are ideal for cooperative learning activities.
- *Margin Notes.* These are brief comments relating to adjacent text material. Some are notes on people, some are light verse, and some take other forms. All are intended to help keep mathematics interesting.
- *On a Tangent.* These are the comments that take off from the text's comment about a person or concept in the same way that a tangent to a curve takes off from a curve. They are longer than the Margin Notes.
- *Exercises.* The exercises are designed to have a number of routine computational-type exercises as well as some applications that go beyond the examples and some exercises that challenge the student's problem solving ability.
- *Written Assignments.* The written assignments include exercises for students to summarize material that has been covered, to write about their personal reactions to material they have studied, to debate some issues, and to do research on historical and cultural developments and applications in mathematics. Numerous teachers have reported on the advantages of having students write to learn mathematics.
- *Chapter Review Exercises.* At the end of each chapter are review exercises. These are directed at measuring the student's mastery of the objectives for the chapter.

- *Bibliography.* A bibliography is included at the end of each chapter. It includes references to books and articles that (1) have been cited in the chapter, (2) have served as background material, of (3) can serve as references for further development.
- *Color.* Extensive use of color helps accent concepts and make the book more attractive.
- *Liberal arts look.* Headings and borders are designed to emphasize visually the location of mathematics within the liberal arts.

It is our viewpoint that the best liberal arts mathematics courses are those in which the instructor is really excited about the material being presented. Students catch that excitement either from the teacher or from a sense of the value of the course to them. We hope that this textbook can be the reference for conveying excitement about knowledge and ways of knowing that should be part of every liberally educated person.

ANCILLARIES

The following supplementary materials are available to assist instruction and learning.

- *Instructor's manual.* This contains a detailed list of objectives for each section. These often expand upon the goals printed at the start of each section in the textbook. It also contains considerations in teaching material in a section, including suggestions for cooperative learning activities.
- *Test bank.* This is available in print and computerized form.
- *Color acetates.* This set contains 150 key figures from the text.
- *Student solutions manual.* This contains worked out solutions to the even-numbered exercises in the text.

ACKNOWLEDGEMENTS

This book is shaped, in part, by the thoughtful suggestions of the following reviewers, to whom we offer our thanks.

H. Rux Allonce
Miami Dade Community College
Jeffrey S. Allbritten
Middle Tennessee State University
George Bradley
Duquesne University
Janis M. Cimperman
St. Cloud State University
Lisa DeMeyer
University of North Carolina—Chapel Hill
Joe S. Evans
Middle Tennessee State University

Marc Franco
South Seattle Community College
Gary Grabner
Slippery Rock University
John S. Haverhals
Bradley University
Evan Innerst
Canada College
Howard Jones
Lansing Community College
Maureen Kelley
North Essex Community College
Thomas Lada
North Carolina State University

Anne F. Landry
Dutchess Community College
Tom Linton
Willamette University
Lee McCormick
Pasadena City College
Janice McFatter
Gulfcoast Community College
Donna Massey
Talahassee Community College
Gael Mericle
Mankato State University
Harald M. Ness
University of Wisconsin–Center
Sunny Norfleet
St. Petersburg Junior College
Stephen K. Prothero
Willamette University
Joe Rappoport
Miami-Dade Community College
Dennis Reissig
Suffolk Community College
Judith Rice
Viollanova University

Stewart M. Robinson
Cleveland State University
Howard L. Rolf
Baylor University
Marsha Self
El Paso Community College–Rio Grande Campus
Henry Mark Smith
University of New Orleans–Loyola University
Raymond F. Smith
Whittier College
Litsa St. Amand
Mesa Community College
Joseph F. Stokes
Western Kentucky University
Hugo S.H. Sun
California State University–Fresno
Richard G. Vinson
University of Southern Alabama
Bruce Williamson
University of Wisconsin–River Falls
Jim Wooland
Florida State University

Jerry Westby, Executive Editor, prompted the writing of this book, saw it through the review process, encouraged us, aided us with his keen sense of what is useful to students and instructors, and served as the monitor of good taste. His personal perspective is of mathematics as one of the liberal arts.

Deanna Quinn, Senior Production Editor, has handled the production of the book from copyediting, through photo and art review, to its finished form with affability, helpfulness, patience, and a higher regard for pedagogical concerns than ease of production. One of us has worked with her before and was delighted to work with her again; the other of us is writing his first book and can't imagine a better production editor.

Elliot Simon, Copyeditor, was thorough and creative in removing grammatical flaws and overreliance on certain words and phrases. Mathematics is sometimes difficult to read. Elliot's work has made it easier by polishing our writing.

Halee Dinsey, Developmental Editor, has provided numerous suggestions and background material for consideration as well as summarizing suggestions for reviewers. In the early stages of writing, Betsy Friedman, Developmental Editor, compiled and analyzed reviewer responses and offered suggestions.

Helen Nance has provided efficient and insightful assistance in all phases of the textbook-preparation process from word-processing early versions of the manuscript to proofreading galleys. Her keen eye for detail and her sense of what looks right have greatly assisted the production process.

To all the people named above we offer our thanks. All of you have played a role in shaping this book.

Thomas J. Miles   Douglas W. Nance

CHAPTER 1

PROBLEM SOLVING AND MATHEMATICAL MODELING

CHAPTER OUTLINE

PROLOGUE

"We're number 1! We're number 1!" The chant rings out as a statement of fact, a statement of hope, or a goal.

During a 1989 Education Summit, the nation's governors, prompted by President George Bush, set the tone and framework for the 1991 National Education Goals Panel Report. This report set six goals for U.S. schools and students. One of these was: "By the year 2000, U.S. students will be first in the world in science and mathematics achievement." This report guided the framing of the Goals 2000: Educate America Act, which President Clinton signed into law on March 31, 1994. One of our nation's goals is thus to become number 1 in student math achievement by the year 2000.

How do we do this? Among other things, we increase the emphasis on problem

solving at all levels. In high schools this means a new curricular emphasis on problem solving, as supported by the National Council of Teachers of Mathematics.

You're in a course looking at mathematics as one of the liberal arts—this is a time to improve your problem-solving skills.

Chapter opening photo: We're number 1! We're number 1!

1.1 INTRODUCTION TO PROBLEM SOLVING

GOALS
1. To describe what problem solving is and what it is not.
2. To identify mathematician George Polya with problem solving, and to describe Polya's heuristic.
3. To describe at least two psychological aspects of problem solving.
4. To describe what an algorithm is, and to give an example of one.

What Is Problem Solving?

The term *problem solving* means different things to different people. Consequently, several definitions exist. Since we are interested in problem solving in mathematics, we will try to formulate a working definition from the perspective of a mathematician.

First, there must be a problem. This means that a situation must exist that requires some solution. Second, the problem solver must be able to apply some principles of mathematics in order to arrive at a solution. This gives us our first definition of problem solving.

DEFINITION 1

Problem solving is the process of applying previously acquired knowledge in order to arrive at a solution to a situation (problem).

Before we modify this definition to be more consistent with the modern-day interpretation of problem solving in mathematics, let's take a brief historical look at problem solving.

Historical Development

Problem solving has been used throughout the development of civilization. In fact, problem solving can be thought of as the essential reason civilization was able to develop. For example, current forms and availability of food, shelter, and clothing are all results of successive generations who solved the problems of providing for their basic needs. More recently, the process of problem solving has been studied with the hope that a better understanding of it will enable more people to be better problem solvers.

The first mathematician to popularize a problem-solving approach was George Polya, in his 1945 book *How to Solve It*. He advocated a four-step process:

Rodin's sculpture The Thinker *represents what is necessary for problem solving—thinking.*

1. *Understand the problem.* What is known? What are the data? What is the condition? Draw a figure, introduce suitable notation. Separate the various parts of the condition.
2. *Devise a plan.* Find the connection between the data and the unknown. Have you seen it before? Do you know a related problem?
3. *Carry out the plan.* Check each step. Is each step correct? Can you prove that each is correct?
4. *Look back.* Examine the solution obtained. Can you check the result? Can you check the argument? Can you derive the result differently? Can you see the result at a glance? Can you use the result, or method, for some other problem?

Polya used the term *heuristics* ("serving to discover") when referring to his problem-solving process. Thus, his often-cited work frequently appears in the following shortened form.

POLYA'S HEURISTIC

1. Understand the problem.
2. Devise a plan.
3. Carry out the plan.
4. Look back.

We will examine more detailed lists later, but every list will contain the elements formulated by Polya.

Psychological Aspects

Before looking more closely at specific problems and the problem-solving process, let's briefly consider psychological aspects of problem solving.

First, when trying to apply a problem-solving process to some situation, many students experience frustration, because most situations considered "problems" do not lend themselves to quick, short solutions. Thus, students must expect to try a variety of approaches they probably have not tried before. This is different from solving mathematical problems.

George Polya

Second, students need to be persistent. Since there may be a variety of reasonable approaches to a problem, students should try alternate approaches if the initial one is unproductive. This means students must persevere even when it appears they are not being successful.

Third, students must be able to read and interpret what a problem says and what a problem requires. Often a student sees some written problem and assumes it will be difficult to solve because it is a "story problem." This sort of negative mental conditioning, which is quite prevalent, must be overcome in order to become a good problem solver.

There is some good news, however. In fact, there is some great news! You can (and probably will) experience the excitement and euphoria that come from being able to solve problems. Once you have been successful at solving a few problems, you will develop a level of confidence that enables you to work on increasingly difficult problems. Your feelings of enhanced self-worth may expand to other areas in which you are working. Furthermore, the skills you hone in solving mathematical problems could become part of your approach to other life situations for many years beyond your formal education.

What Constitutes a Problem?

Let's now consider more closely what we mean by a "problem" within the realm of mathematics. Most mathematicians would agree to the following criteria.

1. A solution to a problem requires the solver to apply previously acquired knowledge.
2. The solution is not something the solver has used before on a similar problem.

This second condition allows us to modify our previous definition of problem solving.

DEFINITION 2

Problem solving is the process of applying previously acquired knowledge to a new and unfamiliar situation in order to arrive at a solution to the situation.

This means that problem solving is a relative process.

To illustrate, consider the following mathematical problem.

Find a number that, when multiplied by 2, is 6.

To a student who knows nothing about division, solving this problem would require the application of previously acquired knowledge to an unfamiliar situation. However, to a more sophisticated student, solving the equation $2x = 6$ would not be considered problem solving because the situation is neither new nor unfamiliar.

What Is Not a Problem?

The second definition of problem solving indicates that what gets solved must be a new and unfamiliar situation. Thus, working a problem in mathematics similar to one (or more) that you have worked previously would not be considered problem solving.

To illustrate this, recall the first time you worked a problem of the following form:

John can shovel the driveway clear of snow in 3 hours, and Mary can do the same job in 4 hours. How long will it take them to clear the driveway if they work together?

You probably tried several different approaches to the problem until you finally arrived at a solution. If you were then to work several similar problems, it would take far less time and original thought because you would already have a method of solution for subsequent problems.

Such a method for solving a problem type is called an *algorithm.* This allows us to make the following generalization.

> If you know an algorithm for solving a problem, then getting a solution is not what is considered to be a problem-solving process.

This generalization is the premise that will be used throughout this chapter.

A consequence of the second definition of problem solving is that finding the solutions to classical coin, age, and rate problems (story problems) of algebra is *not* problem solving once you have worked a few of them. Consistent with this interpretation of problem solving is information contained in the recent *Curriculum and Evaluation Standards for School Mathematics,* published by the National Council of Teachers of Mathematics. While they espouse increased attention to problem solving at all levels and in all areas, they specifically suggest that "word problems by type, such as coin, digit, and work" should receive decreased attention in algebra classes.

WRITTEN ASSIGNMENTS 1.1

1. Another way to answer the question of what constitutes a problem is to state that it must satisfy the following criteria:
   a. The individual must accept the problem.
   b. The individual's initial attempts to solve the problem must fail to work.
   c. The acceptance of the problem must force the individual to try various paths to a solution.
   Discuss the similarities and differences between this definition and the second definition listed in this section. Which do you prefer, and why?
2. Discuss the psychological aspects of problem solving mentioned in this section. Which have you experienced? In general, how long are you willing to work on a problem before you "give up"?

3. Consider the differences between the first and second definitions of this section. List some problem types from your previous mathematics courses that would be considered problems by definition 1 but *not* by definition 2.
4. Reconsider the problem in this section of John and Mary shoveling snow from the driveway of their home. First, solve the problem. Then construct an algorithm for solving similar problems. Construct similar problems and apply your algorithm to find solutions.
5. For each of the five subheadings in this section (What is Problem Solving?, Historical Development, Psychological Aspects, What Constitutes a Problem?, and What Is Not a Problem?), write one sentence summarizing the material in that subsection.

GOAL   To use some common strategies for problem solving.

Several schemes for problem solving have been advocated since Polya's list first appeared in *How to Solve It*. Here are some of the more common strategies.

PROBLEM-SOLVING STRATEGIES

1. Make a picture.
2. Work backwards.
3. Guess and test (trial and error).
4. Find a pattern.
5. Make a table or chart.
6. Collect data.
7. Use a calculating device.
8. Use deductive reasoning.
9. Solve a simpler, analogous problem.
10. Approximate.
11. Determine characteristics of objects.
12. Relate new problems to familiar ones.

We will now illustrate several of these strategies with examples.

EXAMPLE 1.1

Assume you are in a class of 20 students. If every student shakes hands with every other student, how many handshakes will there be?

SOLUTION First, consider reducing the problem to a simpler case (strategy 9). For example, suppose there are only three students: Ann, Bob, and Chris. In this case, Ann shakes the hand of Bob and of Chris, Bob then shakes the hand of Chris. Let's now start a chart of these results.

Number of students	Number of handshakes
3	2 + 1 = 3

Next, let's assume there are four students: Ann, Bob, Chris, and David. Starting with Ann, Ann shakes the hand of Bob, of Chris, and of David (3). Bob then shakes the hand of Chris and of David (2). Finally, Chris shakes the hand of David (1). The chart now becomes

Number of students	Number of handshakes
3	2 + 1 = 3
4	3 + 2 + 1 = 6

We can also draw a picture to help analyze this situation. Arrange the names (letters) in a circular pattern and draw a line between those where a handshake occurs. After Ann shakes hands, we have

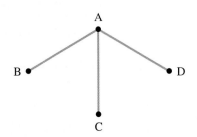

After Bob shakes hands with Chris and David, we get

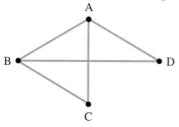

Finally, Chris shakes David's hand to produce

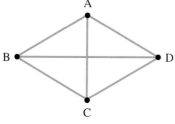

Note that there are six lines in this diagram.

By now you should be able to detect a pattern: For the 20 students in our original problem, there will be $19 + 18 + . . . + 1 = 190$ handshakes.

Note that the techniques we employed in solving this problem included reducing to a simpler problem (strategy 9), making a chart (strategy 5), drawing a picture (strategy 1), and finding a pattern (strategy 4).

YOUR FORMULATION

Let's examine one more case to see if you can detect a pattern.

1. Assume there are five students. Draw a picture to help analyze the situation.
2. How many handshakes are there for the five students? Enter this number in the following chart.

Number of students	Number of handshakes
3	$2 + 1 = 3$
4	$3 + 2 + 1 = 6$
5	

EXAMPLE 1.2

Let's select three distinct digits and use these digits to form a two-digit number and a one-digit number so the product of the two-digit and one-digit numbers is as large as possible. Can we make any generalizations about how to select the two-digit and one-digit numbers?

SOLUTION First, let's try a specific problem. We take the integers 4, 6, and 9, and consider the product of a two-digit number and a one-digit number. The possible products are:

$$4 \times 69, 4 \times 96, 6 \times 49, 6 \times 94, 9 \times 46, 9 \times 64$$

We can immediately eliminate the two-digit numbers where the smaller digit is in the tens position. (Why?) This leaves

$$4 \times 96, \ 6 \times 94, \ 9 \times 64$$

whose products are 384, 564, and 576, respectively. Thus, the correct response is

$$9 \times 64$$

Try this again with the integers 1, 5, and 8. Note that the maximum occurs with the product $8 \times 51 = 408$. The following chart shows the maximum product for each of these two specific problems.

Numbers	Maximum product
4, 6, 9	9×64
1, 5, 8	8×51

Note that the product in both cases is a maximum when the integers are arranged in order from large to small and the largest is the single-digit number.

Does this method always work? Suppose the digits are *a, b,* and *c,* where $a < b < c$. Our "solution" suggests that *c* times the two-digit number *ba,* i.e., $c \times (10b + a)$, is the largest possible product. This product, $10(cb) + ca$, is a maximum because the two largest digits are *c* and *b* and their product is multiplied by 10. The remaining product, *ca,* allows for the largest digit, *c,* to be multiplied by the remaining digit, *a*. Note that our solution never allows for the product *ab*. This general solution applied to the digits 4, 6, and 9 yields

$$
\begin{aligned}
9(64) &= 9 \, (10 \times 6 + 4) \\
&= 10 \, (9 \times 6) + 9 \times 4 \\
&= 10 \, (54) + 36 \\
&= 540 + 36 \\
&= 576
\end{aligned}
$$

If we analyze the methods by which we solved this problem, we find that we did the following:

Used a calculating device (strategy 7)
Reduced to a simpler case (three specific integers—strategy 9)
Used trial and error (tried various products—strategy 3)

EXAMPLE 1.3

Suppose we form a rectangle composed of 30 unit squares. Of all such possible rectangles, let's find the one of minimum perimeter.

SOLUTION First, let's draw a picture to help us visualize the situation. One such rectangle can be formed by placing the squares in a line.

The dimensions of this rectangle are 30×1.

So the perimeter is $30 + 30 + 1 + 1 = 62$. Another rectangle to consider is the following.

2

15

The perimeter of this rectangle is $15 + 15 + 2 + 2 = 34$. Let's now make a chart that lists possible values for appropriate rectangles.

Rectangle dimensions	Perimeter
1×30	62
2×15	34
3×10	26
5×6	22

Because this chart lists all possible rectangles, we can see that the smallest possible perimeter is 22.

Note that in finding the solution to this problem, we drew a picture and made a chart.

EXAMPLE 1.4

At a high school reunion, 100 alumni decided to play an old locker game that involved opening and closing the lockers numbered 1 to 100. Initially, all the lockers were closed. The first person opened all the lockers. The second person then started at the beginning and closed every second locker. The third person then proceeded to change the state of every third locker; that is, this person closed the open ones and opened the closed ones. The fourth and remaining people followed in like manner, each changing the state of the fourth, fifth, etc. lockers. When the last person was finished, which lockers were open?

SOLUTION Let's first consider the problem with only five people and five lockers. We will draw diagrams to illustrate what happens after each person moves down the row of lockers (left to right) as described.

$$
\begin{array}{lllll}
\text{C} & \text{C} & \text{C} & \text{C} & \text{C} \quad \text{initially, all closed} \\
\text{O} & \text{O} & \text{O} & \text{O} & \text{O} \quad \text{after 1st person (all open)} \\
\text{O} & \text{C} & \text{O} & \text{C} & \text{O} \quad \text{after 2nd person} \\
\text{O} & \text{C} & \text{C} & \text{C} & \text{O} \quad \text{after 3rd person} \\
\text{O} & \text{C} & \text{C} & \text{O} & \text{O} \quad \text{after 4th person} \\
\text{O} & \text{C} & \text{C} & \text{O} & \text{C} \quad \text{after 5th person}
\end{array}
$$

At this point, the open lockers are lockers 1 and 4.

Let's next consider 10 people and 10 lockers. We have

$$
\begin{array}{llllllllll}
\text{C} & \text{C} & \text{C} & \text{C} & \text{C} & \text{C} & \text{C} & \text{C} & \text{C} & \text{C} \quad \text{initially} \\
\text{O} & \text{O} & \text{O} & \text{O} & \text{O} & \text{O} & \text{O} & \text{O} & \text{O} & \text{O} \quad \text{after 1st person} \\
\text{O} & \text{C} & \text{O} & \text{C} & \text{O} & \text{C} & \text{O} & \text{C} & \text{O} & \text{C} \quad \text{after 2nd person} \\
\text{O} & \text{C} & \text{C} & \text{C} & \text{O} & \text{O} & \text{O} & \text{C} & \text{C} & \text{C} \quad \text{after 3rd person}
\end{array}
$$

O	C	C	O	O	O	O	O	C	C	after 4th person
O	C	C	O	C	O	O	O	C	O	after 5th person
O	C	C	O	C	C	O	O	C	O	after 6th person
O	C	C	O	C	C	C	O	C	O	after 7th person
O	C	C	O	C	C	C	C	C	O	after 8th person
O	C	C	O	C	C	C	C	O	O	after 9th person
O	C	C	O	C	C	C	C	O	C	after 10th person

At this point lockers 1, 4, and 9 are open. Here's a chart of these two trials:

Number of lockers	Lockers open at end
5	1, 4
10	1, 4, 9

 ## YOUR FORMULATION

Perhaps you have a generalization at this point. If so, try it by considering 20 people and 20 lockers. If not, consider 20 people and 20 lockers and try to find the pattern.

The preceding cases suggest that the open lockers are those whose numbers are perfect squares. If we accept this hypothesis, then for 100 lockers the open ones would be 1, 4, 9, 16, 25, 36, 49, 64, 81, and 100. (See Exercise 4.)

In this section we have applied several strategies for problem solving. The exercises ask you to practice these strategies.

EXERCISES 1.2

1. Consider Example 1.1. Simulate the reduced-cases approach by utilizing students from your class and actually counting the number of handshakes for each group size and recording the results. Discuss what happens with this approach as the group size increases.

2. Consider Example 1.2. Rework the problem by starting with five distinct digits and using them to form a three-digit number and a two-digit number.

3. Consider Example 1.3. Rework the problem for an arbitrary number of unit squares. What conclusions can you reach?

4. Consider Example 1.4. Verify that the solution given is correct and can be generalized to an arbitrary number of lockers. (*Hint:* For each locker number, consider an ordered list of all possible factors of the number. For example, the factors of 24 are 1, 2, 3, 4, 6, 8, 12, 24 and the factors of 25 are 1, 5, 25.)

5. Consider the following figure. Assume the edge of each small square is 1 unit in length. Add squares until the figure has a perimeter of 18. When squares are added, they must meet exactly along at least one edge. What conclusions can you reach? Can you interpret your results geometrically? Make a table that illustrates your results.

6. Find the next three terms in each of the following:

$$3, 8, 13, 18, 23, \ldots$$

$$1, 3, 7, 13, 21, \ldots$$

7. A cashier found that he was often asked to give change for a dollar to people who had made no purchase but wanted a dime or two nickels for a parking meter. He started thinking one day about the number of ways he could make change. If he gave no more than four of any coin, in how many different ways could he give change for a dollar to people who needed a dime or two nickels for a parking meter?

8. These four steps are made of cubes. How many cubes are needed to make 20 steps?

9. Three clubs were meeting after school on the same day: cheerleaders, pep club, and student council. At each meeting not all club members were present, because some students belong to more than one club. Half of the student council members and all of the cheerleaders belong to the pep club. Two students are members of all three clubs. If there are 24 student council members, six cheerleaders, and 40 pep club members, how many students belong only to the pep club?

10. A club of 150 members is holding a Ping-Pong tournament. When a member loses a game, the member is out of the tournament. There are no ties. How many games must be played in order to determine the champion?

11. What is the *least* number of jelly beans you can have so that when they are counted by twos, threes, fours, fives, and sixes, you are always left with one extra, but when counted by sevens, they come out even?

12. A major fast-food chain held a contest to promote sales. With each purchase, a customer was given a card with a whole number less than 100 on it. A $100 prize was given to any person who presented cards whose numbers totaled 100. Here we have several typical cards. Can you find a winning combination? Can you suggest how the contest could be structured so there would be at most 1000 winners throughout the country?

13. Place the whole numbers 1 through 9 in the circles forming the accompanying triangle, with each number appearing exactly once, so that the sum of the numbers on each side of the triangle is 17.

14. Arrange the whole numbers 1 through 9 in a 3 × 3 square array so that the sum of the numbers in each row, column, and diagonal is 15. Each number can be used only once. Show that the number 1 cannot appear in any of the corners.

15. A child has a set of 10 cubical blocks. The lengths of the edges are 1 centimeter, 2 centimeters, 3 centimeters, . . . , 10 centimeters. Using all the cubes, can the child build two towers of the same height by stacking one cube upon another? Why or why not?

16. Prove or disprove the following:

   2 is a factor of $(a - b)(b - c)(c - a)$ for any integer a, b, c.

   (*Hint:* Try a few examples first.)

17. A man's boyhood lasted for $\frac{1}{6}$ of his life, he played soccer for the next $\frac{1}{12}$ of his life, and he married after $\frac{1}{7}$ more of his life. A daughter was born 5 years after his marriage, and the daughter lived $\frac{1}{2}$ as many years as her father did. If the man died 4 years after his daughter, how old was the man when he died?

18. A servant was asked to perform a job that would take 30 days. The servant would be paid 1000 gold coins. The servant replied, "I will happily complete the job, but I would rather be paid 1 copper coin on the first day, 2 copper coins on the second day, 4 on the third day, and so on." (The number of copper coins doubles each day.) The king agreed to the servant's method of payment. If a gold coin is worth 100 copper coins, did the king make the right decision? How much would the servant be paid?

19. If you have a 5-gallon pail and a 9-gallon pail, can you go to a water source and bring back exactly 7 gallons of water? Is it possible to bring back any whole number of gallons, from 1 to 14, with these two containers? How, or why not?

20. One box contains even-numbered chips, one box contains odd-numbered chips, and one box contains a mixture. All three boxes are mislabeled. Can you label the boxes correctly by drawing one chip from one box?

21. Three cylindrical drums of 2-foot diameter are to be securely fastened in the form of a triangle by a steel band. What length of band will be required?

22. Several years ago, Mary put a border of six logs, each 48 inches long, around a blue spruce tree, in the form of a regular hexagon (six equal sides and six equal angles). The trees are much larger now, and Mary wishes to put six new logs around the tree, but the ends of each log need to be 26 inches further out from the center of the tree. What should be the lengths of the new logs?

23. (*The Infinite Forest Problem*) Suppose you have an infinite geoboard (think of this as an infinite checkerboard). On each of the lattice points (points at the corners of the squares) except the one at the origin, there is a tree with a trunk that is only as wide as a line. You are standing on the origin. Is there a straight-line path you can take from the origin that would allow you to walk forever in the forest and not hit a tree?

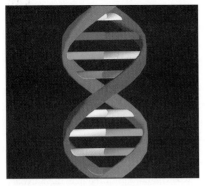

Mathematical modeling helps researchers visualize and understand DNA structure and energetics.

GOALS
1. To describe the mathematical modeling process.
2. To use some mathematical models.
3. To decide which among several mathematical models is most appropriate.

Mathematical structures and ideas often serve as models or descriptions of real-world problems or situations. Using mathematics as a tool to analyze such real-world situations is called *mathematical modeling*. Because employing a mathematical model is one way to apply problem solving to an original real-world problem or situation, this section provides more experience with problem solving as well as introducing mathematical modeling.

Here are some situations that might be modeled.

- Describe our solar system and its operation.
- If a swimming pool is draining so the water level drops 6 inches in the first hour, how long will it take to drain the pool?
- If the counter on a VCR goes from 0 to 400 in 30 minutes, what will the counter register an hour and a half into the tape?
- What is the most efficient route for a mail carrier or school bus?
- What is the probability of winning a particular game of chance?
- Describe the monthly payments that need to be made on a car loan over a 4-year period if $10,000 is borrowed.

The mathematical-modeling process can be described as in the diagram in ▶ Figure 1.1. The process starts with a real-world problem or situation (What is the most efficient route for a mail carrier?). This is converted into a mathematical model by idealizing the situation or making assumptions (e.g., "Suppose all the planets lie in a plane," or "Suppose the mail carrier can go in either direction on every street"). In the mathematical model one employs logic and previously established mathematical results (theorems) to arrive at conclusions. The mathematical conclusions are then interpreted from real-world conclusions. (If the mathematical conclusion is

The Belgian artist René Magritte (1898–1967) in his painting The Blank Signature *had made visual assumptions in presenting a real world situation in this artistic model. The viewer is confronted with interpreting the artistic depiction.*

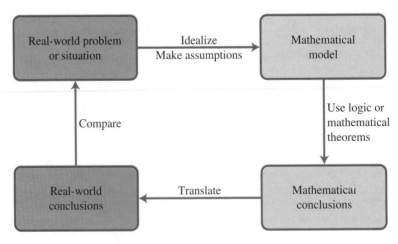

▶ FIGURE 1.1 Mathematical-modeling process

$x = 3$, what does this mean in the real world?) The real-world conclusions are then compared with the real-world problem or situation. Are the conclusions reasonable? Do the conclusions predict accurately what will happen should the real-world problem or situation arise again? If so, the model is ready for use; if not, the model needs to be refined. The logic of the argument is carefully checked and the assumptions are reexamined. Perhaps they oversimplify the situation to the point that conclusions are inaccurate.

As an illustration of the mathematical-modeling process, consider the next example.

EXAMPLE 1.5

A neighbor has a rectangular swimming pool. Tile work was done around the entire edge of the pool. According to the bill, this distance totaled 96 feet. What are the dimensions of the pool?

SOLUTION. There are several rectangular patterns possible for the pool: It could be square, or it could be long and skinny. Suppose we know from looking at it that it's not square and also that it's not real skinny. We use the fact that rectangular pools are often twice as long as they are wide and make an assumption that this is true of this pool. First draw a sketch of the pool.

Let x denote the pool width. Then $2x$ is the pool length. We label the sketch to obtain

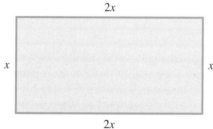

The perimeter (distance around the pool) is $x + 2x + x + 2x = 6x$. Because we know perimeter is 96 feet, we have $6x = 96$. Solving for x we obtain

$$x = \frac{96}{6} = 16$$

We have now reached a mathematical conclusion. We translate this to the real-world conclusion that the pool is 16 feet by 32 feet ($2x$). Notice that this gives a perimeter of $16 + 32 + 16 + 32 = 96$ feet. Next we compare our calculated dimensions of 16 feet by 32 feet with any evidence we have of the true dimensions. (Some comparisons are suggested in Exercises 1–4.) If this other evidence matches our calculations, we'll believe our answer is correct. If the evidence doesn't match our calculation, we'll change our assumption that it's twice as long as it is wide and try again. This points out that the conclusion from mathematical modeling is dependent on the assumptions used to create the model.

The need to refine a model is most often caused by having the initial assumptions being too far from reality. To illustrate this, consider the next example.

Steve is going to move his mattress outside. He would like to move it through his window rather than carrying it down the stairs from his second-floor bedroom. His window opening is 36.5 inches wide and 21 inches high. The mattress is 39 inches wide. Will it fit?

SOLUTION. Without mathematical modeling as an option, we might just open the window, pick up the mattress, and try it. But let's see if we can save some physical work by applying mathematical modeling.

First we draw the diagram.

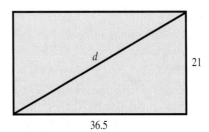

Notice that this is a mathematical model. The widest distance across the window is the diagonal. Call its length d. Then by the Pythagorean theorem (here we use a mathematical result),

$$d^2 = 36.5^2 + 21^2 = 1332.25 + 441 = 1773.25$$
$$d \approx 42.1$$

Next we translate back to the real-world situation and predict that an object a little larger than 42 inches wide will fit through the window. Since the mattress is only 39 inches wide, it should fit. Now we're ready for the physical work.

Can you predict how easy the job will be? When we tried it, the mattress didn't fit! What went wrong? (See if you can figure out what went wrong before reading on.)

We neglected the thickness of the mattress. In terms of mathematical modeling, we made the simplifying assumption that the object we were putting through the window (the mattress) had no thickness. This assumption was too far from reality. The mattress is actually 8 inches thick. If we portray the mattress trying to fit through the window, we get the following diagram.

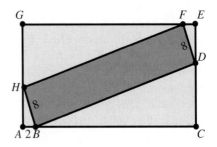

Suppose that we try tipping the mattress so that the length AB is 2 inches. Then by the Pythagorean theorem,

$$(AB)^2 + (AH)^2 = 8^2$$
$$4 + (AH)^2 = 64$$
$$(AH)^2 = 60$$
$$AH = \sqrt{60}$$

Triangles *ABH* and *DEF* are congruent; that is, they have the same size and shape. Therefore, $DE = AH = \sqrt{60}$. In right triangle *BCD, BC* 36.5 − 2 = 34.5 inches, CD = $(21 - \sqrt{60})$ inches, and we hope *BD* is 39 or more inches. By the Pythagorean theorem,

$$BD^2 = BC^2 + CD^2 = 34.5^2 + (21 - \sqrt{60})^2$$
$$BD \approx 36.96$$

We've found that only a mattress 36.96 inches wide or less would fit when *AB* is 2 inches. It can be shown that the widest 8-inch-thick mattress that will fit through this window is a little more than 38.7 inches wide. Therefore, a 39-inch-wide mattress will not fit through.

The initial assumption in Example 1.6 was too far from reality for the model to give an acceptable conclusion. Therefore, the model had to be refined.

In reference to the diagram in Figure 1.1 that describes the mathematical-modeling process, the following observations are true.

- The items in the boxes are nouns; the descriptions by the arrows are verbs.
- The left side describes the real-world part; the right side describes the mathematical part.
- The top half describes the problem or situation; the bottom half describes the conclusions.

With a mathematical model we can describe a real-world situation or solve a real-world problem in a way that allows predictions to be made. A mathematical model also can take the place of a physical model of the same situation.

CLOSE TO HOME   JOHN McPHERSON

"Okee-doke! Let's just double-check. We're 130 feet up and we've got 45 yards of bungee cord, that's uh ... 90 feet. Allow for 30 feet of stretching, that gives us a total of ... 120 feet. Perfect!"

EXERCISES 1.3

In Exercises 1–4, utilize the information in Example 1.5 in addition to that in the respective exercises.

1. A pool supply company has told you that your neighbor purchased swimming pool covers that fit inside their pool, on top of the water and flush against the edges. They fit edge to edge rather than overlapping. They contained 540 square feet of material.
   a. Is the pool really 16 feet by 32 feet?
   b. If you answered yes, support your answer. If you answered no, recalculate the dimensions of the pool.
2. (This question is independent of Exercise 1.) One day you see Carol swimming laps in your neighbor's pool (from one end to the other, parallel to the sides). From observing Carol swimming in a pool that is 25 yards long, you know that she swims

50 yards (one lap) in about 1 minute. In your neighbor's pool she swims a lap (down and back) in about 20 seconds.
   a. Is the pool really 16 feet by 32 feet?
   b. If you answered yes, support your answer. If you answered no, recalculate the dimensions of the pool.
   c. What assumption(s) do you make in stating your answer?
3. (This question is independent of Exercises 1 and 2.) One day you see Carol swimming laps in your neighbor's pool. In order to go the longest possible distance in the water before touching an edge, she swims from one corner diagonally across to the other corner. Because you have seen Carol swim in a pool that is 25 yards long, you know that she swims 50 yards (one lap) in about 1 minute. In your neighbor's pool she swims a lap (corner to opposite corner and back) in about 29 seconds.
   a. Is the pool really 16 feet by 32 feet?

b. If you answered yes, support your answer. If you answered no, recalculate the dimensions of the pool.

c. What assumption(s) do you make in stating your answer?

4. (This question is independent of Exercises 1, 2, and 3.) With your binoculars you can count the number of tiles on one side and on one end (placed vertically on the inside) of the pool, and you find there are 40 on the end and 56 on the side.

a. Is the pool really 16 feet by 32 feet?

b. If you answered yes, support your answer. If you answered no, recalculate the dimensions of the pool.

c. What assumption(s) do you make in stating your answer?

5. The United States population since 1900 is shown in the following table.

Year	Population (in millions)	Year	Population (in millions)
1900	76	1950	151
1910	92	1960	179
1920	106	1970	203
1930	123	1980	227
1940	132	1990	240

Code the years so that 1900 corresponds to $x = 0$, 1910 to $x = 10$, etc. Consider three different mathematical models of the U.S. population. In each of them $P_i(x)$ denotes the population in millions in the year x. For example, $P_1(90) = 220$ means that

according to model 1 the population in the year 1990 is predicted to be 220 million.

(1) $P_1(x) = 1.6x + 76$ (This is a linear model based on the 1900 and 1910 populations.)

(2) $P_2(x) = 1.82x + 76$ (This is a linear model based on the 1900 and 1990 populations.)

(3) $P_3(x) = 1.30x + 123$ (This is a linear model based on the 1980 and 1990 populations.)

a. On a single set of axes, graph the coded years and populations and graph and label the three functions defined above.

b. Which function do you believe best fits all ten data points?

c. Use each function to predict the 1970 population.

d. Use each function to predict the population in the year 2000.

e. Which prediction in part (d) do you believe is most accurate? Why?

6. Consider another mathematical model of the U.S. population (see Exercise 5) given by

$P_4(x) = 1.87x + 68.5$ (This is derived from all ten data points via a technique called linear regression.)

a. On the graph from Exercise 5(a), graph and label this function.

b. How does P_4 fit the data points compared with the other three functions?

c. Use P_4 to predict the 1970 population.

d. Use P_4 to predict the population in the year 2000.

e. Which do you believe is more accurate, the prediction in part (d) here or the "best" prediction in Exercise 5(d)? Why?

WRITTEN ASSIGNMENTS 1.3

1. Describe the mathematical-modeling process.

2. Give an example of a real-world situation or problem (other than one mentioned in this section) that can be modeled mathematically. You need not describe the model itself, just the situation or problem.

CHAPTER 1 REVIEW EXERCISES

1. Describe what problem solving is; i.e., define *problem solving*.

2. Briefly comment on the following statement: "An important goal of problem solving is to find an algorithm that has been used for previous problems."

3. a. Who was the first mathematician to popularize a problem-solving process?

   b. Describe three of the four steps of his four-step process.

4. Describe two psychological aspects of problem solving.

5. a. What is an algorithm?

   b. Give an example of one.

6. Name five strategies for problem solving. If your strategy is not one named in the book or in class, describe it.

7. Find all ways the number 30 can be written as the sum of consecutive integers. Work the same problem for 300 and for 1000. What generalizations can you make?

8. The checkerboard shown on the next page contains one checker. The checker can only move diagonally up the board, along the red squares. In how many ways can this checker reach the square marked A?

9. Draw a diagram that describes the mathematical-modeling process. Label all parts of the diagram.

10. Your neighbor has a swimming pool whose shape is rectangular. The pool distributor said it has a surface area of 648 square feet.

   a. Since many rectangular pools have a length about twice their width, make that assumption and find the dimensions of the pool.

   b. In order to check out this assumption, Jessica paces off the length and width of the pool. She finds that the length is 18 paces and the width 16 paces. Use this information either

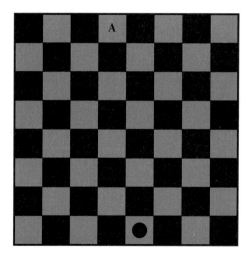

to support the answer you reached in part (a) or to refute it. If you refute it, calculate the dimensions of the pool based on Jessica's pacing.

   c. What assumption(s) do you make in working with part (b)?

11. You have two biological parents, four biological grandparents, eight biological great-grandparents, etc. For convenience, let's say that two generations back you had four grandparents, three generations back you had eight grandparents, etc.

   a. How many grandparents did you have 30 generations back?

   b. What assumption(s) are made in your answer to part (a)?

   c. In about what year did a grandparent of 30 generations back live?

   d. What assumption(s) are made in your answer to part (c)?

12. A residence heated by gas has had monthly gas bills, before tax, of $36.36 and $68.23 for usage of 77 CCF (100 cubic feet) and 162 CCF, respectively. In trying to figure out the billing scheme, Shelley divided $36.36 by 77 to get a cost per CCF of $.472208, and therefore assumes the billing scheme is $.472208 per CCF. Chris suggests that there is a monthly service charge of $7.50, plus a charge for usage. Taking $7.50 away from $36.36 leaves $28.86. This divided by 77 CCF gives a cost per CCF of $.374805, so Chris says the billing scheme is $7.50 plus $.374805 per CCF. John believes the scheme is like the one Chris suggested but with a $10 monthly service charge. He did a calculation like Chris's and suggests that the billing scheme is $10 plus $.342338 per CCF. Which of these models is best, Shelley's, Chris's, or John's? Explain why.

 BIBLIOGRAPHY

Bradley, George. *Problem Solving with Creative Mathematics.* Pacific Grove, CA: Brooks/Cole, 1995.

Polya, George. *How to Solve It: A New Aspect of Mathematical Method.* Garden City, NY: Doubleday, 1957.

———. *Mathematical Discovery: On Understanding, Learning, and Teaching Problem Solving.* New York: Wiley, 1981.

The UMAP Journal

Young, Martha W. "The Goals 2000: Educate America Act," *National Forum, 73* (1993):3–4.

CHAPTER 2

NUMBERS AND NUMERALS

PROLOGUE

The cover story in *Time* magazine of August 9, 1993, read:

SECRETS OF THE MAYA

After a century and a half of research, scientists are finally unraveling the mystery of who the Maya were, how they lived—and why their civilization suddenly collapsed.

Archaeologists have long known that the Maya, who flourished between about A.D. 250

and 900, perfected the most complex writing system in the hemisphere, mastered mathematics and astrological calendars of astonishing accuracy, and built massive pyramids all over Central America.

The latest discovery, announced just this week, underscores how quickly Maya archaeology is changing. Four new Maya sites have been uncovered in the jungle-clad mountains of southern Belize. . . .

Propelled by a series of dramatic discoveries, Mayanism has been transformed over the

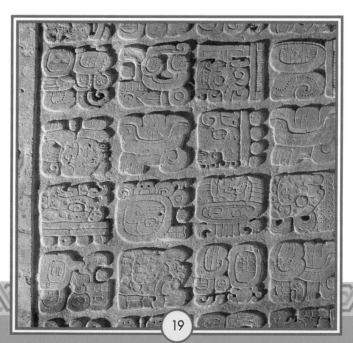

past 30 years from an esoteric academic discipline into one of the hottest fields of scientific inquiry—and the pace of discovery is greater today than ever.

How did the Maya count? How did they record numbers? What mathematics did they master? Knowledge of the Mayan system of numeration is helping archaeologists and anthropologists today learn the secrets of the Maya. If we understand the Mayan system of numeration, we can appreciate the workings of their priests in having developed a numeration system that involves moving through ever-bigger numbers toward the gods.

The Mayan system raises a question of contact between different civilizations. The Hindus in India and the Maya were the first two peoples to employ a symbol for zero as a place holder. Did this idea develop independently in the two civilizations, or was there contact between the civilizations and transmission of the idea?

ANCIENT NUMERATION SYSTEMS

GOALS
1. To describe the early history of counting.
2. To understand the use of numerals of several different civilizations.
3. To describe the origin of our own system of numeration.
4. To explain the development of zero.

Chapter opening photo: A Mayan inscription from Yaxchilan. Numbers are shown by dots, each standing for "one," and upright bars, each standing for "five." The number nine is shown here by one bar and four dots.

Numbers are represented in writing by symbols called *numerals*. The symbols 5 and *V* are both numerals that in different systems, represent the number we call five. As civilizations evolved, numbers were given names and written symbols developed.

The concept of *number* and the representation of numbers by numerals are parts of our heritage. They go back to the earliest peoples and are present in all civilizations. The concept of *number* and the idea of counting are learned relatively early by the developing child.

What is the basic concept of number? How are numbers represented in different civilizations? Are there any new uses for the number concept? These topics are explored in this chapter.

No recorded history exists of the earliest use of numbers or counting. It is quite likely that nonliterate societies had the need and subsequent ability to "keep track" of some items that required number sense. For example, how many children are in my family? This can be answered without stating a number by just naming the children. However, in comparing the sizes of two families, numbers are convenient. How many days does it take to get to the hunting grounds? How many people are in our tribe? With the evolution of a civilization, some form of counting became essential.

It is unclear how people first kept track of "how many," that is, how they counted. Probably some form of matching or marking existed in early societies. For example, a herder may have gathered a pile of pebbles to keep track of how many sheep or goats went out to pasture in the morning. As the animals returned in the evening, pebbles could be removed from the pile. If some pebbles were left, the herder would know to go searching for some missing animals. This method of matching pebbles to animals is an example of a one-to-one correspondence. The pebbles and animals are matched: Each pebble corresponds to one animal, and one animal corresponds to one pebble.

The use of pebbles to count animals is a first level of abstraction. The pebbles are an abstract representation of animals. A second level of abstraction comes when we replace pebbles with tally marks. The tally marks symbolize pebbles, which in turn symbolize animals.

Undoubtedly, nonliterate civilizations gave spoken and written names to numbers, as we do in our civilization today. Our names for numbers go beyond what is the minimum necessary, however. For example, to indicate "two," we have terms

Broom-Hilda

Reprinted by permission: Tribune Media Services.

For Better or For Worse®   **by Lynn Johnston**

like a *team* of horses, a *pair* of shoes, and a *yoke* of oxen—different words for the same number concept. We can mention two possible explanations for the presence of multiple words for the number 2: (1) the idea of "twoness" developed slowly over time, or (2) the society that developed the multiple words did so because it valued such things as horses, shoes, or oxen and consequently applied a value-laden word such as *team, pair,* or *yoke* as a natural expression of the value attached to the objects.

Egyptian Hieroglyphics

One of the first recorded numeral systems is that of Egyptian hieroglyphics. This system dates to as far back as 3400 B.C.E. Egyptian hieroglyphs are made up of pictorial symbols. A dog, for instance, might have been represented by a picture of a dog. Later the picture was simplified into a symbol that looked less like a dog but was easier to write. Here are four basic symbols representing numbers:

1 | 10 ∩   100 ९   1000 ⧟

The symbol for 1 is a single vertical stroke; the symbol for 10 is a heel bone, that for 100, a snare, and that for 1000, a lotus flower. Symbols for 10,000, 100,000, and 1,000,000 also existed but were used less frequently.

These symbols are combined to represent other numbers by adding together the value of all the symbols listed. For example:

2 || 5 ||||| 11 ∩| 30 ∩∩∩   67 ∩∩∩∩∩∩||||||

1,234 ⧟ ९९∩∩∩||||

An Egyptian ruler and his mother are shown on this tomb from about 1475 B.C.E. Food offerings are piled in tiers on the table. Below the table are long-stemmed lotus leaves, hieroglyphic symbols for 1,000. They indicate that the offerings were multiplied by thousands.

TABLE 2.1
Intermediate Combinations of the Elemental Roman Numerals

1	I	10	X	100	C
2	II	20	XX	200	CC
3	III	30	XXX	300	CCC
4	IV	40	XL	400	CD
5	V	50	L	500	D
6	VI	60	LX	600	DC
7	VII	70	LXX	700	DCC
8	VIII	80	LXXX	800	DCCC
9	IX	90	XC	900	CM

We have written these with the symbols for the larger numbers to the left. Often the Egyptians wrote them in the reverse order, and sometimes they wrote them vertically. Ultimately, it doesn't matter in which order they're written since we always evaluate the number by adding the values of the symbols. For example, ∩II and II∩ both denote 12. Notice that this is different than in our system, where 12 and 21 denote different numbers.

This system of Egyptian hieroglyphics is an example of an *additive system* of numeration. In such a system the value of the symbols that represent a number are added together to get the number. Our numeral system is *not* an additive system. For example, 21 is not 2 + 1.

Roman Numerals

You may be familiar with Roman numerals. They appear in the dates on older movies, in the numbering of pages in the preface of books, on some clock faces, as numerals for the Olympics and Super Bowls, and elsewhere. Here are the elemental Roman numerals:

1	I	5	V	10	X	50	L	100	C	500	D	1,000	M

These symbols are combined to form the numbers shown in ● Table 2.1, which are intermediate steps to forming any counting number.

The system of Roman numerals employs a subtraction property, in which if a letter is positioned before a letter of greater value, then the first value is subtracted from the second (larger) value. For example, IV = 5 − 1 = 4, XL = 50 − 10 = 40, and XC = 100 − 10 = 90. Other than this, numbers are represented, as in the Egyptian system, by writing appropriate symbols and adding their values. In other words, the system of Roman numerals is a *modified additive system.* For example, MDCCXLVII represents 1000 + 500 + 200 + 40 + 5 + 2, or 1747. To write 455 (the year the Vandals destroyed Rome) in Roman numerals, we write CD for 400, then L for 50, then V for five to give CDLV.

RULES TO EVALUATE ROMAN NUMERALS

1. If a letter is repeated, repeat its value.
2. If a letter is positioned after another letter of greater value, add both their values.
3. a. If a letter precedes another letter of greater value, subtract its value from the value of the letter of greater value.

b. The only letters of smaller value that precede letters of greater value are I, X, and C.

c. I precedes only V or X.

d. X precedes only L or C.

e. C precedes only D or M.

4. If a letter is positioned between two letters, each of greater value, add the two larger values and then subtract the value of the letter in between.

To illustrate rule 4, consider XIV, whose value is $10 + 5 - 1 = 14$. Rules 3b, c, d, and e limit the arrangement of letters in pairs to "four" and "nine" combinations: IV (4), IX (9), XL (40), XC (90), CD (400), and CM (900). Rule 4 is needed to resolve the seeming contradiction between rules 2 and 3a.

The system of Roman numerals has some advantages and disadvantages when compared with Egyptian hieroglyphics. A look at these advantages and disadvantages helps accent features of the two systems and indicates why different systems of numeration developed. Sometimes a feature of the Roman system can be both an advantage and a disadvantage, depending on your point of view, as illustrated by Table 2.2.

TABLE 2.2
Advantages and Disadvantages of the Roman System vs. Egyptian Hieroglyphics

Advantages	Disadvantages
1. The Roman system has symbols for intermediate numbers, such as V for 5. This permits an economy of symbols. For example, in the Roman system 5 requires only one symbol (V), whereas in Egyptian hieroglyphics 5 requires five symbols: III II	1. Because of the intermediate symbols, the Roman system requires us to memorize more symbols. To represent the counting numbers 1 through 1000 in the Roman system means knowing seven symbols (I, V, X, L, C, D, M), whereas in Egyptian hieroglyphics knowledge of only four symbols is required:
2. The subtractive principle makes the order of appearance of symbols important. For example, IV and VI represent different numbers. Numbers are written with the symbols for larger numbers appearing to the left; e.g., 455 is CDLV, with CD representing 400, L 50, and V 5. In Egyptian hieroglyphics, the symbols can be written in any order. For example, 12 could be written in any of three ways: ∩II I∩I II∩ The order imposed by the Roman system makes it easier to conceptualize the number.	2. In the Egyptian system, the order in which symbols are written is unimportant. This makes it easier to write a number—the symbols can appear in any order.

A building dated with Roman numerals. Can you read the date on it?

The Roman system also has advantages and disadvantages when compared with our Hindu-Arabic system. You are asked to consider these advantages and disadvantages in the exercises at the end of this section.

African Numeration Systems

When Africans speak of their traditional numeration systems, they may mean either a set of number words or a set of standardized gestures that complement the words or substitute for them. Claudia Zaslavsky, in her book *Africa Counts* (p. 37) says:

The number words in some African languages clearly express the finger gestures. In Zulu, the word for six means, literally, "take the thumb," indicating that all the

Finger gestures indicating numbers

	TABLE 2.3		
	Some Igbo Number Words		
1	ofu	7	isa
2	ibua	8	isato
3	ito	9	itenani
4	ino	10	ili
5	ise	11	ili naofu ("10 and 1")
6	isi	12	ili naibua ("10 and 2")

fingers on one hand have been counted and one must start with the thumb of the second hand.

There are many African societies in which the finger gestures have equal status with the spoken numerals and constitute a proper system of numeration which may or may not agree with the spoken number words in its derivation.

Many different systems of numeration exist in Africa. The Igbo people of Nigeria have a culture that dates back at least a thousand years. Some Igbo number words are shown in ● Table 2.3. Notice that this is an additive system, like Egyptian hieroglyphics, since the words for 11 and 12 are both formed from words whose values are added. Although separated in time by at least 2000 years, Egyptian hieroglyphics and the Igbo number words share this additive feature. Still, both have other, distinguishing features.

Hindu-Arabic Numerals

The system of numeration employed in the United States and much of the rest of the world is called Hindu-Arabic. It originated in India with the Hindus and was transmitted to Europe by the Arabs. The oldest-known example of Hindu-Arabic numerals exists on stone columns in India dating from about 250 B.C.E. These numerals were known in Europe about 800 C.E. Some early Hindu-Arabic symbols are shown in ● Table 2.4. A book about these symbols was written by al-Khwārizmī, a Persian mathematician, in 825 C.E. It had the Latinized title *Liber Algorism de Numero Indorum.* The English word *algorithm* comes from "algorism." Another heritage from the same man is *algebra,* from his book *Ilm al-Jabr Wa'l Maqabalah* ("The Science of Reduction and Cancellation").

Prior to the development of the Hindu-Arabic system, there were about the same number of systems of numeration as there were written languages. Today the use of the Hindu-Arabic system cuts across numerous language, cultural, and political groups. Claudia Zaslavsky, in *Africa Counts* (p. 6), says:

"How do Africans count—like us?" I posed this question to two young colleagues, one from Kenya and the other from Tanzania. "How do you count in your language?" They looked at me in surprise, and replied, "Just as you do here in the

United States !" Nowadays school children in most of the world write Hindu numerals (popularly called "Arabic")—truly universal symbols.

Egyptian hieroglyphics, Roman numerals, and the Igbo system are all additive systems. However, none of these three has the property of the Hindu-Arabic system that the place a digit occupies dictates its value. For example, in the symbol 232, the left-most 2 is in the hundreds' place, while the right-most 2 is in the units' place. Such a system is called a *place-value system*. In a place-value system, the position of a symbol determines its value. For example, in the numeral 343, the two 3's have different values—the left 3 represents 300, and the right 3 represents 3.

In a place-value system, it is sometimes necessary to indicate that no symbol appears in a particular place. For example, the fact that 102 and 12 have different meanings is shown by the 0 in the tens' place of 102. Without a symbol like 0, what would we do? Likely, a blank space would be left. For example, 102 might be written as 1 2 to indicate that there are no groups of 10. One system that used a space between symbols was the Babylonian system, dating back to about 3200 or 3400 B.C.E. Such a system could clearly lead to confusion. Does 3 4 mean 34 or 304? How do we write 3004 and 30,004? What about 50 and 500? Clearly, the development of a symbol such as 0, and its use in a place-value system, helped make the Hindu-Arabic system an effective way to represent numbers.

Sets of Hindu numerals have been found that date to before 800 C.E. and have a zero symbol almost identical to the one we use today. The Maya of Central America,

TABLE 2.4
Early Hindu-Arabic Symbols

	1	2	3	4	5	6	7	8	9	0
Mathematical treatise copied at Shiraz in 969 by the mathematician Abd Djalil al-Sidjzy										
Form of numerals used in Europe in 1077										
Form used in Europe in 12th century										
Form used in Europe in early 13th century										
Arithmetical treatise by Ibn al-Banna Al Marrakushi, 14th century										
Arab manuscript dated 1571–72										

SOURCE: Georges Ifrah, *From One to Zero*, New York: Viking Penguin, 1985.

dating back to at least the third or fourth century B.C.E., were the first civilization known to have developed a numeration system with both a symbol for zero and the place-value concept. The Mayan system is discussed in Section 2.2. The Babylonian system, dating back to before 500 B.C.E., utilized the place-value concept but had no symbol for zero.

The use of zero as a number indicating none probably developed after its use as a place holder. Evidence suggests that zero was first recognized as a number by Hindu and Arab mathematicians sometime between 500 and 1100 C.E. Major earlier civilizations such as the Egyptians, Babylonians, Greeks, and Romans apparently had no concept of zero as a number.

In the Hindu-Arabic system, the values of the places, from right to left, are $1 = 10^0$, $10 = 10^1$, $100 = 10^2$, $1000 = 10^3$, and so on. The value of a numeral is found by thinking of it in expanded form. For example, the expanded form is 56 is $(5 \times 10) + 6$ (5 is in the ten's place), or

$$56 = (5 \times 10^1) + (6 \times 10^0)$$

The expanded form of 4032 is

$$4032 = (4 \times 10^3) + (0 \times 10^2) + (3 \times 10^1) + (2 \times 10^0)$$

The development of a place value system of numeration made it easier to represent numbers and made computation much more efficient. Otto Neugebauer, the noted historian of science, describes the significance of this achievement by saying (Neugebauer, p. 5):

> The invention of this place-value notation is undoubtedly one of the most fertile inventions of humanity. It can be properly compared with the invention of the alphabet as contrasted to the use of thousands of picture-signs.

EXERCISES 2.1

1. Write 34 in Egyptian hieroglyphics.
2. Write 306 in Egyptian hieroglyphics.
3. What number is represented by the following Egyptian numeral?

4. What number is represented by the following Egyptian numeral?

5. Write 54 (the year Claudius was poisoned) in Roman numerals.
6. Write 71 (the year the Roman Coliseum was started) in Roman numerals.
7. What number does the Roman numeral CXVII represent? (It is the first year of Hadrian's rule of Rome.)
8. What number does the Roman numeral CXXXVIII represent? (It is the last year of Hadrian's rule of Rome.)

9. What number does the Roman numeral MCMLV represent? (This is the year Albert Einstein died.)
10. What number does the Roman numeral MDCCLXXVI represent? (It is a famous year in U.S. history.)
11. Convert the following Egyptian numeral to a Roman numeral:

12. Convert the Roman numeral MCDXCIV to an Egyptian numeral.
13. List some advantages and disadvantages of our system of numeration compared with Egyptian hieroglyphics.
14. List some advantages and disadvantages of our system of numeration compared with Roman numerals.
15. Write the Igbo number words for 13 and 14.
16. Write the Igbo number words for 15 and 16.
17. Write 51,203 in expanded form.
18. Write 90,786 in expanded form.

1. Read Chapters 1–3 of *Africa Counts* by Claudia Zaslavsky.
   a. Summarize the range of counting systems that exist or have existed in Africa.
   b. If you wrote a book called *Europe Counts,* how might it be the same as and how might it differ from *Africa Counts?*
   c. What does Zaslavsky say is the ultimate basis for the development of numeration in any society? Do you agree? Why or why not?
2. Invent symbols other than Hindu-Arabic symbols for the numbers 0 through 9 (for example, three might be ⦂ or Δ). Use your symbols in a place-value system to write the current year and the approximate date of the origin of Egyptian hieroglyphics (3400 B.C.E.). If it is not obvious, explain what motivated your choice of symbols. (The results of a similar assignment given to two middle school classes are reported by Kim Krusen (see the Bibliography at chapter's end).
3. The *Detroit Free Press* runs a column entitled "New for Young Readers." In the column of April 27, 1992, an 11-year-old reader asked, "Who made the numeral system?" Suppose you're the reporter assigned to answer her question. Do so in 150–200 words.

2.2 ▸ DIFFERENT BASES

GOALS
1. To apply the concept of the base of a system of numeration.
2. To describe and use another civilization's numeration system with a base other than 10.
3. To convert from one base to another.
4. To be familiar with the computer's base-two system.

Our Hindu-Arabic system of numeration has a base of 10. We can make the reliance on 10 apparent by writing 3891 in expanded form:

$$3891 = (3 \times 10^3) + (8 \times 10^2) + (9 \times 10^1) + (1 \times 10^0)$$

Ten is called the *base* of our system of numeration. It is the unit in which we group things: Ten 1's make 10, ten 10's make 100, ten 100's make 1000, etc. It is easy to conjecture that 10 was a natural choice for a base because we humans have 10 fingers. Consequently, people in many civilizations counted by placing items in groups of 10. Egyptian hieroglyphics, Roman numerals, and the Igbo system (see Section 2.1) are all base-10 systems.

The Yuki, American Indians of California, apparently believe that their system's base of 8 is a natural choice, but they focus on the eight gaps between the fingers rather than on the 10 fingers (Ascher, p. 9).

Our society utilizes other groupings, as well: Socks, shoes, mittens, and gloves are grouped by twos; doughnuts and eggs are sold in dozens; canned beverages are sold in 6-packs, 8-packs, 12-packs, and 24-packs.

Several civilizations throughout history have developed numeration systems with bases other than 10. The first numeration systems may have involved a base of 2. The *western tribes of Torres Straits* in Australia employ a base-2 idea by counting (see Conant):

1	urapun	5	okosa okosa urapun
2	okosa	6	okosa okosa okosa
3	okosa urapun	above 6	ras (a lot)
4	okosa okosa		

Eskimos in Greenland have a system that groups by 5's up to 20, then goes on by 20's (see Swain). For example, their description of 27 is 1 man (20), 1 hand (5), other hand 2 (2).

YOUR FORMULATION

Conjecture why 1 man represents 20.

The *Mayan* system, dating back to the third or fourth century B.C.E. (see Sanchez), involved a modified base-20 system. Numerals were written in a vertical column, with the bottom row having the place value 1, the row above it the value 20, the row above that the value 18×20, and the row above that the value 18×20^2. In Hindu-Arabic symbols, the Mayan numeral

$$2$$
$$3$$
$$4$$
$$5$$

would represent

$$(2 \times 18 \times 20^2) + (3 \times 18 \times 20) + (4 \times 20) + (5 \times 1) = 14{,}400 + 1{,}080 + 80 + 5$$
$$= 15{,}565$$

	2	3	4	5
Value	18×20^2	18×20	20	1

In working with systems of numeration with bases other than 10 we have examined numerals by converting them to base 10. If you were, for example, an Eskimo in Greenland, it would be important for you to think numerically in their base-5–20 system rather than converting to and from base 10. For example, if you saw the following collection of animals

you would think "1 man, 1 hand, other hand 2 " rather than "27 = 1 man, 1 hand, other hand 2." Thinking numerically in a different number base is like thinking in another language without translating everything into English. However, the emphasis here is on how numbers are written in other bases rather than on thinking numerically in other bases. Since you are familiar with a base-10 system, we will focus on converting numbers from other bases to base 10, and from base 10 to other bases, in order to understand how numbers are written in other bases.

The following three examples illustrate how numbers are written in bases other than 10 and the conversions of numerals from one base to another. For a number in a base other than 10, the numeral will be written with parentheses around it and a subscript will indicate the base. For example, $(1032)_5$ means the base is 5.

Geographical Distribution of Numeration Systems

The book *Numbers Through the Ages* by Graham Flegg shows the geographical distribution of systems of numeration in the map reproduced here. A "2-count" system refers to base 2, a "4-count" system to base 4, etc. A "5–20-count" system refers to a system with a base of 20 and an intermediate base of 5. Thus, the system of Roman numerals is a 5–10-count system. For Europe and Asia the map shows the numeration systems as they are today. In Africa and the Americas only the counting methods of the original native inhabitants are shown, not the base-10 system of the majority of the current inhabitants.

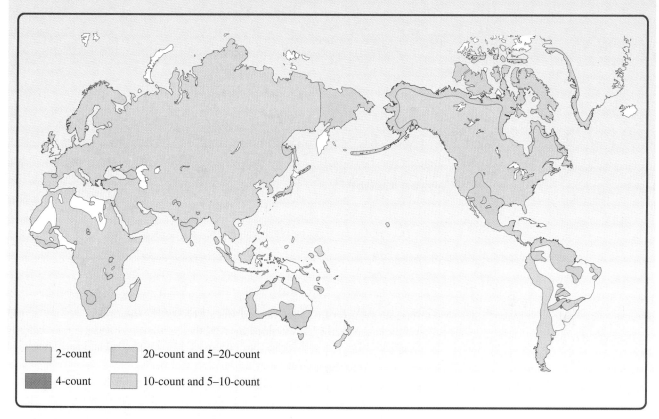

2-count	20-count and 5–20-count
4-count	10-count and 5–10-count

Distribution of counting systems

EXAMPLE 2.1

Convert $(1032)_5$ to base 10.

SOLUTION The place values are indicated below 1032 as follows:

1	0	3	2
5^3	5^2	5^1	5^0
125	25	5	1

Thus

$$(1032)_5 = (1 \times 125) + (0 \times 25) + (3 \times 5) + (2 \times 1) = 125 + 0 + 15 + 2 = 142$$

EXAMPLE 2.2

Convert $(394)_{12}$ to base 10.

Comment. The symbols in base 12 represent our numerals 0, 1, 2, 3, 4, 5, 6, 7, 8, 9, 10, and 11. Since our symbols for ten and eleven require two digits, that creates a problem in base 12. For instance, does $(10)_{12}$ mean ten, or does it mean twelve (1 group of twelve and 0 units)? To resolve this ambiguity, we will use the symbol T for ten and the symbol E for eleven. Thus, $(10)_{12}$ means 12 (one group of twelve and zero units). If we wish to indicate ten in base 12, we write $(T)_{12}$.

SOLUTION The place values are indicated below 394 as follows:

3	9	4
12^2	12^1	12^0
144	12	1

Thus, $(394)_{12} = (3 \times 144) + (9 \times 12) + (4 \times 1) = 432 + 108 + 4 = 544$

EXAMPLE 2.3

Convert 612 to base 5.

SOLUTION In base 5, the place values are as follows:

5^4	5^3	5^2	5^1	5^0
625	125	25	5	1

Here, no groups of 625 are needed since $612 < 625$. How many groups of 125 are needed? When 612 is divided by 125, the quotient is 4 and the remainder is 112 ($612 = 4 \times 125 + 112$), so four groups of 125 are needed and the numeral is

4			
125	25	5	1

How many groups of 25 are needed? We have 112 left. When 112 is divided by 25, the quotient is 4 and the remainder is 12 ($112 = 4 \times 25 + 12$), so four groups of 25 are needed and the numeral is

4	4		
125	25	5	1

For the remaining 12, two groups of 5 and two groups of 1 are needed. Thus, $612 = (4422)_5$. This answer can be checked by noting that

$$(4422)_5 = (4 \times 5^3) + (4 \times 5^2) + (2 \times 5^1) + (2 \times 5^5) = 500 + 100 + 10 + 2 = 612$$

In a base-b place-value system of numeration, exactly b symbols are needed. They are 0, 1, 2, . . . , $b - 1$. For example, in base 10, the ten symbols are 0, 1, 2,

. . . , 9. In base 5, the five symbols are 0, 1, 2, 3, and 4. In base 12, the twelve symbols are 0, 1, 2, . . . , 9, T, and E.

Computer Numeration Systems

Computer circuitry design can be thought of as based on circuits that are either open or closed. If an open circuit corresponds to 1 and a closed circuit corresponds to 0, then a sequence of open and closed circuits corresponds to a sequence of 1's and 0's, which represent the numbers in *base 2*. The binary (base-2) representation of numbers is basic to storing numbers in a computer. Some examples of conversion to and from base 2 follow.

EXAMPLE 2.4

Write 47 in base 2.

SOLUTION In base 2 the place values are as follows:

2^6	2^5	2^4	2^3	2^2	2^1	2^0
64	32	16	8	4	2	1

To write 47, no groups of 64 are needed. One group of 32 is needed, with 15 remaining ($47 = 32 + 15$), so the numeral starts

1					
32	16	8	4	2	1

To represent 15, we use no groups of 16, one group of 8, with 7 remaining. Now the numeral starts

1	0	1			
32	16	8	4	2	1

The remaining 7 is one group of 4, one group of 2, and one group of 1, so the numeral is $(101111)_2$. The answer can be checked by noting that

$$(101111)_2 = (1 \times 2^5) + (0 \times 2^4) + (1 \times 2^3) + (1 \times 2^2) + (1 \times 2^1) + (1 \times 2^0)$$
$$= 32 + 0 + 8 + 4 + 2 + 1 = 47$$

EXAMPLE 2.5

Convert $(1101101)_2$ to base 10.

SOLUTION In expanded form,

$$(1101101)_2 = (1 \times 2^6) + (1 \times 2^5) + (0 \times 2^4) + (1 \times 2^3) + (1 \times 2^2) + (0 \times 2^1) + (1 \times 2^0)$$
$$= 64 + 32 + 0 + 8 + 4 + 0 + 1$$
$$= 109$$

To represent a number in the binary system requires only the two symbols 0 and 1. However, most numbers in base 2 require more places than in base 10.

Some computers use the *hexadecimal* (base-16) system of numeration. It requires a symbol for each of the numbers 0 through 15. One set of hexadecimal

1. For what whole numbers (0, 1, 2, 3, etc.) does it require the same number of places to represent the number in both base 2 and base 10?
2. For what, if any, whole numbers does it require more places to represent the number in base 2 than in base 10?

---

symbols is shown in ⬤ Table 2.5. Example 2.6 shows a conversion from hexadecimal notation to base 10. Example 2.7 shows a conversion from base 10 to hexadecimal notation.

EXAMPLE 2.6

Convert the hexadecimal numeral C3A to base 10.

SOLUTION

	C	3	A
Place value	16^2	16^1	16^0

$$(C3A)_{16} = (C \times 16^2) + (3 \times 16^1) + (A \times 16^0)$$
$$= (12 \times 256) + (3 \times 16) + (10 \times 1)$$
$$= 3130$$

EXAMPLE 2.7

Convert 1946 (the date the first operational electronic digital computer was introduced) to hexadecimal notation.

TABLE 2.5
Three Numeration Systems

Binary (Base 2)	Decimal (Base 10)	Hexadecimal (Base 16)
0001	1	1
0010	2	2
0011	3	3
0100	4	4
0101	5	5
0110	6	6
0111	7	7
1000	8	8
1001	9	9
1010	10	A
1011	11	B
1100	12	C
1101	13	D
1110	14	E
1111	15	F

SOLUTION We look at the places we'll need.

Place value	16^2	16^1	16^0
	256	16	1

Notice that the next place has the value $16^3 = 4096$, so we don't need that place. Since $1946 \div 256 = 7 + \text{remainder}$, we get $1946 = (7 \times 256) + 154$. Therefore, $1946 = (7 \_ \_)_{16}$. Since $154 \div 16 = 9 + \text{remainder}$, we get $154 = (9 \times 16) + 10$. Therefore, $1946 = (79A)_{16}$.

Actual storage of letters and numerals in computer memory involves converting the symbols into a base-2 representation. The ASCII code is one way to assign base-2 numerals to base-10 numerals and text characters.

Application: The Game of Nim

One application of the concept of different bases is the game of Nim. In one version of this game, two players (A and B) manipulate 21 objects (coins, toothpicks, pencils, or whatever). The players take turns. On each turn a player picks up either one or two objects. Whoever picks up the last object loses. Following is a sample game with 21 objects, which are indicated by asterisks.

Start	*********************
A picks up one	********************
B picks up one	*******************
A picks up two	*****************
B picks up two	***************
A picks up one	**************
B picks up two	************
A picks up one	***********
B picks up two	*********
A picks up one	********
B picks up two	******
A picks up two	****
B picks up two	**

If A were to pick up both of the remaining two objects, then A would lose, so A picks up only one. B must pick up the one now remaining, so B loses.

Play this game with an opponent and see if you can create a strategy that enables you to win every time.

To see how the game of Nim is an application of number bases, consider your winning positions in base 3 (this assumes you have developed a winning strategy):

Number Left After Your Turn

Base 10	Base 3
1	1
4	11
7	21
10	101
13	111
16	121
19	201

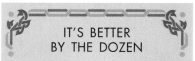

IT'S BETTER BY THE DOZEN

More information on base 12 and advocacy for its use can be obtained from the Dozenal Society of America (DSA), which publishes *The Duodecimal Bulletin* twice a year. One of the DSA publications states the following:

> Mathematicians and philosophers have long been aware that ten is actually a poor base for a number system. Several others have been suggested, with considerable agreement that twelve, with its many factors, would be the most serviceable. Its actual use has been suggested at various times, notably by Herbert Spencer, the British philosopher, and Isaac Pitman, inventor of a system of shorthand. But the hand of tradition is heavy, and few persons . . . have actually tried out this superior system of counting by dozens.

One of the suggestions for the use of base 12 is referred to by Samuel Pepys in his diary. Here's part of his entry for June 9, 1663:

> . . . and then comes Creed and he and I talked about mathematiques and he tells me of a way found out by Mr. Jonas Moore, which he calls Duodecimal arithmetique, which is properly applied to measuring, where all is ordered by inches, which are 12 in a foot.

The DSA can be reached at Dozenal Society of America, Nassau Community College, Garden City, NY 11530-6793. Dues are $12 a year, with a life membership available for $144. More economically inviting but less symbolic is a student membership for $3. Students are cautioned that the root word is *dozen*, not *doze*.

If we look at the other numbers in base 3, we have

Base 10	Base 3	Base 10	Base 3
2	2	11	102
3	10	12	110
5	12	14	112
6	20	15	120
8	22	17	122
9	100	18	200

Notice that the winning positions all have a 1 in the units' place and the nonwinning positions have either a 0 or a 2 in the units' position. Thus, you can produce a winning position by doing whatever is necessary to the current base-3 number to produce a 1 in the units' position. For example, if at your turn the base-3 number is 112, then subtract 1 to obtain 111, which is a winning position.

EXERCISES 2.2

For Exercises 1–9, convert from the indicated base to base 10.

1. $(3401)_5$
2. $(1043)_5$
3. $(7T2)_{12}$
4. $(2E98)_{12}$
5. $(110)_2$
6. $(1010)_2$
7. $(100100)_2$
8. $(45D)_{16}$
9. $(80F9)_{16}$

For Exercises 10–17, convert from base 10 to the indicated base.

10. 84 to base 5
11. 396 to base 5
12. 4123 to base 12
13. 19 to base 2
14. 101 to base 2
15. 3000 to base 12
16. 452 to base 16
17. 4444 to base 16

For Exercises 18–21, convert from one base to another.

18. $(2043)_5$ to base 12
19. $(1T9)_{12}$ to base 5
20. $(101101)_2$ to base 16
21. $(2A3)_{16}$ to base 2
22. Find the base b if $(123)_b = 27$.
23. Find the base b if $(209)_b = 251$.
24. a. Is there any base b for which $(21)_b$ is an even number?
    b. Is there any base b for which $(31)_b$ is an even number?

25. a. Is there any base b for which $(42)_b$ is an odd number?
    b. Is there any base b for which $(52)_b$ is an odd number?
26. a. How many numbers can be represented in base 10 with five or fewer places?
    b. How many numbers can be represented in base 2 with five or fewer places?
27. List all numbers that require the same number of places in base 10 as in base 2.
28. If three places serve to represent a number, how many base-12 numbers can be represented?
29. How many places in the binary system are necessary to write every number through each of the following?
    a. 9   b. 99   c. 999
30. A novelty item is a set of calculating cards as shown. To use them, ask someone to think of a number between 1 and 15, inclusive, then to give you all the cards on which that number appears. For example, if the person's number is 11, the person would give you cards A, B, and D; and if the person's number is 12, the person would give you cards C and D. You "read the person's mind" and report the number thought of by adding together the numbers in the upper left-hand corner of the cards given to you. In these examples, if you are given cards A, B, and D, you add $1 + 2 + 8 = 11$ and report the number as 11; if you are given cards C and D, you add $4 + 8 = 12$ and report the number as 12.

A		B		C		D	
1	9	2	10	4	12	8	12
3	11	3	11	5	13	9	13
5	13	6	14	6	14	10	14
7	15	7	15	7	15	11	15

a. Show why this works by writing numbers on each of the cards in binary notation and noting what happens in the binary addition of the numbers in the upper left-hand corners.

b. Describe the next step in extending this to more than 15 numbers.

31. Suppose the game of Nim is played starting with 22 objects instead of 21. If you want to win, what is your strategy?

32. Suppose you play the game of Nim, starting with 21 objects, with a child and you want to guarantee the other player has the opportunity to win on his/her last move. What is your strategy? Is it possible to guarantee you will lose? Why or why not?

33. We can do arithmetic in other bases just as we do it in base 10. For example, if a child is learning to add in base 3, it is first necessary to learn the addition table for single-digit numbers, just as we do in base 10.

+	0	1	2
0	0	1	2
1	1	2	10
2	2	10	11

To add numbers with more than one digit, proceed as in base 10. For example,

$$\begin{array}{r} 121 \\ + \ 120 \\ \hline 1011 \end{array}$$

$1 + 0 = 10$
$2 + 2 = 11$ (Write the right-most 1 and carry the
$2 + 1 = 10$ left-most 1.)
$1 \text{ (Carried)} + 1 = 2$

Add the following base-3 numbers, showing your work as in the preceding.

a. $10 + 11$
b. $12 + 11$
c. $12 + 22$
d. $210 + 122$
e. $222 + 222$
f. $1201 + 201$
g. $2102 + 1212$

34. a. Construct the addition table for single-digit numbers in base 5.

b. Show your work in adding the following base-5 numbers.
   i. $23 + 11$
   ii. $23 + 42$
   iii. $132 + 404$
   iv. $413 + 424$

WRITTEN ASSIGNMENTS 2.2

1. Some groupings other than base-10 groupings (socks, shoes, and so on) are listed at the start of this section. State some additional examples of groupings other than base-10 groupings.

2. Describe some advantages of a smaller base (2, for example) over a larger base (10).

3. Describe some advantages of a larger base (20, for example) over a smaller base (10).

4. The Babylonian cuneiform system was a place-value system without a symbol for zero. Describe some problems created by the absence of a symbol for zero in a place-value system.

5. Describe one other historical system of numeration. Include a reference to your source; the base of the system; the symbols used; whether the system is additive, place value, or something else; an example of a conversion from a numeral in that system to a Hindu-Arabic numeral, and vice versa: and comments on the historical use of this system.

6. Describe a mechanical device, such as blocks or an odometer, that would help a person *think* in another number base. Try to think in a different base by counting in a base specified by your instructor.

7. Make a case for switching to a base other than 10. State the base you believe we should switch to, and argue persuasively why.

8. Write a paragraph describing why computers use a binary representation of numbers.

9. The prefix "mega-," as in *megabit* and *megabyte,* multiplies the unit following by 1 million. It comes from the Greek word *megas,* which means "great." State two other words, other than units in the metric system, that incorporate the prefix "mega-." Describe for both of the words how the adjective "great" applies.

10. Describe a Nim-like game other than one described in this section. Include a description of a winning strategy, if there is one. One place to look is "All You Need Is Nim" by Dominic Olivastro (see Bibliography).

11. Explain the following:

    Q: Why do mathematicians confuse Halloween and Christmas?

    A: Because oct "31" = dec "25".

2.3 • FRACTIONS AND DECIMALS

GOALS
1. To describe the historical development of fractions and decimals.
2. To work with unit fractions.
3. To compare two fractions to see which is larger.
4. To convert a decimal to a fraction.

As children we learned early about fractions. Phrases like "half a sandwich," "a quarter (of a dollar)," "three quarters of a football game," "a quarter till 10," "two-thirds full," and "an eighth note" became meaningful when we were quite young. In the development of civilizations, the *fraction* concept appeared early and often. However, a systematic approach to fractions often did not develop. What developments have been made in the fraction concept through different civilizations? What prompted these developments?

Unit Fractions

The Rhind papyrus, copied by a scribe named Ahmes around 1650 B.C.E., contains a systematic treatment of unit fractions. A *unit fraction* is a fraction with a *numerator* (top) of 1, such as $\frac{1}{2}$, $\frac{1}{3}$, or $\frac{1}{17}$. The Egyptians of that time relied heavily on unit fractions to do calculations involving fractions. In Writing Exercise 2 you will be asked to speculate on why unit fractions were used extensively. In Egyptian hieroglyphics (see Section 2.1), the fraction $\frac{1}{3}$ was written as ⬭, $\frac{1}{12}$ was written as ⬭, and so on.

● EXAMPLE 2.8

Write the fractions $\frac{1}{4}$ and $\frac{1}{30}$ in Egyptian hieroglyphics.

SOLUTION $\dfrac{1}{4}$ = ⬭ $\dfrac{1}{30}$ = ⬭

● EXAMPLE 2.9

What fractions are represented by the following Egyptian hieroglyphs? ⬭ and ⬭

SOLUTION   ⬭ $= \dfrac{1}{11}$ and ⬭ $= \dfrac{1}{40}$

For a fractional value like $\frac{3}{4}$ that was not itself a unit fraction, the Egyptians wrote a sum of unit fractions. They wrote $\frac{3}{4}$ as $\frac{1}{2} + \frac{1}{4}$. The fraction $\frac{2}{3}$ had a special symbol, ⬭, and $\frac{1}{2}$ was sometimes written as ⬭. Decomposing a fraction into unit fractions is often difficult. Although we could write any fraction *a/b*, where *a* is a counting number (1, 2, 3, and so on), as

$$\frac{1}{b} + \frac{1}{b} + \cdots + \frac{1}{b}$$

with *a* terms, this scheme was often not used. For example, $\frac{2}{35}$ was written as $\frac{1}{30} + \frac{1}{42}$ rather than as $\frac{1}{35} + \frac{1}{35}$. Why this decomposition? No one knows for sure.

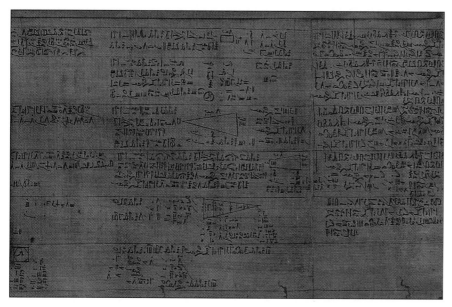

The Rhind papyrus, an Egyptian document dating to about 1650 B.C.E.

Our goal here is to show that the methods for working with fractions we were taught in elementary school are an improvement over the unit-fraction approach employed by several civilizations. We examine how to decompose a fraction into a sum of unit fractions when we are given all but one of the terms of the sum. The next example illustrates this.

EXAMPLE 2.10

Write $2/7$ as $1/4 + x$, where x is a unit fraction.

SOLUTION If $2/7 = 1/4 + x$, then $2/7 - 1/4 = x$. The lowest common denominator is $7 \times 4 = 28$, so

$$x = \frac{2}{7} - \frac{1}{4} = \frac{8}{28} - \frac{7}{28} = \frac{1}{28}$$

Therefore,

$$\frac{2}{7} = \frac{1}{4} + \frac{1}{28}$$

The ancient Egyptians were aided by a table representing fractions of the form $2/n$ as sums of unit fractions. In the Rhind papyrus, such a table for odd values of n from 5 to 101 was included prior to the problem section.

The ancient Greeks also relied on unit fractions. The ancient Romans gave each fraction a special name. They usually kept the *denominator* (bottom part of a fraction) a constant 12. Working with fractions was the main part of instruction in arithmetic in Roman schools.

Our method of writing fractions in the form a/b appears to have originated with the Hindus or with the Greeks. By 1000 C.E., the Arabs had begun using the fraction bar (recall that the ancient Egyptians did not) in the forms

$$\frac{a}{b}, \qquad a-b, \qquad \text{and} \qquad a/b.$$

Although the mathematician Leonardo of Pisa (ca. 1180–1250), known as Fibonacci, regularly put the horizontal bar in fractions, it was not until the sixteenth century that the fraction bar came into general use.

Comparing Fractions

The Hindu Arabic system of numeration and the use of fractions continued to flourish after 1100 C.E., spurred on by an expanded need in areas such as astronomy, business, engineering, government, the military, and navigation. Calculations had to be performed quickly and accurately. In a parallel fashion, as children mature their need for numbers expands from the counting numbers to fractions: half a piece, $\frac{1}{3}$ of a glass of milk, $\frac{1}{4}$ of a ballgame, $\frac{3}{10}$ of a second, $\frac{3}{4}$ of an hour, $\frac{2}{3}$ of a cup in a recipe, and so on. You may recall from your experience that work with fractions is sometimes slow and often susceptible to error.

Comparing two fractions to see which is the larger is frequently difficult. For example, which is larger, $\frac{2}{3}$ or $\frac{3}{4}$? This can be answered by finding a common denominator; 12 is the smallest one. Then

$$\frac{2}{3} = \frac{2}{3} \cdot 1 = \frac{2}{3} \cdot \frac{4}{4} = \frac{8}{12} \qquad \text{and} \qquad \frac{3}{4} = \frac{3}{4} \cdot 1 = \frac{3}{4} \cdot \frac{3}{3} = \frac{9}{12}$$

Since $\frac{9}{12} > \frac{8}{12}$, $\frac{3}{4} > \frac{2}{3}$.

If many comparisons of fractions are to be made, a mechanical approach might be exercised often enough to be remembered. It reduces the comparison of fractions to a comparison of counting numbers by using the fact that, for counting numbers a, b, c, and d, $\frac{a}{b} > \frac{c}{d}$ is equivalent to $ad > bc$. Therefore, $\frac{3}{4} > \frac{2}{3}$ provided $3 \cdot 3 > 4 \cdot 2$. Since $9 > 8$, $\frac{3}{4} > \frac{2}{3}$. The justification for this is explored in Exercise 9.

Probably the easiest method to find which is larger, $\frac{3}{4}$ or $\frac{2}{3}$, is to write both fractions as decimals (this requires a procedure you have seen but that people in 1200 had not): $\frac{2}{3} = .6666 \ldots$ (the three dots indicate that the decimal continues in the same pattern), and $\frac{3}{4} = .75$. From the decimal representations, it is easy to see that $\frac{3}{4} > \frac{2}{3}$.

Development of Decimals

The decimal approach for comparing fractions depends on the use of decimal fractions (like .75), a development the ancient Egyptians, Greeks, and Romans, and people for the subsequent thousand years did not have. The popularization of this concept of decimal fractions, in the late 1500s, is usually credited to the Dutch scientist Simon Stevin. Comparing fractions via decimals is aided greatly by the calculator, which became available in the 1970s.

The only significant improvement in the manipulation of decimal fractions since the time of Stevin is in notation. Stevin wrote 7.843 as

$$\begin{array}{cccc} 0 & 1 & 2 & 3 \\ 7 & 8 & 4 & 3 \end{array} \qquad \text{or} \qquad 7\textcircled{0}8\textcircled{1}4\textcircled{2}3\textcircled{3}$$

Even today the notation for the "decimal point" varies around the world. For 7.843, the English write 7·843, and the French and Germans write 7,843.

The importance of decimal notation is stated by English philosopher and mathematician Alfred North Whitehead (1948, p. 59):

STEVIN'S CARRIAGE WITH SAILS

Simon Stevin was best known to the general population of his day for the invention of a carriage rigged with sails so it could run along the seashore. It carried up to 28 people and could travel faster than a horse.

Before the introduction of the arabic notation, multiplication was difficult, and the division even of integers called into play the highest mathematical faculties. Probably nothing in the modern world could have more astonished a Greek mathematician than to learn that, under the influence of compulsory education, the whole population of Western Europe, from the highest to the lowest, could perform the operation of division for the largest numbers. This fact would have seemed to him a sheer impossibility. . . . Our modern power of easy reckoning with decimal fractions is the most miraculous result of a perfect notation.

Converting from a Decimal to a Fraction

Long division converts a fraction to a decimal, and a calculator simplifies this conversion. For example, $\frac{5}{6}$ is $5 \div 6$, which is $0.833333.$ . . . Converting from a decimal to a fraction often requires more work. The fraction equivalents of some decimals are somewhat recognizable. Are these familiar to you?

$$.5 = \frac{1}{2}$$

$$.25 = \frac{1}{4}$$

$$.75 = \frac{3}{4}$$

$$.3333 . . . = \frac{1}{3}$$

$$.6666 . . . = \frac{2}{3}$$

$$.125 = \frac{1}{8}$$

$$.375 = \frac{3}{8}$$

$$.625 = \frac{5}{8}$$

$$.875 = \frac{7}{8}$$

However, the decimal $.128128128$ . . . is almost certainly not recognizable immediately as a common fraction. The following three examples illustrate a systematic way to convert a decimal to a fraction.

EXAMPLE 2.11

Convert 0.2165 to a fraction.

SOLUTION Let $x = 0.2165$. Multiply both sides of this equation by $10^4 = 10,000$ to get $10,000x = 2165$. (The multiplier 10^4 was chosen to eliminate the decimal number.) So

$$x = \frac{2,165}{10,000} = \frac{433}{2,000}$$

A process like this works on decimal fractions that don't repeat. The next two examples illustrate how to convert repeating decimals to fractions.

EXAMPLE 2.12

Convert .3333 . . . to a fraction.

SOLUTION Let $x = .3333$. . . . Multiply both sides of this equation by 10 to get $10x = 3.333$. . . .

$$10x = 3.3333 \ldots$$
$$x = .3333 \ldots$$

Subtract to get $9x = 3$. Then solve for x:

$$x = \frac{3}{9} = \frac{1}{3}$$

YOUR FORMULATION

To convert a repeating decimal to a fraction we first let x equal the fraction. Then we multiply both sides of the equation by a power of 10. In Example 2.12 this was 10^1, and in Example 2.13 it was 10^3. Describe, in general, what power of 10 to use as a multiplier when converting a repeating decimal to a fraction.

EXAMPLE 2.13

Convert 0.128128128 . . . to a fraction.

SOLUTION Let $x = 0.128128128$. . . . Multiply both sides of this equation by $10^3 = 1000$ to get $1000x = 128.128128128$. . . .

$$1000x = 128.128128128 \ldots$$
$$x = .128128128 \ldots$$

Subtract to get $999x = 128$. Then solve for x:

$$x = \frac{128}{999}$$

Divide this on your calculator to verify that $x = .128128128$. . . .

EXERCISES 2.3

1. List some uses of the fraction concept that children learn while quite young, like "half a sandwich" and "a quarter of a football game."

2. Write each of the following fractions in Egyptian hieroglyphics.

   a. $\frac{1}{2}$ b. $\frac{1}{10}$ c. $\frac{1}{23}$ d. $\frac{1}{40}$

3. What fractions do these Egyptian hieroglyphs represent?

   a.   b.   c.   d.

4. Complete the decomposition of $\frac{2}{99}$ into unit fractions as $\frac{1}{66} + x$.

5. Complete the decomposition of $\frac{2}{5}$ into unit fractions as $\frac{1}{3} + x$.

6. In the Rhind papyrus as shown on page 39, the table decomposing fractions of the form $2/n$ into unit fractions included only odd values of n. Why would it make sense not to include even values of n?

7. Use two different approaches to see which is larger, $\frac{42}{60}$ or $\frac{88}{126}$.

8. Use a calculator to place the fractions $\frac{42}{139}$, $\frac{167}{556}$, $\frac{99}{331}$ in order from smallest to largest.

9. In each of the following pairs, which is larger?

   a. $\dfrac{4}{15}$ or $\dfrac{5}{16}$

   b. $\dfrac{13}{31}$ or $\dfrac{14}{32}$

   c. $\dfrac{21}{132}$ or $\dfrac{22}{133}$

   d. Make a conjecture about comparing the fractions a/b and $(a + 1)/(b + 1)$ when $a < b$.

10. Convert the following decimals to fractions.

    a. 0.5555 . . .
    b. 0.1996
    c. 0.189318931893 . . .
    d. 0.9999 . . .

11. Complete the following steps to show that, for counting numbers a, b, c, and d, $\dfrac{a}{b} > \dfrac{c}{d}$ is equivalent to $ad > bc$.

    a. Find a common denominator for a/b and c/d.
    b. Write a/b and c/d as fractions with a common denominator.
    c. Make a concluding statement.

WRITTEN ASSIGNMENTS 2.3

1. Write about some of your first memories of dealing with fractions. Did these experiences happen in school or out of school? Were they pleasant or unpleasant, exciting or dull? Do you remember any particularly effective ways you were taught about fractions?

2. The ancient Egyptians and Greeks relied heavily on unit fractions to do calculations with fractions. Why do you think this was so?

3. Given the widespread availability of calculators, should instruction in arithmetic operations (addition, subtraction, multiplication, and division) with fractions still be given in elementary school, or should all calculation be done with decimals? Why?

4. Find out more about Simon Stevin and the development of decimals. Consult a book on the history of mathematics (e.g., Boyer and Merzbach, 1991; Eves, 1990; Gittleman, 1975; Katz, 1993), an encyclopedia, or some other reference. Write a one- or two-page paper describing what you have found.

5. A calculator makes it easy to convert fractions to decimals. Describe a difficulty created by working with decimals instead of fractions in doing arithmetic (adding, subtracting, multiplying, or dividing).

6. Some fractions produce terminating decimals ($\frac{1}{2} = .5$); some produce infinite repeating decimals ($\frac{4}{33} = .121212\ldots$).

   a. Make a conjecture about which fractions produce terminating decimals and which produce infinite repeating decimals.
   b. Write an argument that justifies to others in your class your conjecture in part (a).
   c. When a decimal repeats, the length of the repeating part can vary. When $\frac{1}{3}$ is written as the decimal $.3333\ldots$, the length of the repeating part is one digit. When $\frac{4}{33}$ is written as the decimal $.121212\ldots$, the length of the repeating part is two digits. Make a conjecture about the maximum length of the repeating part when the fraction a/b is written as a decimal.
   d. Write an argument that justifies to others in your class your conjecture in part (c).

2.4 NUMBER SYSTEMS

GOALS
1. To describe a need for and the origin of integers and to review arithmetic involving negative integers.
2. To identify rational numbers and their importance to the Pythagoreans and to review arithmetic involving rational numbers.
3. To know and justify that $\sqrt{2}$ is not a rational number.
4. To describe the origins of liberal arts (general education) programs.

So far in this chapter we have approached different kinds of numbers (namely, counting numbers, fractions, and decimals) in the order that a child encounters them

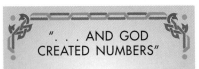
or a civilization deals with them as it develops. From a mathematical viewpoint, a slightly different order of examining number systems is: natural numbers, integers, rational numbers, real numbers, complex numbers.

Natural Numbers

Our starting point in examining number systems is with the counting numbers 1, 2, 3, 4, . . . , just as these are the first numbers for a child or for the earliest form of a civilization. These numbers are called *natural numbers* or counting numbers. Their development predates written history.

Integers

In the natural numbers, addition is always possible. That is, we can add any two natural numbers and get another natural number. However, subtraction is not always possible: $1 - 2$ is not a natural number and $3 - 3$ is not a natural number. In order to get a number system in which subtraction is always possible, the system of integers was created. The system of *integers* consists of the natural numbers, zero, and the negatives of each of the natural numbers as follows:

$$\underbrace{\ldots, -4, -3, -2, -1,}_{\text{Negative integers}} 0, \underbrace{1, 2, 3, 4, \ldots}_{\text{Positive integers}}$$

Note that the subtraction that could not be done in the natural numbers *can* be done in the integers: $1 - 2 = -1$ and $3 - 3 = 0$. Furthermore, if any two integers are added, the sum is again an integer. For example, $2 + 3 = 5$, $2 + (-3) = -1$, $-2 + 3 = 1$, and $-2 + (-3) = -5$. Also, if any two integers are subtracted, the difference is again an integer. For example, $4 - 7 = -3$, $4 - (-7) = 11$, $-4 - 7 = -11$, and $-4 - (-7) = 3$.

Historically, the use of integers was relatively slow to develop. What is a negative number? In particular, what is -2? In a conceptual sense, it is the opposite of 2. A bank balance of $-\$2$ is the opposite of a bank balance of $\$2$: instead of having $\$2$, one owes $\$2$. In a mathematical sense, -2 is something that when added to 2 gives 0. Before negative numbers could be considered, the number zero had to be understood.

It appears that 0 served as a place holder in a place-value system of numeration before it was used as a number to signify none of something (e.g., no apples). Zero was apparently first recognized as a number by Hindu and Arab mathematicians sometime in the sixth century. Major earlier civilizations such as the Egyptians, Babylonians, Greeks, and Romans apparently had no concept of zero as a number. The Chinese did not have a symbol for zero until about the thirteenth century.

Developing the concept of the number zero depends in large part on the need of a society for the concept. Today in the Mende culture, the largest ethnic group in Sierre Leone in West Africa, the word *Gbeogbe* describes "zero" as they need it. Alex Bockarie of the University of Sierra Leone describes the Mende concept of the number zero (p. 210):

Gbeogbe carries several different meanings, ranging from absolutely nothing to not enough to share, depending on the circumstances. To a group of Mende people, the following problem of subtraction of whole numbers was proposed—four leones minus four leones. The answer the respondents gave was *Gbeogbe*, meaning nothing.

Then a similar problem was proposed with larger numbers—200 leones less 199 leones and 50 cents. The common answer was *gbeogbe lo,* meaning nothing remains. Actually 50 cents remained but, relative to the initial money involved, it is almost nothing so it qualifies for *gbeogbe.* They all agreed that nothing remained; even though they admitted that 50 cents remained, it could hardly buy anything.

Gbeogbe is also used to indicate that there is just enough for one. Suppose a person has only a small fraction of kola nut left, and he is asked to share it with someone else. He simply says, "I have nothing left." The concept of *Gbeogbe* is therefore quite vague and very much different from the concept pupils meet in school (and call it zero).

The nature of a society and a society's needs shape the development of mathematical concepts in the same way that cultural values shape other societal beliefs and changes. The concept of zero as we use it mathematically was apparently first developed by the Hindus in the sixth century. Since that time other societies that have come in contact with the concept have come to use it to varying degrees.

Once zero was understood as a number, the door was opened for the development of negative numbers. In the evolution of Western civilization, it appears, this concept was created by the same Hindu and Arab civilizations that gave us the number zero. The seventh century Hindu mathematician Brahmagupta published a work in which he gives the usual rules for computing with negative numbers.

Some commentators (Kline, p. 67; Campbell, p. 96). have suggested that the idea of negative numbers arose in connection with banking in India because the people were so frequently in debt. The use of negative numbers was transmitted by the Arabs from Hindu civilizations to Europe. It caught on slowly. As late as the 1600s, the renowned French mathematician Rene Descartes referred to negative numbers as "false." Morris Kline (p. 69) comments that "it is more difficult to get a truth accepted than to discover it."

Today negative numbers are used in a variety of places. Checkbook balances might be positive or negative. Our calendar marks time from the birth of Christ, with dates B.C.E. (before the common era) being regarded as negative (3400 B.C.E. = −3400) and dates C.E. as positive. Both the Fahrenheit and Centigrade temperature scales have negative temperatures. Elevation above sea level is regarded as positive, while elevation below sea level is negative. The Dow Jones Industrial average is up (positive) or down (negative) some number of points. In football, the yardage on a play might be positive (a gain) or negative (a loss). In all these situations, we can avoid the formal use of negative numbers by substituting descriptive phrases such as "to the good" or "in the hole", "B.C.E." or "C.E.", or "above zero" or "below zero." However, these situations can be dealt with in a common way with negative numbers. The use of negatives rather than descriptive phrases also eases computation.

Although our goal is not to instruct you on operations with integers, the next example reviews some of these operations.

A VOID ZERO

The etymology of the word *zero* reflects its Hindu-Arabic origin. Howard Eves (p. 25) states, "Our word *zero* probably comes from the Latinized form *zephirum* of the Arabic *sifr,* which in turn is a translation of the Hindu *sunya,* meaning 'void' or 'empty.'" Compare with the English word *cipher,* which comes from *sifr.*

EXAMPLE 2.14

Carry out the following calculations.

(a) $2 + (-3)$ (b) $4 + (-2)$ (c) $-3 + (-2)$ (d) $3 - 8$
(e) $4 - (-7)$ (f) $-2 - 3$ (g) $-2 - (-5)$ (h) $-2 - (-1)$
(i) $2(-3)$ (j) $(-4) \cdot 3$ (k) $(-2)(-3)$

SOLUTION (Comments are made with reference to a number line.)

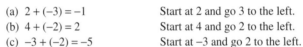

(a) $2 + (-3) = -1$	Start at 2 and go 3 to the left.
(b) $4 + (-2) = 2$	Start at 4 and go 2 to the left.
(c) $-3 + (-2) = -5$	Start at -3 and go 2 to the left.
(d) $3 - 6 = -3$	Start at 3 and go 6 to the left. Subtracting a positive number means going to the left a distance equal to that positive number.
(e) $4 - (-7) = 4 + 7 = 11$	Subtracting a negative number is equivalent to adding the corresponding positive number, so start at 4 and go 7 to the right.
(f) $-2 - 3 = -5$	Start at -2 and go 3 to the left.
(g) $-2 - (-5) = -2 + 5 = 3$	Subtracting -5 is equivalent to adding 5, so start at -2 and go 5 to the right.
(h) $-2 - (-1) = -2 + 1 = -1$	Subtracting -1 is equivalent to adding 1, so start at -2 and go 1 to the right.
(i) $2(-3) = -6$	A positive times a negative is negative.
(j) $(-4) \cdot 3 = -12$	A negative times a positive is negative.
(k) $(-2)(-3) = 6$	A negative times a negative is positive.

Rational Numbers

With the integers, addition, subtraction, and multiplication are always possible. That is, given any two integers, they can be added, subtracted, or multiplied and another integer results. However, division is not always possible. For example, 1 and 2 are integers, but $\frac{1}{2}$ is not an integer. In order to get a number system in which division is almost always possible (division by 0 is always a problem), the system of rational numbers was created.

The root of the word *rational* is "ratio." The number $\frac{1}{2}$ is a ratio of 1 to 2. In general, a *rational number* is a number of the form $\frac{m}{n}$, where m and n are integers and $n \neq 0$.

Why in defining a rational number do we demand that $n \neq 0$? Let's examine the concept of division. Sometimes division of integers produces an integer and sometimes it does not. If m and n are integers, then $\frac{m}{n} = k$ if and only if $m = nk$. For example, $\frac{6}{2} = 3$ since $6 = 2 \cdot 3$. Also, $\frac{1}{2}$ is not a integer since there is no integer k such that $1 = 2k$.

Let's now examine division by zero. What is, say, $\frac{1}{0}$? Call it k. Then $1 = 0 \cdot k$. But $0 \cdot k = 0$ for any number k. Since there is no number k such that $1 = 0 \cdot k$, we say that $\frac{1}{0}$ is undefined. More generally, let m be any nonzero integer and consider $\frac{m}{0}$. This quotient is k if and only if $m = 0 \cdot k$. There is no such number k, so $\frac{m}{0}$ is undefined. If m is zero, then we seek k such that $0 = 0 \cdot k$, and any number k works. This leaves the untenable situation that $\frac{0}{0}$ is anything. Consequently, $\frac{0}{0}$ is also left undefined. Therefore, for any integer m, $\frac{m}{0}$ is undefined.

Note that by the construction of the rational numbers, any integer can be divided by any nonzero integer and the result is a rational number. Furthermore, if we add, subtract, multiply, or divide any two rational numbers, then as long as we do not divide by 0 we get a rational number.

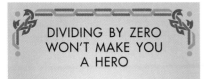

DIVIDING BY ZERO WON'T MAKE YOU A HERO

Nicholas J. Rose, compiler of *Mathematical Maxims and Minims,* warns (p. 162) against division by zero as follows:

Both minus and plus, I'm the same.
There's no one else can make that claim.
Add, subtract me without alarm,
Or multiply—it does not harm,
But—
Mind the rule, be not a hero:
"Thou shalt not divide by zero!"

EXAMPLE 2.15

Find the sum, difference, product, and quotient of $\frac{1}{2}$ and $-\frac{3}{4}$, in that order.

SOLUTION

$$\frac{1}{2} + \frac{-3}{4} = \frac{2}{4} + \frac{-3}{4} = \frac{-1}{4} \qquad \text{is the sum}$$

$$\frac{1}{2} - \frac{-3}{4} = \frac{1}{2} + \frac{3}{4} = \frac{2}{4} + \frac{3}{4} = \frac{5}{4} \qquad \text{is the difference}$$

$$\frac{1}{2} \cdot \frac{-3}{4} = \frac{-3}{8} \qquad \text{is the product}$$

$$\frac{1}{2} \div \frac{-3}{4} = \frac{1}{2} \cdot \frac{4}{-3} = \frac{4}{-6} = \frac{2}{-3} \qquad \text{is the quotient}$$

Real Numbers

We have tried to motivate the mathematical need to develop (a) the integers, in order to make subtraction always possible, and (b) the rational numbers, in order to make division by a nonzero number always possible. Note that subtraction undoes what addition does. For example, if 3 is added to a number n to give $n + 3$, we can get back where we started by subtracting 3 ($n + 3 - 3 = n$). Subtraction is the inverse operation of addition. Division is the inverse operation of multiplication. There is another pair of operations, an operation and its inverse, that we wish to consider.

Recall that natural numbers are used as exponents to indicate repeated multiplication. In particular, if n is a number, then

$$n^1 = n \qquad n^2 = n \cdot n \qquad n^3 = n \cdot n \cdot n \qquad n^4 = n \cdot n \cdot n \cdot n$$

and so forth. The operation of squaring takes a number to its square; for example, $3^2 = 9$. The operation of cubing takes a number to its cube; for example, $4^3 = 64$. Your calculator might have an $\boxed{x^2}$ key that enables you to square any number. The generic name for squaring, cubing, and so forth is *exponentiation*. Addition, subtraction, multiplication, and division are called *binary operations,* since they take a pair of numbers and assign a number. Exponentiation is a *unary operation,* since it takes a single number and assigns another number.

The operation that is the inverse of exponentiation is *extraction of roots.* In particular, taking the square root is the inverse of squaring, and taking the cube root is the inverse of cubing. If n is a number, then $\sqrt{n}$ means a *nonnegative* number such that $(\sqrt{n})^2 = n$. For example, $\sqrt{9} = 3$ since $3^2 = 9$. By convention, we don't write $\sqrt{9} = -3$ even though $(-3)^2 = 9$.

 YOUR FORMULATION

Why has the just-mentioned convention been adopted? In other words, why do we say that $\sqrt{n}$ means a *nonnegative* number such that $(\sqrt{n})^2 = n$? Why do we not say that $\sqrt{9}$ is either 3 or −3?

By $\sqrt[3]{n}$ we mean a number such that $(\sqrt[3]{n})^3 = n$. For example,

$$\sqrt[3]{\frac{8}{27}} = \frac{2}{3} \qquad \text{since} \qquad \left(\frac{2}{3}\right)^3 = \frac{8}{27}$$

and $\sqrt[3]{-8} = -2$ since $(-2)^3 = -8$. As you can see, a cube root can be negative. In the next example, some calculation of square roots is considered.

EXAMPLE 2.16

a. Calculate $\sqrt{16}$.
b. Calculate $\sqrt{17}$ to the nearest tenth with a calculator.
c. Suppose n is a number with $4.3 < \sqrt{n} < 4.4$. What can be said about n?

SOLUTION

a. $\sqrt{16} = 4$ since $4^2 = 16$.
b. $\sqrt{17} \approx 4.1$, using a calculator.
c. If $4.3 < \sqrt{n} < 4.4$, then $(4.3)^2 < n < (4.4)^2$, or $18.49 < n < 19.36$. If n were an integer, then n would be 19. If n is a rational number, then $18.49 < n < 19.36$ is the best that can be said.

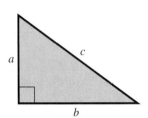

Let's consider the square roots of the first few natural numbers. First, $\sqrt{1} = 1$. Then, $\sqrt{2}$ is not a natural number. Without a calculator, we know it is between 1 and 2 since $1^2 = 1$ and $2^2 = 4$. With a calculator, we can find that $\sqrt{2} \approx 1.41421356$. Why might we be interested in $\sqrt{2}$?

Recall that the Pythagorean theorem says that in a right triangle with legs of lengths a and b and a hypotenuse of length c, $a^2 + b^2 = c^2$. Consider a right triangle with both legs of length 1 and hypotenuse of length c. Then, by the Pythagorean theorem, $1^2 + 1^2 = c^2$, so $1 + 1 = c^2$, or $2 = c^2$, and $c = \sqrt{2}$.

The Pythagorean theorem was known in India, China, and Egypt even before the time of Pythagoras, who lived in the sixth century B.C.E. Consequently, $\sqrt{2}$ has been considered, at least as a length, since before that time. Therefore, $\sqrt{2}$ has been thought about for many more years than has the number zero or negative integers.

The Pythagoreans

Before saying more about $\sqrt{2}$, let's consider the Pythagoreans, a group founded by Pythagoras in the sixth century B.C.E., whose structure has served since then as a model for secret societies. Their mystic religious beliefs blended with their study of philosophy, mathematics, and natural science. Some of the mystic beliefs that Pythagoras ascribed to numbers (5 represents wind) and some of his mathematical knowledge came from his travels to Egypt and China. Although the mysticism of the Pythagoreans is dismissed today, their study of mathematical properties of numbers and their work in geometry were forerunners of the substantial Greek mathematical developments and carry forward to the present.

The Pythagoreans' beliefs were based on the assumption that natural numbers embody the cause of the various properties of matter and humans. The four main subjects studied were arithmetic, music, geometry, and astronomy. This group of subjects became known in the Middle Ages as the *quadrivium*. To these were later added the *trivium* of grammar, logic, and rhetoric. These seven liberal arts were regarded as necessary background for the educated person. This idea of a core of knowledge for all educated people is at the heart of liberal arts or (general education) programs in colleges and universities today.

Aristotle, who lived about 200 years after Pythagoras, described the Pythagorean belief that "all is number" in *Metaphysics,* Book Alpha, 5.985 b:

> [T]he so-called Pythagoreans, the first to be absorbed in mathematics, not only advanced this particular science, but, having been brought up on it, they believed that its principles are the principles of all things. Now, of these principles, numbers are naturally the first. As a result, they seemed to see in numbers . . . many similarities to things as they are and as they come to be: for one sort of modification of numbers, so to speak, is justice; another, soul and mind; still another, opportunity; and so forth. Musical modes and relations, too, they saw in terms of numbers. And all other matters appeared to be ultimately of the nature of numbers; and numbers were for them the primary natures. In view of all this, they took the elements of numbers to be the elements of all things, and the whole heaven to be harmony and number.

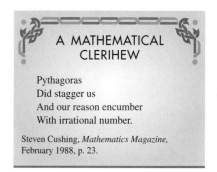

Pythagoras found that ratios of natural numbers were the basis for harmony in music. Shortening a string to $\frac{1}{2}$ of its original length produced a note an octave higher. The ratios $\frac{3}{2}$ and $\frac{4}{3}$ correspond to what are called in music the fifth and the fourth. The Pythagoreans believe that the harmony given by the ratio of natural numbers was important not only in music but in all of nature. (More on the relation between mathematics and music is discussed in Section 13.3.)

Let's now return to $\sqrt{2}$. It occurs in nature as the length of the hypotenuse of a right triangle with both legs of length 1 or the length of a diagonal of a square with side length 1. What ratio of natural numbers represents $\sqrt{2}$? One of the Pythagoreans discovered that $\sqrt{2}$ could not be written as a ratio of natural numbers. In modern terms, he proved that $\sqrt{2}$ is not a rational number or, equivalently, that $\sqrt{2}$ is an irrational number. Since Pythagorean beliefs had as their basis the importance of ratios of natural numbers, this was a heretical discovery. Legend has it that the discoverer was either executed or exiled. Cultural attitudes have changed significantly from those of the Pythagoreans. Now the fact that $\sqrt{2}$ is irrational is basic information in mathematics courses, and the argument that $\sqrt{2}$ is irrational is a model of indirect reasoning.

Other examples of irrational numbers are $\sqrt{3}$, $\sqrt{5}$, and $\sqrt{6}$; and many other irrational numbers exist. (Note that $\sqrt{4} = 2 = \frac{2}{1}$, which is a rational number.) They can, however, be written as decimals, although they are all nonterminating, nonrepeating decimals, unlike $\frac{1}{3} = .3333 \ldots$ and $\frac{4}{33} = .12121212 \ldots$, which are nonterminating but do repeat. For example, $\sqrt{2} = 1.41421356237. \ldots$ Not all irrational numbers are of the form $\sqrt{2}$, $\sqrt{3}$, $\sqrt{5}$, $\sqrt{6}$, etc. For example, the number π is an irrational number (recall that the circumference of a circle with radius r is $2\pi r$ and the area of a circle with radius r is πr^2). There is an infinite number of irrational numbers, and all of them can be written as nonterminating, nonrepeating decimals.

A *real number* is any number that can be written as a decimal. This includes all rational and irrational numbers. The decimal might be terminating, nonterminating and repeating, or nonterminating and nonrepeating. Every rational number is a real number. For example, $\frac{1}{2} = .5$, $-\frac{2}{3} = -0.6666 \ldots$, and $4 = 4.0$. However, there are real numbers that are not rational numbers, e.g., $\sqrt{2}$, $\sqrt{3}$, $\sqrt{5}$, and $\sqrt{6}$.

Consider the following real-number line:

If only integers are identified with points on the line, most of the number line is not covered. If rational numbers are identified with points on the line, the line is densely packed in the sense that between any two rational numbers is another rational number. For example, the number between $\frac{2}{3}$ and $\frac{3}{4}$ is

$$\frac{1}{2}\left(\frac{2}{3}+\frac{3}{4}\right)=\frac{1}{2}\left(\frac{8+9}{12}\right)=\frac{17}{24}.$$

This means that given a rational number q, there is no next rational number, since for any other rational number r, there is a rational number between q and r. This makes the rational numbers different than the integers because given an integer i, there is a next-largest integer $i + 1$ and a next-smallest integer $i - 1$. Although the rational number line is densely packed, there are still holes in it. For example, there are holes at $\sqrt{2}$, $\sqrt{3}$, and $\sqrt{5}$:

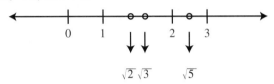

The irrational numbers fill the holes to include every point on the real number line.

In the real numbers we can locate $\sqrt{2}$. In fact, in the real numbers we can extract any root (square root, cube root, and so forth) of any nonnegative number. It wasn't until the seventh century or later that real numbers in the full sense that we know them were considered. In fact, negative numbers were scorned even by prominent mathematicians as late as the mid-1600s. Our description of real numbers as numbers that can be described as decimals would not make sense until Simon Stevin's popularization of decimals in the late 1500s. However, real numbers like $\sqrt{2}$ were understood as lengths by the Pythagoreans of 2500 years ago, and computations with real numbers involving $\sqrt{2}$ can be performed, such as

$$(3\sqrt{2})^2=9\cdot2=18\qquad\text{or}\qquad(\sqrt{2}+1)(\sqrt{2}+4)=2+5\sqrt{2}+4=5\sqrt{2}+6$$

even though the numbers are not written as decimals. Thus, despite the Pythagoreans' partial knowledge of real numbers, a comprehensive understanding of the system of real numbers and the use of the term *real number* are of more recent vintage.

Complex Numbers

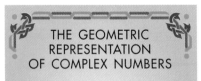

THE GEOMETRIC REPRESENTATION OF COMPLEX NUMBERS

The geometric interpretation of complex numbers was first published by Caspar Wassel (1745–1818), a Norwegian surveyor, in 1797. It was also made use of by Carl Friedrich Gauss (1777–1855), a German regarded as one of the greatest mathematicians of all time, and Jean Argand (1769–1822), a Frenchman whose name ("Argand diagram") often accompanies the geometric representation of a complex number.

In describing what can be done in the set of real numbers, we said we could take a root of any *nonnegative* real number. What about roots of negative numbers? Some are easy. For example, $\sqrt[3]{-8}=-2$, since $(-2)^3=-8$; and $\sqrt[5]{-1}=-1$, since $(-1)^5=-1$. However, consider $\sqrt{-1}$. What is it? Let $i=\sqrt{-1}$. Then $i^2=-1$. But there is no such real number i, since for any real number r, $r^2\geq0$. Consequently, $i=\sqrt{-1}$ is not a real number. A similar argument shows that $\sqrt{-2}$, $\sqrt{-3}$, and $\sqrt{-1.7}$ are not real numbers.

The number i is the prototype *imaginary number*. Any number of the form $r\sqrt{-1}=ri$, where r is a real number, is a pure imaginary number. The collection of numbers of the form $a + bi$, where a and b are real numbers, forms the set of *complex numbers*. They are operated on (added, subtracted, multiplied, divided, raised to powers, have roots taken) under the rules of algebra plus the fact that $i^2=-1$. (We omit details of the calculations.) In the complex numbers, the six operations of addition, subtraction, multiplication, division (but not by zero), exponentiation, and extraction of roots are always possible.

The term *imaginary number* reflects the belief of the first mathematicians to work with imaginary numbers that they aren't real. Gottfried Wilhelm Leibniz

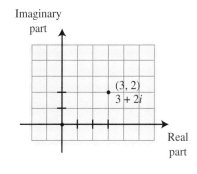

(1646–1716), one of the creators of calculus and a theologian, described imaginary numbers as a sort of amphibian, halfway between existence and nonexistence, resembling in this respect the Holy Spirit of Christianity (Boyer, p. 406). Although imaginary numbers don't represent lengths, they are recognized today as being as real as real numbers, rational numbers, integers, or natural numbers. Just as the integer -1 is needed to solve the equation $x + 1 = 0$, the rational number $\frac{1}{2}$ is needed to solve the equation $2x - 1 = 0$, and the irrational number $\sqrt{2}$ is needed to solve the equation $x^2 - 2 = 0$, the imaginary number $\sqrt{-1}$ is needed to solve the equation $x^2 + 1 = 0$. Today, complex numbers occur in some mathematical calculations, in physics, engineering, and other applications. Some calculators, especially graphing calculators, can do calculations with complex numbers.

Some calculations with imaginary numbers began in the late 1500s. As negative numbers became more accepted, so did roots of negative numbers. Just before 1800, complex numbers were described geometrically by identifying the complex number $a + bi$ with the point (a, b) on the plane. For example, $3 + 2i$ is identified with the point $(3, 2)$.

With this interpretation, complex numbers are operated on via geometric considerations. This geometric representation made people more accepting of complex numbers. People came to understand that imaginary numbers are no more "imaginary" than are negative numbers. Both are abstractions from the natural numbers, although at different levels.

The relationships among number systems is presented schematically in ▶ Figure 2.1.

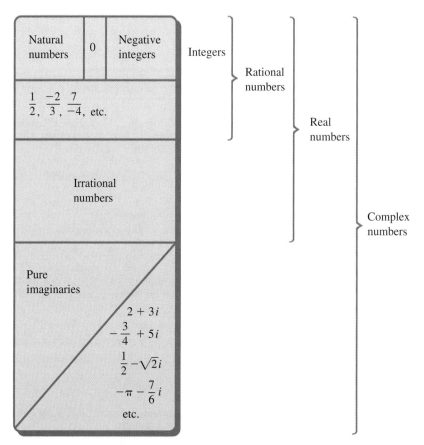

▶ FIGURE 2.1 Relationships among number systems

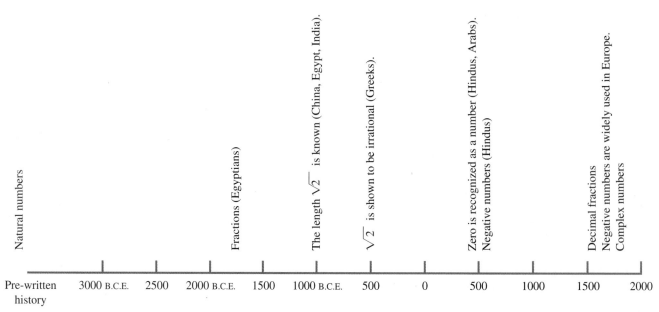

▶ FIGURE 2.2 The evolution of number systems

A timeline showing the historical evolution of number systems is given in ▶ Figure 2.2.

EXERCISES 2.4

1. Give an example of a natural number.
2. Give an example of a number that is not a natural number.
3. Give an example of an integer.
4. Give an example of a number that is not an integer.
5. Do the following computations involving zero.
   a. $1 + 0$ b. $0 + 2$ c. $3 - 0$ d. $0 - 4$
   e. $5 \cdot 0$ f. $0 \cdot 6$ g. $\dfrac{0}{7}$ h. $\dfrac{8}{0}$

6. Negative numbers are used in several situations, such as checkbook balances, calendars, and temperature scales. Name two other places, in addition to those mentioned in this section.
7. Compute the following.
   a. $3 + (-2)$ b. $3 + (-3)$ c. $3 + (-4)$ d. $-2 + 5$
   e. $-2 + 2$ f. $-2 + 1$ g. $-4 + (-3)$ h. $5 - 2$
   i. $5 - 5$ j. $5 - 7$ k. $5 - 0$ l. $5 - (-3)$
   m. $-6 - 2$ n. $-6 - 6$ o. $-6 - 8$ p. $-6 - 0$
   q. $-6 - (-2)$ r. $-6 - (-6)$ s. $-6 - (-7)$ t. $7(-2)$
   u. $(-3)8$ v. $(-4)(-5)$ w. $(-9) \cdot 0$
8. Give an example of a rational number.
9. Give an example of a number that is not a rational number.
10. Give an example, if possible, of a number that satisfies each of the following conditions.

	Natural Number	Integer	Rational Number
a.	yes	yes	yes
b.	yes	yes	no
c.	yes	no	yes
d.	yes	no	no
e.	no	yes	yes
f.	no	yes	no
g.	no	no	yes
h.	no	no	no

In Exercises 11–15, find the (a) sum, (b) difference, (c) product, and (d) quotient, when defined, of the two given rational numbers, in the given order.

11. $\dfrac{2}{3}$ and $\dfrac{3}{4}$ 12. $\dfrac{2}{3}$ and 0

13. $-\dfrac{3}{5}$ and $\dfrac{1}{4}$ 14. 0 and $\dfrac{1}{3}$

15. $-\dfrac{2}{3}$ and 4

16. Perform the following exponentiations.
   a. 2^5 b. $(-4)^2$ c. $(-4)^3$ d. $\left(\dfrac{2}{3}\right)^4$

17. Addition, subtraction, multiplication, and division are examples of binary operations. Give another example of a binary operation.

18. Squaring, cubing, etc. are examples of unary operations. Give another example of a unary operation other than exponentiation.

19. $\sqrt{9} = 3$ and $\sqrt{16} = 4$. What is the next value of n, with $n > 16$, such that $\sqrt{n}$ is an integer?

20. Calculate each of the following to the nearest tenth on a calculator.
   a. $\sqrt{18}$ b. $\sqrt{19}$ c. $\sqrt{20}$ d. $\sqrt{.1}$ e. $\sqrt{.01}$

21. If n is a number with $9.1 < \sqrt{n} < 9.2$, what can be said about n?

22. Give an example of a real number.

23. Give an example of a number that is not a real number.

24. Follow the instructions of Exercise 10, with a fourth column, headed "Real Number." For each line (a through h), create two new lines, one with a yes in column 4 and one with a no in column 4. The first four of 16 parts are given here:

	Natural Number	Integer	Rational Number	Real Number
a.	yes	yes	yes	yes
	yes	yes	yes	no
b.	yes	yes	no	yes
	yes	yes	no	no

25. Calculate the following:
   a. $\sqrt[3]{-27}$
   b. $\sqrt[3]{-64}$
   c. What is the next value of n (going in the negative direction), with $n < -64$, for which $\sqrt[3]{n}$ is an integer?

26. Follow the instruction of Exercise 24, with a fifth column, headed "Complex Number," and with two lines, one for yes and one for no, for each row in Exercise 24. The first four of 32 parts are given here:

Natural Number	Integer	Rational Number	Real Number	Complex Number
yes	yes	yes	yes	yes
yes	yes	yes	yes	no
yes	yes	yes	no	yes
yes	yes	yes	no	no

WRITTEN ASSIGNMENTS 2.4

1. To Leopold Kronecker is attributed the quote "God created the [natural] numbers; all else is the work of men."
   a. Do you agree or disagree with Kronecker's statement? Support your position.
   b. Read more about Kronecker (look in books on the history of mathematics, encyclopedias, or books on set theory). Do you believe his statement reflects a deeply religious view (God set everything in motion by creating the basic building blocks and humans just play with the blocks) or a secular humanist position (God only did the first step; human reason has gone far beyond it)? Support your conclusion.

2. Zero was apparently not recognized as a number until the sixth century. How could this be? Surely in prior civilizations people had no sheep or no money or no children. Don't these qualify as involving zero? Be more specific about what is meant by saying that zero was not recognized as a number until the sixth century. One source is Campbell, pp. 99–100.

3. A Hindu writer named Bhiskara supplies much of our information about Hindu arithmetic. He wrote in the twelfth century (about 600 years after Brahmagupta). One of his works, the *Lilavati*, has an intriguing story told about its title, which is the name of his unlucky daughter whom he hoped to console by immortalizing her in the name of his book.
   a. Look up and describe more about Lilavati. Possible sources are Eves (1990), Gittleman (1975) and Smith (1923).
   b. Comment on the ability of a mathematics book to immortalize someone. If Bhiskara's book had been named *Kathy* or *Lynda* or *Mary*, after a daughter of one of the authors, would she have been immortalized? Could a work of art or literature, a book on a scientific discovery, a book on a discovery in the social sciences, a book on business practices, a book on education, a book on government, or the like immortalize its author or its namesake? Discuss what creative works are lasting and what are temporal.

4. Consider Morris Kline's comment that "it is more difficult to get a truth accepted than to discover it."
   a. Describe at least one of your own experiences in which this was true.
   b. Name historical discoveries besides negative numbers for which it was difficult to get the discovery accepted.

5. This exercise is in the form of a debate.
   a. Argue that the liberal arts (general education) part of a college degree should be eliminated.
   b. Refute the arguments given in (a).

6. If your school is to have a liberal arts (general education) program meant to give all students the background an educated

person should have, should mathematics be required as part of it? If so, what kind of mathematics. If not, why not?

7. Write a one- to three-page report on the mystic beliefs of the Pythagoreans.

8. Without looking in a dictionary or thesaurus, write what the word *cipher* means. Now look it up in a dictionary. Is your own definition the first definition given there? Comment on what you found.

9. Share your thoughts on the following statement: "Imaginary numbers are no more 'imaginary' than are negative numbers."

2.5 CHECK DIGITS

GOALS
1. To find check digits for numbers by three different schemes.
2. To analyze numbers with check digits.
3. To show why an error made by interchanging a 0 and a 9 is not detected under the U.S. Postal Service scheme, and why these are the only single-digit errors not detected in this scheme.

Sometimes when a numeral is written, an error is made in recording it. For example, the telephone number 773–9024 might be mistakenly noted as 772-9024 or the year 1997 might be mistakenly written as 1979. There are many instances where a number serves for identification purposes: Social Security number, telephone number, credit card number, and so on. When an error in recording a number is made, problems usually occur: You can't reach somebody on the telephone, your income tax form is returned because your Social Security number is wrong, someone else's charge gets put on your credit card, etc.

To help detect an error made in recording a number, an extra digit, called a *check digit,* is often added to the number. For example, a major money market fund shows an account number and a check digit in the form shown in ▶ Figure 2.3. Let's now examine several different schemes for assigning check digits.

Fund No.	Account No.	Chk Dgt
95	2306814925	9

▶ FIGURE 2.3 Money market account number including the check digit

U.S. Postal Service Scheme

The United States Postal Service (U.S.P.S.) includes a check digit on each of their money orders. Their scheme involves a 10-digit number followed by a check digit. ▶ Figure 2.4 shows such a money order with the identification scheme labeled. The check digit is the remainder when the 10-digit money order number is divided by 9. For example, when the 10-digit number 0000000012 is divided by 9, the quotient

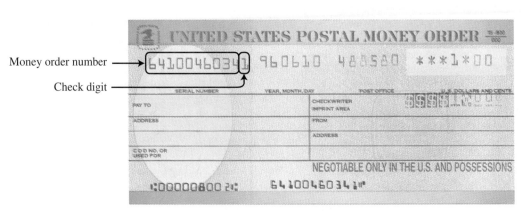

Money order number ⟶

Check digit ⟶

▶ FIGURE 2.4 Postal money order showing the money order number and its check digit (how many times 9 goes into the number) is 1 and the remainder is 3. Consequently, the check digit is 3. When the 10-digit number 0000516772 is divided by 9, the quotient is 57419 and the remainder is 1. Consequently, the check digit is 1. On your calculator, dividing 516772 by 9 will give 57419.111. (Or your calculator may display more than three digits after the decimal point.) The quotient here is 57419 and the remainder, r, is such that

$$\frac{r}{9} \approx .111$$
$$r \approx .999 \qquad \text{or} \qquad r = 1$$

With a calculator, be careful when the 10-digit number has a 9 as the left-most digit (e.g., 9296478533). Many calculators display only 10 or fewer digits, so for $9296478533 \div 9$, such a calculator displays 1032942059, which by the U.S.P.S. scheme may suggest a remainder or check digit of 0. However, if the calculator displays only 10 digits, it has not shown what comes after the decimal point.

 A more exact way to find the remainder on a calculator comes from remembering the meaning of division. To illustrate, consider the problem $17 \div 3$. On a calculator, this yields 5.6666666. If we take the nondecimal part of this answer and multiply by 3, we get $5 \times 3 = 15$. Then we subtract this value from 17 to get the remainder: $17 - 15 = 2$. Another way to write this is

$$17 = (5 \times 3) + 2$$

Let's now apply this technique to a 10-digit number where we want the remainder when it is divided by 9. Suppose the number is 0000516772.

1. Divide by 9 on a calculator:

$$516772 \div 9 = 57419.111$$

2. Multiply the nondecimal part of the result by 9:

$$57419 \times 9 = 516771$$

3. Subtract this result from the original number:

$$516772 - 516771 = 1$$

This gives 1 as the remainder.

To summarize the process:

$$516772 = (57419 \times 9) + 1$$

YOUR FORMULATION

Apply the technique just given to find the remainder when 9296478533 is divided by 9 (the previous example).

CALCULATOR NOTE

Some more sophisticated calculators will have a function key that permits direct calculation of the U.S. Postal Service check digit. It may be called a *mod* key, for "modulus" or "modulo." For example, on the TI-85 calculator, mod (12, 9) when entered gives 3, and mod (516772, 9) when entered gives 1.

To see how the check digit principle is applied, suppose the 10-digit number 0000516772 is an identification number. To this the check digit 1 is appended to give 00005167721. Now suppose the 11-digit number is mistakenly written as 00005187721. When the 10-digit portion of this number is divided by 9, the remainder is 3, not 1. Consequently, the check digit indicates to the postal worker that a mistake has been made.

Are all single-digit errors—i.e., errors where one of the 10 digits is replaced by a different digit—detected by the U.S. Postal Service scheme? Unfortunately, they are not. Consider the number 00005167721 (the check digit 1 has been appended) mistakenly recorded as 00095167721. When the 10-digit part of 00095167721 is divided by 9, the remainder is again 1, so the error is not detected. In general, interchanges of 0 and 9—i.e., a 0 mistakenly written as 9 or a 9 mistakenly written as 0—are not detected by this scheme. These are the only single-digit errors that are not detected by this scheme.

When the U.S. Postal Service scheme detects a single-digit error, it only tells that an error has been made. It does not tell how to correct the error. For example, for the mistakenly written 11-digit ID number 00005187721, the correct number might have been 00005167721 or 00005187701.

What about errors involving more than one digit? See Exercises 27–30 for a discussion of the U.S. Postal Service scheme's ability to deal with one error of this type.

Universal Product Code (UPC)

SAVE RITE

0 11213 92233 1

▶ FIGURE 2.5 UPC bar code

Many products that are packaged prior to being sold have on the package a Universal Product Code (UPC) identification number, which includes a bar code to be read by an optical scanner (see ▶ Figure 2.5).

The UPC number has 12 digits, with the first six digits identifying the manufacturer, the next five identifying the product, and the twelfth being the check digit. Sometimes the check digit is not printed, but it is always included in the bar code. For the bar code in Figure 2.5, the 12-digit UPC number is 011213922331. If the 12-digit number is written as $a_1 a_2 a_3 \ldots a_{12}$, where each of $a_1, a_2, a_3, \ldots, a_{12}$ is a digit, then the check digit a_{12} is chosen so that the remainder is zero when

$$3a_1 + a_2 + 3a_3 + a_4 + 3a_5 + a_6 + 3a_7 + a_8 + 3a_9 + a_{10} + 3a_{11} + a_{12}$$

is divided by 10. For the UPC number in Figure 2.5, we form

$$(3 \cdot 0) + 1 + (3 \cdot 1) + 2 + (3 \cdot 1) + 3 + (3 \cdot 9) + 2 + (3 \cdot 2) + 3 + (3 \cdot 3) + 1$$
$$= 0 + 1 + 3 + 2 + 3 + 3 + 27 + 2 + 6 + 3 + 9 + 1$$
$$= 60$$

Now, $\qquad 60 \div 10 = 6, \quad$ remainder 0

Here's how an error would be detected if the preceding UPC number were mistakenly written with a 7 in place of the 9 to give 011213722331:

$$(3 \cdot 0) + 1 + (3 \cdot 1) + 2 + (3 \cdot 1) + 3 + (3 \cdot 7) + 2 + (3 + 2) + 3 + (3 \cdot 3) + 1$$
$$= 0 + 1 + 3 + 2 + 3 + 3 + 21 + 2 + 6 + 3 + 9 + 1$$
$$= 54$$

And $\qquad 54 \div 10 = 5.4$

The remainder is 4, not 0. Consequently, a mistake has been made. With this scheme, any error involving just one digit will be detected.

EXAMPLE 2.17

The UPC number on a box of envelopes is 04310075050 (note this is only 11 digits) with the check digit in the bar code only. What is the check digit?

SOLUTION Let a_{12} denote the check digit. Then

$$(3 \cdot 0) + (1 \cdot 4) + (3 \cdot 3) + (1 \cdot 1) + (3 \cdot 0) + (1 \cdot 0) + (3 \cdot 7) +$$
$$(1 \cdot 5) + (3 \cdot 0) + (1 \cdot 5) + (3 \cdot 0) + (1 \cdot a_{12})$$
$$= 4 + 9 + 1 + 21 + 5 + 5 + a_{12}$$
$$= 45 + a_{12}$$

If $45 + a_{12}$ is to be a number that has a remainder of 0 when divided by 10, then $45 + a_{12}$ must total to 50: So

$$45 + a_{12} = 50$$
$$a_{12} = 5$$

Thus, the check digit is 5.

As with the U.S. Postal Service scheme, the UPC scheme doesn't tell how to correct errors. Exercise 37 asks for evidence of this.

International Standard Book Number (ISBN)

When a book is published, a number is assigned to it for identification purposes. This number is called an International Standard Book Number (ISBN). The ISBN has 10 digits, as in ▶ Figure 2.6. The first digit tells the language in which the book is written, the second group of digits identifies the publisher, the third group of digits codes title information, and the last digit is the check digit.

The check digit is found by a scheme similar to the UPC scheme. If the 10-digit number is written as $a_1 a_2 a_3 \ldots a_{10}$, where each of $a_1, a_2, a_3, \ldots, a_{10}$ is a digit, then the check digit a_{10} is chosen so that the remainder is zero when

$$10a_1 + 9a_2 + 8a_3 + 7a_4 + 6a_5 + 5a_6 + 4a_7 + 3a_8 + 2a_9 + a_{10}$$

is divided by 11. For the ISBN in Figure 2.6, we form

$$10 \cdot 0 + 9 \cdot 3 + 8 \cdot 1 + 7 \cdot 4 + 6 \cdot 0 + 5 \cdot 2 + 4 \cdot 8 + 3 \cdot 4 + 2 \cdot 1 + 2$$
$$= 0 + 27 + 8 + 28 + 0 + 10 + 32 + 12 + 2 + 2$$
$$= 121$$

Now, $121 \div 11 = 11$, remainder 0

Here's how an error would be detected if this ISBN were mistakenly written with an 8 in place of the 3 to give 0-814-02841-2:

$$10 \cdot 0 + 9 \cdot 8 + 8 \cdot 1 + 7 \cdot 4 + 6 \cdot 0 + 5 \cdot 2 + 4 \cdot 8 + 3 \cdot 4 + 2 \cdot 1 + 2$$
$$= 0 + 72 + 8 + 28 + 0 + 10 + 32 + 12 + 2 + 2$$
$$= 166$$

And $166 \div 11 \approx 15.09$

ISBN 0-314-02841-2

▶ FIGURE 2.6 International Standard Book Number (ISBN)

The remainder here is 1, not 0. Consequently, a mistake has been made. With this scheme, any error involving just one digit will be detected.

A special situation arises with the ISBN scheme if the remainder turns out to be 10. Since 10 is not a single digit, the symbol X is used when the remainder is 10. For example, the book *Lure of the Integers* has ISBN 0-88385-502-X.

YOUR FORMULATION

Why do you think the symbol X is used in an ISBN number, rather than some other symbol, when the remainder is 10?

EXAMPLE 2.18

A book has an ISBN of 1-88s550-04-5, where the s represents a smudge mark. What number should go where the smudge mark is?

SOLUTION

$$10 \cdot 1 + 9 \cdot 8 + 8 \cdot 8 + 7 \cdot s + 6 \cdot 5 + 5 \cdot 5 + 4 \cdot 0 + 3 \cdot 0 + 2 \cdot 4 + 5$$
$$= 10 + 72 + 64 + 7 \cdot s + 30 + 25 + 0 + 0 + 8 + 5$$
$$= 214 + 7 \cdot s$$

We want $214 + 7 \cdot s$ to have a remainder of 0 when divided by 11. In other words, we want $214 + 7 \cdot s$ to be an exact multiple of 11.

$$214 \div 11 = 19.454545 \ldots$$

The first multiple of 11 bigger than 214 is $20 \cdot 11 = 220$. In order to have

$$214 + 7 \cdot s = 220$$
$$7 \cdot s = 6$$
$$s = \frac{6}{7}$$

Here s is not a whole number. So we try the next multiple of 11: $21 \cdot 11 = 231$. In order to have

$$214 + 7 \cdot s = 231$$
$$7 \cdot s = 17$$
$$s = \frac{17}{7}$$

Again, s is not a whole number. But if

$$214 + 7 \cdot s = 242$$

the next multiple of 11, then

$$7 \cdot s = 28$$
$$s = 4$$

Thus, 4 should go where the smudge mark is.

As with the U.S. Postal Service and the UPC schemes, the ISBN scheme doesn't tell how to correct errors. Exercise 38 asks for evidence of this.

Comparison of Schemes

Three different methods of assigning a check digit to a number were described in this section. Each has advantages and disadvantages. All can check whether or not a number has been recorded correctly. None of them tells how to correct any such error.

The U.S. Postal Service scheme detects all single-digit errors except a 0 substituted for a 9 or vice versa. Both the UPC and ISBN schemes improve the U.S.P.S. scheme by detecting all single-digit errors.

The most common type of error people make in recording a number is a single-digit error, such as writing 32 instead of 30. The second most common type of error is interchanging two digits. For example, the year 1997 might be written as 1979 or the zip code 90210 might be written as 20910. Such an error is called a *transposition error*. The ISBN scheme is superior to the UPC scheme in that it detects all transposition errors, whereas the UPC scheme fails to detect some transposition errors. Transposition errors are explored in Exercises 27–36.

Summary

Numerals are symbols that represent numbers. Some systems of numeration are more efficient than others in representing a number and in calculations. Egyptian hieroglyphics, Roman numerals, and Igbo number words are not as efficient as the Hindu-Arabic system. However, when accuracy is at stake, we give up some efficiency by appending a check digit to the way we represent numbers. The U.S. Postal Service scheme, the UPC scheme, and the ISBN scheme are all ways of appending a check digit to help ensure that a number is stated accurately. ● Table 2.6 provides a handy reference for the numeration systems discussed in this chapter.

TABLE 2.6
Numeration Chart

Hindu-Arabic Base Ten	Egyptian (Hieroglyphic)	Roman	Igbo	Mayan	Base Five	Base Twelve	Base Two	Base Sixteen	U.S. Postal Service	UPC	ISBN
1	(glyph)	I	ofu	(glyph)	1	1	1	1	00000000011	000000000017	0000000019
2	(glyph)	II	ibua	(glyph)	2	2	10	2	00000000022	000000000024	0000000027
3	(glyph)	III	ito	(glyph)	3	3	11	3	00000000033	000000000031	0000000035
4	(glyph)	IIII	ino	(glyph)	4	4	100	4	00000000044	000000000048	0000000043
5	(glyph)	V	ise	(glyph)	10	5	101	5	00000000055	000000000055	0000000051
6	(glyph)	VI	isi	(glyph)	11	6	110	6	00000000066	000000000062	000000006X
7	(glyph)	VII	isa	(glyph)	12	7	111	7	00000000077	000000000079	0000000078
8	(glyph)	VIII	isato	(glyph)	13	8	1000	8	00000000088	000000000086	0000000086
9	(glyph)	IX	itenani	(glyph)	14	9	1001	9	00000000090	000000000093	0000000094
10	(glyph)	X	ili	(glyph)	20	T	1010	A	00000000101	000000000109	0000000108
20	(glyph)	XX	ogu	(glyph)	40	18	10100	14	00000000202	000000000208	0000000205
30	(glyph)	XXX	ogunaili	(glyph)	110	26	11110	1E	00000000303	000000000307	0000000302
40	(glyph)	XL	ogunabo	(glyph)	130	34	101000	28	00000000404	000000000406	000000040X
50	(glyph)	L	ogonabonaili	(glyph)	200	42	110010	32	00000000505	000000000505	0000000507
60	(glyph)	LX	oguito	(glyph)	220	50	111100	3C	00000000606	000000000604	0000000604
70	(glyph)	LXX	oguitonali	(glyph)	240	5T	1000110	46	00000000707	000000000703	0000000701
80	(glyph)	LXXX	oguino	(glyph)	310	68	1010000	50	00000000808	000000000802	0000000809
90	(glyph)	XC	oguinonali	(glyph)	330	76	1011010	5A	00000000900	000000000901	0000000906
100	(glyph)	C	nari	(glyph)	400	84	1100100	64	00000001001	000000001007	0000001007
200	(glyph)	CC	narinabo	(glyph)	1300	148	11001000	C8	00000002002	000000002004	0000002003
300	(glyph)	CCC	nari ito	(glyph)	2200	210	100101100	12C	00000003003	000000003001	000000300X
400	(glyph)	CD	nari ino	(glyph)	3100	294	110010000	190	00000004004	000000004008	0000004006
500	(glyph)	D	nari ise	(glyph)	4000	358	111110100	1F4	00000005005	000000005005	0000005002
600	(glyph)	DC	nari isi	(glyph)	4400	420	1001011000	258	00000006006	000000006002	0000006009
700	(glyph)	DCC	nari isa	(glyph)	10300	4T4	1010111100	2BC	00000007007	000000007009	0C00007005
800	(glyph)	DCCC	nari isato	(glyph)	11200	568	1100100000	320	00000008008	000000008006	0000008001
900	(glyph)	CM	nari itenani	(glyph)	12100	630	1110000100	384	00000009000	000000009003	0000009008
1000	(glyph)	M		(glyph)	1300	6E4	1111101000	3E8	00000010001	000000010009	0000010006

1. At the beginning of this section we listed three situations where a number serves for identification purposes (Social Security number, phone number, credit card number). List three other such situations.

In Exercises 2–5, find the check digit by the U.S. Postal Service scheme.

2. 0004561854
3. 0008219466
4. 6178924351
5. 1023876624

In Exercises 6 and 7, determine whether or not the given number (the check digit is appended) has been recorded correctly according to the U.S. Postal Service scheme.

6. 74177611112
7. 27182818282

In Exercises 8 and 9, assume the 11-digit numbers (10 digits plus a check digit following the U.S. Postal Service scheme) are recorded with a smudge mark where the s appears. Find what number should go where the smudge is. Assume that no mistakes except for the smudge mark have occurred.

8. 000002049s4
9. 000019293s5
10. Give an example of a number whose check digit under the U.S. Postal Service scheme is 5.
11. An alternate method of finding the check digit under the U.S. Postal Service scheme is based on the fact that the remainder when a number is divided by 9 is the same as the remainder when the sum of the digits of the number is divided by 9. For example, with the number 0000516772, the sum of its digits is 28. And $28 \div 9 = 3$, remainder 1, so the check digit is 1. This alternate method is useful for hand calculation of the check digit or when the 10-digit number starts with 9 so the quotient requires 10 digits of calculator display. Use this method to find the check digits in Exercises 2–5 and for the number 9901234567.
12. Use the fact stated in Exercise 11 to show why an error made by replacing a 0 with a 9 or replacing a 9 with a 0 is not detected under the U.S. Postal Service scheme and why these are the only errors not detected.

In Exercises 13 and 14 assume either that the UPC number is written correctly or that exactly one single-digit error has been made. Determine which is the case.

13. 044600741607
14. 048011899008

In Exercises 15 and 16, find the check digit given the first 11 UPC digits.

15. 04310075100
16. 02420013101

In Exercises 17 and 18, the digit in a UPC number is smudged in the place where the s is located. Find s.

17. 0412s0958994
18. 07s100072441
19. Under the U.S. Postal Service scheme, the check digit can be any of 0, 1, 2, 3, 4, 5, 6, 7, or 8. What can the check digit be under the UPC scheme?

In Exercises 20 and 21, assume either that the ISBN is written correctly or that exactly one single-digit error has been made. Determine which is the case.

20. 0-12-329870-8
21. 0-7167-5047-3

In Exercises 22 and 23, find the check digit given the first nine ISBN digits.

22. 0-8218-0162
23. 0-87150-160

In Exercises 24 and 25, the digit in an ISBN is smudged in the place where the s is located. Find s.

24. 0-935610-1s-8
25. 0-673-s8039-4
26. Under the U.S. Postal Service scheme, the check digit can be any of 0, 1, 2, 3, 4, 5, 6, 7, or 8. What can the check digit be under the ISBN scheme?
27. When 1997 is written in the U.S. Postal Service scheme with a check digit, it is 00000019978. Suppose a transposition error causes it to be recorded as 00000019798. Does the check digit detect the error?
28. When 90210 is written in the U.S. Postal Service scheme with a check digit, it is 00000902103. Suppose a transposition error causes it to be recorded as 00000209103. Does the check digit detect the error?
29. (*Multiple choice*) Let's examine in general what happens with a transposition error in a two-digit number (such as 96 transposed to 69 or 41 transposed to 14). In general, suppose a number ab is transposed to ba. To get the check digit for ab by the U.S. Postal Service scheme, use the method of Exercise 11. The check digit is the remainder when $a + b$ is divided by 9. The check digit for ba is the remainder when $b + a$ is divided by 9. However, $b + a = a + b$, so the check digit for

ab is the same as the check digit for *ba*. Therefore, a transposition error in a two-digit number is detected by the U.S. Postal Service scheme.

a. Always

b. Sometimes but not always

c. Never

30. (*Multiple choice*) A transposition error of any two digits in a 10-digit number is detected by the U.S. Postal Service scheme.

a. Always

b. Sometimes but not always

c. Never

31. Give an argument to support your answer in Exercise 30.

32. Example 2.17 involves the UPC number 043100750505. Suppose the 4 and 3 are transposed to give 034100750505. Does the check digit detect the error?

33. Suppose the 4 and 1 are transposed in the UPC number 043100750505 of Example 2.17 to give 013400750505. Does the check digit detect the error?

34. (*Multiple choice*) Use the results of Exercises 32 and 33 to answer the following: A transposition error of any two digits in a 12-digit number is detected by the UPC scheme.

a. Always

b. Sometimes but not always

c. Never

35. Describe a general way that two digits in a UPC number can be transposed without having the UPC scheme detect the error.

36. Does the ISBN scheme detect the transposition of two digits?

37. With the UPC scheme, the number 011110070546 is recognized as incorrect. One possible correct number is 011110090546. Find another possible correct number. This shows that the UPC scheme does not tell how to correct an error.

38. With the ISBN scheme, the number 0-675-03784-8 is recognized as incorrect. One possible correct number is 0-675-08784-8. Find another possible correct number by changing just one digit. (This shows that the ISBN scheme does not tell how to correct an error.)

WRITTEN ASSIGNMENTS 2.5

1. A *byte* is a numeral with 8 binary digits (8 bits). Sometimes a byte has a parity check digit attached as a ninth bit. Describe what a parity check digit is. Give some examples to show how it is calculated.

2. Gallian (1991; 1992) discusses check digits on some states' driver's licenses. Does your state add a check digit to a driver's license number? If so, describe it, if possible (Wisconsin does not reveal its scheme). If not, do some research to explain why your state chooses not to do so.

3. At your school, how are student identification numbers assigned? (Social Security numbers and specially assigned numbers are two ways.) Does your identification number include a check digit? If so, how is it calculated and what are its advantages? If not, write a persuasive argument to convince your administration to adopt a check digit scheme.

4. Investigate check digit schemes that allow for the correction of errors.

CHAPTER 2 REVIEW EXERCISES

1. Describe the origins of counting. How old is it and how did it probably first develop?

2. Four basic symbols of Egyptian hieroglyphics were:

$$ 1 \;\; | \quad\quad 10 \;\; \cap \quad\quad 100 \;\; 9 \quad\quad 1000 \;\; \text{} $$

a. Write 231 in Egyptian hieroglyphics.

b. What number is represented by the following Egyptian numeral?

3. a. Write 24 in Roman numerals.

b. What number does the Roman numeral MCMLXXVI represent?

4. a. State two advantages of the Hindu-Arabic system over Roman numerals.

b. State an advantage of Roman numerals over the Hindu-Arabic system.

5. Complete the following sentence: When Africans speak of their traditional numeration systems, they may mean either a set of number words or _____.

6. Describe the origin of Hindu-Arabic numerals. Include some places and approximate dates.

7. a. Describe the distinctions between additive and place-value systems of numeration.

b. Name an additive system of numeration that has been used by a civilization.

c. Name a place-value system of numeration that has been used by a civilization.

8. a. Why is a symbol for zero important in a place-value system of numeration?

 b. Describe more about the use of zero.

9. Write 1985 in expanded form.

10. a. What is the base of our system of numeration?

 b. Name another system of numeration with this same base.

11. What is the base of the system that uses the following numerals in counting? 1, 2, 3, 10, 11, 12, 13, 20, . . .

12. State two examples of grouping—other than base-10 groupings—in use today.

13. a. State an advantage of a base-5 system of numeration over a base-12 system.

 b. State a disadvantage of a base-5 system compared with a base-12 system.

14. Convert the following from the indicated base to base 10.

 a. $(1001001)_2$

 b. $(1023)_5$

 c. $(206)_{12}$

 d. $(314)_{16}$

15. Convert 35 to each of the following bases.

 a. 2   b. 5   c. 12   d. 16

16. Find a base b, if one exists, such that $(12)_b$ is an odd number.

17. Suppose Nim is played with 15 objects.

 a. If your opponent goes first and picks up one object, should you pick up one object or two objects?

 b. If you go first, how many objects should you pick up?

18. If four places can represent a number:

 a. How many base-10 numbers can be represented?

 b. How many base-2 numbers can be represented?

 c. How many base-12 numbers can be represented?

19. What fraction does the following symbol represent in Egyptian hieroglyphics?

$$\overset{\bigcirc}{\text{III}}$$

20. Write $\frac{1}{4}$ in Egyptian hieroglyphics.

21. Describe what unit fractions are. Then name a civilization that used them, and speculate on why.

22. Complete the decomposition of $\frac{2}{7}$ into unit fractions as $\frac{1}{4} + x$.

23. What spurred the development of fractions during the Middle Ages?

24. Which is larger? $\frac{6}{11}$ or $\frac{7}{12}$

25. Place the fractions $\frac{51}{123}$, $\frac{189}{456}$, $\frac{330}{789}$ in order from smallest to largest.

26. Who is generally regarded as the popularizer of decimal fractions, and when did he develop them?

27. a. Is our notation for a decimal point (for example, $2\frac{1}{4} = 2.25$) universal (or at least earthwide)?

 b. If so, when did it become universal? If not, name a country with a different notation, and give an example of that notation.

28. Convert the following decimals to fractions.

 a. 0.65

 b. 0.123123123 . . .

29. Give an example, if one exists, of each of the following. If no example exists, so state.

 a. A complex number that is not a real number

 b. An integer that is not a natural number

 c. An integer that is not a rational number

 d. An irrational number

 e. A natural number

 f. A rational number that is not an integer

 g. A rational number that is not a real number

 h. A number that meets all of the following conditions: natural number, integer, rational number, and real number

30. State a mathematical reason, rather than a societal reason, for extending the natural numbers to another number system.

31. Describe the origins of zero.

32. Describe briefly the history of negative numbers.

33. Describe three places where negative numbers are used.

34. Compute the following.

 a. $-2 + 4$

 b. $5 - 7$

 c. $-3 - (-6)$

 d. $2(-3)$

 e. $(-4)(-5)$

35. Using $\frac{3}{0}$ as an example, explain why division by zero is not defined.

36. Compute and simplify.

 a. $\frac{1}{3} + \frac{2}{5}$

 b. $\frac{3}{4} + -\frac{2}{3}$

 c. $\frac{1}{5} \cdot \frac{2}{-3}$

 d. $-4 \cdot \frac{-2}{3}$

 e. $\frac{3}{8} \div \frac{3}{4}$

 f. $\frac{4}{7} \div 3$

37. Compute.

 a. $(-3)^3$

 b. $\left(\frac{2}{5}\right)^3$

38. Calculate, rounding your answer to the nearest hundredth.

 a. $\sqrt{26}$ c. $\sqrt[3]{2}$

 b. $\sqrt{.02}$ d. $\sqrt[3]{-8}$

39. $\sqrt[3]{1} = 1$ and $\sqrt[3]{8} = 2$. What is the next value of n, with $n > 8$, such that $\sqrt[3]{n}$ is an integer?

40. a. Name an operation on numbers that is a binary operation.

 b. Name an operation on numbers that is a unary operation.

41. Describe the Pythagoreans. About when did they flourish? Where? What were some of their beliefs?

42. Describe the origins of liberal arts (general education) programs.

43. State three situations in which check digits are used.

44. The U.S. Postal Service scheme involves a 10-digit number plus a check digit, which is the remainder when the 10-digit number is divided by 9. Employ this scheme to answer the following.

a. Find the check digit for 2020202020.

b. Has the number 12312312312 (10 digits plus a check digit) been recorded correctly? Why or why not?

c. If *s* indicates a smudge mark, what number should be located at the smudge mark in 123412341*s*3? Assume that no mistakes except the smudge mark have occurred.

45. a. Does the U.S Postal Service scheme detect all transposition errors?

b. If so, explain why. If not, describe all transposition errors it detects, explain why it detects them, and why it does not detect other transposition errors.

46. The UPC check digit is obtained by multiplying the digits of an 11-digit number by 3, 1, 3, 1, etc., adding the resulting

products, and choosing the twelfth digit so that, when it is added to the preceding sum, the remainder upon division by 10 is 0.

a. The check digit for the following UPC code is in the bar code but is not written. Find it.

b. A bottle of Ultra Final Touch has the UPC code 0 11111 58148 7. Has the number been recorded correctly? Why or why not?

c. If *s* indicates a smudge mark in the UPC code on a book of stamps, what number should be where the smudge mark is in 0 15*s*45 16628 8.

47. The ISBN check digit is obtained by multiplying the digits of a nine-digit number by 10, 9, 8, 7, etc., adding the resulting products, and choosing the tenth digit so that, when added to the preceding sum, the remainder upon division by 11 is 0.

a. The ISBN for a book is reported to be 0-471-54391-7. Is the number reported correctly? Why or why not?

b. The first nine digits of an ISBN are 0-06-067768. What is the check digit?

c. If *s* indicates a smudge mark in a book's ISBN, what digit should go where the smudge mark is in 0-8423-*s*691-0?

Aristotle. *Metaphysics.* Translated in Richard Hope, *Aristotle's Metaphysics.* New York: Columbia University Press, 1952.

Ascher, Marcia. *Ethnomathematics: A Multicultural View of Mathematical Ideas.* Pacific Grove, CA: Brooks/Cole, 1991.

Bockarie, Alex. "Mathematics in the Mende Culture: Its General Implication for Mathematics Teaching," *School Science and Mathematics,* April 1993, pp. 208–211.

Boyer, Carl, and Merzbach, Uta. *A History of Mathematics,* 2nd ed. New York: Wiley, 1991.

Campbell, Douglas M. *The Whole Craft of Number.* Boston: Prindle, Weber & Schmidt, 1977.

Colliers Encyclopedia, 1993 ed., s.v. "numerals and systems of numeration."

Conant, Levi Leonard. *The Number Concept.* New York, Macmillan, 1896.

Cushing, Steven. "Four Mathematical Clerihews," *Mathematics Magazine,* February 1988, p. 23.

Eves, Howard. *An Introduction to the History of Mathematics,* 6th ed. Chicago: Saunders, 1990.

Flegg, Graham (ed.). *Numbers Through the Ages.* London: Macmillan Education, 1989.

Gallian, Joseph. "Assigning Driver's License Numbers," *Mathematics Magazine,* February 1991, pp. 13–22.

———. "The Mathematics of Identification Numbers," *College Mathematics Journal,* 1992, p.p. 194–202.

Gittleman, Arthur. *History of Mathematics.* Columbus, OH: Merrill, 1975.

Ifrah, Georges. *From One to Zero.* Harrisonburg, VA: Viking Penguin, 1985.

Katz, Victor J. *A History of Mathematics.* New York: HarperCollins, 1993.

Kline, Morris. *Mathematics: A Cultural Approach.* Reading, MA: Addison-Wesley, 1962.

Krusen, Kim. "A Historical Reconstruction of Our Number System," *Arithmetic Teacher,* March 1991, pp. 46–48.

Lathan, R. C., and Matthews, W, ed. *The Diary of Samuel Pepys.* Berkeley: University of California Press, 1971.

Menninger, Karl. *Number Words and Number Symbols,* Cambridge, MA: M.I.T. Press, 1969.

National Council of Teachers of Mathematics. *Historical Topics for the Mathematics Classroom.* Washington, D.C.: NCTM, 1969.

Neugebauer, Otto. *The Exact Sciences in Antiquity,* 2nd ed. Providence, RI: Brown University Press, 1957.

Olivastro, Dominic. "All You Need Is Nim," *The Sciences,* November/December 1991, pp. 51–53.

Rose, Nicholas J. *Mathematical Maxims and Minims.* Raleigh, NC: Rome Press, 1988.

Sanchez, George I. *Arithmetic in Maya.* Austin, TX, 1961.

Smith, David Eugene. *History of Mathematics,* Vol. I. Chicago: Ginn and Co., 1923.

Swain, Robert L. *Understanding Arithmetic.* New York: Holt, Rinehart and Winston, 1965.

Thompson, J. Eric. *Maya History and Religion.* Norman: University of Oklahoma Press, 1970.

Whitehead, Alfred North. *Introduction to Mathematics.* New York: Oxford University Press, 1948.

Zaslavsky, Claudia. *Africa Counts.* Boston: Prindle, Weber & Schmidt, 1973.

CHAPTER 3

SETS

CHAPTER OUTLINE

Prologue

3.1 Basic Idea and Notation

3.2 Subsets

3.3 Set Operations

3.4 Venn Diagrams

3.5 Set Properties

3.6 Applications of Sets

Review Exercises

Bibliography

PROLOGUE

At least 25 states in recent years have put forward system-wide initiatives, funded by the National Science Foundation, with the goal of improving mathematics and science teaching statewide. A draft of Michigan's guidelines for science and mathematics teacher preparation lists four goals, the first of which is to transform teaching. The second goal is to develop essential supporting knowledge:

Teaching in the manner described in Goal 1 is difficult and demanding, both in terms of the time and energy that it demands of teachers and in terms of the knowledge that they must possess and use. Teachers who can respond to the challenges described above will need deep and integrated knowledge of at least three kinds: *knowledge of science and mathematics, knowledge of students, and knowledge of the pedagogy of science*

and mathematics. Teaching occurs in the intersection of these three domains of knowledge, as represented in the Venn diagram [nearby].

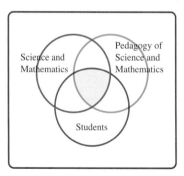

Teaching occurs in the intersection of three domains of knowledge.

This goal statement is meant for the public—teachers, parents, school administrators, legislators, legislative and executive staff members, and taxpayers. You will likely fall in at least one of these categories, if you do not already.

There is an underlying mathematical basis to some of the concepts and some of the language embodied in the goal statement. What is a *Venn diagram?* What is meant by *intersection?* The term *domain* is a specific instance of a broader mathematical concept. In the diagram, the shaded area represents the intersection of three domains of knowledge. There are six other regions in the diagram. How can they be described? Is the diagram an effective way to make a point? Is a picture really worth a thousand words? Is the term *intersection* too technical? Set your mind on these questions.

Chapter opening photo: The Olympic rings are reminiscent of Venn diagrams.

3.1 BASIC IDEA AND NOTATION

GOALS
1. To describe some history of set theory.
2. To use set theory notation.
3. To determine whether or not a given collection of objects is a well-defined set.
4. To determine whether or not a given set is finite and to give examples of infinite sets.

Probably no single concept has had as much influence in mathematics during the last 100 years as the concept of a *set.* The terminology of sets and subsets has spread beyond mathematics to other disciplines, as evidenced by statements such as "A subset of the winners is present." As an abstraction, the concept of *set* is basic to much of today's mathematics.

Intuitively, a *set* is a collection of objects. For example, you could refer to the students in a class as the set of students in a class; buildings on a campus could be referred to as the set of buildings on a campus; and members of the U.S. Senate could be referred to as the set of U.S. senators. In each of these examples, all members of the set share some common property. However, this is not a requirement of a set. For example, you could have a disparate set, such as a cow, a tree, and a book.

History

Georg Cantor (1845–1918), born in St. Petersburg, Russia, is considered the father of *set theory,* a relatively recent area of mathematical study. His early work with infinite sets was controversial and often ridiculed. Some famous mathematicians, including Richard Dedekind, collaborated with Cantor and supported his work. However others, such as Leopold Kronecker, severely criticized it.

Georg Cantor, the father of set theory

Work with set theory is being introduced to students at increasingly early points in their study of mathematics. The late 1950s saw set theory move into junior high schools as part of the "modern mathematics" movement. Today, even elementary school students are being introduced to many of the basic concepts of set theory.

Notation and Terminology

Sets can be described in a variety of ways. Once described, the members of a set are called the *elements of the set.*

Set Description One way to indicate what constitutes membership in a set is to provide a written description of the set members. This is called defining a set by *description*. Here are two examples of set descriptions:

The set of female students admitted to the United States Naval Academy in 1996.
The set of even natural numbers.

Roster Method A second, more frequently used method of indicating members of a set is to list or name the elements between braces {   }. This is called the *roster method*. Here are some examples:

{1, 2, 3, 4}
{George Washington, Abraham Lincoln}
{Canada, United States, Mexico}

Because a set is a collection of objects, the order in which the elements are listed makes no difference. Thus, {1, 3, 4, 2} is the same set as {1, 2, 3, 4}. The roster method is also a useful way to indicate sets of numbers when it would be awkward to list all of the set's elements. For example, the set of positive integers less than or equal to 100 is

$$\{1, 2, 3, \ldots, 100\}$$

The three dots between commas is called an *ellipsis*. The ellipsis means that some elements, following the indicated pattern, have not been listed. One can also indicate elements of an infinite set by this method, such as the set of natural (counting) numbers:

$$\{1, 2, 3, 4, 5, \ldots\}$$

Set-Builder Notation A third method of indicating membership is a set is with *set-builder notation.* The basic form for this involves placing a vertical bar inside the braces:

$$\{ \quad | \quad \}$$

This bar separates the name of an element of a set from a description of the elements in the set. It is read as "such that." For example,

$$\{x \mid x \text{ is a natural number less than } 101\}$$

is read as

"the set of all x such that x is a natural number less than 101"

This is just another way of describing the set

$$\{1, 2, 3, \ldots, 100\}$$

The general format of the set-builder notation is

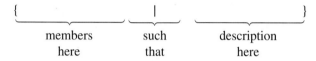

| members here | such that | description here |

The following example shows how the set-builder form can serve to describe sets.

EXAMPLE 3.1

Describe each of the following sets in set-builder notation.

a. The set of even positive integers less than 100
b. The set of all presidents of the United States

SOLUTION The key to translating each of these into set-builder notation is to describe the conditions of membership accurately:

a. $\{x \mid x \text{ is a positive even integer and } x < 100\}$
b. $\{x \mid x \text{ was or is a president of the United States}\}$

The roster and set-builder forms of notation are the more frequently used methods to describe sets. In each of these, we typically label a set with a capital letter, such as

$$A = \{ \quad | \quad \}$$

We then refer to the set thus described as the set A.

Membership in a set is denoted by the symbol $\in$. Thus, if $A = \{1, 2, 3, 4, 5\}$, we could write $2 \in A$, which would be read as "2 is an element of A." If an element is not in a set, we denote this by the symbol $\notin$. Given A as just defined, we could write $6 \notin A$ or $2.5 \notin A$, which would be read as "6 is not an element of A" or "2.5 is not an element of A," respectively.

Well-Defined Sets To be called a set, a collection of objects must be unambiguously determined. A collection of objects is *well-defined* if membership in the collection is clearly determined. For example, the following is well defined:

$$A = \{x \mid x \text{ is a natural number and } x \leq 10\}$$

ON A TANGENT

Russell's Paradox

Bertrand Russell (1872–1970) was one of the great mathematicians of the twentieth century. In his work in set theory he published a paradox known as Russell's paradox or the barber paradox:

> Suppose a barber in a small town has the job of cutting the hair of all those people who don't cut their own hair and only those people who don't cut their own hair. Should the barber cut his own hair?

In set language, we might let C be the set of people whose hair this barber cuts and let b denote the barber. The question then becomes: Is $b \in C$?

Here's another version of Russell's paradox:

> Suppose your job is to give notes to all students in class who do not take their own notes and only those students who do not take their own notes. Do you give notes to yourself?

Russell's paradox tells us that we can't define a set such as C. It gives a deeper meaning to the concept of a well-defined set. Further exploration of issues related to Russell's paradox takes us away from our intuitive development of set theory into issues of axiomatic set theory.

Axiomatic set theory is an area of mathematics developed in the twentieth century that was prompted, in part, by Russell's paradox. A number of colleges and universities offer a course in axiomatic set theory.

The roster method for describing this set produces

$$A = \{1, 2, 3, \ldots, 10\}$$

It is clear exactly which numbers are in this set and which numbers are not in this set. An example of a collection that is not well defined is:

$$B \text{ is the collection of tall people}$$

For the collection B, "tall" is not precise. A better description would be:

$$B = \{x \mid x \text{ is a person and the height of } x \text{ is at least } 6'5''\}$$

Finite vs. Infinite Sets   A *finite set* has finitely many elements; that is, we can count exactly how many elements the set contains, and the counting process terminates with a natural number. Thus, the following are finite sets:

$$A = \{1, 2, 3, \ldots, 10\}$$
$$B = \{100, 200, 300, \ldots, 1,000,000\}$$

and the following are *infinite sets* (not finite):

$$A = \{1, 2, 3, 4, 5, \ldots\}$$
$$B = \{x \mid x \text{ is an even integer}\}$$

Formal work with infinite sets is very abstract and was the basis of most of Cantor's work in set theory. For instance, an attempt is made to explain why there are "different sizes" of infinity.

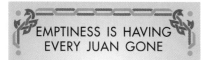

EMPTINESS IS HAVING EVERY JUAN GONE

STUDENT: Is Juan in the empty set?
PROFESSOR: No Juan is in the empty set.

—Nicholas J. Rose, *Mathematical Maxims and Minims* (Raleigh, NC: Rome Press, 1988), p. 142.

The Empty (Null) Set A set that contains no elements is called the *empty,* or *null, set.* It is denoted by the symbol $\varnothing$. (This is a Danish letter and not a computer symbol for zero.) Thus we have

$$\varnothing = \{ \quad \}$$

These symbols can be interchanged. However, most mathematicians simply use $\varnothing$ when referring to the empty set. As an indication of how quickly subtle questions arise when working with set theory, note that $\varnothing$ and $\{\varnothing\}$ are not the same. Can you explain why?

EXERCISES 3.1

For Exercises 1–4, express each set by the roster method.

1. The set of positive integers less than 5
2. The set of living U.S. expresidents
3. The set of U.S. senators from the state where you reside
4. The set of positive integers less than 13 that are divisible by 3

For Exercises 5–8 express each set in set-builder notation.

5. The set of positive integers less than 5
6. The set of living U.S. expresidents
7. The set of U.S. senators from the state where you reside.
8. The set of positive integers less than 13 that are divisible by 3.

For Exercises 9–12, express the sets by the roster method.

9. $\{x \mid x$ is an odd number between 8 and 16$\}$
10. $\{x \mid x$ is one of the two largest states in population in the U.S.$\}$
11. $\{x \mid x$ is one of the two largest states in area in the United States$\}$
12. $\{x \mid x$ is a state beginning with the letter $A\}$

For Exercises 13–16, express the set in set-builder notation.

13. $\{2, 4, 6, 8\}$
14. {George Washington, Thomas Jefferson, Abraham Lincoln, Theodore Roosevelt}

15. $\{a, e, i, o, u\}$
16. a. *(For baseball fans)* {Babe Ruth, Roger Maris}
    b. *(For others)* {Lake Erie, Lake Huron, Lake Michigan, Lake Ontario, Lake Superior}

For Exercises 17–20, (a) determine if the collections are well-defined or not, and (b) justify your answer.

17. The collection of heavy people
18. The collection of short people
19. The collection of all positive integers
20. The collection of all triangles
21. Which sets in Exercises 1 to 20 are infinite?
22. Is the set of grains of sand on earth finite or infinite?

Are the statements in Exercises 23–30 true or false when A is the set of positive integers less than 5 and B is the set of living U.S. expresidents?

23. $1 \in A$
24. $2 \notin A$
25. George Bush $\in$ B
26. Richard Nixon $\notin$ B
27. Madonna $\in$ B
28. Pearl Buck $\in B$
29. Rosa Parks $\notin B$
30. Susan B. Anthony $\notin B$

WRITTEN ASSIGNMENTS 3.1

1. Write a paper on Cantor and his criticism by Kronecker.
2. Sometimes people comment on an organization's inability to fund a project by saying "We only have a finite budget." Is this a proper use of the word *finite?* Explain your answer.
3. Ask three people (none from your class) to give you an example of an infinite set. Record their examples and state whether they are really infinite or are just large finite sets (like the number of grains of sand on the earth). Are there any observations or generalizations you can make from the examples you collected?

3.2 SUBSETS

GOALS
1. To use the concepts of *subset* and *universal set*.
2. To develop, know, and employ the formula for the number of subsets of a set with *n* elements.
3. To know a little about the role of conjectures in mathematics.
4. To describe power in a weighted voting scheme.

Consider the sets $B = \{1, 2, 3, 4\}$ and $A = \{2, 3\}$. Looking at the elements in set A (2 and 3), we note that they are also contained in the set B. This situation, where all the elements of one set are contained in another set, is one of the fundamental relationships in set theory.

Definition and Notation

Let's formalize the property just described with the following definition.

> DEFINITION
>
> Let A and B be sets. Then A is a *subset* of B if and only if every element of A is also an element of B. When this happens, we write
>
> $$A \subseteq B$$

The following examples illustrate some subset relationships.

EXAMPLE 3.2

Let $D = \{1, 2, 3, \ldots, 10\}$. Consider the sets $A = \{2, 4, 6, 8, 10\}$, $B = \{0, 1, 2, 3, \ldots, 11\}$, and $C = \{5\}$. Which are subsets of D? For those that are not subsets, explain why.

SOLUTION Let's examine the elements of the sets A, B, and C.
 Set A:　Since every element of A is also in D, we have $A \subseteq D$.
 Set B:　When looking at set B, note that $0 \in B$ but $0 \notin D$. This means $B \nsubseteq D$ (B is not a subset of D). This conclusion can be made as soon as we find one element that is not in D. It would also be sufficient to point out that since $11 \in B$ but $11 \notin D$, $B \nsubseteq D$.
 Set C:　Set C contains only one element (5). Since $5 \in D$, we have $C \subseteq D$.

EXAMPLE 3.3

Let $D = \{1, 2, 3, \ldots, 10\}$. Let $A = \varnothing$ and $B = \{1, 2, 3, \ldots, 10\}$. Indicate whether or not A and B are subsets of D, and include explanations.

SOLUTION
 Set A:　$A = \varnothing$ is a special case. With a background in formal logic, one can prove that $\varnothing \subseteq D$. For now, we appeal to your intuition by two arguments.

1. Since the statement "Every element contained in A is also contained in D" is true, we use the definition of subset to claim $A \subseteq D$.

2. If it were not true that $\varnothing \subseteq D$, then there would be an element of $\varnothing$ that is not in D; however, there is no such element. We can see from these arguments that $\varnothing \subseteq D$ for any set D.

Set B: $B = \{1, 2, 3, \ldots, 10\}$ is another special case. Here we have B and D containing the same elements. In this case, we could write either $B \subseteq D$ or $D \subseteq B$.

Note that Example 3.3 can be generalized as follows: If S is any set, then $\varnothing \subseteq S$ and $S \subseteq S$. In words, if S is any set, then the empty set is a subset of S and S is a subset of itself. In other words, the empty set is a subset of every set and any set is a subset of itself.

Universal Set

When discussing subset relations, it is often desirable to specify the totality of elements being considered. For example, if you are working with sets of students, you might be thinking in terms of all students in the United States. If you are working with a set of counting numbers, your totality might be the set of integers. We formalize this idea with the following definition.

> ### DEFINITION
>
> The *universal set U* is the set that contains all the elements being considered in a discussion or problem.

To illustrate: If set $A = \{1, 2, 3, \ldots, 10\}$, we could define the universal set to be the natural numbers, $U = \{1, 2, 3, \ldots\}$. But other universal sets could be defined for this problem. We use a universal set to give some frame of reference for the set or sets being considered. Whatever universal set U is defined, we must have $A \subseteq U$ for all sets A in the discussion.

Set Equality

In Example 3.3, we considered the sets $D = \{1, 2, 3, \ldots, 10\}$ and $B = \{1, 2, 3, \ldots, 10\}$. These sets contain exactly the same elements, so we are able to state both $D \subseteq B$ and $B \subseteq D$. When this occurs, we say the sets are equal. This gives us the following definition.

> ### DEFINITION
>
> Sets A and B are *equal* if and only if $A \subseteq B$ and $B \subseteq A$.

This definition merely formalizes a fairly intuitive result: Equal sets contain exactly the same elements.

Much of the early formal work in a course on set theory requires students to prove sets to be equal. Such proofs are typically completed by applying our definition of set equality. Students show both $A \subseteq B$ and $B \subseteq A$ for the sets A and B being considered, and they can then claim $A = B$.

Proper Subsets

At this stage, if $A = \{1, 2, 3\}$ and $B = \{1, 2, 3\}$, we can write both $A \subseteq B$ and $A = B$. We could also write $B \subseteq A$ or $B = A$. Frequently we wish to consider all subsets of a set except the set itself. For example, a teacher might want the class to work together in smaller (than the whole class) groups, or you might wish to apply to a subset of possible employers, but not all of them. To give ourselves the language and symbolism by which to talk about a subset of a set other than the set itself, we set forth the following definition.

DEFINITION

Let A be a subset of B. Then A is a *proper subset* of B if and only if A does not equal B. We write

$$A \subset B$$

Intuitively, this means that for A to be a proper subset of B, A must be a subset of B and B must also contain some element not in A. Thus, if $B = \{1, 2, 3\}$ and $A = \{1, 3\}$, then $A \subset B$.

An Application to Classification

Systems of classification involve the concept of a subset. For example, on a computer a file is considered a subset of a folder. Another example is how biologists classify all animals and plants into related sets, with seven chief types of sets making up the system of scientific classification. Each set is a subset of the previous set: (1) kingdom, (2) phylum, (3) class, (4) order, (5) family, (6) genus, and (7) species. For example, an eastern red squirrel is an element of the species *Tamiasciurus hudsonicus,* which is a subset of the genus *Tamiasciurus,* which is a subset of the family *Sciuridae,* which is a subset of the order *Rodentia,* which is a subset of the class *Mammalia,* which is a subset of the phylum *Chordata,* which is a subset of the kingdom *Animalia.*

Number of Subsets

There are many situations where one might want to know the number of subsets of a set. For example:

1. A student using this book might read none of it, or Chapters 1 and 3, or the first 10 chapters, or some other combination. How many combinations of chapters are there?
2. If a committee has seven members, how many different voting coalitions (alliances) are there?
3. If a family of five is on a trip and wants to take a picture of a landmark with some, all, or none of the family in the picture, how many possibilities are there?
4. If a quartet with four different instruments wants to play something with each possible combination of instruments, how many combinations are possible?

How many combinations of instruments are possible?

For the purpose of this discussion, we assume that the sets being considered have a fixed number of elements, usually denoted by n. To illustrate, suppose $A = \{a, b\}$ and we want to list all subsets of A ($n = 2$ for this problem). The subsets of $\{a, b\}$ are:

$$\varnothing, \{a\}, \{b\}, \{a, b\}$$

Thus A has four subsets.

If A contained only one element, $A = \{a\}$, then the subsets would be $\varnothing, \{a\}$. Therefore, A has two subsets. Let's put these results in the form of a chart:

Number of Elements in A	Number of Subsets of A
1	2
2	4

What happens when $n = 3$? To count the subsets, let $A = \{a, b, c\}$. It is usually desirable to list the subsets in some pattern. Let's start by listing the smallest subsets and then increase their size until we get to $A \subseteq A$. This produces the following:

$$\varnothing \subseteq A \qquad \text{0 element}$$

$$\left.\begin{array}{l} \{a\} \subseteq A \\ \{b\} \subseteq A \\ \{c\} \subseteq A \end{array}\right\} \quad \text{1 element}$$

$$\left.\begin{array}{l} \{a, b\} \subseteq A \\ \{a, c\} \subseteq A \\ \{b, c\} \subseteq A \end{array}\right\} \quad \text{2 elements}$$

$$\{a, b, c\} \subseteq A \quad \text{3 elements}$$

This listing shows that, when A contains three elements, there are eight subsets. Adding this information to the previous chart yields:

Number of Elements in A	Number of Subsets of A
1	2
2	4
3	8

1. If the number of elements in A is 0, so $A = \varnothing$, how many subsets does A have? Enter this information in the chart.

2. Suppose A has four elements, call them $a, b, c,$ and d.

   a. List the subsets of A that contain no elements.
   b. List the subsets of A that contain one element.
   c. List the subsets of A that contain two elements.
   d. List the subsets of A that contain three elements.
   e. List the subsets of A that contain four elements.
   f. How many subsets does a four-element set have? Enter this information in the chart.

3. Generalize from the information contained in the now-complete chart:

   If a set A contains n elements, then A has _____ subsets.

Number of elements in A	Number of subsets of A
0	_____
1	2
2	4
3	8
4	_____

We have the following result:

If A contains n elements, then A has 2^n subsets.

Conjectures and Inductive and Deductive Reasoning

It is interesting to note just how we reached the point where the result in the "Your Formulation" box seemed reasonable. What we did was to examine several small cases and discern a pattern. We then predicted this pattern would hold for other cases, even when it would be extremely tedious to do the counting. Such predictions in mathematics are called *conjectures,* that is, statements you believe to be true. In this particular case, a formal proof that the result holds requires a reasonably sophisticated property. Thus, we merely state the result.

Mathematics employs both inductive and deductive reasoning. *Inductive reasoning* is a process of inferring a general result from the examination of special cases. In looking at the number of subsets a set with n elements has, we looked at special cases for $n = 0, 1, 2, 3,$ and 4; then we inferred a general result (a conjecture) that an n-element set has 2^n subsets. *Deductive reasoning* argues from certain established principles—the undefined terms, definitions, and previously proved theorems of an axiomatic system—to prove a conclusion (conjecture). In mathematics, results are often conjectured based on inductive reasoning and proved deductively. As you might conjecture, much of mathematics consists of proving that conjectures are true.

EXAMPLE 3.4

A student using this book might read none of it, or Chapters 1 and 3, or the first 10 chapters, or some other combination of chapters. How many possible combinations of chapters are there?

SOLUTION Each possible combination of chapters is a subset of the 13-element set of chapters $\{1, 2, \ldots, 13\}$. This set has $2^{13} = 8192$ subsets. Consequently, there are 8192 possibilities.

An Application Involving Weighted Voting

Suppose a small company has three co-owners who have financed the company with $60,000, $60,000, and $30,000, respectively, and who form the board of directors. To reflect the different amounts of money the three people have put up, it is agreed that each will have one vote for every $10,000 invested. Consequently, the first two owners each have six votes and the third owner has three votes. This makes 15 total votes. Suppose they also agree that a simple majority of eight or more votes will win. We show this weighted voting system as follows:

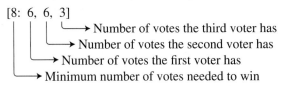

Refer to the three voters as 1, 2, and 3, respectively. There are $2^3 = 8$ possible subsets or coalitions or alliances of the voters: $\varnothing$, $\{1\}$, $\{2\}$, $\{3\}$, $\{1, 2\}$, $\{1, 3\}$, $\{2, 3\}$, $\{1, 2, 3\}$. We separate these into winning and losing coalitions.

Losing Coalitions			Winning Coalitions	
Coalition	*Number of Votes*		*Coalition*	*Number of Votes*
$\varnothing$	0		$\{1, 2\}$	12
$\{1\}$	6		$\{1, 3\}$	9
$\{2\}$	6		$\{2, 3\}$	9
$\{3\}$	3		$\{1, 2, 3\}$	15

Notice that any coalition with two or more voters wins while any coalition with one or fewer voters loses. This is exactly the same as if each voter had only one vote. In other words, voter 3 has exactly as much power as voters 1 and 2 even though voter 3 has only half as many votes as voter 1 or voter 2.

Because this voting scheme gives voter 3 more power than the $30,000 investment warrants, someone suggests another scheme: If each voter had one vote, to win would require two or more votes, which is a two-thirds majority. Therefore, let the voters have their 6, 6, and 3 votes, respectively, but require a two-thirds majority, 10 votes, to win. This is then a [10: 6, 6, 3] weighted voting scheme. Again, separate the eight coalitions into winning and losing coalitions.

Losing Coalitions	Winning Coalitions
$\varnothing$	$\{1, 2\}$
$\{1\}$	$\{1, 2, 3\}$
$\{2\}$	
$\{3\}$	
$\{1, 3\}$	
$\{2, 3\}$	

Notice now that voter 3 has no power: The only way voter 3 can be part of a winning coalition is to join the coalition {1, 2}, which can win even if voter 3 doesn't join it.

Let's make more precise what we mean by power. In the second scenario, voter 3 has no power, in that the only way voter 3 can be part of a winning coalition is to join a coalition that already is a winning coalition. In other words, voter 3 cannot join a losing coalition and turn it into a winning one.

At the opposite end of the spectrum is a weighted voting scheme with a dictator. In this case, the dictator has all the power and the other voters have no power. The other voters cannot join a losing coalition and turn it into a winning coalition while the dictator can change any losing coalition into a winning one by joining it. Notice that a voter need not have all the votes to be a dictator. For example, in the weighted voting scheme [3: 3, 1, 1], voter 1 is a dictator even though voters 2 and 3 each have one vote.

In the foregoing first scenario (the scheme [8: 6,6,3]), we noted that voter 3 had exactly as much power as voters 1 and 2. We define the power index of a voter as follows:

DEFINITION

The *power index* of a voter is the number of losing coalitions that can be turned into winning coalitions by having that voter join.

We apply this to the [8: 6, 6, 3] scheme.

| | If the voter joins the coalition, will it win? | | |
Losing Coalitions	*1*	*2*	*3*
∅	no	no	no
{1}	no	yes	yes
{2}	yes	no	yes
{3}	yes	yes	no
Power index:	2	2	2

Each of the voters has a power index of 2. The total power is $2 + 2 + 2 = 6$, so each voter has $\frac{2}{6} = \frac{1}{3}$ of the power.

This power index, called the *Banzhaf power index,* is due to John F. Banzhaf III, professor of law and legal activism at George Washington University.

EXAMPLE 3.5

Consider the weighted voting scheme [6: 4, 3, 2, 1].

a. Does any voter have no power?
b. Is any voter a dictator?
c. Find the power index for each voter.

SOLUTION A voter will have no power exactly when that voter's power index is 0. A voter will be a dictator exactly when that voter has all the power.

Losing Coalitions	If the given voter joins this coalition, will it win?			
	1	2	3	4
∅	no	no	no	no
{ 1 }	no	yes	yes	no
{ 2 }	yes	no	no	no
{ 3 }	yes	no	no	no
{ 4 }	no	no	no	no
{ 1, 4 }	no	yes	yes	no
{ 2, 3 }	yes	no	no	yes
{ 2, 4 }	yes	no	yes	no
{ 3, 4 }	yes	yes	no	no
Power index: 5		3	3	1 Total = 12

a. There is no voter with no power.

b. No voter is a dictator.

c. The power indices for voters 1, 2, 3, and 4, are, respectively, 5, 3, 3, and 1. Thus voter 1 has $^5/_{12}$ of the power, voter 2 has $^3/_{12} = ^1/_4$ of the power, voter 3 has $^1/_4$, and voter 4 has $^1/_{12}$.

Weighted voting is used in several situations. One is with corporations governed by stockholders. Typically, each share of stock entitles the holder to one vote. In our small-company scenario presented earlier, if each share of stock costs $100 and if a simple majority prevailed, the weighted voting scheme would be [751: 600, 600, 300].

The United Nations Security Council has 15 members, 5 with permanent seats and 10 elected by the U.N. General Assembly for 2-year terms. Decisions on procedural questions are made by an affirmative vote of 9 members, so each member has one vote and equal power. The system is a [9: 1, 1, 1, 1, 1, 1, 1, 1, 1, 1, 1, 1, 1, 1, 1] voting scheme. On all other substantive matters, the affirmative vote of 9 members must include the affirmative vote of all 5 permanent members. This can be described as a weighted voting scheme (see Exercise 20).

Some county and city governments employ weighted voting. These have often been prompted by the idea that if, say, a county has four districts with, say, 40%, 30%, 20%, and 10%, respectively, of the county's population, then the votes should be split 40%, 30%, 20%, and 10% to the four representatives. For example, a [6: 4, 3, 2, 1] scheme might be used. However, that distributes power by a $^5/_{12}, ^3/_{12}, ^3/_{12}, ^1/_{12}$ scheme (see Example 3.5), or about 42%, 25%, 25%, 8% so people in districts 2 and 4 would be upset by having less than their proportional share of power. With weighted voting some voters have more than one vote. So courts have ruled that votes should be distributed so that a voter's power is proportional to the population size of the district represented rather than having the voter's number of votes be proportional to the size of the district.

The U.S. Electoral College, to the extent that states vote as a block, is a weighted voting system. The Banzhaf index and other power indices show a slight bias in favor of the large states relative to their population. Several of the schemes that have been proposed for reforming the Electoral College give too much weight to the smaller states.

Weighted voting is used in a variety of places. Where it is the rule, we should remember that the influence of a voter need not be proportional to that voter's number of votes, as weighted voting can produce. In particular, a voter with twice as many votes as another need not have twice as much influence or power. One way to look at power is by examining certain subsets of the set of all voters.

1. Let $A = \{1, 2, 3, 4, 5\}$, $B = \{1, 3, 5, 7\}$, $C = \{2, 4\}$, and $D = \{1, 2, 3, 4, 5, 6\}$.
   a. Which of B, C, D are subsets of A?
   b. For those that are not subsets of A, explain why.
   c. Write three more subsets of A.
   d. Which of the subsets in part (a) are proper subsets?
2. a. For what sets S is $S \subseteq S$?
   b. For what sets S is $S \not\subseteq S$?
3. a. For what sets S is $\varnothing \subseteq S$?
   b. For what sets S is $\varnothing \not\subseteq S$?
4. Let $A = \{1, 2, 3, \ldots, 10\}$.
   a. Which of the following could be universal sets for A?
      i. B = set of natural numbers
      ii. Z = set of integers = $\{\ldots, -3, -2, -1, 0, 1, 2, 3, \ldots\}$
      iii. E = set of even natural numbers = $\{2, 4, 6, \ldots\}$
      iv. A
   b. Name another universal set for A.
5. If a set has six elements, how many subsets does it have?
6. If a committee has seven members, how many different voting coalitions (alliances) are there? Consider the empty set as a coalition.
7. If a family of five is on a trip and wants to take a picture of Old Faithful with some, all, or none of the family in the picture, how many possibilities are there?
8. A quartet with four different instruments wants to play something (silence does not count) with each possible combination of instruments. How many combinations are possible?
9. If a set has n elements, how many *proper* subsets does it have?
10. A set has 256 subsets. How many elements does the set have?

In Exercises 11–19, consider the given weighted voting scheme and answer the following questions.

   a. Does any voter have no power?
   b. Is any voter a dictator?
   c. Find the power index for each voter.

11. [4: 3, 2, 1]
12. [5: 4, 3, 2]
13. [8: 6, 5, 4]
14. [8: 7, 7, 1]
15. [8: 7, 6, 2]
16. [8: 7, 5, 3]
17. [8: 7, 4, 4]
18. [6: 4, 4, 1, 1]
19. [6: 5, 2, 2, 1]
20. Describe the voting procedure of the U.N. Security Council on a substantive matter as a weighted voting procedure.

1. Make up some other application problems that require knowing how many subsets a given set has. The problems should be like Example 3.4 or Exercises 6–8.
2. What is the difference between a theorem and a conjecture? How are they related?
3. Pretend you are in a political science or history class that is looking at proposals to reform the Electoral College. Write a paper explaining to the others in the class what the Banzhaf index is and why it should be used in considering the reform proposals.

3.3 • SET OPERATIONS

GOALS   1. To find the union, intersection, and difference of two sets.
         2. To find the complement of a set.
         3. To find combinations of unions, intersections, differences, and complements of sets.

The binary operations of addition, subtraction, multiplication, and division combine two numbers to form another number. In a similar way, it is possible to combine two sets to produce yet another set by means of a *binary operation*. This process of starting with two things, doing something to them, and then ending up with one thing of the same kind is not new. For example, what happens when you add two

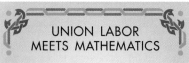
integers, say, 4 + 7? We write 4 + 7 = 11, which means that two things (4 and 7) are combined (4 + 7) and the end result is a single thing (11) of the same kind (an integer). So you see, we have all been performing binary operations with numbers for a long time. We will now develop similar operations for sets.

Union

Suppose $A = \{a, b\}$ and $B = \{c, d, e\}$. Our first set operation is to form a third set by listing all elements contained in either A or B or both A and B. This set would be $C = \{a, b, c, d, e\}$. Here is the more formal definition.

DEFINITION

Let A and B be sets. The *union* of A and B is the set that contains all the elements that are in either A or B or both A and B. We denote this as

$$A \cup B$$

With this definition and notation, we can take the previous sets A and B and write

$$A \cup B = \{a, b, c, d, e\}$$

The next example provides practice in finding the union of two sets.

EXAMPLE 3.6

For each of the following sets, find their union.

a. $A = \{1, 2, 3, 4\}$, $B = \{7, 8, 9\}$
b. $A = \{1, 2, 3, 4\}$, $C = \{3, 4, 5\}$
c. $A = \{1, 2, 3, 4\}$, $D = \{2, 3\}$
d. $A = \{1, 2, 3, 4\}$, $B = \{1, 2, 3, 4\}$,
e. $A = \{1, 2, 3, 4\}$, $C = \varnothing$

SOLUTION Generally, the union of sets is found by first listing all elements of one set and then including any additional elements found in the second set. This produces the following:

a. $A \cup B = \{1, 2, 3, 4, 7, 8, 9\}$
b. $A \cup C = \{1, 2, 3, 4, 5\}$
c. $A \cup D = \{1, 2, 3, 4\}$. Thus, $A \cup D = A$
d. $A \cup B = \{1, 2, 3, 4\}$. This special case has $A = B$.
e. $A \cup C = \{1, 2, 3, 4\}$. This is another special case. Since $\varnothing$ contains no elements, the union of A and $\varnothing$ will merely be the elements of A. Thus, $A \cup \varnothing = A$.

Notice that the union of A and B is a *uniting* of A and B. The union of 50 states produces the United States. The union of two school districts produces a new, united school district.

Intersection

A second method of combining two sets is to list only those elements that belong to both sets. Thus, for $A = \{1, 2, 3, 4\}$ and $B = \{3, 4, 5, 6\}$, the new set is $\{3, 4\}$.

DEFINITION

Let A and B be sets. The *intersection* of A and B is the set that contains all the elements that are in both A and B. We denote this as $A \cap B$

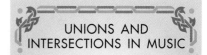

UNIONS AND
INTERSECTIONS IN MUSIC

Dan Tudor Vuza defines an *unending rhythmic canon* in music as a finite nonempty subset of a rhythmic class. Certain classes are defined as the union of sets. These are described in his article "Supplementary Sets and Regular Complementary Unending Canons" in *Perspectives of New Music,* Winter 1992, pp. 184–207. He also counts the number of elements in the intersection of certain pairs of sets.

Given A and B as indicated, we would now write

$$A \cap B = \{3, 4\}$$

or
$$\{1, 2, 3, 4\} \cap \{3, 4, 5, 6\} = \{3, 4\}$$

The next example illustrates how to find the intersection of two sets.

EXAMPLE 3.7

For each of the following sets, find their intersection.

a. $A = \{1, 2, 3, 4\}$, $B = \{7, 8, 9\}$
b. $A = \{1, 2, 3, 4\}$, $C = \{3, 4, 5\}$
c. $A = \{1, 2, 3, 4\}$, $D = \{2, 3\}$
d. $A = \{1, 2, 3, 4\}$, $B = \{1, 2, 3, 4\}$
e. $A = \{1, 2, 3, 4\}$, $C = \varnothing$

SOLUTION Finding an intersection differs from finding a union in that, since the elements must be in both sets, you cannot begin by listing all the elements from one of the sets. Rather, you must determine that an element is in each of the sets before listing it in the intersection.

a. $A \cap B = \{\ \ \}$ (or $\varnothing$)
b. $A \cap C = \{3, 4\}$
c. $A \cap D = \{2, 3\}$. Here we have $D \subseteq A$. Thus, $A \cap D = D$.
d. $A \cap B = \{1, 2, 3, 4\}$. Since $A = B$, $A \cap B = A$.
e. $A \cap C = \varnothing$. Since $C = \varnothing$, there can be no element in both A and $\varnothing$. This leads to the general result of $A \cap \varnothing = \varnothing$ for any set A.

Notice that the intersection of two sets is like the intersection of two streets—it is the part that is common to both of them.

Difference

A third method of combining two sets to produce a third set is to consider all the elements that are in one set and not in the other set. For example, if $A = \{1, 2, 3\}$ and $B = \{2, 3\}$, then the set containing elements in A and not in B is $\{1\}$.

DEFINITION

Let A and B be sets. The *difference* between A and B is the set that contains the elements of A that are not in B. We denote this as

$$A - B$$

3.3 SET OPERATIONS

81

As you can see by the notation, this set operation is analagous to subtracting, or "taking away." In practice, $A - B$ can be found by first listing all elements of A and then deleting from that list any elements that are in B. Using the same sets as listed in Example 3.7, the following example illustrates how to find the difference of two sets.

EXAMPLE 3.8

For each of the following sets, find the indicated difference.

a. $A = \{1, 2, 3, 4\}$, $B = \{7, 8, 9\}$ Find $A - B$.
b. $A = \{1, 2, 3, 4\}$, $C = \{3, 4, 5\}$ Find $A - C$.
c. $A = \{1, 2, 3, 4\}$, $D = \{2, 3\}$ Find $A - D$.
d. $A = \{1, 2, 3, 4\}$, $B = \{1, 2, 3, 4\}$ Find $A - B$.
e. $A = \{1, 2, 3, 4\}$, $C = \varnothing$ Find $A - C$.

SOLUTION We start by listing all elements of the first set, A, and then we delete from the list any elements in the second set.

a. $A - B = \{1, 2, 3, 4\}$
b. $A - C = \{1, 2\}$
c. $A - D = \{1, 4\}$
d. $A - B = \varnothing$. Since $A = B$, everything in A is deleted. Thus, for $A = B$, $A - B = \varnothing$.
e. $A - C = \{1, 2, 3, 4\}$. When the second set is empty, there is nothing to delete. This produces the result $A - \varnothing = A$, which is true for any set A.

Complement

The three previous set operations all combined two sets in some manner to produce a third set. As stated, these are binary operations. We now consider forming a new set from only one set (we suppose that a universal set has been established); this is referred to as a *unary operation*. It parallels the unary number operations of squaring a number.

YOUR FORMULATION

State some other unary operations in mathematics that you know.

Here is the basic concept: Given a set A, define a new set as everything not in A. Next, determine what is not in A. We consider elements that are in the universal set but not in A. To illustrate, suppose $U = \{1, 2, 3, . . .\}$ and A is $A = \{1, 2, 3, 4, 5\}$. If we want a set whose elements are not in A, we would have $\{6, 7, 8, . . .\}$. This could also be written as $U - A$. We can now write the following definition.

> **DEFINITION**
>
> Suppose a universal set has been specified. Let A be a set. Then the *complement* of A is the set of all elements not in A. We denote this as
>
> $$A'$$

As stated, this definition assumes some universal set has been specified. Once this is done, the complement of a set A is relatively easy to find: List all the elements that U contains that are not in A. The following example illustrates how to find complements of sets.

EXAMPLE 3.9

Let the universal set U be defined as $U = \{1, 2, 3, \ldots, 10\}$. Find A' for each of the following.

a. $A = \{2, 4, 6, 8, 10\}$
b. $A = \{1, 2, 3, \ldots, 10\}$
c. $A = \varnothing$

SOLUTION A' consists of the elements not in A.

a. $A' = \{1, 3, 5, 7, 9\}$
b. $A' = \varnothing$. In this case, $A = U$. Since everything in the universe is already in A, the complement of A must be empty. This could be written as $U' = \varnothing$.
c. $A' = \{1, 2, 3, \ldots, 10\}$ (or $A' = U$). This is just the opposite situation from part b. Here A is empty, so everything in the universal set is not in A. This result could also be written as $\varnothing' = U$.

Note that the complement of a set completes the set to the universal set. This is the same sense in which the complement of a 30° angle is a 60° angle (completes it to 90°) or a piece of jewelry complements an outfit (by completing the outfit).

The complement of a set is sometimes easier to describe or work with than the set in which we are interested. For example, consider all the courses at your school except this course; in other words, consider the complement of {this course}. How would you describe this without using the concept of a complement?

Combining Set Operations

Now we will look at ways to combine the set operations just considered. For example, suppose $A = \{a, b, c, d\}$, $B = \{d, e\}$, and $C = \{a, b, e, f\}$. How would you find $A \cap (B \cup C)$? Such problems are solved by first evaluating what is in the parentheses. This produces:

$$A \cap (B \cup C) = \{a, b, c, d\} \cap (\{d, e\} \cup \{a, b, e, f\})$$
$$= \{a, b, c, d\} \cap (\{a, b, d, e, f\})$$
$$= \{a, b, d\}$$

Complements of sets can also be used in conjunction with the other operations. However, a universal set must be defined. We illustrate this in the following example.

EXAMPLE 3.10

Let $U = \{1, 2, 3, \ldots, 10\}$ and

$$A = \{2, 4, 6, 8, 10\}$$

$$B = \{1, 2, 3\}$$

$$C = \{6, 7, 8, 9, 10\}$$

Find each of the following.

a. $A \cup B$
b. $(A \cup B)'$
c. $A' \cup B'$
d. $A' \cap B'$
e. $A' \cap C$

SOLUTION

a. $A \cup B = \{2, 4, 6, 8, 10\} \cup \{1, 2, 3\} = \{1, 2, 3, 4, 6, 8, 10\}$
b. $(A \cup B)' = \{1, 2, 3, 4, 6, 8, 10\}'$ (from part a) $= \{5, 7, 9\}$
c. $A' \cup B' = \{2, 4, 6, 8, 10\}' \cup \{1, 2, 3\}' = \{1, 3, 5, 7, 9\} \cup \{4, 5, 6, 7, 8, 9, 10\}$
   $= \{1, 3, 4, 5, 6, 7, 8, 9, 10\}$

At this point, compare answers to parts b and c. Note that $(A \cup B)' \neq A' \cup B'$. However, there is a relationship between $(A \cup B)'$ and A' and B'. Now consider part d.

d. $A' \cap B' = \{1, 3, 5, 7, 9\} \cap \{4, 5, 6, 7, 8, 9, 10\} = \{5, 7, 9\}$

Here we have the same problem as in part b. For this problem, $(A \cup B)' = A' \cap B'$, which is true in general for sets A and B. This property is examined further in the next section.

e. $A' \cap C = \{1, 3, 5, 7, 9\} \cap \{6, 7, 8, 9, 10\} = \{7, 9\}$

The next section explains how set operations can be studied by means of a geometric approach.

EXERCISES 3.3

For Exercises 1–31, consider the following sets:

$U = \{0, 1, 2, 3, \ldots, 9\}$ Universal set
$S = \{0, 1, 4, 9\}$ Perfect squares in U
$E = \{0, 2, 4, 6, 8\}$ Even integers in U
$O = \{1, 3, 5, 7, 9\}$ Odd integers in U
$P = \{1, 2, 3, \ldots, 9\}$ Positive integers in U

Perform the indicated set operation.

1. $S \cup E$
2. $E \cup O$
3. $O \cup \varnothing$
4. $E \cup E$
5. $S \cap O$
6. $E \cap O$
7. $O \cap P$
8. $S \cap S$
9. $E - S$
10. $O - E$
11. $P - U$
12. $S - \varnothing$
13. S'
14. O'
15. U'
16. $\varnothing'$
17. $(E \cup S)'$
18. $E' \cup S'$
19. $E' \cap S'$
20. $(S \cap O)'$
21. $S' \cap O'$
22. $S' \cup O'$
23. $(P - S)'$
24. $P' - S'$
25. $(S')'$
26. $S \cap (E \cup P)$
27. $S \cup (E \cap P)$
28. $S \cup (E \cup P)$
29. $S \cap (E \cap P)$
30. $(S \cup O) - E$
31. $S \cup (O - E)$

In Exercises 32–45, combine the given sets for any universal set U with any subset A (for example, $\varnothing - A = \varnothing$).

32. $A \cup A$
33. $A \cup U$
34. $A \cup \varnothing$
35. $A \cap A$
36. $A \cap U$
37. $A \cap \varnothing$
38. $A - A$
39. $A - \varnothing$
40. $A - U$
41. $U - A$
42. $A \cup A'$
43. $A \cap A'$
44. $A - A'$
45. $A' - A$

In Exercises 46–49, consider the following sets:
    U = set of letters in the English alphabet
    C = set of consonants
    V = set of vowels = $\{a, e, i, o, u\}$

Describe each of the following sets.

46. $C \cup V$

47. $C \cap V$

48. C'

49. V'

In Exercises 50–56, consider the following sets:

U = states in the United States
R = rural states
S = sunbelt states
M = midwestern states

Describe symbolically each of the following (for example, Urban states = R').

50. Rural states in the sunbelt
51. Rural states in the midwest
52. Urban states in the sunbelt
53. Urban states in the midwest
54. Rural states outside the midwest
55. States that are neither sunbelt nor midwestern states
56. Urban states that are either in the sunbelt or in the midwest

---

WRITTEN ASSIGNMENTS 3.3

1. Union, intersection, and difference are binary operations on sets. What other binary operations in mathematics do you know?
2. Discuss situations outside of mathematics where the term *operation* has a meaning similar to that of binary or unary operation in mathematics.
3. The terms *union, intersection,* and *difference* are suggestive. Come up with an alternate term for each of these three that you believe is more descriptive. Explain why.

3.4 ► VENN DIAGRAMS

GOALS
1. To use Venn diagrams with unions, intersections, differences, and complements of two or three sets.
2. To use Venn diagrams to find the least common multiple or greatest common divisor of two or more numbers.

Sets and set operations can conveniently be presented in diagrams. This method of illustration was introduced by Leonhard Euler (pronounced "oiler") (1707–1783). John Venn (1834–1923) later refined Euler's method to produce what is commonly used today.

The basic idea is to represent the universal set by a rectangle and other sets by circles within the rectangle. Thus, a set A would be illustrated as in ► Figure 3.1, where it is understood that elements of A are contained within the circle that represents A. If $U = \{1, 2, 3, \ldots, 10\}$ and $A = \{2, 4, 6\}$, we would get the illustration in ► Figure 3.2.

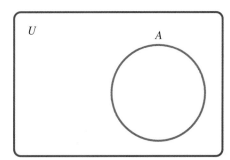

► FIGURE 3.1 Illustration of set A

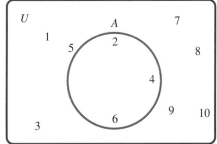

► FIGURE 3.2 Illustration of a universal set and set A

Leonhard Euler: Mathematician Extraordinaire

One of the most prolific mathematicians of all time is **Leonhard Euler.** He published 886 books and articles. These collected works fill over 70 volumes.

Euler was born in Basel, Switzerland, in 1707. He showed signs of genius as a youngster. He was tutored by Johann Bernoulli of the famous Bernoulli family of mathematicians. At the age of 19, Euler was awarded a prize by the French Academy for a paper on the best placement of masts on a ship. He wrote this paper before he had even seen an ocean-going ship.

From 1727 to 1741 Euler held positions at the St. Petersburg (Russia) Academy, an institution that was the dream of Peter the Great. From 1741 to 1766, he held a position at the Berlin Academy under Frederick the Great. In 1766, he returned to St. Petersburg, when he remained until his death in 1783.

Euler published in all branches of mathematics, including algebra, geometry, number theory, and analysis (the calculus branch). He also published in a number of areas of applied mathematics and areas related to mathematics, including astronomy, engineering, acoustics, hydraulics, mechanics, artillery, and the theory of music. Also, he was a masterful writer of textbooks. (Where's Euler when you really need him?)

Leonhard Euler

The creativity and energy of Euler's mathematical activity is even more amazing when it is realized that, during the last 12 years of his life, he was almost totally blind. That seemed not to slow the pace of his mathematical production at all. His ability to do mathematics and write papers on optics while almost totally blind must be related to his tremendous memory and ability to do mental calculations.

Diagrams for Two Sets

Venn diagrams typically serve to describe relationships among two or more sets. Let's consider the various ways two sets could be depicted. In general, A and B could be such that $A \cap B = \varnothing$, $A = B$, $A \subset B$, $B \subset A$, or they have some, but not all, elements in common. Each of these could be shown separately, but convention dictates that sets A and B be illustrated as overlapping, as shown in ▶ Figure 3.3, which allows for each of the possibilities just mentioned. For example, if $U = \{1, 2, 3, \ldots, 10\}$, $A = \{1, 2, 3\}$, and $B = \{4, 5\}$, we would show this by the Venn diagram in ▶ Figure 3.4.

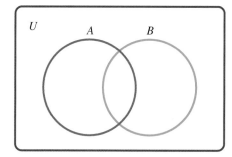

▶ FIGURE 3.3 Illustration of sets *A* and *B* overlapping

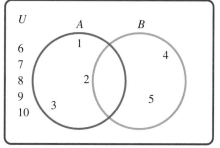

▶ FIGURE 3.4 Illustration of a universal set, set *A*, and set *B*

Using the same universal set, but with $A = \{1, 2, 3, 4\}$ and $B = \{2, 4, 6, 8, 10\}$, we would get the diagram in ▶ Figure 3.5. In this instance, elements in both *A* and *B* are listed in the region common to *A* and *B*. This method allows us to look at the Venn diagram and conclude $A \cap B = \{2, 4\}$. The following example gives you more practice in interpreting Venn diagrams.

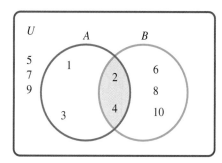

▶ FIGURE 3.5 Illustration of sets *A* and *B* overlapping

EXAMPLE 3.11

Let $U = \{1, 2, 3, \ldots, 10\}$. Using the Venn diagram of ▶ Figure 3.6, find each of the following.

a. $A \cup B$
b. $A \cap B$
c. A'
d. $(A \cup B)'$
e. $A' \cap B'$

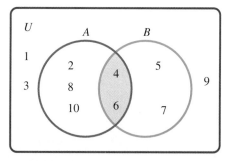

▶ FIGURE 3.6

SOLUTION We could specifically write out *A* and *B* and proceed as in the preceding section; however, many results can be obtained more readily by means of the diagram.

a. $A \cup B = \{2, 4, 5, 6, 7, 8, 10\}$
b. $A \cap B = \{4, 6\}$
c. $A' = \{1, 3, 5, 7, 9\}$
d. $(A \cup B)' = \{1, 3, 9\}$
e. $A' \cap B' = \{1, 3, 9\}$. This one might be a little harder to see. First, A' would contain $\{1, 3, 9, 5, 7\}$ and $B' = \{1, 3, 9, 2, 8, 10\}$. Then, $A' \cap B' = \{1, 3, 9\}$. In other words, these are the elements that are outside of *A* and outside of *B*.

Venn Diagrams and Set Operations

Listing all elements of sets in Venn diagrams becomes tedious and unwieldy if the sets are very large. We solve this problem in two ways: by shading and by numbering.

First, let's consider the method of shading. Rather than writing all the elements of the sets in the circles, as we did in ▶ Figure 3.6, we simply shade to represent the region being considered. Thus, $A \cup B$ would be shown as in ▶ Figure 3.7, and $A \cap B$ would appear as in ▶ Figure 3.8.

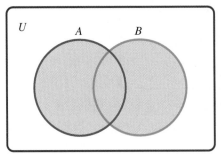

▶ FIGURE 3.7 $A \cup B$

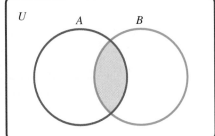

▶ FIGURE 3.8 $A \cap B$

This shading method has a nice, intuitive appeal, but it becomes somewhat limited as diagrams get more complex. We get around this problem by numbering regions of the Venn diagrams. (To avoid confusion, Roman numerals are normally employed.) This is shown in ▶ Figure 3.9. This numbering system may or may not be combined with shading.

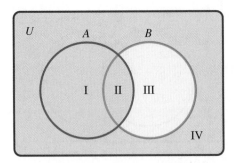

▶ FIGURE 3.9 Regions in Venn diagrams

We can now make statements like "The intersection of A and B is represented by region II." The following example illustrates how regions can help to indicate relations between sets.

EXAMPLE 3.12

Using the Venn diagram of Figure 3.9, indicate which region or regions represent each of the following.

a. $A \cup B$
b. A'
c. $(A \cup B)'$
d. $A' \cap B'$

SOLUTION From the diagram, we get the following results.

a. Regions I, II, and III
b. Regions III and IV
c. Region IV
d. Region IV

Diagrams for Three Sets

Venn diagrams can be extended to illustrate properties and relations involving three sets.

 YOUR FORMULATION

Draw a Venn diagram depicting three sets A, B, and C within a universal set U.

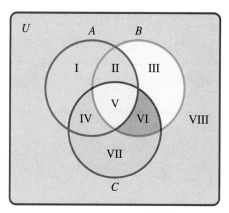

▶ FIGURE 3.10 Venn diagram involving three sets

With three sets, it is usually desirable to label the regions as shown in ▶ Figure 3.10. These eight regions can then be used to identify the desired result(s). For example region V is $A \cap (B \cap C)$.

It can be helpful to combine this labeling with shading for the purpose of identifying which regions satisfy certain problems. To illustrate, consider the next example.

●**EXAMPLE 3.13**

Use shading and labeling to identify the region $A \cap (B \cup C)$.

SOLUTION Borrowing the labeling in Figure 3.10, we first show $B \cup C$ by the blue shading in ▶ Figure 3.11. Then we shade A to produce the diagram in ▶ Figure 3.12. The intersection is seen to be the green-shaded regions in ▶ Figure 3.13 (regions II, IV, and V).

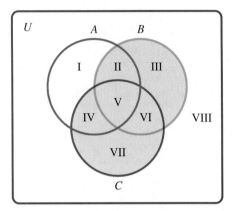

▶ FIGURE 3.11 Shading $B \cup C$

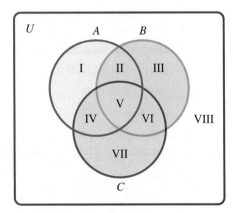

▶ FIGURE 3.12 Shading both A and $B \cup C$

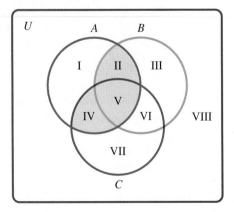

▶ FIGURE 3.13 $A \cap (B \cup C)$

It is an interesting exercise to place elements in the appropriate regions. For one or two sets, this is not difficult; with three sets it gets more complicated. The following example illustrates such an exercise.

EXAMPLE 3.14

Let $U = \{1, 2, 3, \ldots, 20\}$, $A = \{5, 10, 15, 20\}$, $B = \{2, 4, 6, 8, 10, 12, 14, 16, 18, 20\}$, and $C = \{3, 6, 9, 12, 15, 18\}$. Draw a Venn diagram, label the regions, and place each element in the appropriate region.

SOLUTION The labeled Venn diagram is the same as Figure 3.10. Placing elements in their appropriate regions produces the diagram in ► Figure 3.14.

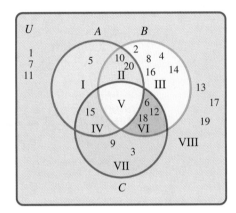

► FIGURE 3.14 Elements in regions

Note that region V, which represents $(A \cap B) \cap C$, is empty.

In Example 3.14, the elements of A are multiples of 5, the elements of B are multiples of 2, and the elements of C are multiples of 3. Elements in $A \cap B$ (regions II and V) are in both A and B, so they are multiples of 5 and multiples of 2; i.e., they are *common multiples* of 5 and 2. The common multiples of 5 and 2 are 10 and 20.

Sometimes we are interested in the *least common multiple* (LCM) of 5 and 2. For example, one way to add 3/5 and 1/2 is to get a common denominator. Calculations are simplified by choosing the lowest common denominator, 10, which is the least common multiple of 5 and 2.

Similarly, to find the least common multiple of 5 and 3 with the Venn diagram of Figure 3.14, we look at the common multiples of 5 and 3. These elements are in $A \cap C$ (regions IV and V). The only element is 15, which is the LCM of 5 and 3. Similarly, LCM $\{3, 2\}$ = smallest number in $B \cap C$ = 6.

To find the least common multiple of 2, 3, and 5, we look at region V, $(A \cap B) \cap C$. To make it nonempty, let $U = \{1, 2, \ldots, 40\}$, $A = \{5, 10, 15, 20, 25, 30, 35, 40\}$, $B = \{2, 4, 6, \ldots, 40\}$, and $C = \{3, 6, 9, \ldots, 39\}$. The corresponding Venn diagram is shown in ► Figure 3.15. The only common multiple of 2, 3, and 5 between 1 and 40 is 30. Therefore, LCM $\{2,3,5\}$ = 30.

Although the Venn diagram helps to visualize the idea of least common multiples, it is not the most efficient way to calculate them. It is one way that sets and Venn diagrams can picture relationships.

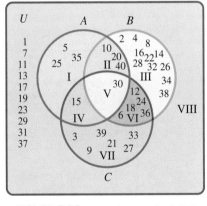

► FIGURE 3.15 Venn diagram for LCM

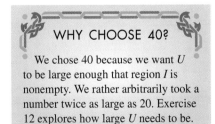

WHY CHOOSE 40?

We chose 40 because we want U to be large enough that region I is nonempty. We rather arbitrarily took a number twice as large as 20. Exercise 12 explores how large U needs to be.

1. Let $U = \{0, 1, 2, 3, \ldots, 9\}$. Using the Venn diagram, list the elements in each of the following.

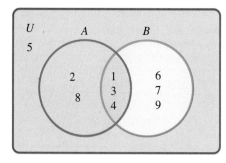

a. $A \cup B$
b. $A \cap B$
c. A'
d. $A - B$
e. $A \cup B'$
f. $(A \cap B)'$
g. (*Multiple choice*) Which one of the following is true?
   i. $A \cap B = \varnothing$
   ii. $A \subset B$
   iii. $B \subset A$
   iv. $A = B$
   v.   None of the above

2. Let $U = \{a, b, c, d, e, f, g\}$. By means of the Venn diagram, list the elements in each of the following.

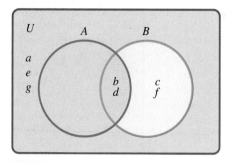

a. $A \cup B$
b. $A \cap B$
c. A'
d. B'
e. $A - B$
f. $B - A$
g. $A' \cap B$
h. $A' \cap B'$
i. (*Multiple choice*) Which one of the following is true?
   i. $A \cap B = \varnothing$
   ii. $A \subset B$
   iii. $B \subset A$
   iv. $A = B$
   v.   None of the above

3. Let $U = \{0, 1, 2, \ldots, 9\}$ $A = \{0, 1, 4, 9\}$, and $B = \{1, 3, 5, 7, 9\}$. Draw a Venn diagram with U, A, and B labeled and each of the 10 elements of U placed appropriately.

4. Let $U = \{a, b, c, d, e, f\}$, $A = \{a, e\}$, and $B = \{b, d\}$. Draw a Venn diagram with U, A, and B labeled and each of the six elements of U placed appropriately.

5. With the aid of the Venn diagram, write each of the following sets in terms of A, B, and set operations (for example: $\{2, 5\} = A - B$ or $A \cap B'$).

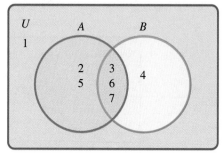

a. $\{1\}$
b. $\{4\}$
c. $\{1, 4\}$
d. $\{1, 2, 5\}$
e. $\{3, 6, 7\}$
f. $\{2, 4, 5\}$
g. $\{1, 2, 4, 5\}$
h. $\{2, 3, 4, 5, 6, 7\}$

6. For each of the following, draw a Venn diagram and shade the indicated set.
   a. $A \cup B$
   b. $A \cap B$
   c. A'
   d. $A - B$
   e. $A' \cup B$
   f. $A' \cap B$
   g. $(A \cup B)'$
   h. $(A \cap B)'$
   i. $A' \cup B'$
   j. $A' \cap B'$

7. Using the Venn diagram, write the set composed of each of the following regions in terms of A, B, and set operations (for example: region III $= B - A$ or $B \cap A'$).

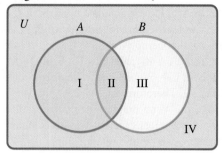

a. Region I
b. Region II
c. Region IV
d. Regions I, II
e. Regions I, III
f. Regions II, III
g. Regions II, IV
h. Regions I, II, III
i. Regions I, II, IV
j. Regions II, III, IV

8. Using shading and the labeling of Figure 3.10, identify the following regions.
   a. $A \cup (B \cap C)$
   b. $(A \cup B) \cap C$
   c. $A \cap B'$
   d. $A \cup (B \cap C)'$
   e. $(A \cup B) \cap (A \cup C)$
   f. $(A \cap C) \cup (B \cap C)$
   g. $A - B$
   h. $A \cup (B' \cup C')$

9. Let $U = \{a, b, c, \ldots, z\}$, $V = \{a, e, i, o, u\}$, $B = \{a, b, c, d, e, f, g, h\}$, $C = \{a, f, g, h\}$. Draw a Venn diagram, label the regions, and place each element in the appropriate region.

10. Given the sets $U = \{1, 2, 3, \ldots, 25\}$, $A = \{3, 6, 9, 12, 15, 18, 21, 24\}$, $B = \{4, 8, 12, 16, 20, 24\}$, and $C = \{6, 12, 18, 24\}$, find:
    a. LCM $\{3, 4\}$
    b. LCM $\{3, 6\}$
    c. LCM $\{4, 6\}$
    d. LCM $\{3, 4, 6\}$

11. Use appropriate sets to find:
    a. LCM $\{2, 3\}$
    b. LCM $\{2, 4\}$
    c. LCM $\{3, 4\}$
    d. LCM $\{2, 3, 4\}$

12. In Figure 3.15 we chose 40 rather than 20 as a stopping element for U.
    a. In that example we were trying to find LCM $\{2, 3, 5\}$. If $U = \{1, 2, \ldots, n\}$, what is the smallest value of n that will make region V nonempty?
    b. In general, if $a, b,$ and c are natural numbers and we wish to find LCM $\{a, b, c\}$ via the Venn diagram approach with a universal set $U = \{1, 2, \ldots, n\}$, what is the smallest value of n that will guarantee that region V is nonempty?

13. Modify the LCM procedure of Example 3.14 and Exercises 10 and 11 to find the greatest common divisor (GCD) of two or three numbers. Illustrate by using this procedure to find:
    a. GCD $\{4, 6\}$
    b. GCD $\{4, 9\}$

14. The Prologue to this chapter showed a Venn diagram representing three domains of knowledge. That Venn diagram is repeated here.

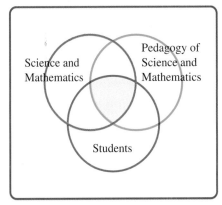

    a. Describe an appropriate universal set.
    b. Describe an element of the set labeled "Science and Mathematics" that is not in either of the other two sets.
    c. The shaded region represents knowledge of all three kinds. There are six other regions in the diagram. Label them and describe what kind of knowledge is represented in each of them.

15. There are four possible ABO blood types, depending on the presence or absence of two antigens called A and B. If only the antigen A is present, the blood is type A. If only antigen B is present, the blood is type B. If both antigens A and B are present, the blood is type AB. If neither antigen is present, the blood is type O. Draw a Venn diagram to illustrate the relation of the four blood types to the two antigens A and B.

16. Another antigen (see Exercise 15) is the Rh factor. People who have this antigen have blood called Rh-positive. Otherwise, a person's blood is Rh-negative. Draw a Venn diagram showing the blood types related to the A, B, and Rh-factor antigens.

<hr/>

WRITTEN ASSIGNMENTS 3.4

1. Write a paper on Leonhard Euler. The "On a Tangent" in this section contains some information about him. Find at least two additional references. The references in the bibliography at the end of the chapter by Boyer (1991), Calinger, (1995), Eves (1990) and Katz (1992) all contain material on Euler. Most encyclopedias and books on the history of mathematics cover Euler.
   a. Include a two- or three-paragraph biography of Euler that mentions at least one fact not in the "On the Tangent."
   b. Include a paragraph on what you consider to be Euler's greatest achievement.
   c. How high do your references place Euler among the great mathematicians? Quote your references in this regard. Given that he was perhaps the most prolific mathematician ever, why is he not regarded as the greatest mathematician of all time?

2. Write a paper on John Venn. Information about him will be much harder to find than information on Euler. If he is so obscure, why did his name stick with Venn diagrams?

3. Much of Euler's mathematical work was done after he was blind. Comment on the use of Venn diagrams by blind people.

SET PROPERTIES

GOALS   1. To employ Venn diagrams to verify properties of sets.
          2. To use Venn diagrams to determine whether or not two sets are equal.

As noted in the previous section, certain general properties seem to hold for sets. For example, when considering the complement of $A \cup B$, we demonstrated in Example 3.12 that $(A \cup B)' = A' \cap B'$. This equality is true for all sets A and B.

In general, how do mathematicians prove that such assertions are true? Venn diagrams certainly help to visualize relations. When the region shaded in $(A \cup B)'$ is the same as the region shaded in $A' \cap B'$, it is natural to conclude that these sets are equal. Unfortunately, this method is not sufficient for a rigorous development of set theory. Much more sophisticated techniques are employed to prove set properties.

The good news is that the Venn diagram is commonly accepted as a way to verify set properties for students who are being introduced to sets. Although this method is not deemed "rigorous" by mathematicians, its intuitive nature makes it acceptable for the beginner, and the Venn diagrams used to demonstrate the truth of a statement usually suggest how to make a truly rigorous proof. In fact, many mathematicians look at Venn diagrams as a means of "verifying" set properties, before they construct a rigorous proof. Our purpose is to employ Venn diagrams to verify several set equalities. Here is the guiding principle for this method.

> If the region in a Venn diagram representing one set is the same as the region representing another set, then the two sets are equal.

We now look at two specific set equalities. Variations of these and related problems appear in the exercises at the end of this section.

DeMorgan's Laws

We previously demonstrated via Venn diagrams that $(A \cup B)' = A' \cap B'$. This property holds in general, and is one of two fundamental properties referred to as *DeMorgan's laws*.

 ON A TANGENT

Who Was Augustus DeMorgan?

Augustus DeMorgan lived during the 1800s and died in 1871. He stated that he was x years old in the year x^2. With a little trial and error you can determine the year of his birth.

    DeMorgan was British and attended Trinity College in London. He could not hold a fellowship at Cambridge or Oxford because he would not take the required religious test, even though he was brought up in the Church of England and his mother hoped he would be a minister. At the age of 22 he was appointed a professor of mathematics at what was later to be known as University College of the University of London. He remained a champion of religious and academic freedom throughout his life.

Let's now verify the second of these laws by means of Venn diagrams. $A \cap B$ is shaded in ▶ Figure 3.16a, and $(A \cap B)'$ is indicated in Figure 3.16b.

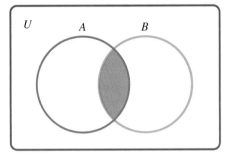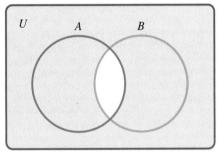

(a) $A \cap B$ (b) $(A \cap B)'$

▶ FIGURE 3.16 $A \cap B$ and $(A \cap B)'$

Next we obtain $A' \cup B'$ by first finding A', then finding B', and finally taking the union of these sets. ▶ Figure 3.17a shows A', Figure 3.17b shows B', and Figure 3.17c shows $A' \cup B'$.

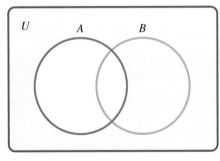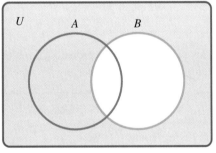

(a) A' (b) B'

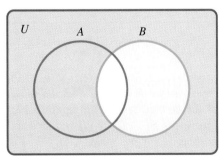

(c) $A' \cup B'$

▶ FIGURE 3.17 Finding $A' \cup B'$

It is sometimes more efficient to label regions and refer to these labels. We could start with the labeling in ▶ Figure 3.18.

1. $A \cap B$ is region II, so what regions would represent $(A \cap B)'$?
2. What regions represent $A' \cup B'$?
3. Does this show that $(A \cap B)' = A' \cup B'$?

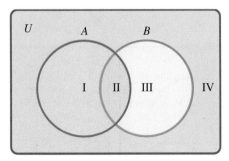

▶ **FIGURE 3.18** Labeled regions

Since the Venn diagram of $(A \cap B)'$ (Figure 3.16b) has the same shaded region as that of $A' \cup B'$ (Figure 3.17c), we conclude $(A \cap B)' = A' \cup B'$.

Distributive Properties

By ordinary arithmetic, the product of one number and the sum of two other numbers, $a(b + c)$, can be written as $ab + ac$:

$$a(b + c) = ab + ac$$

This is the distributive property of multiplication over addition. Two similar results hold for properties of sets.

DISTRIBUTIVE PROPERTIES

Let A, B, and C be sets.

1. $A \cup (B \cap C) = (A \cup B) \cap (A \cup C)$
2. $A \cap (B \cup C) = (A \cap B) \cup (A \cap C)$

We use Venn diagrams to verify the first property. The second one is left for Exercise 2.

The Venn diagram for $A \cup (B \cap C)$ is formed in ▶ Figure 3.19 as follows: A is shaded blue in Figure 3.19a, $B \cap C$ is shaded yellow in Figure 3.19b, and $A \cup (B \cap C)$ is all of the colored regions in Figure 3.19c.

The Venn diagram for $(A \cup B) \cap (A \cup C)$ is found in ▶ Figure 3.20 as follows: $A \cup B$ is shaded blue in Figure 3.20a, $A \cup C$ is shaded yellow in Figure 3.20b, and the green region of Figure 3.20c represents $(A \cup B) \cap (A \cup C)$. Since this is the same region as all of the colored regions in Figure 3.19c, we can conclude $A \cup (B \cap C) = (A \cup B) \cap (A \cup C)$.

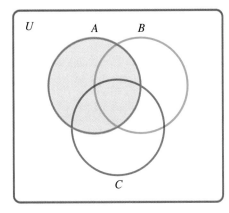

(a) A is shaded

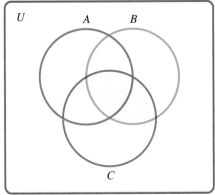

(b) $B \cap C$ is shaded

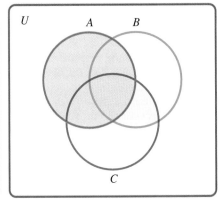

(c) $A \cup (B \cap C)$ is shaded

▶ FIGURE 3.19 Finding $A \cup (B \cap C)$

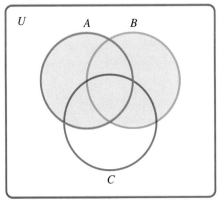

(a) $A \cup B$

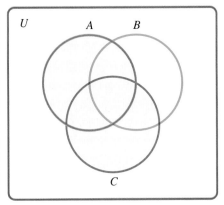

(b) $A \cup C$

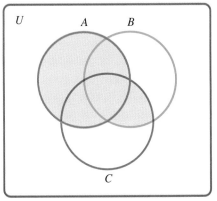

(c) $(A \cup B) \cap (A \cup C)$

▶ FIGURE 3.20 Finding $(A \cup B) \cap (A \cup C)$

For Exercises 1–4, use Venn diagrams to show that the given equality is true.

1. $(A \cup B)' = A' \cap B'$
2. $A \cap (B \cup C) = (A \cap B) \cup (A \cap C)$
3. $(A \cap B) \cap C = A \cap (B \cap C)$
4. $(A \cup B) \cup C = A \cup (B \cup C)$

For Exercises 5–8, use Venn diagrams to show that the given sets are not equal.

5. $A \cap (B \cup C) \neq (A \cap B) \cup C$
6. $A \cup (B \cap C) \neq (A \cup B) \cap C$

7. $(A \cup B)' \neq A' \cup B'$
8. $(A \cap B)' \neq A' \cap B'$

For Exercises 9–14, use Venn diagrams to determine whether or not the given sets are equal.

9. $A - (B \cup C), (A - B) \cup (A - C)$
10. $A - (B \cap C), (A - B) \cap (A - C)$
11. $(A \cup B) - C, (A - C) \cup (B - C)$
12. $(A \cap B) - C, (A - C) \cap (B - C)$
13. $A \cap B', (A \cap B)'$
14. $A \cup B', (A \cup B)'$

1. The two distributive properties for sets are like the distributive property for multiplication over addition of numbers. Give some other properties of numbers and then give the analogous properties for sets. Show with Venn diagrams whether or not the properties of sets are true.

2. One DeMorgan Law is $(A \cup B)' = A' \cap B'$. Write in words what this says. Avoid using: *union, intersection,* or *complement.*

3. One of the distributive properties for sets states that $A \cup (B \cap C) = (A \cup B) \cap (A \cup C)$. Write out in words what this says. Do not use the word *union* or *intersection.*

3.6 APPLICATIONS OF SETS

GOAL   To use Venn diagrams to solve applications involving overlapping sets of data.

Three applications of sets have already been mentioned: analyzing power in weighted voting, using Venn diagrams to describe relationships among sets, and using Venn diagrams to picture least common multiples. In addition, the ability to work with sets and Venn diagrams enables us to discover information about overlapping sets of data, as shown in the following examples.

EXAMPLE 3.15

Data for the nine planets of the sun show that seven have satellites, four have a period of axial rotation of over 20 hours, and two have both satellites and a period of axial rotation of over 20 hours.

a. How many planets have no satellites?

b. How many planets have satellites and a period of axial rotation of not over 20 hours?

c. How many planets have neither satellites nor a period of axial rotation of over 20 hours?

SOLUTION We let S denote the set of planets with satellites; for example, Earth $\in S$. Let R denote the set of planets that have a period of axial rotation of over 20 hours. Also, Earth $\in R$. The Universal set U is the set of all nine planets. In ▶ Figure 3.21, the 2 in the diagram indicates the two planets in S and R.

a. We want to find the number of elements in S'. Since there are nine elements in U and seven in S, there are two elements in S'; i.e., there are two planets that have no satellites.

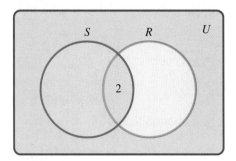

▶ FIGURE 3.21

b. Here we want the number of elements in $S \cap R'$. Since there are seven elements in S and two of them are in $S \cap R$, there are five elements in $S \cap R'$ (the horizontally shaded part of the diagram). Thus, there are five planets that have satellites and a period of axial rotation of not over 20 hours.

c. This question asks for the number of elements in $S' \cap R'$. This is the vertically shaded part of the diagram. Since there are four elements in R and five elements in $S \cap R'$ (the horizontally shaded part), there are nine elements in $S \cup R$. There are only nine planets, so there are no elements in the vertically shaded part. Therefore, there are no planets that have neither satellites nor a period of axial rotation of over 20 hours.

EXAMPLE 3.16

Of the 50 states in the United States, 13 border on Canada, 10 have the Mississippi River as a boundary, and 8 touch one of the Great Lakes. Of the states on the Canadian border, only Minnesota is on the Mississippi River. Of the Great Lakes states, three do not border Canada. Three of the Great Lakes states have the Mississippi River as a boundary. And only Minnesota borders Canada, has the Mississippi River as a boundary, and is a Great Lakes state.

a. Of the states not on the Mississippi, how many are Great Lakes states bordering Canada?
b. How many states do not border Canada?
c. How many states touch neither the Mississippi River nor a Great Lake?

Comment There are at least three ways that we could proceed:

1. Have an extremely impressive recall of U.S. geography.
2. Consult a map of the United States.
3. Construct a Venn diagram.

The intent of this example is to use the third approach. This example is typical of factual information about two or three sets of objects. Often the information comes from a survey—say, of political beliefs—and a map would not be helpful.

SOLUTION Let C denote the set of states bordering Canada, M denote the set of states having the Mississippi River as a boundary, and G denote the set of states on the Great Lakes. In ▶ Figure 3.22, we label the regions from I to VIII.

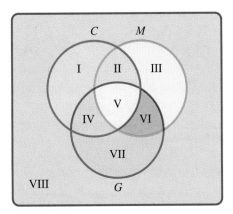

▶ FIGURE 3.22

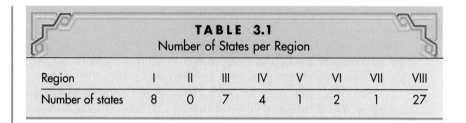

TABLE 3.1
Number of States per Region

Region	I	II	III	IV	V	VI	VII	VIII
Number of states	8	0	7	4	1	2	1	27

We keep track of our counting in ● Table 3.1. We start with the region common to all three sets, region V. It has one element. Next, we focus on regions common to two of the sets. Since, of the states on the Canadian border, only Minnesota has the Mississippi River as a boundary, there is one element in $C \cap M$ and so region II has no elements in it. Of the Great Lakes states, three do not border Canada, so five of the Great Lakes states border Canada (i.e., there are five elements in $G \cap C$). Since one of these states is in region V, four are in region IV. Since three of the Great lakes states have the Mississippi River as a boundary, $G \cap M$ has three elements. Because one of them is in region V, the other two are in region VI. Thirteen states border Canada, and five of them are in regions II, IV, and V; consequently, there are eight states in region I. Since there are 10 states on the Mississippi River and three of them are in regions II, V, or VI, seven states must be in region III. Since there are eight Great Lakes states and seven are in regions IV, V, or VI, there must be one in region VII. Regions I to VII account for 23 states, so there are $50 - 23 = 27$ in region VIII.

A good approach to questions like these is to start with a blank diagram and add numbers to it sequentially: 8 in region I, 0 in region II, and so on. The number of elements in each region is shown in ▶ Figure 3.23 and we can now answer the questions.

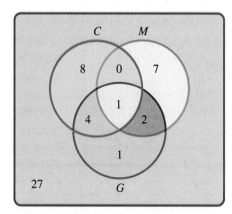

▶ FIGURE 3.23 Number of states per region

a. We want the number of elements in $M' \cap (G \cap C)$. There are four states in region IV.
b. Since 13 states border Canada, $50 - 13 = 37$ states do not. Alternatively, these states are in regions III, VI, VII, or VIII, so there are $7 + 2 + 1 + 27 = 37$ such states.
c. We want the number of elements in $M' \cap G'$ (i.e., the number of elements in region I or VIII). There are $8 + 27 = 35$ such states.

1. Based on the data in Example 3.15 answer each of the following.
   a. How many planets have a period of axial rotation of over 20 hours but do not have satellites?
   b. How many planets have either a satellite or a period of axial rotation of over 20 hours?

2. Of the first 103 chemical elements, 12 have a boiling point below that of water (100°C), 13 have a melting point below that of ice (0°C), and 11 have both a boiling point and a melting point below those of water/ice.
   a. How many elements have a boiling point of at least 100°C?
   b. How many elements have a boiling point at least that of water but a melting point below that of ice?
   c. How many elements have a melting point at least that of ice but a boiling point below that of water?
   d. How many elements have a boiling point at least that of water and a melting point at least that of ice?

3. Mike and Kathy each made a list of favorite courses. Mike's list had nine courses and Kathy's list had seven courses. When they compared their lists, they found that a total of only 12 different courses had been mentioned.
   a. How many courses were on both of their lists?
   b. How many courses were on Mike's list but not on Kathy's list?
   c. How many courses were on Kathy's list but not on Mike's list?

4. There are 37 major league baseball pitchers with more than 250 career victories. Of these, 19 have 300 or more career victories, 28 have 200 or more career losses, and 14 have both 300 or more career victories and 200 or more career losses.
   a. How many major league pitchers have more than 250 but fewer than 300 career victories?
   b. How many major league pitchers have 300 or more career victories and fewer than 200 career lossses?
   c. How many major league pitchers have 200 or more career losses and between 251 and 299, inclusive, career victories?
   d. How many major league pitchers have between 251 and 299, inclusive, career victories and fewer than 200 career losses?

5. Refer to the information in Exercise 4. Of the 37 major league pitchers with more than 250 career victories, 13 have winning percentages of .600 or more; 10 have 300 or more career victories and a winning percentage of .600 or more; 5 have 200 or more career losses and a winning percentage of .600 or more; and 5 have 300 or more career victories, 200 or more career losses, and a winning percentage of .600 or more.
   a. How many major league pitchers have 300 or more career victories, 200 or more career losses, and a winning percentage under .600?
   b. How many major league pitchers have 300 or more career victories, fewer than 200 career losses, and a winning percentage of .600 or more?

   c. How many major league pitchers have between 251 and 299, inclusive, career victories, over 200 career losses, and a winning percentage of .600 or more?
   d. How many major league pitchers have 300 or more career victories and fewer than 200 career looses?
   e. What can be said about the winning percentage of those pitchers in part d?
   f. How many major league pitchers have between 251 and 299, inclusive, career victories and over 200 career losses?
   g. What can be said about the winning percentage of those pitchers in part f?
   h. How many major league pitchers have between 251 and 299, inclusive, career victories, fewer than 200 career losses, and a winning percentage of .600 or more?
   i. How many major league pitchers have between 251 and 299, inclusive, career victories, fewer than 200 career losses, but a winning percentage under .600?

6. On a 50-question final examination, half the questions are multiple choice, half the questions are comprehensive, and 20 questions are open-book. Of the open-book questions, six are multiple choice, seven are comprehensive, and four are comprehensive–multiple choice. Six of the comprehensive questions are closed-book essay questions.
   a. How many of the comprehensive questions are multiple choice?
   b. How many of the multiple choice questions are closed-book and not comprehensive?
   c. How many of the open-book questions are not multiple choice?
   d. How many questions are closed-book, not multiple choice, and not comprehensive?

7. The following data on 1990 wheat, rice, and corn production in 63 selected countries are from *The World Almanac and Book of Facts 1994.*

   Six countries had wheat production of over 20,000 metric tons.
   Four countries had rice production of over 20,000 metric tons.
   Three countries had corn production of over 20,000 metric tons.
   Two countries had both wheat and rice production of over 20,000 metric tons.
   Two countries had both wheat and corn production of over 20,000 metric tons.
   Only China had both rice and corn production of over 20,000 metric tons.
   China had wheat, rice, and corn production of over 20,000 metric tons.

   a. How many countries (of the 63) had only wheat production (wheat production but neither rice nor corn production) of over 20,000 metric tons?
   b. How many countries had only rice production of over 20,000 metric tons?

c. How many countries had only corn production of over 20,000 metric tons?

d. How many countries had no grain (wheat, rice, or corn) production of over 20,000 metric tons?

8. In the 1992 summer Olympic games, 37 countries won gold medals, 44 won silver medals, 54 won bronze medals, 30 won both gold and silver medals, 33 won both gold and bronze medals, 36 won silver and bronze medals, and 28 won gold, silver, and bronze medals.

   a. How many countries won only gold medals?

   b. How many countries won only silver medals?

   c. How many countries won only bronze medals?

   d. How many countries won gold and silver medals but no bronze medals?

   e. How many countries won gold and bronze medals but no silver medals?

   f. How many countries won silver and bronze medals but no gold medals?

9. A group of 47 people is planning to order pizza. Each person is surveyed about three toppings (pepperoni, sausage, and mushrooms). The following data were compiled: 29 like pepperoni, 29 like sausage, 27 like mushrooms, 4 like only pepperoni, 2 like only sausage, 3 like only mushrooms, 10 like all three toppings, 9 like pepperoni and sausage but not mushrooms, 6 like pepperoni and mushrooms but not sausage, and 8 like sausage and mushrooms but not pepperoni.

   a. How many like none of the three toppings?

   b. How many like pepperoni but not sausage?

   c. How many like sausage but not pepperoni?

   d. How many do not like mushrooms?

   e. Eight possible pizzas could be ordered with 0, 1, 2, or 3 of these toppings. If the goal is to order the type that the most people like, what kind should be ordered?

WRITTEN ASSIGNMENTS 3.6

1. Make up a problem like those that appear in this section, based on data from your major or minor or another area in which you are interested.

2. This section discusses a problem-solving technique that is a special case of one of the problem-solving techniques mentioned in Chapter 1.

   a. Of which technique in Chapter 1 is this section's technique a special case?

   b. The technique mentioned in Chapter 1 is more general than the technique of this section. Give an example of a problem that could be solved by the technique mentioned in part a but not by the technique of this section.

CHAPTER 3 REVIEW EXERCISES

1. a. Who is considered the father of set theory?
   b. About when did he do his work?

2. Describe each of the following sets by the roster method.
   a. The set of letters in your last name
   b. {x | x is an ocean bordering the continental United States}

3. Describe each of the following sets in set-builder notation.
   a. The set of even positive integers
   b. {Alabama, Alaska, Arizona, Arkansas}

4. Give an example of a collection of objects that is *not* a well-defined set.

5. Describe each of the following sets as finite or infinite.
   a. The set of molecules of water in all the oceans on Earth
   b. The set of negative integers
   c. The U.S. national debt

6. Give an example of an infinite set other than those listed in Exercise 5.

7. Let S be the set of works written by William Shakespeare. Describe each of the following statements as true or false.

   a. *Hamlet* $\in S$
   b. *Star Wars* $\in S$
   c. *Macbeth* $\notin S$
   d. This book $\notin S$

8. Which symbol(s) are used to represent the set of living humans who are over 200 years old? $\varnothing$, { }, {$\varnothing$}

9. Let $A = \{a, b, c, d, \ldots, z\}$, $V = \{a, e, i, o, u\}$, and $Y = \{y\}$. State whether the following are true or false.

   a. $A \subseteq V$
   b. $A \subset V$
   c. $A \not\subseteq V$
   d. $A \not\subset V$
   e. $V \subseteq A$
   f. $V \subset A$
   g. $V \not\subseteq A$
   h. $V \not\subset A$
   i. $Y \subseteq V$
   j. $Y \subset V$

k. $Y \not\subseteq V$

l. $Y \not\subset V$

10. Let $V = \{a, e, i, o, u\}$.

   a. Which of the following could be universal sets for V?

      i. $\varnothing$

      ii. $I = \{a, e, i\}$

      iii. V

      iv. C = set of all consonants in the alphabet

      v. A = set of all 26 letters of the alphabet

   b. Name another universal set for V.

11. A set has five elements. How many subsets does it have?

12. An art gallery has seven Salvadori Dali paintings to display in a room and it wants to display two or more paintings. How many combinations are possible?

13. Write a paragraph reacting to (supporting or disagreeing with) the following statement. "Since mathematics deals with definite statements like if $2x + 3 = 11$, then $x = 4$, there is little or no room for conjecture in mathematics."

14. Consider the weighted voting scheme [26: 25, 24, 2].

   a. Find the Banzhaf power index for each voter.

   b. Is the power of voter 1 greater than, equal to, or less than the power of voter 2?

   c. Is the power of voter 2 greater than, equal to, or less than the power of voter 3?

15. Give an example of a voting situation that uses weighted voting.

16. *True or false?* In weighted voting, the influence of one's vote is proportional to the number of votes one has.

17. Let $M = \{A, B, C, D, E, F, G\}$ (the set of musical notes), $T = \{C, E\}$ (a major third), and $F = \{C, F\}$ (a perfect fourth). Perform the indicated set operation.

   a. $T \cup F$

   b. $T \cup M$

   c. $T \cap F$

   d. $T \cap M$

   e. $T \cap T$

   f. $T \cup T$

   g. $M - T$

   h. $T - M$

   i. F' (consider M to be the universal set)

   j. $(T \cup F)'$

   k. $(T \cap F)'$

   l. $T' \cup F$

   m. $T \cup F'$

   n. $T' \cup F'$

18. Consider the following sets of colleges or universities: $P = \{\text{public colleges}\}$, $T = \{\text{two-year colleges}\}$, and $E = \{\text{colleges with engineering programs}\}$. Describe symbolically each of the following (for example, $\{\text{private colleges}\} = P'$).

   a. Private two-year colleges

   b. Two-year colleges without engineering programs

   c. Four-year colleges or colleges with engineering programs

19. Let $U = \{1, 2, 3, 4, 5, 6, 7, 8, 9\}$. Use the Venn diagram to find each of the following:

   a. $V \cup T$

   b. $V \cap T$

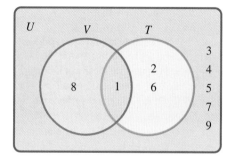

   c. V'

   d. $V - T$

   e. $V \cup T'$

   f. $(V \cap T)'$

20. Let $U = \{a, b, c, d, e, f\}$, $A = \{a, b, c, d\}$, and $B = \{b, d, f\}$. Draw a Venn diagram with U, A, and B labeled and each of the six elements of U placed appropriately.

21. Using the Venn diagram, write each of the following sets in terms of A, B, and set operations.

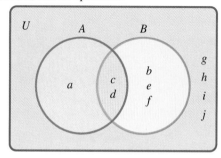

   a. $\{a\}$

   b. $\{c, d\}$

   c. $\{b, e, f\}$

   d. $\{g, h, i, j\}$

   e. $\{a, g, h, i, j\}$

   f. $\{c, d, g, h, i, j\}$

   g. $\{b, e, f, g, h, i, j\}$

   h. $\{a, b, e, f, g, h, i, j\}$

22. For each of the following draw a Venn diagram with a universal set U and three sets A, B, and C. Shade the indicated set.

   a. $A \cap (B \cup C)$

   b. $(A \cap B) \cup C$

   c. $A' \cap B$

   d. $A \cap (B \cup C)'$

   e. $A - (B \cup C)$

   f. $A' - (B \cap C)$

23. Let $U = \{1, 2, 3, \ldots, 16\}$, $O = \{1, 3, 5, 7, 9, 11, 13, 15\}$ (odd numbers), $P = \{2, 3, 5, 7, 11, 13\}$ (prime numbers), and $S = \{1, 4, 9, 16\}$ (squares). Draw a Venn diagram, label the regions, and place each element in the appropriate region.

24. Use appropriate sets to find LCM $\{4, 5, 6\}$. Draw a Venn diagram with numbers appropriately placed.

25. Construct Venn diagrams to determine whether or not the given sets are equal.

   a. $(A \cap B)'$ and $A' \cap B'$

b. $A \cap (B \cup C)'$ and $A - (B \cup C)$

c. $(A \cup B) \cap C$ and $A \cup (B \cap C)$

26. Of the 100 U.S. senators in mid-1993, 56 were Democrats, 5 were female, and 33 had terms ending in 1995. Only Dianne Feinstein met all three of these conditions. Nancy Kassenbaum was the only female Republican senator. Her term ended in 1997. Of the 33 senators whose terms ended in 1995, 20 were Democrats.

a. How many male Democrats had terms not ending in 1995?

b. How many male Republicans had terms not ending in 1995?

c. How many Republicans had terms ending in 1995?

d. Of those whose terms did not end in 1995, how many were Democrats?

 BIBLIOGRAPHY

Boyer, Carl B. (revised by Uta C. Merzbach). *A History of Mathematics,* 2nd ed. New York: Wiley, 1991.

Calinger, Ronald (ed.). *Classics of Mathematics.* Englewood Cliffs, NJ: Prentice-Hall, 1995.

COMAP. *For All Practical Purposes: Introduction to Contemporary Mathematics,* 3rd ed. New York: W.H. Freeman, 1994, Chap. 12.

Eves, Howard. *An Introduction to the History of Mathematics,* 6th ed. Chicago: Saunders College Publishing, 1990.

Johnson, Phillip E. *A History of Set Theory.* Boston: Prindle, Weber & Schmidt, 1972.

Katz, Victor J. *A History of Mathematics.* New York: HarperCollins College Publishers, 1992.

CHAPTER 4

LOGIC

PROLOGUE

What do the fields of archaeology, biology, economics, finance, gerontology, law, mathematics, philosophy psychology, and sociology have in common? They contain different subject matter, although there is some overlap between pairs of them. What they all do have in common is a certain standard of what is rational. They share that commonality of reasoning with many other fields of

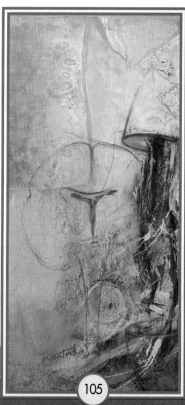

study. In other words, many fields of study share an acceptance of basic principles of logic.

In fact, most, if not all, rational inquiry depends on logic and is judged for its soundness by logic. Academic reasoning, laws, and social conventions all rely on rational behavior. People may disagree because of differences of opinion on the truth of statements in an argument, but there is nearly universal agreement on what constitutes a valid or an invalid agrument. This agreement is housed in the province of logic.

Reasoning takes place in the mind. The accompanying print *States of Mind: A State of the World* by Todd Siler suggests that rational inquiry takes place in our minds and that the state of the world is dependent on our application of the principles of logic.

The principles of valid argument are couched in terms of logic, but reasoning takes place in the mind. That's the subject area of psychology. Psychologists have studied one area of reasoning with a "selection task" developed by P. C. Wason. One form of it follows.

Imagine that you are a police officer on duty. You are patrolling a party where if a person is drinking beer, then the person must be age 21 or older. Before you are four cards with information about four people at the party. On one side of a card is a person's age and on the other side of the card is what the person is drinking. Select the card or cards that you definitely need to turn over to determine whether or not the people are violating the drinking-age rule.

Suppose you are presented with the cards "age 16," "age 22," "drinking beer," and "drinking lemonade." Which cards do you definitely need to turn over to determine which, if any, of these four people are violating the drinking-age rule? (Your answer will be requested in Exercise 32 of Section 4.4.)

Psychologists have found that people reason better with a specific rule like the drinking-age rule than with a more abstract statement like "if *p*, then *q*." This says that people not only reason with the mind, but they do it in a way that goes beyond the mechanical rules of logic.

Nonetheless, at the base of reasoning are the principles of logic. They are based on the states of mind. They affect the state of the world.

4.1 INTRODUCTION

GOALS
1. To give a description of the historical development of the subject of logic, including some approximate dates, the names of some people involved, and some important concepts that have developed.
2. To describe Gödel's incompleteness theorem and its impact on society

Have you ever tried to explain something to a questioning 2-year-old child? Have you ever had a disagreement with your parents in which you tried to present your thoughts in a rational manner but they still wouldn't accept them? Have you ever heard a trial lawyer presenting a case for or against a defendant? These situations all involve the use of logic.

What is logic? At its core, *logic* is the process of combining statements to arrive at conclusions. Viewed as a science, the study of logic requires careful definition of terms and careful analysis of relationships among statements. The result of this study is to determine when conclusions are valid.

Logic is involved in many disciplines. Rules of logic are applied whenever conclusions are made based on facts and relationships. The formal study of logic is conducted in several areas. Philosophy departments frequently offer courses in logic and reasoning. The networks in electronic circuits and computer chips are based on properties studied in logic courses. Lawyers are required to be knowledgeable in logical principles. Why then does a mathematics text include a chapter on logic?

Logic serves in mathematics to establish truth, and it is one of the subdisciplines within mathematics. This chapter will give you a sense of the subdiscipline of logic. In mathematical modeling, one goes from a mathematical model to mathematical conclusions via logic. New mathematics is created by making conjectures. These are then proved by logical arguments. Logic has been called the glue that binds mathematics together. It is the nature of mathematical proof—using logic to establish statements that follow with absolute certainty from the axioms of the system—that separates mathematics from all other disciplines. Mathematics is about truth and proof. Truth is established by proof, and proof relies on logic.

Proofs were probably part of the geometry you studied in high school. Almost all of them go back to the ancient Greeks and in particular to Euclid in around 300 B.C.E. Mathematician Morris Kline puts the connection between Euclidean geometry and logic more strongly (Kline, p. 130):

> By teaching [hu]mankind the principles of correct reasoning Euclidean geometry has influenced thought even in fields where extensive deductive systems could not be or have not thus far been erected. Stated otherwise, Euclidean geometry is the father of the science of logic.

OY! WHAT'S IN A NAME?

If Euler is pronounced "oiler," why isn't Euclid pronounce "oiclid"?

The ancient Greek philosopher Aristotle (384–322 B.C.E.) was one of the first to champion the use of proof to establish truth. He improved the logical rigor of proofs by introducing syllogistic logic. (Here's an example of a syllogism: From the facts that (1) all men are mortal and (2) Socrates is a man, it can logically be argued that Socrates is mortal.) Many call Aristotle "The Father of Logic."

The formalization of logic in written symbols rather than words started with the German mathematician Gottfried Wilhelm Leibniz (1646–1716). He sought a method of formal reasoning with exact rules that would increase the precision and rationality of everyday life.

Leibniz's work was converted into "symbolic logic" by the English mathematician and logician George Boole (1815–1864). Boole developed a system of writing logical statements in symbols so they can be manipulated much like expressions in algebra. He used 1 to represent the universal set, 0 to represent the empty set, and $1 - x$ to represent the complement of a set x. His algebra, now called Boolean algebra, also has applications in engineering problems, such as the design of electrical switching circuits, and in computer science in the design of circuits for computers. Boole's work in this area is contained in the books *The Mathematical Analysis of Logic* (1847) and *An Investigation of the Laws of Thought* (1854). In the latter he stated, "The object of science . . . is the knowledge of laws and relations." He used his algebra of symbolic statements to discover and verify laws and relations.

Gottlob Frege (1848–1925), regarded as the founder of modern mathematical logic, is another person who developed Leibniz's ideas and also extended Boole's

Leibniz: The Last Universal Scholar

Gottfried Wilhelm Leibniz was a German child prodigy who by the age of 12 had taught himself Latin and Greek, by age 18 had completed his bachelor's degree, and by age 20 had essentially completed all of the requirements for a doctor of law degree—which was denied him, ostensibly because of his youth. He shortly thereafter received his doctorate in jurisprudence from another university. He went to work in diplomatic service for the Elector of Mainz, the head of one of the many small states that then comprised Germany.

Leibniz is regarded as perhaps the last universal scholar. He was well versed in law, philosophy, theology, and mathematics. His two notable mathematical achievements are the development of calculus (Isaac Newton is the other developer) and his pioneering work in what is now called symbolic logic. Leibniz was also a great builder of mathematical notation. He was probably second only to Leonhard Euler in that regard. He also helped popularize complex numbers and intertwined a binary system (base two) with his theology.

His beliefs embodied a very optimistic view of the world. It was this view that was satirized by the French author Voltaire in *Candide* (1759).

Gottfried Wilhelm Leibniz (1646–1716)

work. He built logic on explicit axioms (assumptions). In looking at the relation between mathematics and philosophy, he once said, "every good mathematician is at least half a philosopher and every good philosopher is at least half a mathematician" (quoted in Calinger, 1995, p. 646).

Another pioneer in symbolic logic was Giuseppe Peano (1858–1932). He saw his contributions in this area as responding to Leibniz's call for a universal scientific language. He is also known for his work in the axiomatization of mathematics, including the Peano postulates for the natural numbers.

Cantor's developments in set theory led to showing that almost all mathematical objects, including numbers, could be defined in terms of sets. This meant that the axioms for numbers, geometry, and other branches of mathematics could all be stated in terms of axioms about sets. This led to attempts to put all of mathematics on a more rigorous footing. In 1900 David Hilbert, probably the greatest mathematician of his time, focused part of this effort by suggesting that one of the problems that mathematicians should attack is to show that the axioms of arithmetic are consistent; i.e., in using the axioms of arithmetic, one must not be able to prove, via principles of logic, that a given statement is both true and false. This was problem 2 in a list of 23 problems presented by Hilbert in an address to the International Congress of Mathematicians.

A decade after Hilbert's problems were stated, Bertrand Russell (1872–1970) and Alfred North Whitehead (1861–1947) published the first volume of the three-volume *Principia Mathematicia,* an attempt to develop the fundamental properties of numbers (arithmetic) from a precise set of axioms. However, it did not solve Hilbert's second problem.

The solution to Hilbert's second problem was a surprise. In 1931 Kurt Gödel (1907–1978) showed that in any system that contains ordinary arithmetic and uses

David Hilbert, the great man of mathematics at the turn of the twentieth century.

ON A TANGENT

George Boole: Pure Mathematician

George Boole attended a local elementary school in his home town in England and briefly attended a commercial school. His favorite subject was the classics. By age 14 he knew Latin, Greek, French, and German. At age 15 his father's business as a cobbler had serious problems, so Boole went to work. He taught elementary school and at age 20 opened his own elementary school in his native Lincoln, England. During these times (the 1830s), Boole's interest in mathematics blossomed.

Boole's wife was Mary Everest, the niece of Sir George Everest, after whom Mt. Everest is named. They had five daughters.

It is ironic that Georg Cantor, who in his pioneering work on set theory had problems with the opposition to it by Leopold Kronecker, did not accept Boole's work. However, a majority of Boole's contemporaries accepted his work and recognized its importance. Bertrand Russell, a noted twentieth century philosopher and mathematician, thought of Boole's work as the prototype of what mathematics is, and credited him as being the discoverer of "pure mathematics," in Boole's work called *The Laws of Thought.*

Bertrand Russell: Mathematician and Philosopher

Bertrand Russell was one of the most profound and influential intellectuals of the twentieth century. He was a mathematician and philosopher. Substantial parts of his work were in logic. He wrote *A Critical Exposition of the Philosophy of Leibniz* in 1900, corresponded with Frege, and met Peano. His three-volume *Principia Mathematica,* published in 1910, 1912, and 1913, has been very influential with logicians. He maintained that mathematics can be derived from a small number of self-evident logical principles. Among his awards was the 1950 Nobel prize for literature "as a defender of humanity and freedom of thought."

The relations between some mathematicians have been less than cordial. However, a strong statement about the character of Gottlob Frege and the relationship between him and Bertrand Russell is made in the following letter from Russell to a publisher.

I should be most pleased if you would publish the correspondence between Frege and myself, and I am grateful to you for suggesting this. As I think about acts of integrity and grace, I realize that there is nothing in my knowledge to compare with Frege's dedication to truth. His entire life's work was on the verge of completion, much of his work had been ignored to the benefit of men infinitely less capable, his second volume was about to be published, and upon finding that his fundamental assumption was in error, he responded with intellectual pleasure clearly submerging any feelings of personal disappointment. It was almost superhuman and a telling indication of that of which men are capable if their dedication is to creative work and knowledge instead of cruder efforts to dominate and be known.

Bertrand Russell

---

the ordinary axioms about sets, there are statements that can be proved neither true nor false within the system. This rendered Hilbert's hope of rigorously proving all true statements in arithmetic an impossibility.

Consider the implications of this for other disciplines. If in arithmetic there are statements that can neither be proved true nor be proved false, must this not also be the case in other disciplines? This mathematical result had a profound impact on the intellectual community. It was similar to the impact on the Pythagoreans of the discovery that $\sqrt{2}$ is irrational. Mathematics again had influenced society.

Gödel's result emphasized that in an axiomatic system, not only is truth dependent on the collection of axioms, but also no matter what collection of axioms is used there will be some statements whose truth values cannot be determined. In set theory, the continuum hypothesis is such a statement for typical axioms.

Kurt Gödel and His Axiomatic Analysis of the Constitution

Kurt Gödel was born in Brunn, Moravia (now Brno, Czech Republic). As a child, he was characterized by his curiosity. His family called him Herr Warum (Mr. Why). He received his doctorate at the University of Vienna and taught there until 1938, when, in the face of Nazi domination of eastern Europe, he emigrated to the United States. In the United States he was affiliated with the Institute for Advanced Study in Princeton, New Jersey.

In 1948 Gödel decided to become a U.S. citizen. John Casti (1990, pp. 373–374) tells the story.

[I]n his characteristically thorough way, he began a detailed study of the U.S. Constitution in preparation for the citizenship examination. On the day before the exam Gödel called his friend, the noted economist Oskar Morgenstern, saying with great excitement and consternation that he had discovered a logical flaw in the Constitution, a loophole by which the United States could be transformed into a dictatorship. Morgenstern, who along with Einstein was to serve as one of Gödel's witnesses at the examination the next day, told him that the possibility he had uncovered was extremely hypothetical and remote. He further cautioned Gödel not to bring the matter up the next day at the interview with the judge.

The following morning Einstein, Morgenstern, and Gödel drove down to the federal courthouse in the New Jersey state capital of Trenton, where the citizenship examination was to take place. As legend has it, Einstein and Morgenstern regaled Gödel with stories and jokes on the trip from Princeton to Trenton in order to take his mind off the upcoming test. At the interview itself the judge was suitably impressed by the sterling character and public personas of Gödel's witnesses, and broke with tradition by inviting them to sit in during the exam. The judge began by saying to Gödel, "Up to now you have held German citizenship." Gödel corrected this slight affront, noting that he was Austrian. Unfazed, the judge continued, "Anyhow, it was under an evil dictatorship . . . but fortunately, that's not possible in America." With the magic word *dictatorship* out of the bag, Gödel was not to be denied, crying out, "On the contrary, I know that can happen. And I can prove it!" By all accounts it took the efforts of not only Einstein and Morgenstern but also the judge to calm Gödel down and prevent him from going into a detailed and lengthy discourse about his "discovery."

Such is the mind of the man who analyzed the axiom set of the U.S. Constitution in the same way he analyzed the axioms of arithmetic.

Kurt Gödel (second from right) receives the first Albert Einstein Award for Achievement in the Natural Sciences from Einstein (left). Also pictured are Julian Schwinger (right), a co-recipient of the award, and Lewis T. Strauss (second from left), president of the board of trustees of the Institute for Advanced Study.

Because of this and other results, Gödel is recognized as one of the most outstanding logicians of all times. Astronomer John Barrow calls him (1992, p. 117) "the most famous logician of all time," and mathematical historian Ronald Calinger describes him (1995, p. 744) as "the leading mathematical logician of the twentieth century." Mathematical historians Carl Boyer and Uta Merzbach (1991, p. 611) say that Gödel's theorem is "sometimes regarded as the most decisive result in mathematical logic."

What is logic? Let's answer this time with the words of some of the people we've met.

> To discover truths is the task of all sciences; it falls to logic to discern the laws of truth.
> — Frege, 1918

> Logic is the theory of pure concepts; it includes set theory as a proper part.
> — Gödel, 1971 and 1975

Why should you study logic? First, if your purpose in this course is to learn about mathematics, you must have some understanding of the process by which mathematical systems are built. Second, an understanding of logical principles will enable you to organize your thoughts better and to present your arguments more convincingly. Finally, you will be better at analyzing others' arguments. Is the arguer's reasoning logically correct? Is an advertisement making misleading statements? What is the difference between a valid and an invalid conclusion? The goal of this chapter is to provide the tools for answering such questions.

WRITTEN ASSIGNMENTS 4.1

1. a. Write a two- or three-page paper on Hilbert and his second problem. Include a brief biography of Hilbert, a description of the work on Hilbert's second problem, and some of its spin-offs, and a description of its current status. Two references are Campbell and Higgins, 1984, Vol. I, pp. 273–278, 300, and Eves, 1990.

   b. Write a three- to five-page paper on Hilbert's problems (23 of them). What is their current status? Two references are Campbell and Higgins, 1984, Vol. I, pp. 273–278, 300–304, and Eves, 1990.

2. (*Library research*) Write two more facts about each of the following.
   a. Aristotle
   b. Leibniz
   c. Boole
   d. Frege
   e. Peano
   f. Russell

3. a. Contact a mathematician and get her or his reaction to Frege's belief that every good mathematician is at least half a philosopher and every good philosopher is at least half a mathematician.

   b. Do the same for a philosopher.

4. We commented that Gödel's incompleteness theorem had an impact on society similar to the impact on the Pythagoreans of the discovery that

$$\sqrt{2}$$

is irrational. Write a three- to five-page paper describing each of these discoveries, their impact on society, and the relations between the impacts. Assume that your audience is a class just starting this course, one of whose goals is to look at the impact of mathematics on society.

⬤ **4.2** ⬤ TERMINOLOGY AND NOTATION

GOALS
1. To give an example of a statement and, given a phrase or sentence, to tell whether or not it is a statement.
2. Given a statement, to write it in symbolic form using the connectives ∧, ∨, or ~.
3. To write an English translation of a symbolic statement involving ∧, ∨, or ~.
4. To consider conjunction, disjunction, and negation as operations on statements, and to relate these operations to the set operations of intersection, union, and complementation.

The study of any area of mathematics requires knowledge of basic terminology and agreement as to the notation for representing various concepts. We continue our standard approach of starting with terminology and notation for logic, even though it is studied as part of other disciplines.

Statements

The fundamental unit in a study of logic is a statement.

> DEFINITION
> ──●──
>
> A *statement* is a sentence that can be judged as either true or false.

Some examples:

There are seven days in a week.
There are eight days in a week.
$3 + 4 = 7$

Not all sentences are statements in the previously defined sense. For example, the following sentences fail to satisfy the definition of a statement.

Sue has an interesting personality.
Mathematics is a fascinating subject.
$x + 3 = 7$

What is common to each of these sentences is that there is no universally agreed-upon answer.

For the remainder of this chapter, we use the term *statement* in the defined sense. That is, a *statement* is a sentence that can be judged as either true or false. There may be times when we want to include as statements sentences such as "Sue has an interesting personality" or "All teenagers want a car." When we do, it will be under the assumption that a truth value has been assigned. Example 4.1 provides additional practice in recognizing statements.

For each of the following sentences, indicate whether it is or is not a statement. Explain each answer.

a. Telephones are a nuisance.
b. A telephone is a communication device that was not available in 1900.
c. The Berlin Wall came down in 1990.
d. Life as a university student is great.

SOLUTION In some respects, analysis of these sentences requires you to decide whether or not they are opinions. Since opinions may vary about issues, it is not always possible to determine whether an opinion is true or false.

a. Not a statement. Some people may consider the telephone a nuisance, others may find it an invaluable asset.
b. A statement. It is false. As you may know, Alexander Graham Bell patented the telephone in 1876.
c. A statement. It is true.
d. Not a statement. Although many students would agree with this sentence, many others would disagree, feeling themselves overwhelmed by a combination of social, psychological, and financial demands.

We now introduce some notation to accompany the concept of a statement. We will use a single letter to denote a statement. Thus, for the statements "This book has a blue cover" and "Boise is the capital of Idaho," we write

p : This book has a blue cover
q : Boise is the capital of Idaho

We can then talk about statements p and q. It is customary, though not essential, to use letters such as p, q, r, and s to denote statements.

Connectives

Statements can be combined to form compound statements by the use of *connectives*. The two connectives considered here are "and" and "or." Formally, we have the following notations.

NOTATION

Let p and q be statements. The compound statement "p and q" is denoted

$$p \wedge q$$

NOTATION

Let p and q be statements. The compound statement "p or q" is denoted

$$p \vee q$$

The statement "*p* and *q*" ($p \wedge q$) is called the *conjunction* of *p* and *q*, and the statement "*p* or *q*" ($p \vee q$) is called the *disjunction* of *p* and *q*.

Notice the similarity of this notation with set notation. The statement "*p* and *q*" is denoted $p \wedge q$, and the intersection of sets *A* and *B* is denoted $A \cap B$; the statement "*p* or *q*" is denoted $p \vee q$, and the union of sets *A* and *B* (*A* or *B*) is denoted $A \cup B$.

The comparison of the logic connectives "and" and "or" with the set operations intersection and union goes beyond notation to basic meaning. Define the statements:

$p(x) : x$ is in *A*
$q(x) : x$ is in *B*

Then, $\qquad A \cap B = \{x : x \text{ is in } A \text{ and } x \text{ is in } B\} = \{x : p(x) \wedge q(x)\}$

and $\qquad A \cup B = \{x : x \text{ is in } A \text{ or } x \text{ is in } B\} = \{x : p(x) \vee q(x)\}$

Notice also that the use of "or" in logic, what is called the inclusive use of "or," is the same as its use in set theory with union; namely, $p \vee q$ means *p* or *q* (or both *p* and *q*). The inclusive or is sometimes written in ordinary usage as "and/or." For example, the statement "Cindy's father's name is George or her mother's name is Argie" might be written "Cindy's father name is George and/or her mother's name is Argie" to emphasize that it is intended to be true if both parts are true.

Notice also that the connectives "and" and "or" are binary operations on statements: They take any two statements *p* and *q* and produce the new statements $p \wedge q$ and $p \vee q$. As such, they are comparable to $+, -, \times, \div$ on numbers and to $\cap$ and $\cup$ on sets.

When statements are represented by letters and combined via the symbols "$\wedge$" and "$\vee$," we say the statements are written in *symbolic form*. This is illustrated in the next example.

LET US JOIN TOGETHER

The word *conjunction* comes from the Latin *con-*, meaning "together with," and *iunct-*, the past participial stem of *iungere*, meaning "to join." The Indo-European root word *yeug-*, meaning "to join," was transformed into the native English word *yoke*, which is a device for joining two animals. The *dis-* in *disjunction* comes from the Latin prefix *dis-*, meaning "apart" or "away."

EXAMPLE 4.2

Express the following compound statement in symbolic form.

Jim is hiking the Appalachian Trail and Donn lives in Knoxville, TN.

Comment on Instructions: When the instructions in this book say to write a statement "in symbolic form," we mean that each of the labeled parts should be free of connectives. That is, each of the labeled statements should not involve the connective "and" or "or" or any connectives we introduce later. We will refer to these "free-of-connectives" statements as *simple statements*. Thus, a simple statement is one free of connectives; and when we assign labels *p, q, r*, etc. to statements, we intend that they be simple statements.

SOLUTION We first isolate and label the simple statements. This produces

p : Jim is hiking the Appalachian Trail
q : Donn lives in Knoxville, TN

We then use these designations with the connective "$\wedge$" to write the symbolic form as

$$p \wedge q$$

The following example provides additional practice in writing the symbolic form of compound statements.

EXAMPLE 4.3

Consider the following statements:

p : Helen drives a Buick
q : Mike plays basketball
r : Jack is a golfer

Write each of the following in symbolic form:

a. Helen drives a Buick and Jack is a golfer
b. Mike plays basketball or Jack is a golfer

SOLUTION These are written symbolically by replacing each statement with its designated letter, replacing "and" with "$\wedge$," and replacing "or" with "$\vee$". We then have

a. $p \wedge r$
b. $q \vee r$

Negation

The *negation* of a statement is a statement with the opposite truth value of the original statement. If p is a statement, "not p" is its negation. We denote this "$\sim p$."

NOTATION

Let p be a statement. The statement "not p" is denoted $\sim p$.

It is not difficult to write the negation of a simple statement. If the statement is "Richard is 51 years old," a negation is "Richard is not 51 years old." It is possible to write in a somewhat more stilted manner and say, "It is not the case that Richard is 51 years old." However, the shorter version is more common.

When considering the truth value of a negation, we note that the truth values are reversed from those of the original statement. To illustrate, let p be the statement

p : Sacramento is in California

This is a true statement; its negation, $\sim p$, is thus false.

Negation is called a *unary operation* on statements: It takes any statement p to another statement $\sim p$.

 YOUR FORMULATION

1. The connectives "and" and "or" are binary operations on statements. Is the connective "not" a binary operation on statements? Why or why not?
2. State an arithmetic operation that the connective "not" is like.
3. What set operation is the connective "not" like? In particular, if A is a set and we define a statement $p(x)$ by

$p(x)$: x is in A

then name $\{x : \sim p(x)\}$.

Multiple Connectives

We conclude this section by considering how statements can be related by more than one connective and/or by negation. We have all heard conversations containing compound statements such as "John is going home and Mary is staying here or Joe will be here on Saturday." Symbolically, we have the statements

p : John is going home
q : Mary is staying here
r : Joe will be here on Saturday

The compound statement can be written as

$$p \wedge q \vee r$$

At this point, it is not clear what the compound statement means. Symbolically, you could have either

$$(p \wedge q) \vee r \quad \text{or} \quad p \wedge (q \vee r)$$

These groupings produce different meanings for the original compound statement. Typically, such ambiguities are avoided by adding commas when writing. If the statement is meant to say

$$(p \wedge q) \vee r$$

we write "John is going home and Mary is staying here, or Joe will be here on Saturday." The expression

$$p \wedge (q \vee r)$$

represents "John is going home, and Mary is staying here or Joe will be here on Saturday." This convention groups statements that are separated by commas. Although some texts assign different priorities to "$\wedge$" and "$\vee$," we believe it is best always to use parentheses to avoid possible misinterpretation.

Negation can also be used with connectives. With negation it is important to add parentheses for clarification of meaning. For example, consider the sentence "It is not the case that Ohio State won the Rose Bowl and Notre Dame won the Orange Bowl." If we let

p : Ohio State won the Rose Bowl
q : Notre Dame won the Orange Bowl

two possible symbolic forms would be

$$\sim (p \wedge q) \quad \text{and} \quad (\sim p) \vee q$$

The parentheses certainly clarify the meaning of the statements. Which one of these does the English sentence mean? The sentence is vague. It could be clarified by saying either, "It is not the case that both Ohio State won the Rose Bowl and Notre Dame won the Orange Bowl," which would be written as $\sim(p \wedge q)$, or "Notre Dame won the Orange Bowl and Ohio State did not win the Rose Bowl," which would be written as $q \wedge (\sim p)$ and would be equivalent to $(\sim p) \wedge q$. For a sentence of the form "It is not the case that . . . ," we consider the entire statement after "that" (the entire ". . ." part) negated.

In statements involving negation and connectives, parentheses help us to be precise. They help eliminate ambiguity. However, their use can get unwieldy. For example, consider the statement $(\sim p) \wedge (\sim q)$. We can eliminate two sets of parentheses by writing it as $\sim p \wedge \sim q$ and agreeing that the operation of negation takes

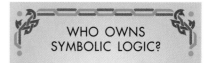

precedence over the operations of $\land$ or $\lor$. This is similar to the use of negation with numbers or the use of complement with sets. This convention allows us to write the statement $(\sim p) \land q$ of the previous paragraph as $\sim p \land q$.

Our final examples in this section provide practice in translating between written form and symbolic form when using negation and multiple connectives.

EXAMPLE 4.4

Consider the statements

p : Rosa is a lawyer
q : Rosa drives a blue car

Write an English translation for each of the following symbolic statements:

a. $p \land (q \lor r)$
b. $(\sim p \land q) \lor r$

SOLUTION Using commas to separate simple statements, we get:

a. Rosa is a lawyer, and Rosa drives a blue car or Rosa is taking night classes.
b. Rosa is not a lawyer and Rosa drives a blue car, or Rosa is taking night classes.

EXAMPLE 4.5

Consider the statements

p : Rosa is a lawyer
q : Rosa drives a blue car
r : Rosa is taking night classes

Provide symbolic forms for each of the following compound statements.

a. Rosa is a lawyer who drives a blue car, and is not taking night classes.
b. Rosa does not drive a blue car or Rosa is a lawyer, and Rosa is taking night classes.

SOLUTION Taking commas to indicate grouping and using the convention on priority of operations we obtain

a. $(p \land q) \land (\sim r)$
b. $(\sim q \lor p) \land r$

EXERCISES 4.2

In Exercises 1–3, indicate whether what is written is a statement.

1. Valentine's Day is February 14.
2. Valentine's Day is February 16.
3. Martha Washington
4. William Shakespeare is the world's greatest playwright.
5. Is Hillary Clinton the First Lady of the United States?

In Exercises 6–14 write the symbolic form of the given statement in terms of simple statements.

6. Cindy is shopping and Mike is at basketball practice.
7. Steve is going to Paul's house or Paul is coming to Steve's house.
8. Eugene is not the capital of Oregon.

9. The party is on Friday evening or it is on Saturday noon, and we should bring candy.
10. The sun is shining and it is 80°, or I am dreaming.
11. This is not the case: The New York Giants played in the 1992 Super Bowl and the Los Angeles Rams played in the 1992 Super Bowl.
12. You will be home by midnight or you will not go out tomorrow night.
13. Vote for Smith for governor and vote for Brown for lieutenant governor, or don't vote.
14. Study this material, or don't expect a passing grade and don't expect extra help.

For Exercises 15–22, consider the statements

p : Suriname is in South America
q : Nigeria is in Africa
r : The United States is in Europe

Write an English translation for each of the following symbolic statements.

15. $\sim r$
16. $p \wedge q$
17. $q \vee p$
18. $r \wedge \sim p$
19. $p \wedge (q \vee r)$
20. $\sim p \wedge (q \vee r)$
21. $(\sim p \vee \sim q) \wedge r$
22. $(p \vee q) \wedge (\sim r)$

WRITTEN ASSIGNMENTS 4.2

1. Consider some famous quotations you know, or look in a book of quotations. Isolate the simple statements and write the symbolic form of some quotations.
2. Listen to several people talk. Isolate simple statement and write the symbolic form of some compound statements.
3. Find some appropriate sentences in this chapter, isolate simple statements, and write the symbolic form of the sentences.
4. In which form do you think ambiguity is reduced—symbolic form or ordinary English? Give some examples to support your answer.

5. a. Describe a binary operation and give examples of binary operations on
   i. Numbers
   ii. Sets
   iii. Statements
   b. Describe a unary operation and give examples of unary operations on
   i. Numbers
   ii. Sets
   iii. Statements

4.3 TRUTH TABLES

GOALS
1. To construct a truth table for a statement composed of one, two, or three simple statements and the connectives ∧, ∨, or ∼.
2. To construct a truth table for a statement composed of two simple statements and the connectives ∧, ∨, ∼, and XOR.

p	q
T	T
T	F
F	T
F	F

▶ FIGURE 4.1 Truth values for p and q

Symbolic logic permits a convenient tabular form for determining the truth of compound statements. A *truth table* is formed by creating one column per statement and then considering all possible combinations of values of true or false for each statement. To illustrate, suppose a compound statement involves two simple statements p and q. The start of a truth table would be as shown in ▶ Figure 4.1. This table is a symbolic method of considering all possibilities of combining truth values for p and q. Note that they can both be true, both be false, or one true and the other false. Subsequent columns can now be added to analyze the truth value of compound statements.

Before looking at truth tables involving two simple statements, let's briefly consider a truth table for one simple statement. If p is a simple statement, the truth table for $\sim p$ is as shown in ▶ Figure 4.2. All possible values for p are listed (T, F) and then the column for $\sim p$ is formed by listing the appropriate value.

p	$\sim p$
T	F
F	T

▶ FIGURE 4.2 Truth table for $\sim p$

Truth Tables for Two Statements

p	q	p ∧ q
T	T	T
T	F	F
F	T	F
F	F	F

▶ FIGURE 4.3 Truth table for $p \wedge q$

Each example in this subsection involves exactly two statements, p and q. Thus, each truth table consists of four lines (2×2). We *always* assign initial values as shown in Figure 4.1.

Let's now consider the truth tables for "and" ($\wedge$) and "or" ($\vee$). If p and q represent two simple statements, then $p \wedge q$ (read "p and q") is true when and only when both p is true and q is true. In all other cases, $p \wedge q$ is false. This is shown in ▶ Figure 4.3. Notice that we have adopted the convention of always putting the compound statement being considered in the right-hand column and then shading it.

To see how rows of this truth table relate to specific statements, suppose p and q are the following true statements:

p : Annette lives on the fourth floor
q : Deanna lives on the second floor

The compound statement

Annette lives on the third floor and Deanna lives on the second floor

is then false. This is represented by line 3 of the truth table. The compound statement

Annette lives on the second floor and Deanna lives on the fourth floor

is also false. (Two wrongs don't make a right). This is represented by line 4 of the truth table.

In contrast, the statement $p \vee q$ (read "p or q") is true whenever at least one of the simple statements is true. This is shown in ▶ Figure 4.4. We again illustrate how this truth table relates to specific statements by considering true statements p and q:

p	q	p ∨ q
T	T	T
T	F	T
F	T	T
F	F	F

▶ FIGURE 4.4 Truth table for $p \vee q$

p : Annette lives on the fourth floor
q : Deanna lives on the second floor

The compound statement

Annette lives on the third floor or Deanna lives on the second floor

is now true, as can be seen by row 3 of the truth table for $p \vee q$. Recall that the "or" we are using is the inclusive "or"—that is, $p \vee q$ is true if p is true or q is true or if both p and q are true. Sometimes "or" is meant in an *exclusive* sense, for example, in statements such as "She drank coffee or she drank tea," which are true if either part is true but not if both parts are true. When "or" is used in this exclusive sense, this statement means that she drank coffee or she drank tea, but she did not drink both coffee and tea. We will use "or" and $\vee$ only in the inclusive sense. The exclusive use of "or" is discussed later in this section.

In the next example, we look at truth tables for more complex statements involving two simple statements.

p	q	p ∨ q
T	T	T
T	F	T
F	T	T
F	F	F

p	q	p ∨ q	~(p ∨ q)
T	T	T	F
T	F	T	F
F	T	T	F
F	F	F	T

▶ FIGURE 4.5 Negation of $(p \vee q)$

EXAMPLE 4.6

Make a truth table for the compound statement $\sim(p \vee q)$.

SOLUTION This truth table will resemble the one in Figure 4.4, but with one column added for the negation of $(p \vee q)$. We start by constructing the same table shown in Figure 4.4. Next we negate the truth values in the final column of this table to get the table in ▶ Figure 4.5. The interpretation of this truth table is that $\sim(p \vee q)$ is true only when both p and q are false, as shown in row 4.

EXAMPLE 4.7

Make a truth table for the compound statement $\sim p \wedge \sim q$.

SOLUTION We start with the four possible truth values for p and q, as shown in Figure 4.1 (p goes two trues and two falses, q alternates true and false), then add a column for $\sim p$, then add a column for $\sim q$, then combine these last two columns using $\wedge$ to give ► Figure 4.6.

p	q	$\sim p$	$\sim q$	$\sim p \wedge \sim q$
T	T	F	F	F
T	F	F	T	F
F	T	T	F	F
F	F	T	T	T

► **FIGURE 4.6** Truth table for $\sim p \wedge \sim q$

As in the preceding example, the compound statement is true only when both p and q are false.

The preceding two truth tables give a symbolic verification that $\sim(p \vee q)$ and $(\sim p) \wedge (\sim q)$ have identical truth values for all assignments of truth values to p and q. An application of this result is that the statement "It's not the case that I have a 4.0 grade point average or I have a red sports car" has identical truth values to the statement "I don't have a 4.0 grade point average and I don't have a red sports car."

EXAMPLE 4.8

Construct a truth table for $\sim p \vee (p \wedge q)$.

SOLUTION Start with the basic configuration for two statements p and q: p has two trues then two falses, q alternates true, false, true, false (Figure 4.1). Successively, add columns for $\sim p$, $p \wedge q$, and $\sim p \vee (p \wedge q)$. ► Figure 4.7 presents this truth table.

p	q	$\sim p$	$p \wedge q$	$\sim p \vee (p \wedge q)$
T	T	F	T	T
T	F	F	F	F
F	T	T	F	T
F	F	T	F	T

► **FIGURE 4.7** Truth table for $\sim p \vee (p \wedge q)$

EXAMPLE 4.9

Construct a truth table for $\sim(p \vee q) \wedge q$.

SOLUTION State with the basic configuration for two statements p and q and add successively the columns as shown in ► Figure 4.8 In this case our truth table

> **YOURS TRULY**
>
> *True* is a native English word that originally meant "loyal" or "trustworthy," as in "true friend." In logic, a truth table can be your true friend.

p	q	$p \vee q$	$\sim(p \vee q)$	$\sim(p \vee q) \wedge q$
T	T	T	F	F
T	F	T	F	F
F	T	T	F	F
F	F	F	T	F

▶ FIGURE 4.8 Truth table for $\sim(p \vee q) \wedge q$

reveals that the compound statement $\sim(p \vee q) \wedge q$ can never be true. Such a statement which always takes on the truth value false is called a *contradiction*.

Exclusive or

The connective "or" that we've been using is called the "inclusive or" since the statement "p or q" is true when and only when p is true or q is true or both p and q are true; i.e., "p or q" being true includes the case in which both p and q are true. Sometimes we join two statements with the conjunction "or" and mean that the compound statement is true if and only if the first statement is true or the second statement is true, but not both of them are true. For example, the statement "I will study math for two hours or I will study English for two hours" could mean "I will study math for two hours or I will study English for two hours, but I'm not going to do both." If so, this is the exclusive use of "or." We denote it XOR (for "exclusive or"). Thus p XOR q is true if and only if p is true or q is true, but not both.

 YOUR FORMULATION

Construct the truth table for p XOR q.

Truth Tables for Three Statements

A truth table for three simple statements involves twice as many combinations of values of true and false as a truth table for only two simple statements, since for each combination of true and false values for the first two statements there are two choices (true and false) for the third statement. There are $4 \times 2 = 2^3 = 8$ such combinations. We list these possibilities in ▶ Figure 4.9: This pattern of truth values is fairly easy to remember. For the first statement (p) we put four trues then four falses; for the second statement (q) we put two trues, two falses, two trues, and two falses; and for the third statement (r) we alternate true, false, true, false, etc. Although a different listing order could be technically correct, we encourage all students to use this same form. This allows for a quick check of results by merely looking at the last column.

The remaining examples in this section consist of finding truth tables for compound statements involving three simple statements.

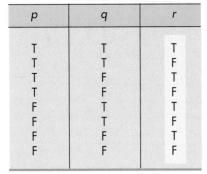

p	q	r
T	T	T
T	T	F
T	F	T
T	F	F
F	T	T
F	T	F
F	F	T
F	F	F

▶ FIGURE 4.9 Combinations of truth values for p, q, and r

EXAMPLE 4.10

Construct a truth table for the statement $\sim p \vee (q \wedge r)$.

SOLUTION As with two statements, we start with the basic format for three statements (Figure 4.9), then create a column for $\sim p$, then a column for $q \wedge r$. The final column is the desired column. The completed table is shown in ▶ Figure 4.10.

p	q	r	$\sim p$	$(q \wedge r)$	$(\sim p) \vee (q \wedge r)$
T	T	T	F	T	T
T	T	F	F	F	F
T	F	T	F	F	F
T	F	F	F	F	F
F	T	T	T	T	T
F	T	F	T	F	T
F	F	T	T	F	T
F	F	F	T	F	T

▶ FIGURE 4.10 Truth table for $(\sim p) \vee (q \wedge r)$

EXAMPLE 4.11

Construct a truth table for $p \wedge (q \vee r)$.

SOLUTION The table for this statement is shown in ▶ Figure 4.11.

p	q	r	$(q \vee r)$	$p \wedge (q \vee r)$
T	T	T	T	T
T	T	F	T	T
T	F	T	T	T
T	F	F	F	F
F	T	T	T	F
F	T	F	T	F
F	F	T	T	F
F	F	F	F	F

▶ FIGURE 4.11 Truth table for $p \wedge (q \vee r)$

EXAMPLE 4.12

Construct a truth table for $(p \wedge q) \vee (p \wedge r)$.

SOLUTION The table for this expression is shown in ▶ Figure 4.12.

p	q	r	$(p \wedge q)$	$(p \wedge r)$	$(p \wedge q) \vee (p \wedge r)$
T	T	T	T	T	T
T	T	F	T	F	T
T	F	T	F	T	T
T	F	F	F	F	F
F	T	T	F	F	F
F	T	F	F	F	F
F	F	T	F	F	F
F	F	F	F	F	F

▶ FIGURE 4.12 Truth table for $(p \wedge q) \vee (p \wedge r)$

If you compare the right-hand columns in the truth tables for $p \land (q \lor r)$ and $(p \land q) \lor (p \land r)$, you will see that they are identical. This shows that these expressions have identical truth values for all assignments of truth values to p, q, and r. This means that the statement "You will make $25,000, and you can start on June 15 or you can start on July 1" has exactly the same truth values as the statement "You will make $25,000 and you can start on June 15, or you will make $25,000 and you can start on July 1" for each combination of truth values of the three simple statements involved. In other words, those two statements are equivalent. Furthermore, *any* two sentences in these forms will have the same truth values for each combination of the truth values of the parts.

EXERCISES 4.3

1. Can the truth values for p and q be abbreviated to the following form? Why or why not?

p	q
T	F
F	T

2. In a truth table, answer each of the following as: (i) T, (ii) F, or (iii) more information is needed.
   a. If the truth value of statement p is T, then the truth value of $\sim p$ is _____ .
   b. If the truth value of $\sim p$ is F, then the truth value of p is _____ .
   c. If the truth value of p is T, then the truth value of q is _____ .

For Exercises 3–11, construct a truth table for the given statement. When possible, make an interpretation of this truth table as in Examples 4.6 and 4.7.

3. $\sim(p \land q)$
4. $\sim p \lor \sim q$
5. $\sim q \lor (p \land q)$
6. $\sim(p \land q) \lor q$
7. $(p \lor q) \lor \sim q$
8. $(\sim p \lor q) \land r$
9. $\sim p \land (q \land r)$
10. $(p \lor q) \land (p \lor r)$
11. $(p \lor q) \land r$

12. For which of the following pairs of statements are the truth values identical?
    a. p XOR q, $(p \lor q) \land \sim(p \land q)$
    b. p XOR q, $p \lor q$
    c. $p \lor q$, $(p$ XOR $q) \lor \sim p$

13. In Exercises 15–22 of Exercises 4.2, p is true, q is true, and r is false. Find the truth value for the statement of each of those exercises.

WRITTEN ASSIGNMENTS 4.3

1. In looking at the truth tables for the examples or exercises in this section, are there any places in the truth tables that surprise you? That is, are there any combinations of truth values that surprise you? Describe them, and state why you find them surprising. Are you comfortable with them now, or are they still puzzling?

2. Construct truth tables for the statements you found in Written Assignments 1, 2, and 3 of Written Assignments 4.2. Use your tables to describe how you would show that these statements are true (false).

3. Find a statement in a book, magazine, or newspaper or on radio or television that involves two or more statements combined by "and," "or," or negation.
   a. Write the symbolic form for it.
   b. Construct a truth table for it.
   c. What are the ways you could show that the statement is false?

4. Write two or three paragraphs comparing and contrasting the construction of the symbolic form of a statement and the diagramming of a statement in an English class.

4.4 CONDITIONAL SENTENCES

GOALS
1. To work with conditional sentences.
2. To work with biconditional sentences.
3. To know the relations among a statement, its converse, its inverse, and its contrapositive.

Probably the most fundamental concept in the study of logic is a *conditional sentence;* it is the basis for constructing valid arguments. You have used conditional sentences most of your life, for instance:

If you mow the lawn, then you can use the car Saturday night.

Terminology and Notation

Consider the sentence "If you understand this section, then you will get an A on the next test." Let's symbolize this sentence as follows:

p : You understand this section
q : You will get an A on the next test

We could rewrite the sentence symbolically as

If p, then q.

This is an example of a conditional sentence. More specifically, we have the following definition.

DEFINITION

Let p and q be statements. A *conditional sentence* is a sentence of the form "If p, then q." We denote this

$$p \rightarrow q$$

which is read as "*p* implies *q*." Statement p is the *hypothesis* and q is the *conclusion.*

With suitable definition for statements p and q, each of the following has the form "If p, then q," or $p \rightarrow q$.

If I fall in love, [then] it will be forever.
If you have a 4.0 grade point average, then you will be valedictorian.

Conditional sentences are also called *implications.*

Truth Tables

When is a conditional sentence true? Let's take the sentence

If you mow the lawn, then you can use the car Saturday night.

We define statements p and q as

p : You mow the lawn
q : You can use the car Saturday night

p	q	$p \to q$
T	T	
T	F	
F	T	
F	F	

▶ FIGURE 4.13 Incomplete truth table for $p \to q$

As statements, p and q can be either true or false, so a truth table for $p \to q$ has four rows, as shown in ▶ Figure 4.13.

We must now decide what values are appropriate for the column $p \to q$. Let's first consider row 1, where p and q are both true. Given p and q as last defined, this means you did mow the lawn and you did get to use the car. Thus, you would believe that the following sentence is true:

If you mow the lawn, then you can use the car Saturday night.

The partially completed truth table for $p \to q$ is then as shown in ▶ Figure 4.14.

p	q	$p \to q$
T	T	T
T	F	
F	T	
F	F	

▶ FIGURE 4.14 Partial truth table for $p \to q$

What happens when p is true and q is false—that is, when you mow the lawn but do not get to use the car on Saturday night? In this case, you would believe the sentence is false. In other words, you would believe that you had not been told the truth. (Can you imagine your reaction after mowing the lawn and then being told you couldn't use the car?) At this point, the partially completed truth table is as shown in ▶ Figure 4.15.

What about the third row? The third row has p false and q true: You didn't mow the lawn but you get to use the car anyway (such as when it rains and you are unable to mow because the grass is wet). What value would then be assigned to the sentence? In this instance, the sentence is still true. There's no reason to believe it's false. The partially completed truth table than becomes as shown in ▶ Figure 4.16.

p	q	$p \to q$
T	T	T
T	F	F
F	T	
F	F	

▶ FIGURE 4.15 Partial truth table for $p \to q$

Finally, let's consider row 4. When p and q are both false, what of the sentence $p \to q$? In our example, you don't mow the lawn and you don't get to use the car. How does this affect the truth of the conditional sentence? Your expectation is not violated. In this instance, we agree that the implication $p \to q$ shall be assigned the value true. The complete truth table for $p \to q$ is then as shown in ▶ Figure 4.17.

The third and fourth lines of the truth table are perhaps the hardest with which to feel comfortable. Consider this mathematical example:

If $x = 2$, then $x^2 = 4$.

We define statements p and q as

p: $x = 2$
q: $x^2 = 4$

p	q	$p \to q$
T	T	T
T	F	F
F	T	T
F	F	

▶ FIGURE 4.16 Partial truth table for $p \to q$

In line 3 of the table in Figure 4.17, p is false and q is true. Can you think of a value of x that makes p false and q true? $x =$ _____. This happens when, and only when, $x = -2$. In this case, it seems reasonable to say that the conditional statement "If $x = 2$, then $x^2 = 4$" is true. Having x be -2 does not make the statement false. Enter this observation in ▶ Figure 4.18. In line 4 of the table in Figure 4.17, p and q are both false. Suppose x is 3. Then $x^2 = 9$, so p and q are both false. However, this example doesn't challenge the truth of the statement "If $x = 2$, then $x^2 = 4$." Enter this observation in Figure 4.18.

x	$p : x = 2$	$q : x^2 = 4$	$p \to q$
-2	F	T	
3	F	F	

▶ FIGURE 4.18 Some truth values for $p \to q$

p	q	$p \to q$
T	T	T
T	F	F
F	T	T
F	F	T

▶ FIGURE 4.17 Complete truth table for $p \to q$

Let's reconsider the truth values of the conditional statement $p \to q$ by thinking of how to show that it is false. Consider the statement,

If I get A's on two tests, then I will get an A in the course.

What would have to be done to show that this statement is false? The only way is for you to get A's on two tests but not to get an A in the course. In other words, the only way to make $p \rightarrow q$ false is for p to be true and q to be false. Each of the other three arrangements of truth values allows $p \rightarrow q$ to be true.

At this point, you should be carefully note the following.

Just as a one-way street has traffic going in only one direction, so a conditional sentence p $\rightarrow$ q *has thought going in only one direction from the hypothesis* p *to the conclusion* q.

RESULT
___

A conditional sentence, $p \rightarrow q$, is false when p is true and q is false. In all other cases, $p \rightarrow q$ is true.

The next example provides practice in working with truth tables and conditional sentences.

EXAMPLE 4.13

Write the following conditional sentence symbolically. Then construct a truth table and give an illustration of when the sentence would be false.

Class will be cancelled on Friday provided our basketball team wins on Thursday.

SOLUTION This sentence does not have an explicit "if–then" form. However, it may be rewritten as

If our basketball team wins on Thursday, then class will be cancelled on Friday.

We can then define

p : Our basketball team wins on Thursday
q : Class will be cancelled on Friday

and write the symbolic form $p \rightarrow q$. The truth table for this is the same as that shown in Figure 4.17. The only time this sentence would be false is when the basketball team did win but class was not cancelled. (Again, imagine your reaction.)

It is possible to construct truth tables for variations of the basic conditional sentence $p \rightarrow q$. In writing such statements it is convenient to establish another convention regarding the precedence of operations. It is customarily agreed that negation ($\sim$) takes precedence over implication ($\rightarrow$). For example, $\sim p \rightarrow q$ means $(\sim p) \rightarrow q$ rather than $\sim(p \rightarrow q)$. The next example illustrates such a situation.

EXAMPLE 4.14

Construct a truth table for $\sim p \rightarrow q$.

SOLUTION This table requires a column for $\sim p$. Once that column has been established, the column for $\sim p \rightarrow q$ is found by remembering that a conditional sentence is always true except when the hypothesis is true and the conclusion is false. The truth table for $\sim p \rightarrow q$ is shown in ▶ Figure 4.19.

p	q	$\sim p$	$\sim p \rightarrow q$
T	T	F	T
T	F	F	T
F	T	T	T
F	F	T	F

▶ FIGURE 4.19 Truth table for $\sim p \rightarrow q$

Biconditional Sentences

A second fundamental concept in the study of logic is that of a *biconditional sentence,* for which we have the following definition.

DEFINITION

Let p and q be statements. The *biconditional sentence, p if and only if q,* is the sentence $p \rightarrow q$ and $q \rightarrow p$. This is denoted

$$p \leftrightarrow q \qquad \text{or} \qquad p \text{ iff } q$$

A biconditional sentence is merely two conditional sentences joined by "and." Note that $p \leftrightarrow q$ is just $(p \rightarrow q) \wedge (q \rightarrow p)$.

 YOUR FORMULATION

1. Use the definition of the biconditional sentence $p \leftrightarrow q$ to construct a truth table for it.
2. For what truth values of p and q is the biconditional sentence $p \leftrightarrow q$ true?

How does a biconditional sentence differ from a conditional sentence? To answer this question, ask yourself whether a friend has ever said something like the following:

If you wash it afterwards, then you may borrow my sweater.

You will know this is a conditional sentence and would be logically true if you borrowed the sweater and didn't wash it afterwards ($p \rightarrow q$ is true for p false and q true). Do you think this is really what your friend meant? Probably what should have been said is

You may borrow my sweater if and only if you wash it afterwards.

This means that either you both borrow it and wash it or you both don't borrow it and don't wash it. It is not possible for you to do one of these things and not the other. (*Authors' note:* We do *not* recommend using formal logic on friends or parents unless they are understanding and tolerant. Sometimes emotional arguments carry the day. Use with caution!)

You can now construct truth tables for more complex biconditional statements. The next example provides one such instance.

EXAMPLE 4.15

Construct a truth table of $p \leftrightarrow (q \rightarrow r)$.

SOLUTION Since there are three statements, we need eight rows. A column for $q \rightarrow r$ is listed. The column for $p \leftrightarrow (q \rightarrow r)$ is formed by considering

$$[p \rightarrow (q \rightarrow r)] \wedge [(q \rightarrow r) \rightarrow p]$$

or by remembering that a biconditional statement is true exactly when the two parts have the same truth value. This is shown in ▶ Figure 4.20.

p	q	r	$q \rightarrow r$	$p \leftrightarrow (q \rightarrow r)$
T	T	T	T	T
T	T	F	F	F
T	F	T	T	T
T	F	F	T	T
F	T	T	T	F
F	T	F	F	T
F	F	T	T	F
F	F	F	T	F

▶ FIGURE 4.20 Truth table for $p \leftrightarrow (q \rightarrow r)$

Contrapositive

The biconditionals just studied were really combinations of two conditional sentences. We next look at three variations of a single conditional sentence. The first of these is the *contrapositive sentence*.

DEFINITION

Let $p \rightarrow q$ be a conditional sentence. The *contrapositive* of $p \rightarrow q$ is the sentence $\sim q \rightarrow \sim p$.

To illustrate a contrapositive, let's reconsider the sentence

If the basketball team wins on Thursday, then class will be cancelled on Friday.

With p and q as previously defined, the contrapositive of this is

If class is not cancelled on Friday, then the basketball team did not win on Thursday.

Let's now look at the truth table for $\sim q \rightarrow \sim p$, as shown in ▶ Figure 4.21. Note that the columns for $p \rightarrow q$ and $\sim q \rightarrow \sim p$ are identical. This property will be examined in more detail in Section 4.6.

In practice, contrapositives are frequent. For example, consider the statement

If you aren't here by 8:00, then you can't go with us.

If we define p and q as

p : You can go with us
q : You will be here by 8:00

p	q	$p \rightarrow q$	$\sim p$	$\sim q$	$\sim q \rightarrow \sim p$
T	T	T	F	F	T
T	F	F	F	T	F
F	T	T	T	F	T
F	F	T	T	T	T

▶ FIGURE 4.21 Truth table for $\sim q \rightarrow \sim p$

then the previous statement is written symbolically as $\sim q \rightarrow \sim p$. Our observation with Figure 4.21 was that $\sim q \rightarrow \sim p$ has identical truth values with $p \rightarrow q$. Therefore the preceding statement has identical truth values to the statement

If you can go with us, then you will be here by 8:00.

The wording can be made a little smoother by saying

If you want to go with us, then you need to be here by 8:00.

Converse and Inverse

Two other variations of the conditional sentence $p \rightarrow q$ exist, the *converse* and the *inverse*.

DEFINITION

Let $p \rightarrow q$ be a conditional sentence. The *converse* of $p \rightarrow q$ is $q \rightarrow p$. The *inverse* of $p \rightarrow q$ is $\sim p \rightarrow \sim q$.

Truth tables for both variations are shown in ▶ Figure 4.22. Note that neither of the last two columns is identical to the column for $p \rightarrow q$. Consequently, a statement and its converse have different truth values. Also, a statement and its inverse have

p	q	$p \rightarrow q$	$q \rightarrow p$	$\sim p \rightarrow \sim q$
T	T	T	T	T
T	F	F	T	T
F	T	T	F	F
F	F	T	T	T

▶ FIGURE 4.22 Truth table for $q \rightarrow p$ and $\sim p \rightarrow \sim q$

A conditional statement p → q *is like traffic on one side of the boulevard going from* p *to* q *while the converse* q → p *is like traffic on the other side going from* q *to* p.

different truth values. However, the converse and the inverse of a conditional statement have the same truth values. This must be the case since the inverse is the contrapositive of the converse. We now look at one example that provides practice in writing all three, given a conditional sentence.

EXAMPLE 4.16

Write the converse, the inverse, and the contrapositive, given the following true conditional sentence:

If x has a factor of 4, then x has a factor of 2.

Which of them are true statements?

SOLUTION We begin by defining p and q:

p : x has a factor of 4
q : x has a factor of 2

Our sentence is then $p \to q$ and the symbolic converse is $q \to p$. This would be written as

If x has a factor of 2, then x has a factor of 4.

This is not true, because, for instance, 10 is an integer with a factor of 2 but not a factor of 4. The inverse, $\sim p \to \sim q$, is

If x does not have a factor of 4, then x does not have a factor of 2.

Again, this can be seen to be false by considering the case where $x = 10$. Finally, the contrapositive is $\sim q \to \sim p$. This is written as

If x does not have a factor of 2, then x does not have a factor of 4.

We know this to be true by considering the truth table of Figure 4.21.

The following exercises provide practice in working with variations of conditional sentences. In the succeeding sections we will see how to use conditional sentences to construct valid arguments.

> **CONDITIONAL VERSE**
>
> Conditional sentences
>   Now let us rehearse.
> If you pull up your mind,
>   Then we will converse.
>
> But the converse reverses
>   The implication we find.
> If we will converse,
>   Then you'll pull up your mind.
>
> If we won't converse,
>   Then your mind isn't along.
> Changes the original statement
>   To the contrapositive song.
>
> If you don't bring your mind,
>   Then we won't converse.
> Negate hypothesis and conclusion.
>   There, I've said it inverse.

EXERCISES 4.4

In each of Exercises 1–6, construct a truth table for the given statement.

1. $q \to p$
2. $\sim p \to \sim q$
3. $(p \wedge q) \to p$
4. $p \to (p \vee q)$
5. $p \to \sim p$
6. $(p \wedge \sim p) \to q$

In each of Exercises 7–12, write the conditional sentence symbolically, construct a truth table, and give an illustration of when the sentence would be false.

7. You can do anything with children if you only play with them.
8. When a subject becomes totally obsolete, we make it a required course.

9. Don't talk unless you can improve the silence. (*an old Vermont proverb*)
10. If children live with approval, they learn to live with themselves.
11. If you don't know where you are going, you will probably end up somewhere else. (*quoted from Lawrence J. Peter*)
12. If you don't say anything, you won't be called on to repeat it.
13. From listening to people talk, write some conditional statements, then write their symbolic form.
14. Find some sentences in this chapter that are conditional statements, then write their symbolic forms.
15. Compare the truth table for $p \to q$ with that for $\sim p \vee q$.

In each of Exercises 16–19, construct a truth table for the given statement.

16. $(p \rightarrow q) \leftrightarrow (\sim p \vee q)$
17. $p \rightarrow (q \leftrightarrow r)$
18. $(p \wedge q) \rightarrow r$
19. $(p \rightarrow q) \leftrightarrow (q \rightarrow p)$
20. Given the true conditional sentence

   If $A \cup B \subseteq C$, then $A \subseteq C$

   write the converse, the inverse, and the contrapositive. Which of these are true statements?
21. Given the true conditional sentence

   If x is a nonzero real number, then x^2 is a nonzero real number.

   write the converse, the inverse, and the contrapositive. Which of these are true statements?
22–27. For each of the statements in Exercises 1–6 write the converse, the inverse, and the contrapositive.
28. Write some conditional statements that people sometimes make when what they really mean is the converse or the inverse.
29. Statements of the form "p only if q," such as "Proceed to Section 4.5 only if you understand Section 4,4," mean the same as "$p \rightarrow q$." Note that the biconditional "p if and only if q" means "p if q and p only if q"; i.e., $p \leftrightarrow q$ means $(q \rightarrow p) \wedge (p \rightarrow q)$.
   a. Construct a truth table for "p only if q."
   b. For each of the following sentences, identify and label the simple statements and write the sentence in symbolic form.
      i. Enter this gate only if you know geometry. (*Note:* Plato's Academy had over the gate the inscription "Let no man ignorant of geometry enter.")
      ii. We will go only if the probability of snow is less than 20%.
   c. *True or false.*
      i. A number x is positive if x^2 is positive.
      ii. A number x is positive only if x^2 is positive.
      iii. A number x is positive if and only if x^2 is positive.
      iv. A quarter note gets one beat if music is in $\frac{3}{4}$ time.
      v. A quarter note gets one beat only if music is in $\frac{3}{4}$ time.
      vi. A quarter note gets one beat if and only if music is in $\frac{3}{4}$ time.
   d. Give an example of a statement p and a statement q so that all three of the following are true: p if q, p only if q, p if and only if q.
30. Sometimes one condition is necessary or sufficient for another condition. For example, in Illinois, being age 16 is a necessary condition for getting a driver's license but it it not

sufficient—one must also pass a test. At a certain store showing your driver's license is a sufficient condition for cashing a check but it is not necessary—you can show a credit card instead. The statement "p is a necessary condition for q" means "$q \rightarrow p$." The statement "p is a sufficient condition for q" means "$p \rightarrow q$."
   a. For each of the following sentences, identify and label the simple statements and write the sentence in symbolic form.
      i. Having $A \subseteq B$ is a necessary condition for having $A = B$.
      ii. Having x be an integer is a sufficient condition for having x be a rational number.
      iii. Having $p \rightarrow q$ and $q \rightarrow p$ are necessary and sufficient conditions for having $p \leftrightarrow q$.
   b. In each of the following, give an example other than one in part a of a statement p and a statement q such that
      i. p is a necessary and sufficient condition for q.
      ii. p is a necessary but not sufficient condition for q.
      iii. p is a sufficient but not necessary condition for q.
      iv. p is neither a necessary nor a sufficient condition for q.
   c. For each of the following pairs of conditions, tell whether p is a necessary condition for q, a sufficient condition for q, or both a necessary and sufficient condition for q.
      i. $p : A = B$
         $q : A \subseteq B$
      ii. $p : x = 0$
          $q : x^2 = 0$
      iii. $p : x = 2$
           $q : x^2 = 4$
      iv. $p :$ Pat is in Atlanta, Georgia
          $q :$ Pat is in the United States of America
      v. $p :$ He was a president of the United States
         $q :$ His face is carved on Mt. Rushmore
31. Suppose the truth value for $(p \vee \sim q) \rightarrow r$ is:
   a. T when p is T and q is T. What must the truth value of r be?
   b. F when p is F and r is F. What must the truth value of q be?
   c. F when q is T and r is F. What must the truth value of p be?
   d. T when p is T. What must the truth value of r be?
32. Provide a solution to the example of a Wason test in the Prologue to this chapter.
33. The philosopher and mathematician Rene Descartes is noted for saying "I think, therefore I am." This could be represented symbolically by $p \rightarrow q$, where

   $p :$ I think

   $q :$ I am

A story is told that Descartes was attending a party at which the hostess asked him if he'd like more wine. He answered, "I think not" and disappeared. What is wrong with the *logic* of the punchline?

1. The authors tried to convince you of the reasonableness of the assignment of truth values to a conditional sentence by considering the sentences "If you mow the lawn, then you can use the car Saturday night," "If $x = 2$, then $x^2 = 4$," and "If I get A's on two tests, then I will get an A in the course." Construct another conditional sentence to argue for the reasonableness of the assignment of truth values to a conditional sentence. Present your argument in a form similar to what is done in the subsection "Truth Tables," through Figure 4.17.

2. Definitions are really biconditional statements. However, definitions are often stated just as conditional statements such as the following:

   Definition: If $p \rightarrow q$ is a conditional statement, then its converse is $q \rightarrow p$.

   Comment on this practice. Is it confusing? Would you argue that all definitions should be stated as "if and only if" statements? Why or why not?

3. The text cautions you about using formal logic on friends or parents unless they are understanding and tolerant. Describe a real or fictional instance in which such a use of formal logic creates a problem.

4. Describe an instance in which a conditional sentence is used but the converse of the sentence is really intended.

4.5 VALID ARGUMENTS

GOALS
1. To work with the concept of a tautology.
2. To use the *modus ponens* form of argument.
3. To recognize and know the validity of reasoning by the converse, the inverse, or the contrapositive.
4. To write an argument symbolically and use a truth table to determine its validity.

Normal conversation often includes drawing conclusions. The practice of law is heavily dependent on the ability to arrive at valid conclusions from stated assumptions. Reasoning involves drawing conclusions. When are these conclusions warranted? We will now see how a knowledge of symbolic logic can enable you to determine when an argument is valid.

Basic Form of a Valid Argument

Consider the following sequence of statements:

> If the snow is 6 inches deep, then we can go skiing.
> The snow is 6 inches deep.
> _____
> Therefore, we can go skiing.

This is the basic form of an argument. More specifically, a symbolic argument has the following form:

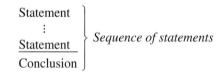

Statement
⋮
Statement
Conclusion
} *Sequence of statements*

Statements given before the conclusion are called *premises.* In our example, we have

If the snow is 6 inches deep,
then we can go skiing. ⎫
                        ⎬ *Premises*
The snow is 6 inches deep. ⎭

Therefore we can go skiing. *Conclusion*

In this case, one of the premises is a conditional sentence. This is one of the most frequently used forms of an argument. Symbolically, this becomes

p : The snow is 6 inches deep
q : We can go skiing

The argument can then be stated as

$$p \rightarrow q$$
$$\underline{p}$$
$$q$$

When is such an argument valid? That is, under what circumstances does a conclusion logically follow from the premises? The argument is saying essentially that if we have the premises, then we have the conclusion. In other words, the argument is the conditional sentence

[premises] → conclusion

If this conditional sentence is always true, then we have a *valid argument;* otherwise, it is an *invalid argument.*

Think about a valid argument from the opposite point of view. What would lead you to say an argument is invalid? It would be when all the premises are true but the conclusion is false. However, this makes the conjunction of the premises true and the conclusion false. This is the one and only situation that makes

"[premises] → conclusion"

false. In other words, if this conditional sentence is false, the argument is invalid; and if this conditional sentence is always true, the argument is valid.

An additional definition is now appropriate.

ETYMOLOGY OF *TAUTOLOGY*

Tautology comes from the Greek roots *tauto-,* meaning "same," and *logos,* meaning "speech, reasoning, discourse, proposition" (it's a root word for *geology, psychology,* and *sociology*). A tautology is a proposition that has the same truth value (true) no matter what the truth of its parts.

DEFINITION

A *tautology* is a statement that is always true.

Thus, a valid argument is one in which the conditional sentence

"[premises] → conclusion"

is a tautology.

Tautologies are most easily seen by looking at truth tables. To illustrate, consider the truth table for $p \vee (\sim p \vee q)$ shown in ▶ Figure 4.23. This statement is always true, no matter what values are assigned to p and q; therefore, it is a tautology.

p	q	$\sim p$	$\sim p \vee q$	$\sim p \vee (\sim p \vee q)$
T	T	F	T	T
T	F	F	F	T
F	T	T	T	T
F	F	T	T	T

▶ FIGURE 4.23 Truth table for $p \vee (\sim p \vee q)$

YOUR FORMULATION

Reconsider our original example, which was last written as

$$p \rightarrow q$$
$$\underline{p}$$
$$q$$

The form means that the two premises are joined by "and," so the premises become the compound statement

$$(p \rightarrow q) \wedge p$$

and the argument can be written as

$$[(p \rightarrow q) \wedge p] \rightarrow q$$

1. Complete the truth table for this argument in ▶ Figure 4.24.

p	q	$p \rightarrow q$	$(p \rightarrow q) \wedge p$	$[(p \rightarrow q) \wedge p] \rightarrow q$
T	T			
T	F			
F	T			
F	F			

▶ FIGURE 4.24 Truth table for $[(p \rightarrow q) \wedge p] \rightarrow q$

2. Is the argument of our original example a tautology? Why or why not?

This form is the most frequently used method of constructing a valid argument. That is:

1. A conditional sentence is given.
2. The hypothesis is stated to be true.
3. The conclusion is deduced to be true.

In logic, this form is called the *law of detachment*, or *modus ponens*.
Let's now consider the following logical argument.

If $3 + 5 = 8$, then 9 is an even number
$\underline{3 + 5 = 8}$
9 is an even number

As before, we symbolize this as

$p : 3 + 5 = 8$
$q : 9$ is an even number

and rewrite the argument as

$$\begin{array}{c} p \rightarrow q \\ \underline{p} \\ q \end{array}$$

This form can be rewritten as $[(p \rightarrow q) \wedge p] \rightarrow q$. As we found in Figure 4.24, this is a tautology. Therefore, it is a valid argument.

Here is the significance of this example: *It is possible to have a valid argument that contains a false conclusion.* This particular aspect of valid arguments is discussed later in this section.

Using Truth Tables

Truth tables provide a convenient means of analyzing logical arguments. The general method for using truth tables is to do the following:

1. Convert the argument to a symbolic statement.
2. Construct a truth table.
3. Determine whether or not you have a tautology.

This method is illustrated in the next example.

EXAMPLE 4.17

Use a truth table to determine whether or not the following is a valid argument.

If you study for 2 hours every day, you will get an A on the test.
You got a B on the test.
Therefore, you did not study for 2 hours every day.

SOLUTION We define statements p and q as

$p :$ You study for 2 hours every day
$q :$ You will get an A on the test

and then write the symbolic argument as

$$\begin{array}{c} p \rightarrow q \\ \underline{\sim q} \\ \sim p \end{array}$$

We next consider the truth table found for the conditional sentence in which the premises are the hypothesis $p \rightarrow q$ and $\sim q$ and the conclusion is $\sim p$,

$$[(p \rightarrow q) \wedge \sim q \rightarrow \sim p$$

This is given in ▶ Figure 4.25, which shows that this conditional statement is a tautology, so the argument is valid. (This should come as no surprise, since a knowledge of symbolic logic is unnecessary to see that the conclusion was warranted from the premises.)

p	q	$p \to q$	$\sim q$	$(p \to q) \wedge \sim q$	$\sim p$	$[(p \to q) \wedge \sim q] \to \sim p$
T	T	T	F	F	F	T
T	F	F	T	F	F	T
F	T	T	F	F	T	T
F	F	T	T	T	T	T

▶ FIGURE 4.25 Truth table for $[(p \to q) \wedge \sim q] \to \sim p$

The form of reasoning analyzed in Example 4.17 is called *reasoning from the contrapositive.*

Invalid Arguments

Not all arguments are valid. So let's look at some invalid arguments. Consider the following:

If it rains today, the swimming pool will be closed.
The swimming pool is closed.
Therefore, it rained today.

We identify statements p and q:

p : It rains today
q : The swimming pool will be closed

and write the symbolic argument as

$$p \to q$$
$$\frac{q}{p}$$

At this point, we construct the truth table for $[(p \to q) \wedge q] \to p$ (▶ Figure 4.26). It shows that this statement is not a tautology, so the argument is invalid.

p	q	$p \to q$	$(p \to q) \wedge q$	$[(p \to q) \wedge q] \to p$
T	T	T	T	T
T	F	F	F	T
F	T	T	T	F
F	F	T	F	T

▶ FIGURE 4.26 Truth table for $[(p \to q) \wedge q] \to p$

This form of reasoning is called *reasoning from the converse.* It is invalid. Never use it!

Now let's look at the original argument. It is not valid because of the line in the truth table where p is false while q is true (in other words, the pool is closed but it did not rain). This line in the truth table is an instance where the conjunction of the premises is true but the conclusion is false. This is *the* situation we call an invalid argument. In terms of rain and pools, it happens when the pool was closed for some other reason (cold weather, malfunctioning equipment, or so on).

The next example illustrates another invalid argument.

EXAMPLE 4.18

Use a truth table to determine if the following argument is valid or invalid.

If an integer has a factor of 4, then it has a factor of 2.
10 does not have a factor of 4.
Therefore, 10 does not have a factor of 2.

SOLUTION This is symbolized by

p : x has a factor of 4
q : x has a factor of 2

and the argument is written as

$$\begin{array}{c} p \rightarrow q \\ \underline{\sim p} \\ \sim q \end{array}$$

We construct the truth table for $[(p \rightarrow q) \wedge \sim p] \rightarrow \sim q$, as shown in ► Figure 4.27. This shows that the argument is invalid because we do not have a tautology. This confirms that the conclusion does not follow from the premises. Clearly, 10 does have a factor of 2.

p	q	$p \rightarrow q$	$\sim p$	$(p \rightarrow q) \wedge \sim p$	$[(p \rightarrow q) \wedge \sim p] \rightarrow \sim q$
T	T	T	F	F	T
T	F	F	F	F	T
F	T	T	T	T	F
F	F	T	T	T	T

► FIGURE 4.27 Truth table for $[(p \rightarrow q) \wedge \sim p] \rightarrow \sim q$

The form of reasoning demonstrated in Example 4.18 is called *reasoning from the inverse*. It is invalid. Don't ever use it!

Arguments and Truth

The validity of an argument has nothing to do with the truth of premises or conclusions. A valid argument can be constructed that contains a false conclusion. For example, let's reconsider the following argument:

$$\begin{array}{c} \text{If } 3 + 5 = 8, \text{ then 9 is an even number} \\ \underline{3 + 5 = 8} \\ \text{9 is an even number} \end{array}$$

Symbolically, this becomes $[(p \rightarrow q) \wedge p] \rightarrow q$. Since this is a tautology, the argument is valid even though 9 is not an even number. How is this possible? It happens because, when examining the validity of an argument, you merely decide whether or not the conclusion follows from the premises. (In this case, it does.) It is not necessary to decide whether or not the premises are true. A symbolic analysis of this argument produces the following:

$$\begin{array}{ll} p \rightarrow q & (p \rightarrow q \text{ is false}) \\ \underline{p} & (p \text{ is true}) \\ q & (q \text{ is false}) \end{array}$$

Another way to construct a valid argument with a false conclusion is to have two false premises, as shown in the next example.

EXAMPLE 4.19

Construct a truth table to show that the following argument is valid.

If 3 + 5 = 7, then 9 is an even integer.
3 + 5 = 7.
Therefore, 9 is an even integer.

SOLUTION Symbolically, we define

$p : 3 + 5 = 7$
$q : 9$ is an even integer

and rewrite the argument as

$$[(p \rightarrow q) \wedge p] \rightarrow q$$

Recall, this is a tautology, as shown in Figure 4.24. Hence it is a valid argument (though somewhat ridiculous). However, both of the premises $p \rightarrow q$ and p are false. Again, this shows that a valid argument can contain a false conclusion.

What is the point of the last two examples? Do people actually use valid arguments with false conclusions? Let's consider a slight variation of a television commercial.

If more doctors prescribe aspirin than any other painkiller, then you should take aspirin.

What conclusions can the consumer make? If we symbolize this as

p : More doctors prescribe aspirin than any other painkiller
q : You should take aspirin

we have the conditional sentence $p \rightarrow q$. The advertisers hope you will assume p is true and think

$$\begin{array}{l} p \rightarrow q \\ \underline{p} \\ q \end{array}$$

which is a valid argument whether or not q is true. Here we have a case of a valid argument that can have a false conclusion.

In this commercial, it may be the case that very few doctors prescribe aspirin (fewer than for some other painkiller). In this case, the symbolic form becomes

$$\begin{array}{l} p \rightarrow q \\ \underline{\sim p} \\ q \end{array}$$

p	q	$p \rightarrow q$	$\sim p$	$(p \rightarrow q) \wedge \sim p$	$[(p \rightarrow q) \wedge \sim p] \rightarrow q$
T	T	T	F	F	T
T	F	F	F	F	T
F	T	T	T	T	T
F	F	T	T	T	F

▶ FIGURE 4.28 Truth table for $[(p \rightarrow q) \wedge \sim p] \rightarrow q$

If we look at the truth table for $[(p \rightarrow q) \wedge \sim p] \rightarrow q$ shown in ▶ Figure 4.28, we see this is not a tautology; hence, the argument is invalid. If the television viewer believes that very few doctors prescribe aspirin (the viewer believes that p is false or $\sim p$ is true), then the advertiser has implicitly constructed an invalid argument.

EXERCISES 4.5

1. Which of the following statements are tautologies?
   a. p
   b. $p \vee \sim p$
   c. $p \vee q$
   d. $p \wedge q$
   e. $[(p \rightarrow q) \wedge p] \rightarrow q$
2. In Exercises 4.4, Exercises 1–6, which are tautologies?
3. Using *modus ponens,* what conclusion can be drawn from the following statements?

   If ever I should leave you, it wouldn't be in springtime.
   I am leaving you.

4. Using *modus ponens* with the premise "If $A \cap B = A,$ then $A \subseteq B$," what must the other premise be in order to reach the conclusion "$A \subseteq B$"?
5. Using *modus ponens* with the premise "The government is a lawbreaker," what must the other premise (due to Justice Louis D. Brandeis) be in order to reach the conclusion "The government breeds contempt for the law"?

In each of Exercises 6–12, label the argument as one of the following types:
   a. Valid—*modus ponens*
   b. Valid—reasoning from contrapositive (like Example 4.17)
   c. Invalid—reasoning from converse (like Figure 4.26)
   d. Invalid—reasoning from inverse (like Example 4.18

6. If my house is furnished by an interior designer, it will look nice. My house looks nice. Therefore, it was furnished by an interior designer.
7. "If there had been a good rap group around in those days, I might have chosen a career in music instead of politics" (*Richard Nixon*). He chose a career in politics, not music. Therefore, there was not a good rap group around in those days.

8. If you haven't taken a year of high school mathematics, then you can't take this course. You have taken a year of high school mathematics. Therefore, you can take this course.
9. If this section were well written, I could answer Exercise 8 correctly. I answered Exercise 8 correctly. Therefore, this section is well written.
10. If one is sick, then one should go to the doctor. I am not sick. Therefore, I will not go to the doctor.
11. According to Abigail VanBuren (Dear Abby), if a man gives a woman an engagement ring and 4 hours later they break up, then the woman should return the ring. A man gave my cousin an engagement ring and 4 hours later they broke up. Therefore, my cousin should return the ring.
12. One can join AARP only if one is over 50. Tom belongs to AARP. Therefore, Tom is over 50.

In each of Exercises 13–19, write the symbolic form of the argument, then use a truth table to determine if the argument is valid or invalid.

13. Either it is Friday or I don't have to go to work. I have to go to work. Therefore, it is not Friday.
14. Either it is Friday or I don't have to go to work. I have to go to work. Therefore, it is Friday.
15. Either it is Friday or I don't have to go to work. I don't have to go to work. Therefore, it is Friday.
16. If $x = 3$, then $x^2 = 9$. Therefore, if $x^2 \neq 9$, then $x \neq 3$.
17. If you have Devil Squares or Figaroos, you have a Little Debbie product. You don't have a Little Debbie product. Therefore, you don't have Devil Squares.
18. If you have Devil Squares or Figaroos, you have a Little Debbie product. You have Figaroos. Therefore, you have a Little Debbie product.
19. If you are on the Pacific Ocean or the Atlantic Ocean, then you are on one of the world's two largest oceans. You are not

on one of the world's two largest oceans. Therefore, you are not on the Pacific Ocean.

In Exercises 20–25, state whether the given condition will
  a. Always give a true conclusion
  b. Always give a false conclusion
  c. Sometimes give a true conclusion and sometimes give a false conclusion

20. A valid argument
21. An argument with all premises true
22. An argument with all premises false
23. A valid argument with all premises true
24. A valid argument with all premises false
25. A valid argument with some premises true and some premises false
26. For each of Exercises 20–25 for which you answered c, give an example of an argument satisfying the given condition for which the conclusion is true and a second example of an argument satisfying the given condition for which the conclusion is false.

WRITTEN ASSIGNMENTS 4.5

1. State some examples of arguments you have seen in writing or have heard that are invalid arguments. Use truth tables to show that they are invalid.
2. Do you think most people understand the distinction between validity and truth? Why do you believe thus?
3. Suppose you want to explain to a person of junior high school age that reasoning by the converse is not valid. A typical junior high school student has not seen a truth table. How do you make the explanation?

In the following projects, when you write a paper, assume that your audience is a group of readers of this book.

4. One of the functions of a mathematician is to create new mathematics. This involves conjecturing new mathematical results, then proving that they are true. These true statements are proved by the reasoning in this chapter and are called theorems.
  a. Some statements exist as conjectures for a number of years before being proved as theorems. One of the most interesting of these is the Four Color Theorem. Write a two- to four-page paper on the Four Color Theorem from its conjecture in the 1850s to its proof in 1976 by Appel and Haken. One reference is Campbell and Higgins (1984), Vol. I, pp. 154–173. Include at least one paragraph commenting on whether or not what Appel and Haken did was really a proof.
  b. A conjecture that had not been proven as of 1996 is Goldbach's conjecture:

> Every even integer greater than 4 is the sum of two odd prime numbers (for example, 6 = 3 + 3 and 8 = 5 + 3).

For this conjecture, write a one- or two-page paper that includes comments on the history of the conjecture, its current status (for what positive integers is it current known to be true), and the possibility of proof being found in the near future.

  c. In 1931 the logician Kurt Gödel proved that, in the common logical systems of mathematics, there exist statements that can be proved neither true nor false. Gödel's theorem has made a substantial impact on intellectual thought. (See Section 4.1) Write a two- or three-page paper on Gödel, Gödel's theorem, and its impact on society. One reference is Campbell and Higgins (1984), Vol. II, pp. 262–275.

5. a. Write a one-page review of the article "Are Logic and Mathematics Identical?" by Leon Henkin in *Science,* Vol. 138 (November 16, 1962), pp. 788–794. An adapted version of it is in Campbell and Higgins (1984), Vol. II, p. 223.
  b. Write a one-page review of the article "Proof" by Philip J. Davis and Rueben Hersh in Davis and Hersh (1981), pp. 147–152, and reprinted in Campbell and Higgins (1984), pp. 248–250.
  c. Write a two- to four-page paper comparing Henkin's view with that of Davis and Hersh on the similarities of mathematics and logic.

6. Computer programs have been written that will prove theorems. In fact, most, if not all, theorems of high school geometry can be proved by such a program. Write a two- to four-page paper about automated theorem proving or machine proofs.

7. One way of acquiring knowledge and, more importantly, of establishing knowledge is through deductive reasoning. For example, in this section we argued via deductive reasoning, that the form of the following argument is valid:

$$\begin{array}{c} p \to q \\ \sim q \\ \hline \sim p \end{array}$$

Deductive reasoning relies on valid arguments to acquire or establish knowledge. What other ways are there of acquiring knowledge besides deductive reasoning? That is, what other ways are there of knowing?

4.6 ● EQUIVALENT STATEMENTS

GOALS
1. To use truth tables to determine whether or not a pair of statements are logically equivalent.
2. To use DeMorgan's Laws and the fact that $p \rightarrow q \equiv \sim p \vee q$ to write a statement in an alternate form.
3. To know and use the analogies between logically equivalent statements and set theory properties.

You may have noticed that some logic statements have identical truth tables. For example, reconsider the truth tables for $p \rightarrow q$ and $\sim q \rightarrow \sim p$, shown in ▶ Figure 4.29, and note that the values of true and false are identical in both columns.

p	q	$\sim p$	$\sim q$	$p \rightarrow q$	$\sim q \rightarrow \sim p$
T	T	F	F	T	T
T	F	F	T	F	F
F	T	T	F	T	T
F	F	T	T	T	T

▶ FIGURE 4.29 Truth table for $p \rightarrow q$ and $\sim q \rightarrow \sim p$

When it is the case that two expressions have this property, it means they are interchangeable as far as their logical values are concerned. The following definition formalizes this property.

> DEFINITION
>
> ————●————
>
> Let s and t be two logical statements. We say that s is *logically equivalent to* t if and only if s and t have identical truth tables. When s and t are logically equivalent, we write
>
> $$s \equiv t$$

Notice that s is logically equivalent to t ($s \equiv t$) if and only if $s \leftrightarrow t$. With this definition in mind, we now investigate several equivalent statements.

DeMorgan's Laws

Let p and q be statements. The following can be shown to be true.

1. $\sim(p \wedge q) \equiv \sim p \vee \sim q$
2. $\sim(p \vee q) \equiv \sim p \wedge \sim q$

These are known as *DeMorgan's Laws*.

Before looking at the truth tables for these statements, let's consider a specific example. Suppose p and q are the following statements.

p : My test grade was A
q : I went home last weekend

A translation of $\sim(p \wedge q)$ then becomes

It was not the case that my test grade was A and I went home last weekend.

Is there another way to say the same thing? Consider the following sentence.

Either my test grade was not an A or I did not go home last weekend.

These two sentences have the same meaning. The second form is written symbolically as $\sim p \vee \sim q$.

Let's now consider the truth tables for $\sim(p \wedge q)$ and $\sim p \vee \sim q$, as shown in ▶ Figure 4.30.

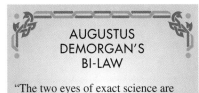
p	q	$\sim p$	$\sim q$	$p \wedge q$	$\sim(p \wedge q)$	$\sim p \vee \sim q$
T	T	F	F	T	F	F
T	F	F	T	F	T	T
F	T	T	F	F	T	T
F	F	T	T	F	T	T

▶ FIGURE 4.30 Truth table for $\sim(p \wedge q)$ and $\sim p \vee \sim q$

Since the truth tables are identical, we can write

$$\sim(p \wedge q) \equiv \sim p \vee \sim q$$

Truth tables for $\sim(p \vee q)$ and $\sim p \wedge \sim q$ are deferred to Exercise 1. However, they are identical, so we also have

$$\sim(p \vee q) \equiv \sim p \wedge \sim q$$

Before looking at other logically equivalent statements, let's briefly reconsider a property from set theory. Recall, if A and B are sets, we have

$$(A \cap B)' = A' \cup B'$$

Compare this to the logic result

$$\sim(p \wedge q) \equiv \sim p \vee \sim q$$

If we form the following associations,

Complement with negation ($\sim$)
Intersection with and ($\wedge$)
Union with or ($\vee$)

we see these are analogous results. Given these associations, we can also note the analogy between

$$\sim(p \vee q) \equiv \sim p \wedge \sim q \qquad \text{and} \qquad (A \cup B)' = A' \cap B'$$

The parallel development between logic and set theory does not end here. Set theoretic companions to other logical equivalences are shown later in this section.

Implication

As was said earlier, the most fundamental concept in the study of logic and constructing valid arguments is the implication, or conditional sentence, $p \rightarrow q$. Showing that a conditional sentence is true ($[p \rightarrow q) \wedge p] \rightarrow q$) permeates the develop-

p	q	$p \rightarrow q$
T	T	T
T	F	F
F	T	T
F	F	T

▶ FIGURE 4.31 Truth table for $p \rightarrow q$

ment of all areas of mathematics. Because the conditional sentence is so important, it is useful to have some logically equivalent forms. We now look at two such forms.

Let p and q be two statements. For reference, the truth table for $p \rightarrow q$ is shown in ▶ Figure 4.31.

 YOUR FORMULATION

1. Construct the truth table for $\sim p \vee q$ by completing ▶ Figure 4.32.

p	q	$\sim p$	$\sim p \vee q$
T	T		
T	F		
F	T		
F	F		

▶ FIGURE 4.32 Truth table for $\sim p \vee q$

2. Since the truth tables for $p \rightarrow q$ and $\sim p \vee q$ are identical, what can we conclude?

The significance to mathematicians of the observation that $p \rightarrow q \equiv \sim p \vee q$ is that, if they need to prove that some conditional sentence $p \rightarrow q$ is true, it is sufficient to show that either p is false or q is true. The significance to anybody is that a conditional statement can be converted to a disjunction, or vice versa. For example, the statement in the cartoon below is equivalent to "I didn't know there was math involved or I wouldn't have taken algebra."

"I wouldn't have taken Algebra if I had known there was math involved."
© 1993. Reprinted courtesy of Bunny Hoest and *Parade* magazine.

As a second equivalence for $p \to q$, let's reconsider the contrapositive $\sim q \to \sim p$. Figure 4.29 shows the truth tables for these expressions are identical. Hence, we can write

$$p \to q \equiv \sim q \to \sim p$$

This particular logical identity is very important to mathematicians. Frequently, the proof of a conditional sentence in its stated form is not obvious. However, if the problem is rewritten as

$$\sim q \to \sim p$$

this conditional sentence may be easier to prove. Since

$$\sim q \to \sim p \equiv p \to q$$

a proof of the contrapositive is equivalent to proving the original form, $p \to q$.

To illustrate, let's consider a short proof. The goal here is that you understand the logic involved in this proof, not the result being proved. Specifically, consider the following statement:

If x^2 is an odd integer, then x is an odd integer.

Define p and q:

$p : x^2$ is an odd integer
$q : x$ is an odd integer

We then have the conditional sentence

$$p \to q$$

A typical start to a proof of this is to assume p is true and then deduce that q is true. This can be shown as follows:

$$\left.\begin{array}{c} \text{Assume } p \\ \vdots \\ \text{Therefore } q \end{array}\right\} \text{Mathematical reasons go here}$$

The problem is that it is not easy to show that x is odd when you're working directly from the assumption that x^2 is odd. On the other hand, consider the contrapositive,

$$\sim q \to \sim p$$

This is the conditional sentence

If x is an even integer, x^2 is an even integer.

A fairly short proof is then:

Assume x is an even integer. Then $x = 2m$ for some integer m.
Therefore, $x^2 = (2m)^2 = 4m^2 = 2(2m^2)$.
Thus, x^2 is an even integer.

Symbolically, this is

$$\left.\begin{array}{c} \text{Assume } \sim q \\ \vdots \\ \text{Therefore } \sim p \end{array}\right\} \text{Reasons go here}$$

At this stage, we have proven $\sim q \to \sim p$ is true. Since this is logically equivalent to $p \to q$, we may claim that $p \to q$ has been proven.

Again, our focus is not on the specific parts of this proof, but on the logic involved in the proof. Remember that working with logically equivalent forms can often simplify a proof.

The equivalence of a statement and its contrapositive can also serve to rephrase conditional statements. For example, the statement "If you don't understand this, then you won't get 100 on the next test" is logically equivalent to the statement "If you get 100 on the next test, then you understand this."

Distributive Properties

Our final look at logically equivalent statements is to show that the *distributive property* holds for "and" and "or." These properties are:

1. $p \land (q \lor r) \equiv (p \land q) \lor (p \land r)$
2. $p \lor (q \land r) \equiv (p \lor q) \land (p \lor r)$

Note the similarity to the distributive property for sets.

$$A \cap (B \cup C) = (A \cap B) \cup (A \cap C)$$

$$A \cup (B \cap C) = (A \cup B) \cap (A \cup C)$$

and to the properties of real numbers

$$a(b + c) = a \cdot b + a \cdot c$$

such as

$$2(3 + 4) = 2 \cdot 3 + 2 \cdot 4$$

▸ ▶ CAUTION

Notice that for real numbers, the "other" distributive property, $a + (bc) = (a + b)(a + c)$, does *not* hold. For example, if $a = 2$, $b = 3$, and $c = 4$, then $a + bc = 2 + 3 \cdot 4 = 2 + 12 = 14$, whereas $(a + b)(a + c) = (2 + 3)(2 + 4) = 5 \cdot 6 = 30$. Consequently, the pattern for logical statements and sets does *not* carry over completely to numbers.

Truth tables for both $p \land (q \lor r)$ and $(p \land q) \lor (p \land r)$ are shown in ▶ Figure 4.33.

The respective columns are identical, so we write

$$p \land (q \lor r) \equiv (p \land q) \lor (p \land r)$$

p	q	r	$q \lor r$	$p \land q$	$p \land r$	$p \land (q \lor r)$	$(p \land q) \lor (p \land r)$
T	T	T	T	T	T	T	T
T	T	F	T	T	F	T	T
T	F	T	T	F	T	T	T
T	F	F	F	F	F	F	F
F	T	T	T	F	F	F	F
F	T	F	T	F	F	F	F
F	F	T	T	F	F	F	F
F	F	F	F	F	F	F	F

▶ FIGURE 4.33 Truth table for $p \land (q \lor r)$ and $(p \land q) \lor (p \land r)$

We defer to Exercise 6 constructing a truth table to show that

$$p \vee (q \wedge r) \equiv (p \vee q) \wedge (p \vee r)$$

An ordinary statement that can be changed via one of the distributive properties is "Ed is my cousin, and his wife's name is Marian or Mary Ann." This is logically equivalent to "Ed is my cousin and his wife's name is Marian, or Ed is my cousin and his wife's name is Mary Ann."

Expressions That Are Not Equivalent

Truth tables can also show that certain expressions are not equivalent. This is related to the material presented earlier when discussing invalid arguments.

Consider the conditional sentence $p \rightarrow q$. Sometimes an invalid argument is constructed by considering the converse, $q \rightarrow p$. For example, let p and q be

p : I receive a paycheck today
q : It is Friday

and consider the conditional sentence

If I receive a paycheck today, then it is Friday.

What can be deduced from the fact that it is Friday? Some might think, "Oh, I get paid today." Basically, they are thinking that $q \rightarrow p$ is logically equivalent to $p \rightarrow q$. ▶ Figure 4.34 shows that they are not equivalent. (For example, you might get paid on alternate Fridays.)

The next example allows you to examine one more situation in which errors in logic are sometimes made.

p	q	$p \rightarrow q$	$q \rightarrow p$
T	T	T	T
T	F	F	T
F	T	T	F
F	F	T	T

▶ FIGURE 4.34 Truth table for $p \rightarrow q$ and $q \rightarrow p$

EXAMPLE 4.20

Use truth tables to show that $p \rightarrow q$ is not logically equivalent to $\sim p \rightarrow \sim q$.

SOLUTION Truth tables for these expressions are shown in ▶ Figure 4.35. The columns for $p \rightarrow q$ and $\sim p \rightarrow \sim q$ show that these expressions are not equivalent.

p	q	$\sim p$	$\sim q$	$p \rightarrow q$	$\sim p \rightarrow \sim q$
T	T	F	F	T	T
T	F	F	T	F	F
F	T	T	F	T	F
F	F	T	T	T	T

▶ FIGURE 4.35 Truth table for $p \rightarrow q$ and $\sim p \rightarrow \sim q$

Again let's define p and q as

p : I receive a paycheck today
q : It is Friday

and consider the conditional sentence $p \rightarrow q$

If I receive a paycheck today, it is Friday.

What can be concluded from the fact that a person did not receive a paycheck? Nothing. It could be Friday or it could be some other day of the week, and the conditional sentence would still be true.

Some of this section's exercises illustrate what can be done with symbolic logic. In the abstract, manipulating symbols and constructing truth tables become routine. When possible, you should consider how equivalent statements can be used in both oral and written communication.

EXERCISES 4.6

In each of Exercises 1–6, use truth tables to determine whether or not the given pair of statements are logically equivalent.

1. $\sim(p \vee q)$, $(\sim p) \wedge (\sim q)$
2. $\sim(p \vee q)$, $(\sim p) \vee (\sim q)$
3. $p \rightarrow q$, $p \vee \sim q$
4. $p \leftrightarrow q$, $(p \rightarrow q) \wedge (q \rightarrow p)$
5. $p \wedge (q \vee r)$, $(p \wedge q) \vee r$
6. $p \vee (q \wedge r)$, $(p \vee q) \wedge (p \vee r)$

In each of Exercises 7–10, write the given statement in an alternate form using one of DeMorgan's Laws.

7. It's not the case that Amy is single and does not claim any dependents.
8. Carol itemizes deductions and does not claim any adjustments to income or tax credits.
9. It's not the case that Ben is not 65 or older or blind.
10. Dan's unearned income was not $0 or the total of that income plus his earned income was $3, 250 or less.
11. For each of Exercises 1–6 in which the given pair of statements are logically equivalent, write the analogous property from set theory.
12. In this section we proved "If x^2 is an odd integer, then x is an odd integer" by letting p be "x^2 is an odd integer" and q be "x is an odd integer" and showing that $\sim q \rightarrow \sim p$. Consider the following argument for that statement.

Suppose x is an odd integer. Then $x = 2m + 1$ for some integer m (for example, if $x = 5$, let $m = 2$). Therefore, $x^2 = (2m + 1)^2 = 4m^2 + 4m + 1 = 2(2m^2 + 2m) + 1$. Thus x^2 is an odd integer.

Is this a valid argument for the conditional sentence "If x^2 is an odd integer, then x is an odd integer"? Why or why not?

In each of Exercises 13–16, write the given statement in an alternate form taking the fact that $p \rightarrow q$ is logically equivalent to $\sim p \vee q$.

13. If you don't go to other men's funerals, they won't go to yours. (*quoted from Clarence Day*)
14. It is not 3 A.M. or I am Leibnitz.
15. If one man offers you democracy and another offers you a bag of grain, then at some stage of starvation you will prefer the grain to the vote. (*paraphrase of Bertrand Russell*)
16. This course is not trivial or I am really a slow learner.

In each of the Exercises 17–18, change the given statement to a logically equivalent *conditional* statement.

17. It's not the case that the suspect was both in Los Angeles at 9:00 P.M. and in San Francisco at 9:30 P.M.
18. The painting is not both a Cubist and a Renaissance painting.

WRITTEN ASSIGNMENTS 4.6

1. Lewis Carroll, author of *Alice's Adventures in Wonderland,* also wrote about logic in *Symbolic Logic.* Report on several of his logical puzzles.
2. We have maintained that the single most fundamental concept in the study of logic is the conditional sentence $p \rightarrow q$—If p, then q. If–then statements are important in a variety of situations and circumstances. Expand on each of the following ideas.
   a. A first-grader learning to read once asked the teacher what the biggest word in the English language was. Helen Strode answered, "If." Do you agree with her? Why or why not?

   b. A multiple-choice aptitude test has one question:

   A cow has four legs. If we call a tail a leg, then how many legs does a cow have?

   Pick your answer before reading what follows.
   If you answered 3, you are destined to go into politics, because politicians can do anything with numbers. If you answered 4, you are destined to be a hard scientist such as a chemist, physicist, or engineer, because you take the facts as you know them and are not influenced by flights of fancy. If

you answered 5, you are destined to be a mathematician, because you are willing to take a hypothesis (the "if" part) and see what follows from it. What, in your mathematical experience, illustrates the idea of taking a hypothesis and seeing what follows from it?

c. Games of strategy—like tic-tac-toe, athletic contests, and debates—are often analyzed by a player's saying "If I move here, what will my opponent do?" Therefore, conditional statements are basic in games of strategy.

d. In creative endeavors like painting, sculpting, composing music, and writing literature, the creator thinks, "If I do this, what will be the result?" For example, "If I paint an orange sky, what will that convey?" Therefore, conditional statements are basic in creative endeavors.

3. How are conditional statements involved in the following? (See Written Assignments 2c and 2d for some examples.)

a. Business
b. Education
c. Engineering
d. Health Sciences
e. Law
f. Social Sciences

4. (*Library research*) Augustus DeMorgan is described briefly in Section 3.5. Write a one- to three-page paper about him.

5. Find a conditional sentence in a song or a poem. Write it in a logically equivalent, but different, form. Try to write a portion of the song or poem with at least one conditional sentence written in a logically equivalent form and with a rhyme maintained.

6. Describe some situations (like the Friday-paycheck situation) in which people sometimes construct an invalid argument by using the converse.

CHAPTER 4 REVIEW EXERCISES

1. Write a one-page description of the historical development of the subject of logic. Include some approximate dates and the names of at least three people involved in the development.

2. Indicate whether or not each of the following is a statement, from the point of view of logic.
   a. *Schindler's List* won an Academy Award for best movie.
   b. *King Kong* was a great movie.
   c. *Gone With the Wind* was the first movie ever made.

3. Write the symbolic form of each of the following statements.
   a. Chapter 4 is on logic and Chapter 3 is on sets.
   b. "I love the game, but I don't need that kind of turmoil." (*George Bush, on whether he was interested in being commissioner of baseball*)
   c. "If Thomas Jefferson was alive today, I would appoint him secretary of state." (*Bill Clinton*)
   d. This is night or we're having an eclipse.
   e. She has a journalism major or she has a journalism minor, and she writes well.
   f. It is not the case that the Los Angeles Dodgers will win the 2001 World Series and the New York Mets will win the 2001 World Series.
   g. If this is the weekend, then today is Saturday or today is Sunday.
   h. Speak only if called on.
   i. Getting a C is a sufficient condition for passing this course.

4. Given the following statements:
   h : The Hindu–Arabic numeration system has base 10.
   r : Roman numerals use base 10.
   s : Some numeration systems have a base of 20.
   Write an English translation for each of the following symbolic statements.
   a. $\sim r$
   b. $h \wedge s$
   c. $h \vee r$

   d. $\sim h \wedge r$
   e. $h \vee \sim r$
   f. $h \wedge r \wedge s$

5. Construct a truth table for each of the following statements.
   a. $\sim(p \wedge \sim q)$
   b. $p \vee \sim q$
   c. $\sim p \vee (p \wedge q)$
   d. $\sim p$ XOR q
   e. $p \rightarrow \sim q$
   f. $(p \wedge q) \rightarrow q$
   g. $q \rightarrow (p \vee q)$
   h. $(\sim p \rightarrow q) \leftrightarrow (\sim q \rightarrow p)$
   i. $[(p \vee q) \wedge r] \leftrightarrow [p \vee (q \wedge r)]$

6. Via a truth table, answer each of the following as (i) T, (ii) F, or (iii) more information is needed
   a. If the truth value of $\sim p$ is F, the truth value of p is _____.
   b. If the truth value of p is T, the truth value of $p \vee q$ is _____.
   c. If the truth value of p is T, the truth value of $p \wedge q$ is _____.
   d. If the truth value of p is T, the truth value of p XOR q is _____.
   e. If the truth value of p is T, the truth value of $p \rightarrow q$ is _____.
   f. If the truth value of p is F, the truth value of $p \rightarrow q$ is _____.
   g. If the truth value of $p \rightarrow q$ is T, the truth value of $q \rightarrow p$ is _____.
   h. If the truth value of $p \rightarrow q$ is T, the truth value of $\sim p \rightarrow \sim q$ is _____.
   i. If the truth value of $p \rightarrow q$ is T, the truth value of $\sim q \rightarrow \sim p$ is _____.

7. a. Write the following sentence symbolically: "If I had a hammer, I'd hammer in the morning."

b. Construct a truth table for it.

c. Give an illustration of when the sentence would be false.

8. Given the true sentence "If x is a negative real number, then x^2 is a positive real number." Write its
   a. Contrapositive
   b. Converse
   c. Inverse
   d. Which of the statements in parts a, b, and c are true?

9. Give an example of a tautology that is an English sentence rather than a statement in symbolic form.

10. Which of the following statements are tautologies? Leave evidence of your reasoning.
    a. $p \rightarrow q$
    b. $(p \vee q) \vee (p \wedge q)$
    c. $[(p \rightarrow q) \wedge \sim q] \rightarrow \sim p$

11. Using *modus ponens,* what conclusion can be drawn from the following statements?

    If guns are outlawed, only outlaws will have guns.
    State X outlawed guns.

12. Label each of the following arguments as one of the following types:
    i.   Valid—*modus ponens*
    ii. Valid—reasoning from the contrapositive
    iii. Invalid—reasoning from the converse
    iv. Invalid—reasoning from the inverse
    a. If I study too much, I don't do well.
       I didn't do well.
       Therefore, I studied too much.
    b. If there is a traffic jam, they will be late.
       There was no traffic jam.
       Therefore, they will not be late.
    c. If the standards of Goals 2000 are met, high school graduates should know about dances and dancers prior to the twentieth century.
       Most high school graduates don't know about any dances or dancers prior to the twentieth century.
       Therefore, the standards of Goals 2000 are not being met.
    d. If the standards of Goals 2000 are met, high school graduates should know how to operate a calculator.
       High school graduates know how to operate a calculator.
       Therefore, the standards of Goals 2000 are being met.

13. In each of the following, write the symbolic form of the argument, then use a truth table to determine if the argument is valid or invalid.
    a. If a high school graduate meets the standards of Goals 2000, he knows about the function of Amnesty International and NATO.
       Ed is a high school graduate who doesn't know about the function of Amnesty International but does know about the function of NATO.
       Therefore, Ed doesn't meet the standards of Goals 2000.
    b. If you do a crossword puzzle or lease an Acura, then you demonstrate your intelligence.

You don't do the crossword puzzle and you don't lease an Acura.
Therefore, you don't demonstrate your intelligence.

14. State whether the given conditions will
    i.   Always give a true conclusion
    ii. Always give a false conclusion
    iii. Sometimes give a true conclusion and sometimes give a false conclusion
    a. An invalid argument
    b. An invalid argument with all premises false
    c. A valid argument with all premises true

15. a. Give an example of a conditional sentence.
    b. Give an example of a valid argument with a true conclusion.
    c. Give an example of a valid argument with a false conclusion.

16. In each part, use truth tables to determine whether or not the given pair of statements are logically equivalent.
    a. $\sim(p \vee q)$, $\sim p \vee \sim q$
    b. $p \rightarrow q$, $p \wedge q$
    c. $p \rightarrow (q \wedge r)$, $(\sim p \vee q) \wedge (\sim p \vee r)$

17. In each of the following, write the given statement in an alternate form using one of DeMorgan's Laws.
    a. It's not the case that she is at least 25 and at least 7 years a citizen of the United States. (*a qualification for a U.S. Representative*)
    b. He was not young and he was not wealthy.

18. The statements $\sim(p \vee q)$ and $\sim p \wedge \sim q$ are logically equivalent. Write the analogous property from set theory.

19. The sets $A \cap (B \cup C)$ and $(A \cap B) \cup (A \cap C)$ are equal. Write the corresponding logically equivalent statements.

20. In this chapter, we proved "If x^2 is an odd integer, then x is an odd integer" by letting p be "x^2 is an odd integer" and q be "x is an odd integer" and showing $\sim q \rightarrow \sim p$. Consider the following argument for that statement.

    Suppose x is an odd integer. Then x ends in 1, 3, 5, 7, or 9, so x^2 ends in 1, 9, 5, 9, or 1, respectively. Therefore, x^2 is odd.

    Is this a valid argument that if x^2 is an odd integer, then x is an odd integer? Why or why not?

21. Write each of the following statements in an alternate form taking the fact that $p \rightarrow q$ is logically equivalent to $\sim p \vee q$.
    a. If the shoe fits, wear it.
    b. The steak is either not done or I'm a poor judge of meat.

22. For each of the following statements tell whether it is
    A. Always true
    N. Never true
    S. Sometimes true and sometimes false
    a. A valid argument leads to a true conclusion.
    b. An invalid argument leads to a false conclusion.
    c. An argument with a true conclusion is valid.
    d. An argument with a false conclusion is invalid.

Barrow, John D. *Pi in the Sky.* New York: Little, Brown, 1992.

Boyer, Carl B. and Merzbach, Uta C. *A History of Mathematics,* 2nd ed. New York: Wiley, 1991.

Calinger, Ronald (ed.). *Classics of Mathematics.* Englewood Cliffs, NJ: Prentice-Hall, 1995.

Campbell, Douglas M., and Higgins, John C. (ed.). *Mathematics: People, Problems, Results.* Belmont, CA: Wadsworth, 1984.

Casti, John L. *Searching for Certainty: What Scientists Can Know About the Future.* New York: William Morrow, 1990.

COMPAC. *For All Practical Purposes,* 3rd ed. New York: W.H. Freeman, 1994.

Davis, Phillip J., and Hersh, Reuben. *The Mathematical Experience.* Boston: Birkhauser, 1981.

Eves, Howard. *An Introduction to the History of Mathematics,* 6th ed. Chicago: Saunders College, 1990.

Kline, Morris. *Mathematics: A Cultural Approach.* Reading, MA: Addison-Wesley, 1962.

Stillwell, John. *Mathematics and Its History.* New York: Springer-Verlag, 1989.

CHAPTER 5

ALGEBRA

PROLOGUE

"So, you're studying algebra."
"Yes."
"Have they found x yet?" (chuckle, chuckle)

How long has this joke been told? Some of the earliest-known writing was of mathematical problems in what is known as the *Rhind Papyrus*, dating to about 1650 B.C.E. Here's one of the problems in the *Rhind Papyrus*: "A quantity,

its $\frac{2}{3}$, its $\frac{1}{2}$, and its $\frac{1}{7}$, added together, becomes 33. What is the quantity?" In other words, find x. But the question did not say, "Find x." in fact, it was close to two millennia later before such symbolism was adopted.

Solving linear equations for "the quantity" is a skill that was known over 4000 years ago. Solving some quadratic equations (remember the

quadratic formula?) is a skill known almost as long. Solving a third-degree equation with a formula similar to the quadratic formula has been known for less than 500 years. Solving a fourth-degree equation has been known for a little over 450 years. Progress seems to be picking up. What about a fifth- or sixth-degree equation? Equations of how big a degree can be solved now?

Surprise! It has been proven that there is no formula like the quadratic formula that will solve a general equation of degree five or larger. No matter how much time and money are thrown at finding a solution, one cannot be found. What happened to the power of mathematics to explain everything? Attitudes needed to change.

Equations of degree five or bigger *can* be solved by approximation techniques rather than exact formulas. These approximation techniques are behind the "solver" procedure on graphing calculators.

The world is interested in answers: Go find the solutions. The mathematician is interested in the process: Develop some notation and ask how we can find *x*. Society sits back and observes: What can mathematics do? How powerful is it? Have they found *x*? Have they found a way to find *x*? Can we always find *x*?

5.1 GENERALIZED ARITHMETIC AND APPLICATIONS OF LINEAR EQUATIONS

GOALS
1. To describe algebra as generalized arithmetic and to do arithmetic operations on polynomials.
2. To describe some of the historical developments in algebra.
3. To identify and give examples of linear equations in one variable.
4. To solve application problems using linear equations.

Just what is algebra? We expect that you have had at least one algebra course in high school. You perhaps recall working with *x*'s and being told to solve equations. However, you might have a difficult time giving a succinct explanation of what algebra is. Many people have offered explanations and/or definitions for algebra. Although most of these are accurate and acceptable, there is no universally accepted "best" definition. If we look at several such definitions, certain terms emerge as essential. These include *variable, equation, solving,* and *generalized arithmetic.*

Knowing that any attempt at a rigorous definition of algebra would be incomplete, we offer the following description.

Algebra is a system that generalizes arithmetic. It is characterized by:

1. The use of variables to represent numbers
2. The analysis of methods for solving equations

Let's now proceed to examine some aspects of algebra.

Concept of a Variable

Students in elementary schools spend most of their arithmetic development time working with numbers. They learn what numbers represent and how to operate with them. At some point in the middle school years, students are introduced to the concept of a *variable* (usually *n* or *x*), that is, a letter serving to represent some number. For example:

Let *n* represent some number. Write an expression for 2 more than *n*.

Students would then be expected to produce an answer such as $n + 2$. However, it has become customary to use the letter *x* to denote the variable when only one letter is required. Thus, students might be asked to work algebraic statements such as the following:

$$\text{Solve for } x: \quad x + 3 = 5$$

$$\text{Combine:} \quad \frac{x}{3} + \frac{2}{x}$$

Being able to work with variables rather than just numbers is a major transition that students make when beginning the study of algebra. But when did this transition occur historically? Two of the oldest extant records of mathematical activity are called the *Rhind Papyrus* and the *Moscow Papyrus.* They come from Egyptian civilizations dating to about 1850 B.C.E. Both sets of mathematical problems contain some problems that we would classify as algebra. For example, problem 35 of the *Moscow Papyrus* asks the reader to find the size of a scoop for which it takes $3\frac{1}{3}$ trips to fill a 1-hekat measure. The problem is recognizable as the type we encounter in algebra, although the solution technique of the Egyptians was different than the one you were taught in algebra.

Historians of mathematics have described three stages in the development of algebraic notation. First, there is *rhetorical algebra,* in which everything is written

ON A TANGENT

How Many Were Going to St. Ives?

Problem 79 of the *Rhind Papyrus* contains the following set of data:

Estate	
Houses	7
Cats	49
Mice	343
Heads of wheat	2,401
Hekat measures	16,807
	19,607

The numbers above the line are the first five powers of 7: $7^1 = 7$, $7^2 = 49$, $7^3 = 7 \cdot 49 = 343$, $7^4 = 7 \cdot 343 = 2,401$, and $7^5 = 7 \cdot 2,401 = 16,807$. It has been suggested that this problem was similar to that of the Old English children's rhyme:

As I was going to St. Ives
I met a man with seven wives;
Every wife had seven sacks;
Every sack had seven cats;
Every cat had seven kits.
Kits, cats, sacks, and wives,
How many were going to St. Ives?

out in words. The problem from the *Moscow Papyrus* is an example of rhetorical algebra. In fact, all of the algebra done by the ancient Egyptians was rhetorical algebra. The second stage has been called *syncopated algebra,* in which some abbreviations are adopted for the variables. Finally, comes *symbolic algebra,* in which symbols like *x* or *n* represent variables. Symbolic algebra first appeared in western Europe in the 1500s but did not become widespread until the mid-1600s. Therefore, much of the algebraic symbolism with which you are familiar is less than 400 years old. That may seem rather old to you, but measured against the extent of human history it is quite young. Imagine doing algebra without symbols!

Generalized Arithmetic

Earlier in this section we described algebra as generalized arithmetic. Let's examine this idea in more detail.

Arithmetic includes the study of integers (. . . −3, −2, −1, 0, 1, 2, 3, . . .) and the basic operations of addition, subtraction, multiplication, and division with integers. We also examine the concept of factors for integers; for example, since 15 = 3 · 5, we say 15 has factors of 3 and 5. Thus, the concepts of *prime* and *composite numbers* are developed. Recall, that a positive integer greater than 1 is considered *prime* provided it has no divisors except itself and 1. A *composite number* is a positive integer that does have a divisor other than itself and 1. Thus, 7, 19, 23, 41 are examples of prime numbers, while 10, 15, 25, and 39 are examples of composite numbers.

To see how algebra generalizes the study of arithmetic, let's think of polynomials (expressions of the form $a_0 + a_1x + a_2x^2 + . . . + a_nx^n$, where *n* is a nonnegative integer and $a_0, a_1, a_2, \ldots, a_n$ are real numbers) as the "integers" of algebra. With this relationship, note the similarities indicated in ● Table 5.1.

If one polynomial can be written as the product of two other polymials, it is composite. For example, since

$$x^2 + 3x + 2 = (x + 1)(x + 2)$$

then $x^2 + 3x + 2$ is composite and $(x + 1)$ and $(x + 2)$ are factors of $x^2 + 3x + 2$. If a polynomial cannot be written as a product of factors other than $\pm$itself and ± 1, it is prime. Thus, the following are prime:

$$x^2 + x + 2 \qquad \text{and} \qquad -x^2 - x - 2 = (-1)(x^2 + x + 2)$$

(A detailed study of factors of polynomials requires an examination of what kinds of numbers are permitted as coefficients of terms. For the sake of this discussion, we assume coefficients are integers.) Being able to find factors of expressions is a significant part of high school algebra.

One particularly effective way to illustrate how algebra generalizes arithmetic is to analyze the addition of rational numbers (numbers that are ratios of integers). From arithmetic, consider

$$\frac{2}{5} + \frac{1}{3}$$

This would typically be evaluated as follows:

1. Get a common denominator: $\qquad 3 \cdot 5 = 15$

2. Rewrite the fractions using the common denominator:

$$\frac{2}{5} \cdot \frac{3}{3} + \frac{1}{3} \cdot \frac{5}{5} = \frac{6}{15} + \frac{5}{15}$$

3. Add the numerators: $\qquad \dfrac{6}{15} + \dfrac{5}{15} = \dfrac{11}{15}$

Using the analogy of polynomials as integers and the ratio of polynomials as the rational numbers of algebra, let's consider the problem

$$\frac{x}{x + 1} + \frac{x - 1}{x - 2}$$

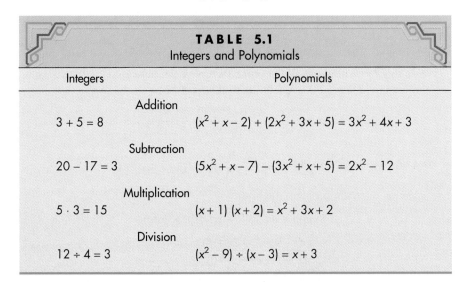

TABLE 5.1
Integers and Polynomials

Integers	Polynomials
Addition	
$3 + 5 = 8$	$(x^2 + x - 2) + (2x^2 + 3x + 5) = 3x^2 + 4x + 3$
Subtraction	
$20 - 17 = 3$	$(5x^2 + x - 7) - (3x^2 + x + 5) = 2x^2 - 12$
Multiplication	
$5 \cdot 3 = 15$	$(x + 1)(x + 2) = x^2 + 3x + 2$
Division	
$12 \div 4 = 3$	$(x^2 - 9) \div (x - 3) = x + 3$

TABLE 5.2
Arithmetic versus Algebra

Arithmetic	Algebra

Problem

$$\frac{2}{5} + \frac{1}{3} \qquad\qquad \frac{x}{x+1} + \frac{x-1}{x-2}$$

Get a common denominator

$$5 \cdot 3 \qquad\qquad (x+1)(x-2)$$

Rewrite the fractions

$$\frac{2}{5}\left(\frac{3}{3}\right) + \frac{1}{3}\left(\frac{5}{5}\right) \qquad\qquad \frac{x}{x+1}\left(\frac{x-2}{x-2}\right) + \frac{x-1}{x-2}\left(\frac{x+1}{x+1}\right)$$

Simplify

$$\frac{2 \cdot 3}{15} + \frac{1 \cdot 5}{15} = \frac{6}{15} + \frac{5}{15} \qquad\qquad \frac{x(x-2)}{(x+1)(x-2)} + \frac{(x-1)(x+1)}{(x-2)(x+1)}$$

$$= \frac{x^2 - 2x}{(x+1)(x-2)} + \frac{x^2 - 1}{(x-2)(x+1)}$$

Add numerators:

$$\frac{11}{15} \qquad\qquad \frac{2x^2 - 2x - 1}{x^2 - x - 2}$$

Via development parallel to the arithmetic problem, we have the comparison shown in Table 5.2.

Reducing fractions and division are two other examples to illustrate the parallel development of algebra and arithmetic. The following example illustrates reducing fractions; division will be included in Exercises 9–11.

EXAMPLE 5.1

Recall from arithmetic a problem such as the following:

$$\text{Simplify } \frac{60}{90}$$

We do this by writing out all the prime factors of the numerator and the denominator and reducing 60/90. A solution to this could be

$$\frac{60}{90} = \frac{6 \cdot 10}{3 \cdot 30} = \frac{2 \cdot 2 \cdot 3 \cdot 5}{2 \cdot 3 \cdot 3 \cdot 5}$$

We then "divide out" to get $\dfrac{2 \cdot 2 \cdot 3 \cdot 5}{2 \cdot 3 \cdot 3 \cdot 5} = \dfrac{2}{3}$

In algebra, a similar problem might be:

Simplify the following expression:

$$\frac{(x^2 - 9)(x^2 + 2x + 1)}{(x+1)(x^2 - x - 6)}$$

Finding factors for the numerator and denominator, we get

$$\frac{(x^2 - 9)(x^2 + 2x + 1)}{(x + 1)(x^2 - x - 6)} = \frac{(x - 3)(x + 3)(x + 1)^2}{(x + 1)(x - 3)(x + 2)}$$

We could then "divide out" to obtain

$$\frac{(x - 3)(x + 3)(x + 1)(x + 1)}{(x + 1)(x - 3)(x + 2)} = \frac{(x + 3)(x + 1)}{(x + 2)}$$

Finding the factors of this example is secondary. The point is how the solution in algebra parallels the arithmetic problem.

Applications of Linear Equations

Recall that a *linear equation in one variable* is an equation that can be written in the form $ax + b = 0$, where a and b are real numbers and $a \neq 0$. Such equations are called linear because graphs of them are lines. An essential feature is that a linear equation has no x^2, x^3, etc. term. For example, the following are linear equations in one variable:

$$2x - 6 = 0$$

$$3x + 5 = 2$$

$$-6x + 2x = 3 - x$$

whereas the following are not linear equations (although each has one variable):

$$2x^2 - 6 = 0$$

$$3x^2 + 5 = 2$$

$$x + 2 = \frac{1}{x}$$

YOUR FORMULATION

1. Solve the linear equation $2x - 6 = 0$.
2. Write the equation $2x - 6 = 0$ in rhetorical algebra; i.e., write or state the equation without any symbol for the variable.
3. Solve the equation $2x - 6 = 0$ in rhetorical algebra; i.e., solve the equation without writing or stating a symbol for the variable.
4. Comment on the advantage, if any, of symbolic algebra over rhetorical algebra.

Many situations in life can be translated into mathematical terms that result in linear equations. We will now investigate several such problems.

Writing the Equation   Problems translatable into mathematical sentences typically appear in written form. Perhaps you recall your reaction to "story problems" from a previous course. Often, students have developed an aversion to working such problems. We will attempt to show you how certain routine steps can be followed

THE FAR SIDE   By GARY LARSON

Hell's library

to help you translate written problems into mathematical sentences. Let's consider a typical one:

A box of computer disks was marked down 15% and offered for sale at $20.36. What was the original price for the box of disks?

Problems like this require the use of a variable, which should be identified and labeled accordingly. For example:

Let x denote the original cost of the disks.

Further information from the problem should then be stated as clearly as possible. For our problem we get:

Let $.15x$ denote the reduction in price.
Then, $x - .15x$ is the sale price.

We next write an equation based on what has been identified. Since the sale price is $20.36, we assume x is in dollars and have

$$x - .15x = 20.36$$

At this point, we use techniques of solving equations to obtain

$$x - .15x = 20.36$$
$$.85x = 20.36$$
$$x = \frac{20.36}{.85} = 23.95$$

So our answer is:

The original price for the box of disks was $23.95.

Further examples of interpreting written expressions follow. We will write a mathematical equation for each one.

EXAMPLE 5.2

The sum of twice a number and 7 is 19.

SOLUTION In this problem, let n represent the number. We then have

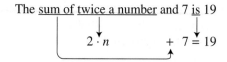

The sum of twice a number and 7 is 19

$$2 \cdot n + 7 = 19$$

EXAMPLE 5.3

Three times a number divided by 5 is 9.

SOLUTION Again, let n denote the number. We then have

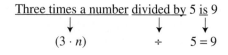

Three times a number divided by 5 is 9

$$(3 \cdot n) \div 5 = 9$$

Summary of Steps for Writing an Equation Let's now look at mathematical expressions that are contained within somewhat more elaborate problems. We recommend you approach such problems as follows.

1. Read slowly all the way through the problem to get an idea of what is to be done.
2. Start reading again, stopping for some note taking or sketching.
3. Make a sketch (if possible).
4. Choose a variable; write what it represents; label the sketch accordingly.
5. Write a mathematical statement from the problem.
6. Solve the equation.
7. Answer the question.
8. Check your answer.

Notice how these steps are followed in the ensuing examples.

More Examples

EXAMPLE 5.4

Rectangular swimming pools are often twice as long as they are wide. If the perimeter of such a pool is 96 feet, what are the dimensions of the pool?

SOLUTION First, read slowly through the problem. A sketch is possible, so we get the following:
(Notice that the length is approximately twice the width. Sketches should always reasonably reflect the physical setting described in the problem.)

Choose a variable:

Let x denote the pool width.
Then $2x$ is the pool length.

We label the sketch:

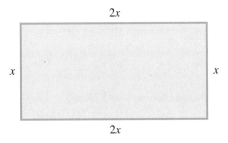

At this point, an equation can be written. From the sketch, we determine the perimeter as

$$x + 2x + x + 2x = 6x$$

The perimeter is known to be 96 feet. Thus, we have

$$6x = 96 \text{ ft}$$

Now we solve for x to obtain

$$x = \frac{96}{6} \text{ ft}$$

$$x = 16 \text{ ft}$$

The question requests the dimensions, so our answer is

Width = 16 ft

Length = 32 ft

We then check the answer. If the pool is 16 ft by 32 ft, the perimeter is 16 ft + 32 ft + 16 ft + 32 ft, which equals 96 ft.

EXAMPLE 5.5

A person who itemizes expenses on the federal income tax return, is single, and earns over $18,550 but not over $44,900 owes tax of $2,782.50 + 28% of the amount over $18,550. If such a person paid $5,667.90 in income tax, what was the person's taxable income?

SOLUTION We let x be the person's taxable income, in dollars. Since the tax is $2,782.50 + 28% of the amount over $18,550, next we need to know the amount over $18,550, which is $x - 18,550$. The person's tax, in dollars, is

$$2,782.50 + .28(x - 18,550)$$

Since we're told that this is $5,667.90, we have

$$2,782.50 + .28(x - 18,550) = 5,667.90$$

We solve the equation:

$$2,782.50 + .28x - 5,194 = 5,667.90$$

$$.28x - 2,411.50 = 5,667.90$$

$$.28x = 5,667.90 + 2,411.50$$

$$.28x = 8,079.40$$

$$x = \frac{8,079.40}{.28} = 28,855$$

Thus, the person's taxable income was $28,855.

We check this answer by figuring the tax on an income of $28,855. The amount over $18,550 is $10,305. We find 28% of this, which is $2,885.40, then add on $2,782.50 to get a tax of $5,667.90, as stated.

The preceding examples have emphasized the translation of written problems into mathematical sentences. Unfortunately, many students in a first algebra course are unable to do this well. But students can—and must—learn good translation techniques from fairly easy examples so they can then apply these techniques to more difficult problems.

1. Which of the following numbers are composite? 11, 21, 31, 41, 51
2. Find a prime number larger than 60.
3. List all the even prime numbers.

For Exercises 4 and 5, perform the indicated operations.

4. $\dfrac{2x + 3}{x - 1} + \dfrac{x - 4}{x + 2}$

5. $\dfrac{5x - 2}{x + 3} + \dfrac{3x + 4}{x - 4}$

For Exercises 6–8, simplify the expressions.

6. $\dfrac{x^2 - 4}{x^2 + 5x + 6}$

7. $\dfrac{(x^2 - x - 2)(x + 3)}{x^2 + x - 6}$

8. $\dfrac{(x^2 + 2x + 1)(x^2 - 2x - 8)}{(x^2 - 16)(x^2 + 6x + 5)}$

For Exercises 9–11, carry out the indicated divisions and simplify. Recall that

$$\frac{2}{9} \div \frac{4}{3} = \frac{\cancel{2}}{\cancel{9}_3} \cdot \frac{\cancel{3}}{\cancel{4}_2} = \frac{1}{6}$$

9. $\dfrac{x}{x^2 + 3x + 2} \div \dfrac{x^2 + 3x}{x + 2}$

10. $\dfrac{x^2 - 1}{x^2 - x - 6} \div \dfrac{x^2 - 2x + 1}{x^2 + 6x + 8}$

11. $\dfrac{x^2 + 3x + 2}{(x - 3)\,(x^2 + 4x)} \div \dfrac{x + 1}{x^2 - 3x}$

12. Give an example of a linear equation in one variable.
13. Give an example of an equation that is not a linear equation.
14. Redbird Pet Food Center advertised 10% off on a canine maintenance diet.
    a. If the total before tax came to $12.51, what was the original price?
    b. If the original price was $8.50, what is the sale price?
15. A clothing store advertises slacks at 30–50% off the regular price. Suppose the sale price on a pair of slacks is $19.90.
    a. If this price is 30% off, what was the original price?
    b. If this price is 50% off, what was the original price?
    c. If the original price was $30.00, what is the percentage discount?
16. The sum of 3 and twice a number is 17. Find the number.
17. Four times a number divided by 3 is 8. Find the number.
18. If a rectangular swimming pool is twice as long as it is wide and the perimeter is 108 ft, what are the dimensions of the pool?

19. If a rectangular swimming pool is twice as long as it is wide and the perimeter is 120 ft, what are the dimensions of the pool?
20. The ancient Greeks discovered a number called the *golden ratio*, which is exactly $(1 + \sqrt{5})/2$ and approximately 1.6. People through the years have been intrigued with this number for both its mathematical properties and its aesthetic properties.
    a. Calculate the golden ratio to three decimal places.
    b. If a rectangular picture frame has the ratio of its long side to its short side of 1.6 (about the golden ratio) and its perimeter is 52 inches, find the dimensions of the picture frame.
21. A person who itemizes expenses on the federal income tax return, is single, and earns not over $18,500 owes tax of 15% of taxable income. If such a taxpayer pays $1,851.30 in tax, what is the person's taxable income?
22. A person who itemizes expenses on the federal income tax return, is single, and earns over $44,900 but not over $93,130 owes tax of $10,160.50 + 33% of the amount over $44,900. If such a taxpayer pays $12,539.80 in tax, what is the person's taxable income?
23. To convert tempertures from Celsius degrees to Fahrenheit degrees, take nine-fifths of the Celsius degrees, then add 32; i.e., $F = (9/5)C + 32$. For example, if the temperature is 30°C, the temperature in degrees Fahrenheit is $(9/5)(30) + 32 = 54 + 32 = 86$.
    a. Find the temperature in degrees Fahrenheit if it is 25°C.
    b. Find the temperature in degrees Celsius if it is 70°F.
    c. Find the temperature in degrees Celsius at freezing (32°F).
    d. Find the temperature in degrees Celsius at which water boils (212°F).
    e. Find the temperature at which degrees Celsius and degrees Fahrenheit are the same.
24. The Kelvin scale is a temperature scale in which absolute zero, −273.16° on the Celsius (centigrade) scale, is 0 Kelvin. Thus, the degrees Kelvin are equivalent to degrees Celsius plus 273.16; i.e., $K = C + 273.16$.
    a. The freezing point of water is 0°C. Find the freezing point of water on the Kelvin scale.
    b. Find the temperature in degrees Celsius if it is 100 K.
    c. Find the temperature in degrees Celsius if it is 300 K.
25. Some states have a 6% sales tax. If a purchase (item price plus sales tax) in such a state totals $6.89, what is the price of the item before tax?
26. An electric utility charges residential customers $0.067797 per kilowatt-hour (kWh) for the first 510 kilowatt-hours and $0.083097 after that.
    a. If a person uses 927 kWh, what is the charge?
    b. If a person's bill is $103.22, how many kilowatt-hours, to the nearest kilowatt-hour, were used?
    c. Suppose there is a 4% tax on the amount figured. What is the amount due, with tax, if 842 kWh is used?

d. Suppose a person's bill, with 4% tax, is $46.16. How many kilowatt-hours, to the nearest kilowatt-hour, were used?

27. Some credit cards permit the card holder to get a cash advance with the card. One credit card company has a transaction fee for a cash advance of 2.5% of the cash advance taken, with the exception that the minimum fee is $2.00 and the maximum fee is $10.00.
    a. What range of cash advances yield transaction fees of $2.00?
    b. What range of cash advances yield transaction fees of $10.00?

28. A national multiple-choice test was scored by taking the number of correct answers plus one-fifth the number of blank answers. The test had 60 questions. Suppose a student who answered 30 questions (and left 30 blank) was notified she had a score of 31. How many questions did she answer correctly?

29. Here is a mathematical trick. Ask someone to do the following steps. You can then predict the answer. To see how and why it works, at each step write the algebraic expression for the number that results.
    a. Pick a number.
    b. Double it.
    c. Add 10.
    d. Divide by 2.
    e. Subtract the number you started with.
    f. Your answer is _____.

30. In the mathematical trick of Exercise 29, change step c to add a number of your choice, say, k. Leave all other steps as in Exercise 29. What is the new answer? Why?

31. The historical material on the development of algebra presented problem 35 from the *Moscow Papyrus:*

    Find the size of a scoop for which it takes $3\frac{1}{3}$ trips to fill a 1-hekat measure.

    Solve it.

1. Ask some students in their first junior high or high school algebra class what algebra is. Report their results, and discuss how their descriptions fit your understanding of what algebra is and the description presented in this section.

2. From your previous experience with algebra, what is a concept you found difficult or confusing? Do you still regard it as difficult or confusing?

3. This question is for people who have studied both algebra and geometry. Which did you like better? Why?

4. In your experience, which of the following holds? Offer a guess as to why this so.
   a. Boys usually do better in algebra than geometry, while girls usually do better in geometry than algebra.
   b. The reverse of part a is true.
   c. Neither part a nor part b is true.

5. Assume a high school teacher in your area is the person referred to in Example 5.5. Write a note to that person stating what her or his taxable income was and explaining why your answer is correct.

6. Find or make up an application problem that can be solved by means of a linear equation. Describe the problem, and give its solution.

7. The solutions to Examples 5.4 and 5.5 can be thought of as mathematical modeling. Fit each step of their solutions to a part of the mathematical modeling diagram in Section 1.3 (a box or an arrow).

8. (*Library research*) Write a two- to four-page paper on the golden ratio. Cite at least one reference other than this textbook. Assume that the audience for your paper is your class.

5.2 LINEAR AND QUADRATIC FUNCTIONS AND QUADRATIC EQUATIONS

GOALS
1. To review functions of one variable and the *f(x)* notation.
2. To give examples of functions of more than one variable, and to use function notation for them.
3. To distinguish between linear and nonlinear functions.
4. To review quadratic functions, and to examine some of their applications.

Definitions

One concept that unifies a number of algebraic ideas is that of a *function*. The idea underlying a function is that two (or more) variables are related in some manner. For example, for the equation

$$y = 3x + 1$$

we could say y "is a function" of x because each choice of a value for the variable x produces a unique value for the variable y. An everyday example is the cost of a long-distance phone call, which is a function of the amount of time you talk. Here then is a common definition for a function relating two variables.

> ### DEFINITION
>
> A *function between variables x and y* is a rule relating x and y such that for each value of x, there is exactly one corresponding value of y.

Let's consider some equations that may or may not produce examples of functions.

EXAMPLE 5.6

Consider the equation $y = 3x + 1$. If $x = 1$, we get

$$y = 3(1) + 1$$
$$= 3 + 1$$
$$= 4$$

In other words, when x is assigned the value 1, y is always equal to 4. Since this same process produces a unique value for y for each choice of x, we say y is a function of x.

EXAMPLE 5.7

Consider the equation $y = x^2$. When $x = 2$, we get

$$y = 2^2$$

Furthermore, when $x = -2$, we get

$$y = (-2)^2$$
$$= 4$$

Note that we obtain the same y value for different choices of x. This still satisfies our definition, however, because each choice of x produces a single value for y.

EXAMPLE 5.8

Consider the equation $y^2 = x$. For the value $x = 4$, what is y? One person might note that since $2^2 = 4$, $y = 2$ would solve the equation $y^2 = 4$. However, another person might note that $(-2)^2 = 4$ and claim $y = -2$ would solve the equation

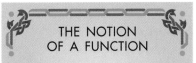

THE NOTION OF A FUNCTION

That flower of modern mathematical thought—the notion of a function.

—Thomas J. McCormack, "On the Nature of Scientific Law and Scientific Explanation," *Monist*, 1899–1900, p. 555.

Nature herself exhibits to us measurable and observable quantities in definite mathematical dependence; the conception of a function is suggested by all the processes of nature where we observe natural phenomena varying according to distance or to time. Nearly all the "known" functions have presented themselves in the attempt to solve geometrical, mechanical, or physical problems.

—J. T. Merz, *A History of European Thought in the Nineteenth Century*, London, 1903, p. 696.

$y^2 = 4$. Both would be correct. In this case, for the value $x = 4$, y could be either 2 or –2, so y is *not* a function of x.

Functional Notation

A fairly standard method for denoting a functional relationship is to replace the y in an equation with $f(x)$ (which is read "f of x"). Thus,

$$y = 3x + 1$$

would be written as

$$f(x) = 3x + 1$$

This notation reinforces the concept that a value is assigned to x and then a resulting value is obtained. Using this notation, for the value $x = 2$ we would write

$$f(2) = 3 \cdot 2 + 1$$
$$= 6 + 1$$
$$= 7$$

or just

$$f(2) = 7$$

EXAMPLE 5.9

Consider the function $f(x) = x^2 + x - 3$. Let's evaluate the following:

a. $f(1)$ b. $f(0)$ c. $f(-4)$

SOLUTIONS The notation $f(1)$ means find the value of the expression when 1 is substituted for x. Thus, we get

a. The notation $f(1)$ means find the value of the expression when 1 is substituted for x. Thus, we get

$$f(1) = 1^2 + 1 - 3$$
$$= 1 + 1 - 3$$
$$= 2 - 3$$
$$= -1$$

b. Similarly, $f(0)$ yields

$$f(0) = 0^2 + 0 - 3$$
$$= 0 + 0 - 3$$
$$= 0 - 3$$
$$= -3$$

c. Finally, $f(-4)$ gives

$$f(-4) = (-4)^2 + (-4) - 3$$
$$= 16 - 4 - 3$$
$$= 12 - 3$$
$$= 9$$

To function is to perform.

ON A TANGENT

Euler and Mathematical Notation

Whereas Leibniz (1646–1716, see Sections 2.4 and 4.1) was the first person to use the word *function*, (many people also credit him with being the first to use the function concept), the notation $f(x)$ to indicate the value of the function f at x was one of the many notational inventions of Euler's (1707–1783; see Sections 3.4 and 4.3).

Mathematical notations are what they are today more because of Euler than because of any other person in history. His notational inventions include:

$f(x)$

i for the imaginary number $\sqrt{-1}$

e for the number that is the base for natural logarithms (on your calculator, above the $\boxed{\text{LN}}$ key might be an $\boxed{e^x}$ key.)

π was invented by others but popularized by Euler

If you still lack a clear understanding of what a function is, consider the illustration representing a machine or "function box" with an opening for numbers to be put into the box and an opening for numbers to come out of the box. The work of the machine is to perform the operations indicated by the function under consideration. Thus, for $f(x) = x^2 + x - 3$, put in the number 1 and the resulting output would be the number −1. In general, the function box can be thought of as having input x and output $f(x)$. In fact, a calculator is such a function box for a variety of functions. For example, if you enter 4 and hit the $\boxed{\sqrt{\ }}$ key, out comes 2; if you enter 4 and hit the $\boxed{x^2}$ key, out comes 16; and if you enter 4 and hit the $\boxed{1/x}$ or $\boxed{x^{-1}}$ key, out comes 0.25. The $\boxed{\sqrt{\ }}$ key performs the function given by $f(x) = \sqrt{x}$; the $\boxed{x^2}$ key performs the function given by $f(x) = x^2$; and the $\boxed{1/x}$ key performs the function given by $f(x) = 1/x$.

Function Machine

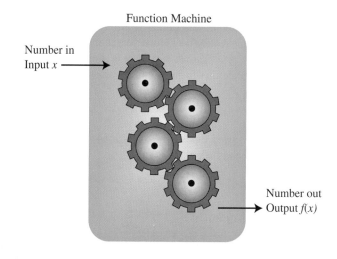

Number in
Input x

Number out
Output $f(x)$

A calculator contains function keys.

Single-Variable Functions

Our early examples all considered y as a function of x or some rule denoted by $f(x) = \ldots$. Note that each of these examples used only one variable for assigning values. When this happens, we say we have a *function of a single variable*.

Although all our early examples use x as the variable for which values are substituted, this does not have to be the case. Consider, for example, the function given by

$$f(t) = t^2 + t - 3$$

The variable here is t, but it is the same function as

$$f(x) = x^2 + x - 3$$

For instance, in each case we find $f(5)$ as

$$f(5) = 5^2 + 5 - 3$$

As soon as a value is substituted for the variable, the choice of the variable letter becomes immaterial.

Multiple-Variable Functions

Many instances arise where more than one variable is necessary to describe a functional relationship. Consider, for example, the formula for computing interest:

$$I = Prt$$

where I represents interest earned, P is the principal, r is the rate of interest, and t is the time. In this case, I is a function of P, r, and t. Thus, we could write

$$f(P, r, t) = Prt$$

A second example with which you may be familiar is the equation relating distance, rate, and time:

$$d = rt$$

Again, since distance depends on both the rate of motion and the time, we could write

$$f(r, t) = rt$$

We will not attempt to extend the formal definition of function to include functions of more than one variable at this point. You may think focusing on functions of one variable is somewhat restrictive. However, that is far from true. We could develop all of trigonometry, precalculus, and what is typically covered in a full year of differential and integral calculus from functions of a single variable.

Linear Functions

One type of function that is used extensively is a *linear function*. A linear function of one variable is given by the following rule:

$$f(x) = ax + b, \qquad \text{where } a \text{ and } b \text{ are real numbers}$$

Such a function is called linear because its graph is a line. For example, the following functions are all linear functions:

$$f(x) = 2x + 3$$

$$g(x) = -4x$$
$$h(x) = \sqrt{2}x + \pi$$

while the following functions are not linear functions:

$$f(x) = x^2$$
$$g(x) = \sqrt{x} + 2$$
$$h(x) = \frac{x+1}{x+2}$$

To this point we have dealt with linear functions, their graphs, and applications of linear functions. Recall, these functions are characterized by the property that their graphs are lines. Algebra encompasses many other functions; generally, these other functions are referred to as *nonlinear functions*. As you might expect, a nonlinear function is a function whose graph is not a line. We will now look at a particular class of nonlinear functions.

Quadratic Functions

Recall, linear functions are functions which can be written as

$$y = ax + b$$

When graphed, such lines have slope a and y intercept b. A *quadratic function* is a function that can be written as

$$y = ax^2 + bx + c$$

where $a \neq 0$. Notice the term containing x^2. This causes the function to be nonlinear, for without it (i.e., when $a = 0$) the function would be linear.

THE ETYMOLOGY
OF *QUADRATIC*

The word *quadratic* is from the Latin *quadratum,* meaning "square." The Romans followed the Greek model to consider the quantity s^2 as the area of a square of side length s.

Parabolas

Quadratic functions produce parabolas when graphed. For example, $y = x^2$ produces the graph shown in ▶ Figure 5.1, and $y = -x^2 + 2x - 1$ produces the graph shown in ▶ Figure 5.2. In general, when the coefficient a of $ax^2 + bx + c$ is positive, the

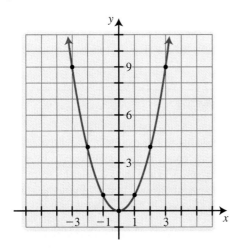

▶ FIGURE 5.1 Graph of $y = x^2$

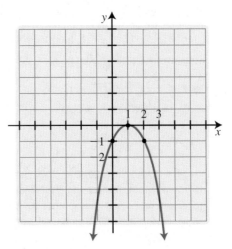

▶ FIGURE 5.2 Graph of $y = -x^2 + 2x - 1$

Gateway Arch in St. Louis. A parabola is the gateway to the graph of nonlinear functions.

graph opens upward; and when the coefficient a is negative, the graph opens downward.

> ▶ ▶ ▶ ▶ 0000
> ### CALCULATOR NOTE
> Quadratic functions are just one type of function that can easily be graphed on a graphing calculator. For example, to get the graph of $y = x^2$ that looks like Figure 5.1, set the range so the x-values go from -6 to 6 (like those shown on the x-axis of Figure 5.1) and the y-values go from -1 to 11. The function is $y(x) = x^2$. The calculator will display the same graph as in Figure 5.1.

The vertex of a parabola with $a > 0$ is the lowest point on the curve. For $a < 0$, the vertex is the highest point on the curve. As you perhaps recall, the x-coordinate of the vertex is $x = -b/2a$. Knowing this value can be an aid in graphing.

Zeros of Quadratic Functions

It is often important to know where (or even if) a quadratic function crosses the x-axis. Since the x-axis is the line $y = 0$, values of x where a curve crosses the x-axis are called *zeros of the function*. For example, if $y = x^2 - 4$, then when $x = 2$, $y = 0$, and when $x = -2$, $y = 0$. Thus, 2 and -2 are zeros of $y = x^2 - 4$, as illustrated in ▶ Figure 5.3. As you can see, zeros of a quadratic indicate where the parabola crosses the x-axis. If we now switch to functional notation and write the quadratic as $f(x) = x^2 - 4$, we have $f(2) = 0$ and $f(-2) = 0$.

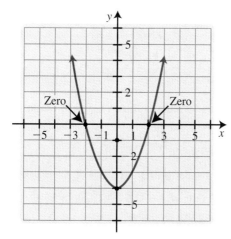

▶ FIGURE 5.3 Zeros of a quadratic function

 YOUR FORMULATION

Do all parabolas cross the x-axis?

1. In Figure 5.3, the graph crosses the x-axis twice. Sketch the graph of a parabola that opens downward and crosses the x-axis twice. In this case, there are two distinct values, x_1 and x_2, for x such that $f(x_1) = 0$ and $f(x_2) = 0$.
2. Sketch the graph of a parabola that opens upward and touches the x-axis exactly once. This graph represents a quadratic function with one distinct zero.
3. Repeat step 2 for a parabola that opens downward.
4. Sketch the graph of a parabola that opens upward and doesn't intersect the x-axis. This graph represents a quadratic function with no real zeroes.
5. Repeat step 4 for a parabola that opens downward.
6. Could a parabola cross the x-axis more than twice?
7. Summarize what you have found in steps 1–6: A parabola can cross the x-axis _____ times.

> The quadratic function $f(x) = ax^2 + bx + c$ may have 0, 1, or 2 real zeros.

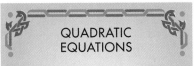
The Quadratic Formula

Perhaps you recall working with or hearing about the quadratic formula. Simply stated, the quadratic formula is a formula that produces the x-coordinates of zeros of a quadratic function.

QUADRATIC FORMULA

For $f(x) = ax^2 + bx + c$,

$$x = \frac{-b \pm \sqrt{b^2 - 4ac}}{2a}$$

produces the values such that

$$f(x) = 0.$$

Finding zeros via this formula is illustrated in the following example.

EXAMPLE 5.10

Use the quadratic formula to find the zeros of $f(x) = x^2 - x - 6$.

SOLUTION In this example, $a = 1$, $b = -1$, and $c = -6$. Substituting these values into the quadratic formula produces

$$x = \frac{-(-1) \pm \sqrt{(-1)^2 - 4(1)(-6)}}{2(1)}$$

$$= \frac{1 \pm \sqrt{1 + 24}}{2}$$

$$= \frac{1 \pm \sqrt{25}}{2}$$

$$= \frac{1 \pm 5}{2}$$

If $+5$ is chosen, we have

$$x = \frac{1 + 5}{2} = 3$$

If -5 is chosen, we have

$$x = \frac{1 - 5}{2} = -2$$

Going back to the original function, we get

$$f(3) = 3^2 - 3 - 6 = 9 - 9 = 0$$

$$f(-2) = (-2)^2 - (-2) - 6 = 4 + 2 - 6 = 0$$

The graph of $f(x) = x^2 - x - 6$ is shown in ▶ Figure 5.4.

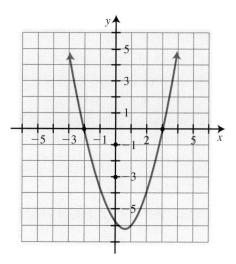

▶ FIGURE 5.4 Graph of $f(x) = x^2 - x - 6$

In the quadratic formula, the expression $b^2 - 4ac$ is the *discriminant*. The value of the discriminant is related to the zeros of the function. If $b^2 - 4ac > 0$, then there are two distinct real zeros:

$$x = \frac{-b + \sqrt{b^2 - 4ac}}{2a} \quad \text{and} \quad x = \frac{-b - \sqrt{b^2 - 4ac}}{2a}$$

When the discriminant equals zero, the vertex is on the *x*-axis and there is only one real zero for the function: $x = -b/2a$. For example, recall $f(x) = x^2 - 2x + 1$ had only one zero. We can now see this by looking at $b^2 - 4ac$. Since $a = 1$, $b = -2$, and $c = 1$, we have $b^2 - 4ac = (-2)^2 - 4(1)(1) = 4 - 4 = 0$.

If $b^2 - 4ac < 0$, there are no real zeros. This means the parabola is either entirely above or entirely below the *x*-axis. To illustrate, let's look again at $f(x) = x^2 + 1$. Here we have $a = 1$, $b = 0$, and $c = 1$. The discriminant is $b^2 - 4ac = 0^2 - 4(1)(1) = -4$. The graph is entirely above the *x*-axis.

Applications of Quadratic Functions

Many realistic problems can be solved by the properties of quadratics just developed. Since quadratics have a vertex that yields a maximum or minimum value for a function, any situation that can be described by a quadratic function can accordingly be found to have a maximum or minimum value. In business, maximums can be found for profit functions and minimums found for cost functions. When studying materials for use in building objects, you may wish to minimize the cost of the object or to maximize the area of the object for a fixed perimeter of material. If a quadratic function can describe the physical situation, the desired maximum or minimum can then be found. The following two examples illustrate applications of quadratic functions.

EXAMPLE 5.11

A charter bus company charges an organization $3200 for a trip for 40 or fewer people ($80 per person if there are 40 people).

a. For each additional person over 40, the company reduces the per-person charge by $1. For example, if there are 41 people, the cost is $79 per person.

For what number of people will the company's revenue be a maximum? What is this maximum revenue?

SOLUTION Let x be the number of people on the trip in excess of 40. The revenue, in dollars, is

$$R = \underset{\substack{\text{Number of}\\\text{people who go}}}{(40 + x)} \cdot \underset{\substack{\text{Charge}\\\text{per person}}}{(80 - x)}$$

$$R = 3200 + 40x - x^2$$

$$R = -x^2 + 40x + 3200$$

The graph of this revenue function is a parabola that opens downward (since $a < 0$). The x-coordinate of the vertex is

$$-\frac{b}{2a} = -\frac{40}{2(-1)} = 20$$

Thus, the bus company's revenue is a maximum when $x = 20$, which is when the number of people is 60. The maximum revenue is $(40 + 20)\ (80 - 20) = 60 \cdot 60 = 3600$, or \$3,600.

b. Suppose that for each additional person the company reduces the per-person charge by \$2. Again, answer the two questions in part a.

SOLUTION Again, let x be the number of people who take the trip in excess of 40. Now the revenue in dollars is

$$R = (40 + x)\ (80 - 2x) = 3200 - 2x^2$$

The graph of this revenue function is a parabola that opens downward. The x-coordinate of the vertex is $-b/2a = -0/-4 = 0$. Thus, the bus company's revenue is a maximum when $x = 0$, which is when the number of people is 40. The maximum revenue is $\$(40)(80) = \$3,200$.

● EXAMPLE 5.12

The distance it takes to stop a car depends on the speed of the car. Let d be the distance, in feet, it takes to stop a car with brakes that meet minimum standards, and let x be the speed of the car, in miles per hour.

a. A good relation between x and d is given by

$$d = .071x^2$$

Note that braking distance is a quadratic function of the speed of the car.
  i. What is the braking distance if the car is going 30 mph?

SOLUTION $\qquad\qquad d = .071(30^2) = 63.9$

The braking distance is 63.9 ft.
  ii. What is the braking distance if the car is going 60 mph?

SOLUTION $\qquad\qquad d = .071(60^2) = 255.6$

The braking distance is 255.6 ft.
  iii. In general, what happens to the braking distance if the speed of the car doubles?

SOLUTION If the first speed is x mph, the braking distance is $d_1 = .071x^2$. If the second speed is the first speed doubled, it is $2x$ and the braking distance is

$$d_2 = .071(2x)^2 = .071(4x^2) = 4(.071x^2) = 4d_1$$

Thus, if the speed of the car doubles, the braking distance is quadrupled. Note that this happened in the special case in subparts i and ii.

b. If we take into account the reaction time of the driver before the brakes are applied and assume it is half a second, then the braking distance is

$$d = .071x^2 + .73x$$

Let's examine the same three questions as in part a.
   i. What is the braking distance if the car is going 30 mph?

SOLUTION $\qquad d = .071(30^2) + .73(30) = 85.8$

The braking distance is 85.8 ft.
   ii. What is the braking distance if the car is going 60 mph?

SOLUTION $\qquad d = .071(60^2) + .73(60) = 299.4$

The braking distance is 299.4 ft.
   iii. In general, what happens to the braking distance if the speed of the car doubles?

SOLUTION If the first speed is x mph, the braking distance is $d_1 = .071x^2 + .73x$. If the second speed is the first speed doubled, it is $2x$ and the braking distance is

$$d_2 = .071(2x)^2 + .73(2x) = .071(4x^2) + 1.46x = .284x^2 + 1.46x$$

Note that this is between

$$2d_1 = (.142x^2 + 1.46x) \qquad \text{and} \qquad 4d_1 = (.284x^2 + 2.92x)$$

c. A driver wants to be able to stop within 300 feet of the flashing of a signal (300 ft is the length of a football field). If the driver's reaction time is assumed to be half a second, what is the maximum speed the driver can be going and still stop?

SOLUTION $\qquad d = .071x^2 + .73x$

$$300 = .071x^2 + .73x$$

$$0 = .071x^2 + .73x - 300$$

$$x = \frac{-.73 \pm \sqrt{(.73)^2 - 4(.071)(-300)}}{2(.071)}$$

If we accept the minus sign before the square root, x will be negative, contrary to x as a speed. Thus, we take the plus sign before the square root to get $x = 60.065$. Therefore, the maximum speed the driver can go and still stop is 60 mph.

As a check, we let $x = 60$ in the braking-distance formula and find that $d = .071(60^2) + .73(60) = 299.4$. The driver has .6 ft = 7.2 in. to spare.

1. Which of the following equations define y as a function of x?
   a. $y = 2x + 3$
   b. $y = 2x^2 - 5$
   c. $y = x^2$
2. If $f(x) = x^2 - x + 2$, find
   a. $f(1)$ b. $f(3)$ c. $f(0)$ d. $f(-2)$
3. If $f(t) = 2t^2 - 3t + 4$, find
   a. $f(1)$ b. $f(2)$ c. $f(0)$ d. $f(-1)$
4. *True or false?* Every equation in x and y defines y as a function of x.
5. Give an example of an equation in x and y that does not define y as a function of x.

In Exercises 6–8, some inputs and outputs for a function are given; try to guess the function. For example, $f(1) = 2$ and $f(2) = 4$. By noting that the output is twice the input, we see that one possibility for f is $f(x) = 2x$. Another possibility is that $f(x) = x^2 - x + 2$.

6. State an example of a function f such that $f(1) = 3$.
7. State an example of a function f such that $f(0) = 0$ and $f(2) = 6$.
8. State an example of a function f such the $f(1) = 1$, $f(2) = 4$, and $f(3) = 9$.
9. Define the term *function of a single variable*.
10. Two examples of functions of more than one variable are distance and interest.
    a. Describe another function of two variables.
    b. Describe another function of three variables.
11. Denote distance d as a function of rate r and time t by $d = f(r, t) = rt$. Find each of the following:
    a. $f(2,3)$ b. $f(55,2)$
12. Which, if any, of the equations in Exercise 1 that define y as a function of x define y as a *linear* function of x?
13. Give an example of a linear function.
14. Give an example of a function that is not a linear function.

For each of Exercises 15–23, find the vertex and the zeros and sketch the graph.

15. $y = x^2 + 2x - 8$ 16. $y = 3x^2 + 6x$
17. $y = 5x^2 - x - 2$ 18. $y = 2x^2 + 3$
19. $y = 9x^2$ 20. $y = 1 - 2x - 3x^2$
21. $y = -2 + 3x - 4x^2$ 22. $y = -7 - 3x^2$
23. $y = -3x - 7x^2$

24. In a quadratic equation, what is implied by having each of the following?
    a. A positive discriminant
    b. A negative discriminant
    c. A zero discriminant
25. Give an example of a function that is not a quadratic function.
26. A catering service charges $300 for up to 30 people ($10 per person if 30 people are present). For each additional person, the catering service reduces the per-person charge by 10¢. For example, for 31 people, the charge is $9.90 per person for a total charge of $306.90. For what number of people will the catering service's revenue be a maximum? What is the maximum revenue?
27. Suppose a car is going 55 mph.
    a. What is the braking distance (ignoring reaction time)?
    b. What is the braking distance if the driver's reaction time is half a second?
28. If we ignore reaction time, what happens to the braking distance if the speed of the car triples?
29. A driver wants to be able to stop within 100 feet of the flashing of a signal. If the driver's reaction time is assumed to be half a second, what is the maximum speed the driver can be going and still stop?
30. In Illinois, all cars must be able to stop in under 30 feet when traveling 20 miles per hour. Example 5.12 used two different models to find the distance it takes a car to stop, one involving reaction time and one without reaction time. Does the Illinois law consider reaction time? Support your answer.
31. A dog owner has available 30 feet of fencing with which to fence in a dog pen. One side of the pen is against a building and does not need to be fenced. Assume the pen is rectangular in shape. What should be its dimensions in order to have a maximum area?
32. Consider the dog-pen problem of the preceding exercise.
    a. If the assumption that the pen is rectangular is removed, can a pen with a larger area be built?
    b. If so, what are the shape, dimensions, and area of such a pen. If not, why not?
33. The box in the subsection entitled "Zeros of Quadratic Functions" says that a quadratic function may have 0, 1, or 2 real zeros. How many real zeros can a linear function have?

1. The word *function* is used in ordinary conversation in a variety of ways—for example: "He doesn't function well under pressure" and "What is the function of the button?"
   a. How do these uses of the word *function* compare with the mathematical use of *function* described earlier in this section?
   b. Write another statement involving the word *function* in yet another context. How does it compare with the mathematical use of *function*?

2. In this section, the idea of a calculator as a function box is described and keys such as $\boxed{\sqrt{}}$, $\boxed{x^2}$, and $\boxed{x^{-1}}$ are

referred to as function keys. Are the keys $+$, $-$, $\times$, and $\div$ function keys? Why or why not?

3. Write one to three pages on the origin of the function concept. See Eves (1990), NCTM (1969), or other books on the history of mathematics.

4. A graphing calculator will graph a quadratic function. Experiment with this on some examples or exercises that you have graphed. Given this ability of a graphing calculator, do you think the graphing of quadratic functions by hand should be de-emphasized? Support your position.

5. In the word *quadratic, quad-* suggests "four." Why is *quadratic* used when the biggest exponent is 2?

6. Several applications of quadratic functions are given in Examples 5.11 and 5.12 and Exercises 26–31. State and solve another application problem involving a quadratic function. Your problem should differ more from those already given than just changing numbers in a previous problem or changing "bus company" to "plane company."

5.3 EXPONENTIAL FUNCTIONS

GOALS
1. To identify and give examples of exponential functions.
2. To graph the function $f(x) = a^x$.
3. To graph the function $f(x) = e^x$.
4. To work with applications of exponential functions.

The last section dealt with one type of nonlinear function, the quadratic function, whose graph is a parabola, together with some applications. However, there are many other types of nonlinear functions. In this section, we will describe one such function, the exponential function. This was chosen because it has some very interesting applications. Most calculators have function keys for this function.

Exponential Functions

An exponential function is a function where the variable (x) is in the exponent. Recall that when working with exponents, in an expression such as 2^3, 2 is the base and 3 is the exponent. We then extended this to include expressions such as x^2, where the base (x) is a variable and the exponent (2) is a constant. We are now going to consider expressions such as 2^x. This expression has a constant base (2) and a variable exponent (x). A specific example of an exponential function is

$$y = 2^x$$

In general, exponential functions have the form

$$y = a^x$$

where a is a positive constant other than 1. Functions of this type have many uses, for example, computing interest, obtaining information about population growth, and modeling radioactive decay. Some of these applications are developed later.

 YOUR FORMULATION

In the description of an exponential function we said that a is not 1. Why? What happens if $a = 1$?

Let's now take a closer look at the exponential function $y = 2^x$. When computing values for coordinates of points on the graph of this function, we substitute values for x and then compute y. To illustrate, for $x = 1$ we have $y = 2^1 = 2$. Thus, the point $(1, 2)$ is on the graph of $y = 2^x$. Coordinates of some other points can be found as follows:

x	y	Point
2	$2^2 = 4$	(2, 4)
3	$2^3 = 8$	(3, 8)
4	$2^4 = 16$	(4, 16)

When plotted, these points appear as follows:

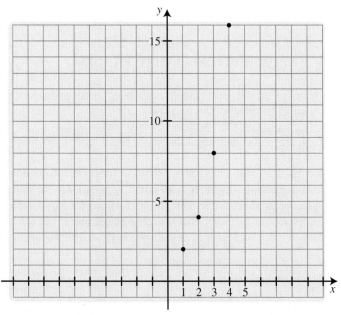

They can be connected to produce part of the curve:

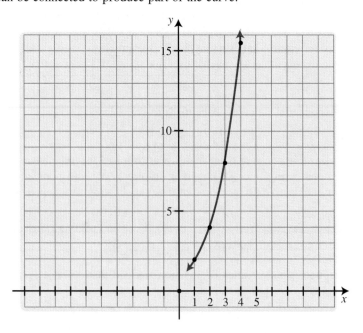

For $x = 0$, we have $y = 2^0$. Recall that any base (except 0) to the zero power has value 1. Thus, $(0, 1)$ is on the graph.

What happens for negative values of x? For example, what is 2^{-1}? When working with exponents, a negative exponent can be changed to positive by putting the expression with the negative exponent in the other part of the fraction.

That is, move the expression from the numerator to the denominator or from the denominator to the numerator. Thus 2^{-1} becomes $1/(2^1)$. We then have for $x = -1$ and $y = 2^x$,

$$y = 2^{-1} = \frac{1}{(2^1)} = \frac{1}{2}$$

This gives us the point $(-1, \frac{1}{2})$ on the graph of $y = 2x$. With zero and negative values of x, we get the following:

x	y	Point
0	$2^0 = 1$	$(0, 1)$
-1	$2^{-1} = \dfrac{1}{2}$	$(-1, \frac{1}{2})$
-2	$2^{-2} = \dfrac{1}{(2^2)}$	$(-2, \frac{1}{4})$
-3	$(2^{-3} = \dfrac{1}{(2^3)}$	$(-3, \frac{1}{4})$

Plotting these additional points along with those previously plotted, we get the following graph:

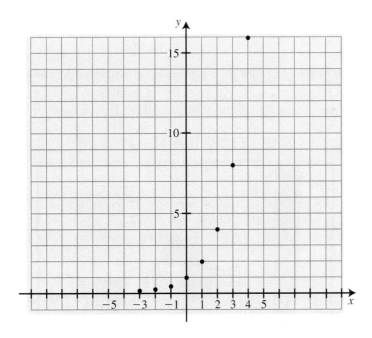

We connect these points with a smooth curve and extend in a natural manner to produce the curve shown in ▶ Figure 5.5.

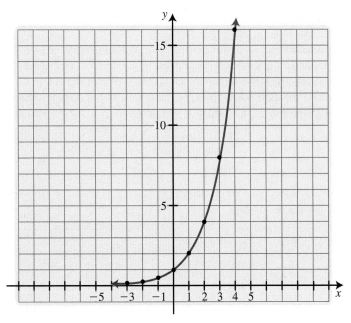

▶ FIGURE 5.5 Graph of $y = 2^x$

Calculator Use

Computing y-values for exponential functions can be simplified by using the $\boxed{y^x}$ key on a calculator. For example, the value of 2^3 is obtained by the following key stroke sequence.

$$\boxed{2} \qquad \boxed{y^x} \qquad \boxed{3} \qquad \boxed{=}$$

The display reads 8, which is the value of 2^3. Note that some calculators have an $\boxed{x^y}$ key instead of a $\boxed{y^x}$ key. Experiment with some values you know ($2^3 = 8$, $3^2 = 9$) to become familiar with your calculator. We said a little earlier that 0^0 is not 1. In fact, 0^0 is not defined. Try finding it on a calculator. Do you get an error message?

THE LOCKHORNS by Bunny Hoest & John Reiner

"WE'RE AT CHAPTER II TO THE THIRD POWER."

A Special Exponential Function

One primary use of exponential functions is for computing interest. Of particular importance is the computation of interest compounded continuously. However, before we look at this application, we will examine an especially significant exponential function given by $f(x) = e^x$.

An important irrational number, usually denoted e, arises in the study of mathematics. In practice, this irrational number has a value between 2 and 3, with a typical approximation given as

$$e \approx 2.718$$

This number comes up so frequently it has a special function key on many calculators, the $\boxed{e^x}$ key, which is usually paired with the $\boxed{LN}$ key. To see what approximation your calculator has for e, enter the sequence

$$\boxed{1} \qquad \boxed{e^x}$$

The display will be close to 2.718281828. This is the calculator value of e^1. The following example gives further practice in using the $\boxed{e^x}$ key. If your calculator has no $\boxed{e^x}$ key, use the $\boxed{y^x}$ key with $y = 2.71828$. If your calculator has neither key, try to borrow one for the remainder of this section and the next section.

EXAMPLE 5.13

Plot points and sketch the graph of the function $y = e^x$.

SOLUTION As before, we compute coordinates using a few integer values. The points in the chart at the top of the next page are plotted and connected to produce the graph in ▶ Figure 5.6.

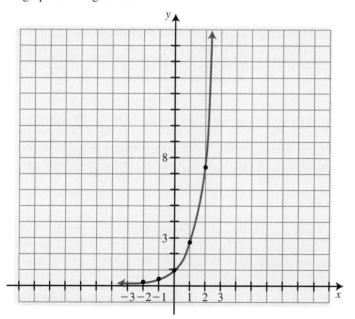

▶ FIGURE 5.6 Graph of $y = e^x$

x	Keystroke Sequence	Display
1	[1] [e^x]	2.718 . . .
2	[2] [e^x]	7.389 . . .
3	[3] [e^x]	20.085 . . .
0	[0] [e^x]	1
-1	[1] [+/-] [e^x]	0.367 . . .
-2	[2] [+/-] [e^x]	0.135 . . .

Applications

We will now examine several applications of exponential functions: compound interest, inflation, population growth, continuous compounding, half-life, radioactive decay, and the Richter scale.

Compound Interest   The following example shows how compounding works.

•EXAMPLE 5.14

Suppose $1000 is invested for 3 years at 6% interest compounded annually. After 1 year the interest earned is 6% of $1000, which is .06 × $1000 = $60. At the end of 1 year the original investment has increased in value to $1060 (the original $1000 plus $60 interest). During the second year the interest earned is 6% of $1060 which is $63.60. Therefore, at the end of the second year, the value of the investment is $1060 + $63.60 = $1123.60. Cover up the following material and try to determine the value of the investment at the end of 3 years.

SOLUTION The interest earned during the third year is 6% of $1123.60, which is $67.416, or approximately $67.42. Therefore, at the end of 3 years, the value of the investment is $1123.60 + $67.42 = $1191.02.

The solution steps for this example are straightforward, but it would take a long time and be very tedious if we wanted the value of the investment after 50 years. We will soon develop a method to shortcut this process. First, however, consider an example in which interest is compounded over a shorter period of time.

•EXAMPLE 5.15

Suppose $1000 is invested at $5\frac{1}{4}$% compounded quarterly. [Quarterly compounding means that, at the end of each quarter (3 months), the interest earned for that quarter is credited to the account.] During the first year, how much interest is earned?

SOLUTION The rate is $5\frac{1}{4}$%, so the interest would be $5\frac{1}{4}$% of $1000 = $52.50 if it were for the entire year. However, since the interest is figured at the end of one quarter, it is $\frac{1}{4}$ as much, or $\frac{1}{4}$($52.50) = $13.125. If the bank is generous, the value at the end of one quarter is $1000 + $13.13 = $1013.13. During the second quarter the interest is $\frac{1}{4}(.0525)(\$1013.13)$, which rounds to $13.30. After two quarters the investment's value is $1013.13 + \$13.30 = \$1026.43$. During the third quarter the interest is $\frac{1}{4}(.0525)(\$1026.43)$, which rounds to $13.47. After three quarters the investment's value is $1026.43 + \$13.47 = \$1039.90$. During the fourth quarter the interest is $\frac{1}{4}(.0525)(\$1039.90)$, which rounds to $13.65. Therefore, after four quarters, or 1 year, the investment's value is $1039.90 + \$13.65 = \$1053.55$.

In this example, note (1) how compounding over a period of time less than a year is handled, and (2) that the value at the end of the year is greater than the value if the investment had simply earned $5\frac{1}{4}\%$ for the year, or $52.50, for a total value of $1052.50.

To develop a method to shortcut the process, assume an amount P (for principal) is invested at an interest rate i (in the preceding examples, i was .06 or .0525) compounded m times per year (with quarterly compounding, $m = 4$; with annual compounding, $m = 1$) for n periods of time (in the first example, $n = 3$; in the second example, $n = 4$). Let A_n denote the value of the investment at the end of n periods of time.

After one period of time:

$$A_1 = \underset{\text{Principal}}{P} + \underset{\text{Interest}}{P\left(\frac{i}{m}\right)} = P\left(1 + \frac{i}{m}\right)$$

After two periods of time:

$$A_2 = \underset{\substack{\text{Amount after} \\ \text{1 period}}}{\left[P\left(1 + \frac{i}{m}\right)\right]} + \underset{\text{Interest}}{\left[P\left(1 + \frac{i}{m}\right)\right] \cdot \frac{i}{m}}$$

$$= \left[P\left(1 + \frac{i}{m}\right)\right]\left(1 + \frac{i}{m}\right)$$

$$= P\left(1 + \frac{i}{m}\right)^2$$

After three periods of time:

$$A_3 = \underset{\substack{\text{Amount after} \\ \text{2 periods}}}{\left[P\left(1 + \frac{i}{m}\right)^2\right]} + \underset{\text{Interest}}{\left[P\left(1 + \frac{i}{m}\right)^2\right] \cdot \frac{i}{m}}$$

$$= \left[P\left(1 + \frac{i}{m}\right)^2\right]\left(1 + \frac{i}{m}\right)$$

$$= P\left(1 + \frac{i}{m}\right)^3$$

Continuing in this manner, we get

COMPOUND-AMOUNT FORMULA

$$A_n = P\left(1 + \frac{i}{m}\right)^n$$

EXAMPLE 5.16

Use the compound-amount formula to calculate the value of the investment in Example 5.14.

SOLUTION $A_3 = \$1000\left(1 + \dfrac{.06}{1}\right)^3 = \$1000(1.06)^3 \approx \$1191.02.$

EXAMPLE 5.17

Calculate the value of the investment in Example 5.14 after 50 years; i.e., calculate the value of $1000 at 6% compounded annually after 50 years.

SOLUTION

$$A_{50} = \$1000\left(1 + \frac{.06}{1}\right)^{50} = \$1000(1.06)^{50} \approx \$18{,}420.15$$

EXAMPLE 5.18

Use the compound-amount formula to calculate the value of the investment in Example 5.15; i.e., calculate the value of $1000 at $5\frac{1}{4}$% compounded quarterly for 1 year.

SOLUTION

$$A_4 = \$1000\left(1 + \frac{.0525}{4}\right)^4 = \$1000(1.013125)^4 \approx \$1053.54$$

4 quarters in a year

Keep as many decimal places here as you can. Keep the number stored in your calculator—you need not write it down.

Note that this result is off by $.01 (1¢) from our previous result. Can you explain why? (See Exercise 11.)

EXAMPLE 5.19

Calculate the value of the investment in Example 5.15 after 50 years; i.e., calculate the value of $1000 at $5\frac{1}{4}$% compounded quarterly for 50 years.

SOLUTION There are 50 · 4 = 200 quarters.

$$A_{200} = \$1000\left(1 + \frac{.0525}{4}\right)^{200} \approx \$13{,}570.83$$

The $1000 has grown to $13,570.83.

Note that the compound-amount formula, $A_n = P(1 + i/m)^n$, gives the compound amount A_n as an exponential function of n (the variable is in the exponent) when P, i, and m are fixed. In other words, when the principal, the interest rate, and the frequency of compounding are fixed, the compound amount increases exponentially with time (the number of compounding periods).

To see the power of this exponential growth, suppose $1000 is invested at 6% compounded semiannually. ● Table 5.3 shows the compound amount after various periods of time. ▶ Figure 5.7 illustrates the same information graphically.

TABLE 5.3	
$1000 Compounded Semiannually at 6%	
Number of Years, x	Compounded Amount: $1000\left(1 + \dfrac{.06}{2}\right)^{2x}$
5	$ 1,343.92
10	1,806.11
15	2,427.26
20	3,262.04
25	4,383.91
30	5,891.60
35	7,917.82
40	10,640.89

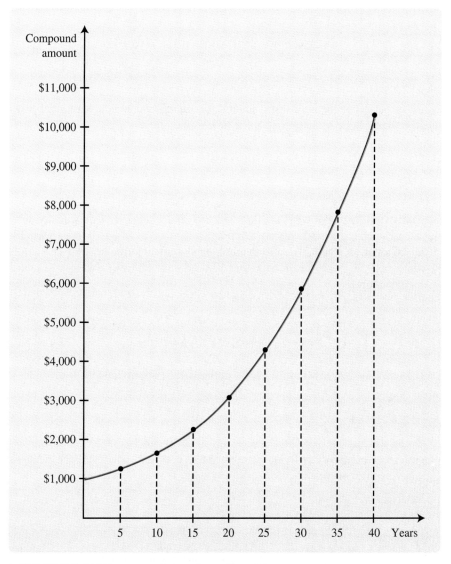

▶ FIGURE 5.7 $1000 compounded semiannually at 6%

One lesson from this is that if at age 25 you invested $1000 at 6% compounded semiannually, it would be worth $10,640.89 by age 65.

Inflation The purchasing power of a dollar shrinks every year because of inflation. Another way to view inflation is to think of the amount an item costs as increasing each year in the same manner that the value of an investment grows with compound interest. The inflation rate is like an interest rate: Think of a 4% inflation rate as an interest rate of 4%. In our treatment of inflation, we will compound annually.

EXAMPLE 5.20

If inflation is 4% per year and an item costs $10 today, what will it cost in 5 years?

SOLUTION Use the formula for compound amount, $A_n = P(1 + i/m)^n$. Here, $n = 5$ (5 years), $P = \$10$ (today's cost), $i = .04$ (4%), and $m = 1$ (compounding once a year).

$$A_5 = \$10\left(1 + \frac{.04}{1}\right)^5 = \$10(1.04)^5 \approx \$12.17$$

We can also use the compound-amount formula in connection with inflation to find the future value of an item at some time in the future.

EXAMPLE 5.21

If inflation is 3% per year, what will be the value of a dollar in 10 years?

SOLUTION This asks what a dollar will buy in 10 years, or, equivalently, how much something would cost now in order for it to cost $1 in 10 years with 3% annual inflation. Again, we apply the compound-amount formula, $A_n = P(1 + i/m)^n$, with $n = 10$, $A_n = \$1$ (the amount after 10 years), $i = .03$ (3%), and $m = 1$ (annual compounding).

$$\$1 = P\left(1 + \frac{.03}{1}\right)^{10}$$

$$P = \frac{\$1}{(1.03)^{10}} \approx \$0.74 \qquad \text{Solve for } P$$

Therefore, in 10 years the value of a dollar will be approximately 74¢.

EXAMPLE 5.22

If a person just out of college starts a job at a salary of $25,000 a year and every year receives a cost-of-living raise of 3.5% per year, what will the person's salary be after 40 years?

SOLUTION

$$A_n = P\left(1 + \frac{i}{m}\right)^n$$

$$A_{40} = \$25,000\left(1 + \frac{.035}{1}\right)^{40} = \$25,000(1.035)^{40} \approx \$98,981.49$$

The effect on a salary of a fixed-percentage salary increase is calculated in the same way as the effect of a cost-of-living increase on a salary. The following example illustrates this.

EXAMPLE 5.23

A school district is negotiating a contract with its teachers. Proposal A calls for a 6% increase each year for 3 years, while proposal B calls for a 3% increase every 6 months for 3 years. Do proposals A and B cost the school district the same amount? If not, which costs more?

SOLUTION For both proposals, we apply the formula for the compound amount,

$$A_n = P\left(1 + \frac{i}{m}\right)^n$$

where P denotes the present collective salary amount paid to teachers.
For proposal A:

$$A_3 = P\left(1 + \frac{.06}{1}\right)^3 = P(1.06)^3 = 1.191016P$$

For proposal B: The compounding occurs semiannually, so $n = 6$ (six 6–month periods) and $i/m = .03$, so

$$A_6 = P(1 + .03)^6 = P(1.03)^6 \approx 1.194052P$$

Because of the power of compounding, proposal B costs more. The preceding argument shows that the amount paid at the end of 3 years is more under proposal B. A similar argument shows the same thing to be true at the end of 1 year and 2 years.

Population Growth   The compound-amount formula, $A_n = P(1 + i/m)^n$ can be thought of as applying to a population with a population size of P and a growth rate of i. For population problems, we take m to be 1 and n the number of time periods over which the growth is compounded. For example, suppose a population of people (or wolves or bacteria) has a constant growth rate of 2% a year. What is the population after 15 years? In this situation, $m = 1$, $i = .02$, and $n = 15$. The following examples illustrate the application of the compound-amount formula to population growth. The nearby graph shows the growth of the U.S. population. Try to picture an exponential curve that closely models this population growth.

EXAMPLE 5.24

The 1970 U.S. population was 203,302,031. If the population grew at the rate of 2% a year, what would the population have been in 1980?

SOLUTION Using the compound-amount formula, $A_n = P(1 + i/m)^n$,

$$A_{10} = 203{,}302{,}031\left(1 + \frac{.02}{1}\right)^{10} \approx 247{,}824{,}041$$

(Actually, the 1980 U.S. population was 226,545,805. Consequently the growth rate from 1970 to 1980 was less than 2% a year.)

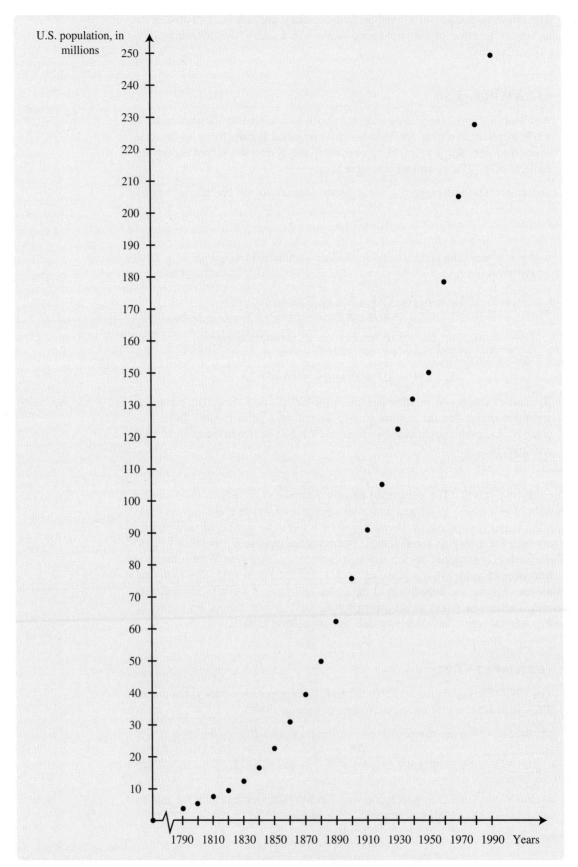

U.S. population, in millions

Years

Source: U.S. Bureau of the Census, *Current Population Reports,* P25-1045 and P25-1112.

a. If the U.S. population grew at the rate of 1% a year from 1970 to 1980, what would the population have been in 1980?

b. Based on the answers to part a and Example 5.24 what can be said about the growth rate from 1970 to 1980?

Continuous Compounding   Let's return to the situation where money, or any population, is being compounded and consider what happens under more and more frequent compounding: annually, semiannually, quarterly, monthly, and so on. Clearly, the compound amount increases. Does it grow without bound, or is there a limit to the growth?

Recall that the formula for the compound amount is

$$A_n = P\left(1 + \frac{i}{m}\right)^n$$

In order to help answer our question, suppose 1 unit (think of $1) is invested at an interest rate of 100% ($i = 1.00$). (This interest rate is much higher than one could expect to earn, but it has certain mathematical advantages in answering the question.) ● Table 5.4 shows the effect of more and more frequent compounding. The length of time the money is invested will hold constant at 1 year.

Note that the compound amount continues to grow with more frequent compounding but that the additions to the compound amount are smaller and smaller (going from annual to semiannual compounding adds .25000 to the compound amount, while going from semiannual to quarterly compounding adds only .19141 to the compound amount). It can be shown that the value of $(1 + 1/n)^n$ as n gets larger and larger gets closer and closer to the constant named e, whose value to five decimal places is 2.71828. This is referred to earlier in this section. Thus, the compound amount does not grow without bound; it reaches a limit which, in the example we're discussing, is 2.71828.

TABLE 5.4			
Value of 1 Unit at 100% under More Frequent Compounding			
Frequency of Compounding	m	n	$A_n = P\left(1 + \frac{1}{m}\right)^n$
Annually	1	1	$\left(1 + \frac{1}{1}\right)^1 = 2.00000$
Semiannually	2	2	$\left(1 + \frac{1}{2}\right)^2 = 2.25000$
Quarterly	4	4	$\left(1 + \frac{1}{4}\right)^4 \approx 2.44141$
Monthly	12	12	$\left(1 + \frac{1}{12}\right)^{12} \approx 2.61304$
Daily	365	365	$\left(1 + \frac{1}{365}\right)^{365} \approx 2.71457$
Hourly	8760	8760	$\left(1 + \frac{1}{8760}\right)^{8760} \approx 2.71813$

The limiting amount (2.71828 in our example) is called the result of *continuous compounding*. It can be shown that, if an amount P (P dollars or a population of size P) is compounded continuously at an interest rate i for n years, the compound amount is given by

$$A = Pe^{in}$$

If we apply this formula to our example, $P = \$1$, $i = 1.00$ (100%), and $n = 1$, so

$$A = \$1e^{1 \cdot 1} = \$e \approx \$2.72$$

CALCULATOR NOTE

If you have an $\boxed{e^x}$ key on your calculator, let $x = 1$ and calculate $e = e^1$. You should get some part of 2.718281828. For whatever calculator you have, verify that $e^2 \approx 7.389$.

EXAMPLE 5.25

Calculate the value of $1000 after 3 years at 6% compounded continuously.

SOLUTION $A = \$1000e^{(.06)3} = \$1000e^{.18} \approx \$1197.22$

Compare this result with Example 5.14 to see that continuous compounding generates $6.20 more than annual compounding over 3 years.

EXAMPLE 5.26

Calculate the value of $1000 at 6% compounded continuously for 50 years.

SOLUTION $A = \$1000e^{(.06)\ 50} = \$1000e^3 \approx \$20{,}085.54$

Note: This calculation was done using an $\boxed{e^x}$ key. If it is calculated as $\$1000(2.71828)^3$, the result is $20,085.50, which is less accurate because the calculator with an $\boxed{e^x}$ key uses a value for e that is more accurate than 2.71828.

Compare this result with Example 5.17 to see that continuous compounding generates $1665.39 more than annual compounding over 50 years.

Half-Life   Our next example of an exponential function is called a *half-life function*. The basic form for this is

$$y = \left(\frac{1}{2}\right)^x$$

The distinguishing feature of this exponential function is that the base is a number less than 1. To see how values are computed, consider the following:

$$x = 1, \qquad y = \left(\frac{1}{2}\right)^1 = \frac{1}{2}$$

$$x = 2, \qquad y = \left(\frac{1}{2}\right)^2 = \frac{1}{2} \cdot \frac{1}{2} = \frac{1}{4}$$

$$x = 3, \qquad y = \left(\frac{1}{2}\right)^3 = \frac{1}{2} \cdot \frac{1}{2} \cdot \frac{1}{2} = \frac{1}{8}$$

For $x = 0$, we get
$$y = \left(\frac{1}{2}\right)^0 = 1$$

For $x = -1$, we have
$$y = \left(\frac{1}{2}\right)^{-1} = 2^1 = 2$$

For other negative values of x, we have

$$x = -2 \qquad y = \left(\frac{1}{2}\right)^{-2} = 2^2 = 4$$

$$x = -3, \qquad y = \left(\frac{1}{2}\right)^{-3} = 2^3 = 8$$

When these points are plotted and connected, we obtain the graph nearby. Note one important characteristic of this graph: As x gets larger, the value of y decreases.

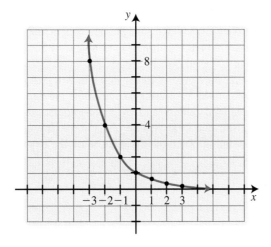

Exponential functions like this one with a base less than 1 are sometimes written in a different form, as follows:

$$y = \left(\frac{1}{2}\right)^x = (2^{-1})^x = 2^{-x}$$

This alternate form has a base greater than 1 but has a negative sign in the exponent. The graphs of such functions have shapes like that in the graph above.

Radioactive Decay   A number of substances in nature decay so that the rate of decay is proportional to the amount present at a given time. It can be shown (via calculus) that the quantity Q of such a substance that is present after t units of time (t seconds, t years, or so on) is given by

$$Q = Q_0 e^{kt}$$

where Q_0 is the amount present at $t = 0$ and k is a negative constant that depends on the particular substance. For example, if we start with Q_0 grams of barium-137, the quantity Q, in grams, that is present after t minutes is given by $Q = Q_0 e^{-.27t}$.

EXAMPLE 5.27

a. If you start with 2.00 grams of barium-137, how much will be present after 5 minutes?

SOLUTION $Q_0 = 2.00$ g and $t = 5$, so

$$Q = (2.00 \text{ g})e^{-.27(5)} = (2.00 \text{ g})e^{-1.35} \approx .52 \text{ g}$$

About .52 g is left after 5 minutes.

b. If one starts with 2.00 grams of barium-137, how much will be present after 2.55 minutes?

SOLUTION $Q = (2.00 \text{ g})e^{-.27(2.55)} = (2.00 \text{ g})e^{-.6885} \approx 1.00 \text{ g}$

About 1.00 g will be present.

The time it takes for such a substance to decay to half its original amount—i.e., the value of t so that $Q = \frac{1}{2}Q_0$—is called the half-life of the substance. Note that Example 5.27b verifies that the half-life of barium-137 is 2.55 minutes.

Radiocarbon dating is based on the decay of a naturally occurring carbon isotope. It can help to determine the age of organic matter to within 100 years. Radiocarbon dating is employed by geologists to determine the age of such materials as wood, bone, and fossilized pollen, and by archaeologists to date artifacts from ancient civilizations. Both the Dead Sea Scrolls and the Shroud of Turin have been dated via radiocarbon dating. Such dating is based on the exponential function

$$Q = Q_0 e^{kt}.$$

MOTHER GOOSE & GRIMM

Reprinted by permission: Tribune Media Services.

Radiocarbon dating is a positive application of radioactive decay. However, negative features are also associated with it. Some radioactive materials involved in the production of nuclear energy have very long half-lives. For example, the half-life for plutonium-239 (Pu-239) is 24,000 years. Consequently, for such radioactive materials, a sizable portion is left with the potential of contaminating the environment after many years. Pu-239 is highly toxic. The exact amount that is left after a given number of years can be determined via the formula for radioactive decay. This is explored in Exercise 34.

The Richter Scale   The traditional system for measuring the strength of an earthquake is the Richter scale. It was developed by Charles F. Richter, an American seismologist, in 1935. The first known device to sense an earthquake (a model shown in the following photo) was designed by the Chinese scholar Chang Heng about 130 C.E. The direction of an earthquake was indicated by a ball that rolled out of an appropriate dragon's mouth.

Today seismologists have much more sophisticated devices for detecting earthquakes, but the words that roll off a reporter's tongue are perhaps as mysterious as the mechanism that caused the dragon to spew the ball out of a particular dragon's mouth. It is common for newspaper or magazine accounts of an earthquake to say something like:

> The Richter scale measures the amount of force released by an earthquake. Each whole number represents a tenfold increase in a quake's power. Thus:
>
> $$6 = 10 \text{ times stronger than } 5$$
> $$7 = 100 \text{ times stronger than } 5$$
> $$8 = 1000 \text{ times stronger than } 5$$

Initially, we will be loose in interpreting the word *power;* however, you should generally be very careful in your interpretation.

Let x denote the Richter scale measurement of an earthquake and let y denote the power of the earthquake. Then x and y are related by $y = 10^x$. Note that when x is 5, y is 10^5; when x is 6, $y = 10^6$, which is 10 times the value when x is 5; when x is 7, $y = 10^7$, which is 100 times the value when x is 5 ($10^7 = 10^2 \cdot 10^5 = 100 \cdot 10^5$); and when x is 8, $y = 10^8$, which is 1000 times the value when x is 5 ($10^8 = 10^3 \cdot 10^5 = 1000 \cdot 10^5$). Note, also, that the power of an earthquake is an exponential function of its Richter scale measurement.

What does *power* mean? E. S. Oberhofer (in "Different Magnitude Differences," *The Physics Teacher,* May 1991, pp. 273–274) refers to B. A. Bolt (*Earthquakes.* New York: W. H. Freeman, 1988, p. 112) in stating that *power* is a "measure of the maximum deflection that would be produced on a 'standard' seismograph at a standard distance from the earthquake, and as such, for each factor of 10 increase in the amplitude of the earthquake waves, the magnitude goes up one unit." Oberhofer points out that the usually accepted expression for the energy E, in joules, being released during an earthquake with Richter scale measurement x is given by

$$E = 25{,}000 \cdot 10^{1.5x}$$

Note that energy is an exponential function of x, the Richter scale measurement. However, when x is 5, $E = 25{,}000 \cdot 10^{7.5}$, when x is 6, $E = 25{,}000 \cdot 10^9$, which is $10^{1.5}$ (≈ 32) times the value when x is 5. Therefore, the "power" of an earthquake is something different from the energy released by it.

EXAMPLE 5.28

The strongest earthquakes so far recorded were located in the Pacific Ocean near the Colombia–Ecuador border in 1906 and in Japan in 1933. Both measured 8.9 on the Richter scale. The next-strongest earthquake measured 8.4 and was located in Valdez, Alaska, in 1964.

a. The strongest recorded earthquakes were how many times as powerful as the next strongest?
b. How many times as much energy did they expend?

SOLUTION

a. Let y_S denote the power of the strongest earthquakes and y_A denote the power of the 1964 Alaska earthquake. Then

$$\frac{y_S}{y_A} = \frac{10^{8.9}}{10^{8.4}} = 10^{0.5} \approx 3.2$$

Therefore, the strongest recorded earthquakes were about 3.2 times as strong as the next-strongest quake.

b. Let E_S denote the energy, in joules, released by the strongest quakes and E_A denote the energy, in joules, released by the 1964 Alaska quake. Then

$$\frac{E_S}{E_A} = \frac{25{,}000 \cdot 10^{1.5(8.9)}}{25{,}000 \cdot 10^{1.5(8.4)}} = \frac{10^{13.35}}{10^{12.6}} = 10^{.75} \approx 5.6$$

Therefore, the strongest recorded earthquakes expended about 5.6 times as much energy as did the next-strongest quake.

The mathematics you learn in this book might not be earth shaking, but it is involved in reporting on earth shakings. We hope this doesn't make you tremble.

1. Which of the following are exponential functions?
   a. $y = 2^2$
   b. $y = 3^x$
   c. $y = x^2$
2. Graph $y = 3^x$ and $y = (1/3)^x$.
3. Sketch the general shape of the graph(s) of $y = a^x$, where $a > 0$.
4. Exponential functions have the form $y = a^x$, where a is a positive constant. What is a problem that arises if a is negative?
5. Use a calculator to find the following.
   a. 4^6 b. 6^4 c. 4^{-3} d. $(-3)^4$
   e. $4^{1/3}$ f. $(1/3)^4$ g. $(1/3)^{-4}$ h. -3^4
6. Find the value of e^x for $x =$ each of the following.
   a. 2.5   b. 1.5   c. 0.5   d. −0.5   e. −1.5
7. Plot points and sketch the function $y = e^{-x}$.
8. (*Multiple choice*) For the function $y = a^x$, where a is a positive constant, the statement "As x increases, y increases" is true for which of the following?
   a. Any positive value of a
   b. Some, but not all, positive values of a
   c. No positive values of a
9. For what values of x is a^x (where a is positive)
   a. Positive
   b. Zero
   c. Negative
10. The quadratic function $y = ax^2 + bx + c$ has its maximum or minimum value when $x = -b/2a$. What can be said about the maximum and minimum values of an exponential function $y = a^x$?
11. Explain why the amount in Example 5.18 is off by 1¢ from the earlier-calculated amount. (Instead of a penny for your thoughts, this is your thoughts for a penny.)

In Exercises 12–15, find the compound amount under the given conditions.

12. $2000 is invested at $5\frac{1}{2}$% compounded annually for 5 years.
13. $4000 is invested at $5\frac{3}{4}$% compounded semiannually for 10 years.
14. $6000 is invested at 6% compounded quarterly for 3 years.
15. $8000 is invested at 9% compounded monthly for 7 years.

For Exercises 16–17, use the compound-amount formula to solve for P if A_n, i, m, and n are given.

16. If money is invested at $5\frac{1}{4}$% compounded quarterly for 5 years, how much should be invested in order to have $10,000 at the end of 5 years? This amount is called the *present value* of $10,000.
17. Find the present value of $15,000 if money is invested for 3 years at 7% compounded semiannually.
18. If inflation is 5% per year and an item costs $100 today, what will it cost in 4 years?

19. If inflation is 4% per year, what will be the value of a dollar in 20 years?
20. If inflation is 3% per year, what will be the value of a dollar in the year 2020?
21. If a person starts a job at a salary of $30,000 and receives a cost-of-living increase of 3% per year, what will the person's salary be after 35 years?
22. In Example 5.23, assume the average salary per teacher is $35,000 and that there are 40 teachers. Over a 3-year contract, how much more in total would the school district pay under proposal B than under proposal A?
23. The world population was estimated at 5,234,000,000 in 1989.
    a. In 1989 the growth rate was 1.8% per year. If that rate continued to the year 2000, what would the population be in 2000?
    b. In 1988 the growth rate was 1.7% per year. If that rate had continued from 1989 to 2000, what would the world population be in 2000?
24. The population of Australia in 1988 was estimated to be 16,800,000, with an average annual rate of increase of 0.8%. If that rate continues to the year 2000, what would be Australia's population in 2000?

In Exercises 25–29, find the compound amount under the given conditions.

25. $1000 is invested at 4% compounded continuously for 9 years.
26. $3000 is invested at 5% compounded continuously for 4 years.
27. $5000 is invested at $5\frac{1}{4}$% compounded continuously for 6 years.
28. $7000 is invested at $6\frac{1}{2}$% compounded continuously for 2 years.
29. $9000 is invested at 6% compounded continuously for 8 years.

In Exercises 30 and 31, use the compound-amount formula with continuous compounding to find P if A, i, and n are given.

30. If money is invested at $5\frac{1}{4}$% compounded continuously for 5 years, how much should be invested in order to have $10,000 at the end of 5 years? This amount is the present value of $10,000.
31. Find the present value of $15,000 if money is invested for 3 years at 7% compounded continuously.
32. Table 5.3 and Figure 5.7 show the value of $1000 invested at 6% compounded semiannually for various numbers of years.
    a. Calculate the value of $1000 invested at 6% compounded continuously for 5, 10, 15, 20, 25, 30, 35, and 40 years. Enter these amounts as a third column in Table 5.3.
    b. Illustrate the information in part a graphically by putting it on a graph with Figure 5.7.

5.3 EXPONENTIAL FUNCTIONS

33. For gallium-68, the formula for radioactive decay is given by

$$Q = Q_0 e^{-.01t}$$

where Q is the amount present after t minutes and Q_0 is the original amount present.

   a. If 1.50 g is present originally, how much is present after 10 minutes?

   b. If 1.50 g is present originally, how much is present after 1 hour?

   c. If 1.50 g is present originally, how much is present after 1 hour?

   d. Which of the following is closest to the half-life of gallium-68? 10 minutes, 1 hour, a day

   e. Suppose that, after a quantity of gallium-68 has decayed for 20 minutes, 2.00 g is present. How much was present originally?

   f. Graph this radioactive decay function using $Q_0 = 1.50$.

34. In our description of radioactive decay we mentioned that some radioactive materials used to produce nuclear energy have very long half-lives. For example, the half-life of plutonium-239 is 24,000 years.

   a. The formula for the quantity Q of Pu-239 that is present after t years is given by $Q = Q_0 e^{kt}$, where Q_0 is the original amount present. Which is the best choice for k? −2888, 2888, −0.00002888, 0.00002888

   b. If 450 g (slightly less than 1 pound) of Pu-239 is discarded today, how much of it will be left after 100 years?

35. a. The 1906 San Francisco earthquake, measuring 8.3 on the Richter scale, was how many times as powerful as the 1989 San Francisco earthquake (during the World Series), measuring 7.1 on the Richter scale?

   b. How many times as much energy did it expend?

36. (*Multiple choice*) An earthquake measuring 6.2 on the Richter scale is fairly mild. If another quake were twice as strong, what would it measure on the Richter scale?

$$2.0, \quad 3.1, \quad 6.2, \quad 6.5, \quad 8.2, \quad 12.4$$

WRITTEN ASSIGNMENTS 5.3

1. This section dealt with a particular kind of nonlinear function, namely, an exponential function. The previous section dealt in part with another type of nonlinear function, namely, a quadratic function. Write a section of material comparable to the first part of this section dealing with yet another type of nonlinear function, i.e., other than an exponential or quadratic function.

2. The term *exponential growth* carries with it the idea of very rapid growth. Illustrate how rapid by comparing the growth of an exponential function with other functions that grow, such as $f(x) = x$ and $f(x) = x^2$.

3. The notation x^2 for $x \cdot x$ and x^3 for $x \cdot x \cdot x$ probably seems very natural to you. However, its original use is usually credited to René Descartes in about 1637. What was used, if anything, prior to that time to indicate a repeated product $x \cdot x \cdot \cdots \cdot x$? Capsule 89 in the Thirty-First Yearbook of the National Council of Teachers of Mathematics, entitled *Historical Topics for the Mathematics Classroom,* is one place to start collecting information.

4. The number e has an interesting development. Write a paper that contains details of this development. Consult a book on the history of mathematics as your reference.

5. Interview some of your mathematics faculty about whether or not they believe there are any mathematicians active today who are comparable to Euler in the amount or significance of the mathematics they do. Report your results.

6. Euler's work during the last 17 years of his life was done while he was almost totally blind. Describe what particular problems you believe a person would have in doing mathematics while unable to see.

7. Find out what local bank interest rates are.

   a. Compare two or more of them.

   b. Do you think a typical investor would find it difficult to compare interest rates (for example, 6% compounded annually versus $5\frac{1}{2}\%$ compounded quarterly)? Why or why not?

   c. If you answered yes to part b, would you support legislation to standardize calculation or reporting of interest rates? Why or why not?

8. Recount some stories you have heard parents, grandparents, or other "old-timers" tell about prices or salaries when they were younger. Make (and state) some assumptions about the rate of inflation, and calculate what some of those prices or salaries would be today just due to inflation.

9. Criticize the population growth models described in this section. What assumptions are made in them that might be unrealistic?

10. Write several paragraphs describing continuous compounding. Assume your audience is someone with whom you graduated from high school but who probably doesn't know what continuous compounding is.

11. a. Do any of your textbooks refer to radioactive decay, half-life, or radiocarbon dating? If so, cite the book(s) and describe what is said.

   b. Find an article in a newspaper or magazine published in the last 5 years that deals with radioactive decay, half-life, or radiocarbon dating. Report on the article.

12. One problem with nuclear energy is the disposal of nuclear wastes, such as plutonium-239. Find a newspaper article, magazine article, or textbook passage that deals with this. Cite your reference, and either report on a calculation done with the radioactive decay formula, $Q = Q_0 e^{kt}$, or do such a calculation with information reported in the article.

13. Find a newspaper article, magazine article, or textbook passage that deals with the Richter scale. Does it make any explicit mention of an exponential function? Does it add anything to your understanding of the Richter scale beyond what this section has said? If so, what?

 ## 5.4 FURTHER DEVELOPMENTS IN ALGEBRA

GOALS
1. To describe briefly the history of algebra from the ninth century through the work of Galois in the nineteenth century.
2. To solve linear inequalities in one variable.
3. To describe an application of linear inequalities.
4. To describe a development in algebra prompted by the work on solving the general fifth-degree equation.

In this chapter algebra has been considered as generalized arithmetic and the basic functions—linear, quadratic, and exponential—have been considered. What else is there to algebra? How has it developed, and where else is it applied? We will briefly examine these questions in this section.

Historical Interlude

Arithmetic and algebra as generalized arithmetic are two prime examples of the importance of good notation. As two mental exercises to appreciate this point, consider multiplying the numbers written in Roman numerals as CCLVIII times CCCIX without converting to Hindu-Arabic numerals and consider solving the equation "two times a number plus nine is twenty-one" without writing a name like *x* for the number. The word *equation* has as its root the word *equal*. Let's consider the symbol for equality.

The first use of the symbol "=" to indicate that two quantities are equal was made by Robert Recorde, an English mathematician, in his 1557 book *The Whetstone of Witte*. Recorde said: "To avoid the tedious repetition of these words—is equal to—I will set, as I do often in work use, a pair of parallels or [twin] lines of one length, thus ═══, because no two things can be more equal." However, it took a century or more for Recorde's notation to become the predominant one.

The word *whetstone* in the title of Recorde's book prompts a look at developments in algebra. It will take us a while to get back to whetstone, but we need to lay some background to whet your appetite.

The word *algebra* comes from the Arabic word *al-jabr* (see Section 2.1 and this chapter's etymology), from the title of a book written by al-Khwārizmī in about 825. The term *al-jabr* can be translated as "restoring." It refers to moving a term that is subtracted from one side of an equation to where it is added, or restored. In the introduction to his book, the Muslim al-Khwārizmī described why he wrote it:

That fondness for science, by which God has distinguished the Iman al-Ma'mun, the Commander of the Faithful, . . . has encouraged me to compose a short work on calculating by *al-jabr* . . . confining it to what is easiest and most useful in arithmetic, such as men constantly require in cases of inheritance, legacies, partition, lawsuits, and trade, and in all their dealings with one another, or where the measuring of lands, the digging of canals, geometrical computation, and other objects of various sorts and kinds are concerned.

There are two things in this quote that are characteristic not only of al-Khwārizmī but also of much of mathematical and other intellectual development in the Middle Ages in the West: a deep respect for religious beliefs, and the development of mathematics for practical purposes. In the latter regard, the attitude is much more like the Babylonian and Egyptian attitudes than the attitude of the intervening Greeks.

Islamic mathematicians continued their development of algebra for the next several centuries. One of the developers was the Persian poet, astronomer, and mathematician known in the West as Omar Khayyam (1048–1131), who is best known for his *Rubaiyat.* A significant part of his work in algebra was the geometrical solution of cubic (third-degree) equations. (We've discussed the solution of quadratic equations in Section 5.2.) Since at his time all numbers were considered positive, he dealt with cubic equations that had positive coefficients and positive solutions. In his *Algebra* he describes his view of algebra:

By the help of God and with His precious assistance, I say that Algebra is a scientific art. The objects with which it deals are absolute numbers and measurable quantities which, though themselves unknown, are related to "things" which are known, whereby the determination of the unknown quantities is possible. . . . What one searches for in the algebraic art are the relations which lead from the known to the unknown. . . . The perfection of this art consists in knowledge of the scientific method by which one determines numerical and geometric unknowns.

While intellectual activity blossomed in the Islamic world, the Middle Ages found little such activity in Europe. In this relative wasteland appeared Leonardo of Pisa (Italy) (c. 1170–1240), also known as Fibonacci (son of Bonaccio). His background says much about the spread of mathematics in the period in which he lived. His father was a merchant in Pisa who did substantial business in North Africa. Fibonacci spent much of his early life in North Africa learning Arabic and studying mathematics under Islamic teachers. He later traveled extensively in the Mediterranean area, probably doing business for his father. During those travels he learned computation with Hindu-Arabic numerals and perhaps some Chinese mathematics. The increased business activity of the time in Europe prompted travel such as Fibonacci's. This, in turn, helped transmit the mathematical developments of the Islamic world to Europe.

In 1202 Fibonacci published *Liber abbaci,* or *Book of the Abacus,* whose title is misleading, since it is not about the abacus but about algebraic methods and problems in which the use of Hindu-Arabic numerals is described. Its content reflects the influence of the algebra of al-Khwārizmī and other Islamic mathematicians. In fact, it made little, if any, advance over the algebra of the Islamic mathematicians, but it did serve to introduce their methods to Europe. Fibonacci was also one of the first to employ the horizontal bar as we do in fractions (see Section 2.3). He also made several original contributions to mathematics.

The *Liber abbaci* introduced the Hindu-Arabic symbols (essentially the numerals we use) into a Europe where most people used Roman numerals. He also explained the principle of place value (which is not present in Roman numerals) and employed the Hindu-Arabic place-value system in arithmetical operations. Then he applied this computation to commercial problems, such as interest, profit margins, money changing, and conversion of weights and measures. Many merchants of Pisa and Florence adopted these new computational methods. In fact, they were more responsive to their use than were academic scholars.

A book published probably a little later in the 1200s than the *Liber abbaci* had more of an emphasis on doing algebra as generalized arithmetic than on doing it geometrically, as Omar Khayyam and other Islamic mathematicians had been doing. Algebra also started to move from rhetorical algebra to symbolic algebra. The book was entitled *De numeris datis* (*On Given Numbers*). It was written by Jordanus de Nemore, about whom very little is known, although he is recognized as one of the best mathematicians of the Middle Ages. It is believed that he taught in Paris about 1220. The translator of *De numeris datis,* Barnabas Hughes, makes an intriguing conjecture about Jordanus and recalls Hypatia, the first significant woman mathematician (see Section 6.1):

> The only explanation [of why there is no known biographical information] that appealed to me was that the name is a pseudonym. But why a *nom de plume*? Could it be that Jordanus was really a woman? Shades of Hypatia! Thirteenth century women were good for writing poems, songs, and prayers; but science?

Whether Jorandus was a woman or not is unknown, but the comments by Hughes reflect societal attitudes of the Middle Ages and before. This is why the mathematicians mentioned up until this time are men, except for Hypatia and perhaps Jordanus.

The period of time from the fourteenth through the sixteenth centuries is usually referred to as the Renaissance, denoting a rebirth of arts and letters in Europe. This cultural movement was fueled by increased trade between Europe and places in the East. New technology in shipbuilding allowed merchants to remain at home and send their agents "on the road" to transact business. The largely barter economy was replaced by a need for the home office to deal with credit and interest calculations. Double-entry bookkeeping began as a way to keep track of the transactions. The merchants needed new and faster tools for calculating, problem solving, and mathematical modeling. During this period the Italians were the European leaders in commerce, in the Renaissance, and in developments in algebra. We now have the stage set to wind back to Robert Recorde's "whetstone."

The Italian word for "unknown" was *cosa*, which literally meant "thing." By the late fourteenth century algebra had moved into Germany, where it was known as the Art of the Coss. (Some wit has said that algebra in the United States is the art of the cuss.) *Coss* was the German form of the Italian *cosa*. People versed in algebra were called Cossists. One of the Cossists, Christoff Rudolff, wrote the first comprehensive German algebra book in 1525, called simply *Coss*.

Robert Recorde's "whetstone" was a play on the German *coss* (the *Whetstone of Witte* was written in 1557), based on the idea that *cos* is Latin for "whetstone." We mentioned Recorde's book because it contains the first use of our "=" sign. Enroute to describing the reason for using the word "whetstone," we have viewed algebra from al-jabr in 825 done rhetorically to a movement into symbolic algebra by the end of the Renaissance; we have seen the cultural influences of religious beliefs, commerce, beliefs about women's roles, and the Renaissance; we have looked at a change in the geometric locus of mathematical activity from the Islamic world to Western Europe; and we've briefly considered contributions by three people: al-Khwārizmī, Omar Khayyam, and Fibonacci.

The equal sign, "=" is used to form equations. In your previous experience with algebra, you probably remember solving equations. Many real-world problems or situations give rise to an equation as a type of mathematical model.

Finding the zeros of a function requires solving an equation. In this book, discussion of solving equations is included in earlier sections of this chapter. Historically, the problem of solving an equation goes back to some of the earliest-recorded

ROBERT RECORDE REMEMBERED

Robert Recorde studied at Oxford and Cambridge and practiced medicine. He served in several civil service positions, but he was most successful in writing mathematics textbooks. His books were written in the form of a dialogue between master and student, in which each step in a particular computational process was carefully explained. Do you understand what we mean? *The Whetstone of Witte* was published in 1557, about a year before he died in prison. The authors of this textbook are relieved to know that he was jailed for political or religious reasons or because of troubles related to his last position as Surveyor of the Mines and Monies of Ireland rather than for writing a mathematics textbook.

1. Give an example of a real-world problem or situation that is modeled by an equation.
2. Write the equation. Be sure to identify what your variable represents.
3. If possible, solve the equation.

---

mathematics, found on the *Rhind Papyrus* and *Moscow Papyrus* dating to about 1850 B.C.E. We will focus for a while on solving polynomial equations. A *polynomial equation of degree n* is an equation of the form $a_0 + a_1x + a_2x^2 + \ldots + a_nx^n = 0$, where the coefficients $a_0, a_1, a_2, \cdots, a_n$ are real numbers, n is a positive integer, and $a_n \neq 0$. (Why is this last condition imposed?) For example, the equation $1 - 2x + 3x^2 = 0$ is a polynomial equation of degree 2, and the equation $4/3 - x^5 = 0$ is a polynomial equation of degree 5.

Methods for solving the general first-degree equation $ax + b = 0$ as $x = -b/a$ were known back to the ancient Babylonians and Egyptians and in ancient Chinese mathematics, although negative solutions were problematic.

What about solutions to quadratic equations? To solve the general quadratic equation $ax^2 + bx + c = 0$ we can use the quadratic formula,

$$x = \frac{-b \pm \sqrt{b^2 - 4ac}}{2a}$$

The ancient Babylonians and Chinese could solve quadratic equations, although, again, negative solutions and complex solutions were disregarded. However, neither the ancient Babylonians nor the Chinese made explicit use of the quadratic formula.

What about solutions to cubic (third-degree) equations? Recall that a third-degree equation has the form $ax^3 + bx^2 + cx + d = 0$. Solving such an equation is a problem that Fibonacci worked on. He found ways to classify and solve geometrically all cubic equations, disregarding negative and complex solutions. Although he solved them geometrically (1202), he came up short on finding ways to solve them algebraically. He wrote about finding an algebraic solution to a cubic equation: "Neither we, nor any of those who are concerned with algebra, have been able to solve this equation—perhaps others who follow us will be able to fill the gap." The gap was filled, but not for over 300 years. Who filled it is a matter of dispute, but the names of the Italians Girolamo Cardano, Scipione del Ferro, Lodovico Ferrari, Antonio Maria Fior, and Niccolo Tartaglia are part of the story. It's not a matter of dispute that filling the gap was significant. The mathematical historian John Stillwell (1985) considers the solution of cubic equations in the early sixteenth century as the first clear advance in mathematics since the time of the Greeks.

Cardano's 1845 book, *Ars Magna*, included the first published algebraic solution to the cubic equation. It contained not only the solution but considered negative solutions. In it Cardano also used complex numbers in connection with solutions to a quadratic equation.

The solution to the general equation of the fourth degree $ax^4 + bx^3 + cx^2 + dx + e = 0$ was found by Cardano's student Ferrari and was also included in *Ars Magna*. The time from the first special-case solution of the third-degree equation in the early 1500s to the general solution of the fourth-degree equation in 1545 was quite short. This gave the hope that the solution of the fifth-degree equation would follow very quickly.

Fibonacci Sequence

In *Liber abbaci* Fibonacci poses and solves a problem known as "The Rabbit Problem." Here is his description of the problem.

> How many pairs of rabbits can be bred from one pair in 1 year? A man has one pair of rabbits at a certain place entirely surrounded by a wall. We wish to know how many pairs can be bred from it in 1 year, if the nature of these rabbits is such that they breed every month one other pair and begin to breed in the second month after their birth.

Let's solve the problem.

1. We start with one pair of rabbits. In the first month that pair produces a pair, so at the end of 1 month there are two pairs of rabbits. We'll keep track of this in a table.

End of Month	0	1
Number of Pairs	1	2

How many pairs will there be after two months? _____

2. At the end of 3 months there will be five pairs. How many pairs will there be after 4 months? Put your answers to parts 1 and 2 in the following table.

End of Month	0	1	2	3	4
Number of Pairs	1	2	_____	5	_____

3. Continue the table to find how many pairs there will be at the end of a year. _____

The sequence of numbers giving the number of pairs

$$1, 2, 3, 5, 8, 13, 21, 34, \ldots$$

is called the Fibonnaci sequence. Often an additional 1 is added at the start of the sequence to make the sequence

$$1, 1, 2, 3, 5, 8, 13, 21, 34, \ldots.$$

Notice, as did Fibonacci, that after the first two terms of the sequence, each term is obtained by adding the preceding two terms: $3 = 1 + 2, 5 = 2 + 3$, etc. This sequence has many interesting and wide-reaching properties unsuspected by Fibonacci. For example, on a pineapple or pine cone, the number of spirals winding clockwise and the number of spirals winding counterclockwise are consecutive terms of the Fibonacci sequence. In fact, there is such a richness of ideas related to the Fibonacci sequence that there is a mathematics journal, *The Fibonacci Quarterly,* devoted entirely to them.

© 1982 by Sidney Harris—*American Scientist* magazine.

The number of spirals in each direction are consecutive terms of the Fibonacci sequence.

Intrigue to the Third Power

The truth about integrity in publishing the solution of the cubic equation seems as elusive as truth in some trials, but here is a common version of the story: Between 1500 and 1515 Scipione del Ferro, a professor of mathematics, solved algebraically a particular type of cubic equation that can be written in the form $x^3 + cx - d = 0$, where c and d are positive numbers. But the academic community of that day was not a publish-or-perish world, so del Ferro did not publish or even publicly announce his result. However, sometime before he died in 1526 he showed his solution to his student Antonio Maria Fior. Fior also did not publish the solution.

Another mathematician, Niccolo Tartaglia ("Tartaglia" was actually a nickname, meaning "stutterer," because of a childhood injury when his home city was sacked by the French), was more vocal. He announced that he could solve algebraically a cubic equation of the form $x^3 + bx^2 - d = 0$, where b and d are positive numbers. In 1535 Fior, knowing that he could solve for sure a certain type of cubic equation, challenged Tartaglia to a public contest of mathematical problems, a practice that was common among academics of that time. Each submitted a set of problems for the other to solve. Fior's submission consisted of 30 problems, each dealing with the type of cubic equation he could solve. One night not long before the contest, perhaps the night before, Tartaglia figured out how to solve the cubic equation that Fior was submitting. Fior could not solve a number of the problems that Tartaglia submitted, so Tartaglia was declared the winner.

Word of the contest and Tartaglia's efforts spread rapidly across Italy (at least as rapidly as was possible in a world without e-mail, television, radio, or next-day mail). Girolamo Cardano, a physician and mathematics teacher, asked Tartaglia to show him the solution. Initially Tartaglia refused, but after many requests he finally relented and on a visit from him to Cardano in 1539, he delivered to Cardano, in a poem, his method for solving an equation of the form $x^3 + cx - d = 0$. His technique is far from obvious from the poem. Cardano subsequently set about working on the problem of algebraically solving a cubic equation, probably with the assistance of his servant and student Ferrari. Over a several-year period Cardano managed to solve algebraically *all* the cases of the cubic. Meanwhile, Cardano heard reports of del Ferro's earlier solution, so he and Ferrari located and inspected del Ferro's papers (del Ferro had died over 13 years earlier). This showed that del Ferro had discovered the solution first.

Although Cardano had originally been told the solution to one case of the cubic equation by Tartaglia, it is understandable that Cardano might have reasoned that since Tartaglia's solution was predated by del Ferro's solution, since del Ferro was now dead, since several years had passed, and since Cardano had discovered how to solve algebraically *all* cases of the cubic equation, it was acceptable for him to publish the solution to the cubic. He did so in his 1545 book, *Ars Magna, sive de Regulis Algebraicis* (*The Great Art, or On the Rules of Algebra*), with the following acknowledgment:

> In our own days Scipione del Ferro . . . has solved the case of the cube and first power equal to a constant [$x^3 + cx = d$ or $x^3 + cx - d = 0$], a very elegant and admirable accomplishment. Since this art surpasses all human subtlety and the perspicuity of mortal talent and is a truly celestial gift and a very clear test of the capacity of men's minds, whoever applies himself to it will believe that there is nothing that he cannot understand. In emulation of him, my friend Niccolo Tartaglia . . . , wanting not to be outdone, solved the same case when he got into a contest with his [del Ferro's] pupil, Antonio Maria Fior, and, moved by my many entreaties, gave it to me. . . . Having received Tartaglia's solution and seeking a proof of it, I came to understand that there were a great many other things that could also be had. Pursuing this thought and with increased confidence, I discovered these others, partly by myself and partly through Lodovico Ferrari, formerly my pupil.

Tartaglia and many others have accused Cardano of dishonesty. Tartaglia claimed that Cardano had promised not to publish the solution. Ferrari came to Cardano's defense, stating that he was present as a servant at the time and that no such promise had been made. Ferrari and Targalia traded insults and mathematical challenges in a series of 12 printed pamphlets and in a public contest in 1548 (Ferrari probably won). Cardano, in a later book, places Tartaglia with people who he "cannot understand by what impertinence they have managed to get themselves into the ranks of the learned."

Many of the best mathematicians of the sixteenth, seventeenth, and eighteenth centuries worked on finding a solution to the general fifth-degree equation, but without success. The problem was not rigorously solved until 1824, and the solution was negative. The young Norwegian mathematician Niels Henrik Abel (1802–1829) showed that the general fifth-degree equation is not "solvable by radicals"; that is, there is no solution like the quadratic formula that uses the coefficients of the equation and a finite sequence of the algebraic operations addition, subtraction, multiplication, division, and extraction of roots.

Abel's work inspired an even younger French mathematician, Evariste Galois (1811–1832), to investigate when an equation is solvable by radicals. (See the On a Tangent on the next page.) This led to what is now called abstract algebra and, within that broader field, to group theory and field theory.

The results of Abel and Galois showed that for any equation of degree five or higher, there is no formula analogous to the quadratic formula to give the solution in terms of the coefficients of the equation. These results completed one branch of algebra, namely, solving polynomial equations by means of formulas like the quadratic formula. However, it prompted the development of ways to approximate the solutions to an equation. A number of methods have been found, and this continues to be an area of active mathematical research.

Inequalities

We've just discussed some history of finding methods to solve polynomial *equations*. Section 5.2 discusses methods of solving linear and quadratic *equations*. Many quantitative comparisons deal with *inequality* rather than equality. Work with equations can be extended to work with inequalities.

The basic symbol for an inequality is "<." For example, $x < 2$ is read "x is less than 2." This can be combined with equality to produce the symbol "≤." Thus, $x \le 2$ is read "x is less than or equal to 2." These symbols can go in either direction. When faced the other way, we have ">" and "≥." These are read "greater than" and "greater than or equal to," respectively. Thus, $x \ge 10$ is read "x is greater than or equal to 10." These are summarized in ● Table 5.5.

When and by whom were the inequality symbols "<" and ">" introduced? They were first used by another English mathematician, Thomas Harriot (1560–1621), roughly half a century after the "=" sign had been introduced. However, it was a number of years before these symbols became widely used.

THE SECRET LIFE OF
THOMAS HARRIOT

Thomas Harriot led a secretive life, but one in company with the explorer Sir Walter Raleigh, the dramatist Christopher Marlowe, and political activist Guy Fawkes. That he was secretive may not be surprising when it is noted that Raleigh was executed on an old treason charge when he returned from an ill-fated expedition, Marlowe was tried for atheism and murdered in a tavern brawl, and Fawkes was hanged for his role in an unsuccessful plot to blow up King James I and the Parliament on November 5, 1605 (England observes Guy Fawkes Day each November 5).

Solving Inequalities in One Variable

Problems involving inequalities do not always come in a form as simple as $x < 2$. Frequently you will encounter expressions such as

$$2x + 1 > 3$$

You will then be asked to find the values of x that satisfy the inequality. In this example, if $x = 3$, we have

$$2 \cdot 3 + 1 > 3$$

$$6 + 1 > 3$$

$$7 > 3 \quad \text{True}$$

ON A TANGENT

Evariste Galois: The Dueling Mathematician

Evariste Galois was born near Paris in 1811 and died 20 years and 7 months later, from wounds received in a duel, in 1832. His short, tragic life and outstanding mathematical achievements make him one of the most romantic people in the history of mathematics.

He was the middle of three children born to well-educated parents. His childhood seems to have been happy. In 1823 he entered a boarding school in Paris. By 1827 he found that many textbooks did not meet his needs, so he began reading original sources. (This would be analogous—Galois was 15 when he did this—to your switching from reading this book or your high school algebra book to reading the publications of the leading research mathematicians in the world.) In the five years between then and his death in 1832, he progressed in mathematics at a rate that is among the top of all time. In 1829 (at age 17), one of his papers appeared in a major mathematical journal. His future looked bright.

From the middle of 1829 on, a number of setbacks hit Galois. He twice failed the entrance examination to the École Polytechnique. His father committed suicide. A paper he submitted for publication was not published, perhaps because the noted mathematician who was dealing with it lost it. Although he was admitted to another school, the École Normale, he was expelled a little over a year later for writing a letter berating the school's director. During this time he became politically active in the revolution against the Bourbon monarchy. He was arrested twice and jailed because of his political activity. He was killed in a duel about a month after his second release from confinement.

The cause of the duel has some uncertainty to it. It is known that Galois had a broken love affair. It appears that Galois may have felt a duty to protect her honor because of sorrows someone else had caused her. In letters he wrote to friends on the night before the duel, he mentioned an "infamous coquette."

A fictionalized account has him lured into the duel by a police agent as a way to dispose of a political activist. In any case, Galois dueled on May 30, 1832.

Evariste Galois

During the time of Galois' setbacks, he continued a vigorous program of mathematical activity dealing with solving equations. On the night before his duel, he scratched out a lengthy letter to a friend, presenting in a comprehensive manner his principal mathematical results. He hoped that "there will be men who will find it profitable to decipher all this mess." In the duel Galois was shot, and he died the next day. Mathematicians have been able to "decipher all this mess" and Galois' name lives on in a branch of algebra called "Galois theory."

But for $x = -2$, you get

$$2(-2) + 1 > 3$$

$$-4 + 1 > 3$$

$$-3 > 3 \qquad \text{False}$$

Formal methods of working with inequalities exist. They are very similar to methods of working with equations. You may add or subtract the same quantity to both sides of an inequality, and you may multiply or divide both sides of an inequality by the same *positive* number. Based on these properties, we can solve the inequality $2x + 1 > 3$ as follows:

$$2x + 1 > 3$$

$$2x > 2 \qquad \text{Subtract 1}$$

$$x > 1 \qquad \text{Divide by 2}$$

YOUR FORMULATION

What happens when we multiply or divide both sides of an inequality by a negative number?

a. Consider the true inequality $2 < 3$. Multiply both sides by -4. Insert the appropriate inequality symbol in the following: -8 _____ -12.

b. Consider the true inequality $-2 < 4$. Divide both sides by -2 and insert the appropriate inequality symbol in the following 1 _____ -2.

c. Based on the examples in parts a and b, conjecture about what happens when both sides of an inequality are multiplied or divided by a negative number.

d. Multiplying both sides of an inequality by a negative number, say, -4, can be regarded as a two-step process: First multiply by -1, then multiply by 4. Similarly, dividing both sides of an inequality by a negative number, say, -2, can be regarded as a two-step process: First multiply by -1, then divide by 2. Since we know the effect of multiplying or dividing both sides of an inequality by a positive number (what is the effect?), we will know the effect of multiplying or dividing both sides of an inequality by a negative number if we know the effect of multiplying both sides by -1. Picture numbers a and b on a number line, with $a < b$. In the number lines nearby, consider three cases: (i) a and b are both positive, (ii) a is negative and b is positive, and (iii) a and b are both negative. In the number line on the left, sketch in a and b. In the number line on the right, locate $(-1)a$ and $(-1)b$ for the choice of a and b on the left. What do you conclude about the relationship between $(-1)a$ and $(-1)b$?

e. Based on what you've found in part d, is your conjecture in part c correct? If so, state it again. If not, develop a correct statement.

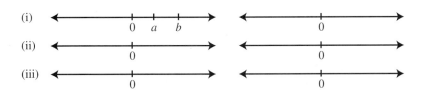

What you found in the Your Formulation is that if both sides of an inequality are multiplied by a number or divided by a number, the sense of the inequality remains unchanged only if the number is positive. When a negative number is involved, you must reverse the inequality. Thus,

$$-2x \le 6$$

when divided by -2 becomes $\qquad x \ge -3$

Here's another explanation why this works: We first use addition and subtraction properties as follows:

$$-2x \le 6$$

$$0 \le 6 + 2x \qquad \text{Add } 2x$$

$$-6 \le 2x \qquad \text{Subtract } 6$$

Now divide by 2 to obtain $\qquad -3 \le x$

which is the same as $\qquad x \ge -3$

The properties for solving inequalities are summarized in ● Table 5.6. [These properties could be combined into only three statements (see Written Assignment 8). However, listing them separately may aid students who are not familiar with solving inequalities.]

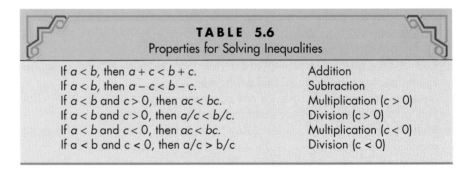

TABLE 5.6
Properties for Solving Inequalities

If $a < b$, then $a + c < b + c$.	Addition
If $a < b$, then $a - c < b - c$.	Subtraction
If $a < b$ and $c > 0$, then $ac < bc$.	Multiplication $(c > 0)$
If $a < b$ and $c > 0$, then $a/c < b/c$.	Division $(c > 0)$
If $a < b$ and $c < 0$, then $ac < bc$.	Multiplication $(c < 0)$
If $a < b$ and $c < 0$, then $a/c > b/c$	Division $(c < 0)$

The next examples show how these properties can help to solve inequalities.

●**EXAMPLE 5.29**

Solve the inequality $3x - 2 \le 7$.

SOLUTION Using properties from Table 5.6, we get

$$3x - 2 \le 7$$

$$3x \le 9 \qquad \text{Add } 2$$

$$x \le 3 \qquad \text{Divide by } 3$$

●**EXAMPLE 5.30**

Solve the inequality $5x + 4 > 2x + 10$.

SOLUTION This problem has terms containing the variable on both sides. It is usually easier to choose to add or subtract variables in such a way as to end up with a positive coefficient. With this in mind, we get

$$5x + 4 > 2x + 10$$

$$3x + 4 > 10 \qquad \text{Subtract } 2x$$

$$3x > 6 \qquad \text{Subtract } 4$$

$$x > 2 \qquad \text{Divide by } 3$$

EXAMPLE 5.31

Solve the inequality $-4x + 2 \leq 5$ by two different methods.

SOLUTION

a. We will leave $-4x$ on the left side. This yields

$$-4x + 2 \leq 5$$

$$-4x \leq 3 \qquad \text{Subtract } 2$$

$$x \geq -\frac{3}{4} \qquad \text{Divide by } -4$$

b. We will first add $4x$ to both sides to avoid having to divide by a negative number.

$$-4x + 2 \leq 5$$

$$2 \leq 5 + 4x \qquad \text{Add } 4x$$

$$-3 \leq 4x \qquad \text{Subtract } 5$$

$$-\frac{3}{4} \leq x \qquad \text{Divide by } 4$$

As you can see, both methods produce the same result.

Application

A teacher grades on a percentage basis according to the following scale:

100–90.0	A
89.9–80.0	B
79.9–70.0	C
69.9–60.0	D

a. Assume the grade in the course is based on four tests with equal weights. Ann's scores on tests 1, 2, and 3 are 94, 91, and 87, respectively. What can Ann get on the fourth test and still receive at least a B?

Let x be Ann's score on test 4. Ann wants

$$\frac{94 + 91 + 87 + x}{4} \geq 80.0$$

$$\frac{272 + x}{4} \geq 80.0$$

$$272 + x \geq 320.0 \qquad \text{Multiply by } 4$$

$$x \geq 48.0 \qquad \text{Subtract } 272$$

A score of 48 or higher will earn Ann at least a B.

b. Assume, again, that the grade in the course is based on four tests with equal weight. Brad has scores of 57, 48, and 72 on the first three tests, respectively. He wonders what he can do to get his grade up to a C.

Let x be Brad's grade on the fourth test. He wants

$$\frac{57 + 48 + 72 + x}{4} \geq 70.0$$

$$\frac{177 + x}{4} \geq 70.0$$

$$177 + x \geq 280.0 \qquad \text{Multiply by 4}$$

$$x \geq 103.0 \qquad \text{Subtract 177}$$

To get a C, Brad needs to repeat the course.

c. Suppose the grade in the course is based on four exams, with the fourth exam having twice the weight of the other three. In other words, the fourth exam score is counted twice as two of five scores. Now what does Ann need to earn, after scores of 94, 91, and 87 on the first three exams, to get at least a B?

Let x be Ann's score on test 4.

$$\frac{94 + 91 + 87 + 2x}{5} \geq 80.0$$

$$272 + 2x \geq 400.0 \qquad \text{Multiply by 5}$$

$$2x \geq 128.0 \qquad \text{Subtract 272}$$

$$x \geq 64.0 \qquad \text{Divide by 2}$$

Ann needs a score of 64 or higher.

Linear Inequalities in Two Variables

A linear equation in two variables is an equation such as $2x + 3y = 4$. In general, a linear equation in two variables is an equation of the form $ax + by = c$, where a, b, and c are real numbers. Similarly, a *linear inequality in two variables* is an inequality of the form

$$ax + by \leq c$$

Whereas, the graph of a linear equation in two variables is a line, the graph of a linear inequality in two variables is a half-plane (points on one side of a line).

Linear equalities and inequalities are useful in a fairly recent area of mathematical activity called *linear programming*. Linear programming is a method of solving problems that require a maximum or minimum value to be found in which the variables satisfy several conditions defined by linear equations or linear inequalities. Linear programming is used by large businesses, government agencies, nutritional and medical researchers, and numerous others.

The following is an example of the kind of problem that can be solved with linear programming. A bicycle manufacturing company produces two kinds of bicycles, a mountain bike and a 10-speed. Production of the mountain bike requires 2 hours on machine A and 4 hours on machine B. Production of the 10-speed bicycle requires 3 hours on machine A and 2 hours on machine B. Machine A can operate at most 12 hours per day; machine B can operate at most 16 hours per day. If the manufacturing company makes a profit of $12 on a mountain bike and $10 on

Emmy Noether: Premier Woman Mathematician

One of the people to establish significant results in the area of abstract algebra was Emmy Noether (1882–1935). She was perhaps the most creative woman mathematician of all time. She succeeded in her mathematics research in the face of two cultural factors of her time that made success—in fact, existence—difficult: She was female and she was Jewish.

She grew up in the town of Erlangen in southern Germany and graduated from a girls' high school. Three years later, in 1900, she passed a test to qualify as a teacher of French and English. About this time she became interested in mathematics and in pursuing university studies. However, most universities at the time did not admit women students. In 1900 she was one of two women allowed to audit, but not take for credit, classes at the Univesity of Erlangen. In 1904 the University of Erlangen permitted women to register and she became a regular student. In 1908 she received her doctorate, specializing in abstract algebra.

She lived with her family in Erlangen for seven years until in 1915 David Hilbert, the most influential mathematician of the time, invited her to the University of Göttingen (Germany), the leading university in mathematics at that time. Although as a woman she was not allowed to teach courses or to receive a salary, Hilbert arranged for her to teach courses given under his name. Hilbert also spoke to the university senate in favor of appointing her to a position of *Privatdozent:* "I do not see that the sex of the candidate is an argument against her admission as *Privatdozent.* After all, the Senate is not a bathhouse." Hilbert had much more influence in the mathematical community than he had on Noether's appointment. It was not until changes in Germany after World War I that she finally became a *Privatdozent* at Göttingen in 1919.

Noether continued to teach at Göttingen until 1933, when the Nazis were in power and attacking both Jews and nonconforming mathematicians. In that year the university dismissed her. The Nazi machine effectively dismantled the premier mathematical institution in the world of the time and caused the locus of mathematical activity to be redistributed around the world. Noether accepted a visiting professorship at Bryn Mawr College in Pennsylvania. She died suddenly in 1935, after only about two years in the United States.

Her work through the years in abstract algebra established her as one of the significant mathematicians of the twentieth century and as probably the principal woman mathematician of all time.

Emmy Noether

a 10-speed, how many of each should be manufactured each day in order to maximize profit?

Although this problem has only two variables (the number of mountain bikes and the number of 10-speed bikes), real-world linear programming problems often have hundreds or thousands of variables and hundreds or thousands of linear equations or inequalities. Research continues on the effectiveness of known methods of linear programming and the development of new methods. At the heart of these developments are ideas from algebra regarding the solving of linear equations.

Abstract Algebra

In showing that for any equation of degree five or higher there is no formula analogous to the quadratic formula (see earlier in this section), Abel and Galois developed methods of argument that were applied elsewhere. Their methods joined with ideas from geometry and a part of mathematics called number theory in the first half of the nineteenth century to open the doors to a new direction in algebra called abstract (or modern) algebra. It involves dealing with a collection of objects, like numbers, sets, logic statements, or polynomials, and operations, like addition or multiplication, on those objects. Abstract algebra has many subbranches and is an area of active mathematical research.

EXERCISES 5.4

1. For each of the following values of x, state whether the inequality $3x - 1 < 4$ is true or false.
   a. -4? b. -3 c. 1 d. 4/3 e. 5/3 f. $\sqrt{3}$ g. 3
2. For each of the following values of x, state whether the inequality $5x + 2 \geq 6$ is true or false.
   a. -6 b. -5 c. 4/5 d. 5/4 e. $\sqrt{2}$
3. In each of the following, place the appropriate symbol: $<$, $\leq$, $>$, or $\geq$. If the symbol depends on the value of c, explain why.
   a. $2 + c$ _____ $5 + c$
   b. $c + 7$ _____ $c + 4$
   c. If $a < b$, then $a - 3$ _____ $b - 3$.
   d. If $a > b$ then $a/4$ _____ $b/4$.
   e. $3c$ _____ $8c$
   f. $(1/2)c$ _____ $(1/3)c$
   g. If $a < b$, then $-2a$ _____ $-2b$.
   h. If $a < b$, then $a/-3$ _____ $b/-3$.

In Exercises 4–13, solve the inequality.

4. $x + 2 < 3$
5. $x - 3 \leq 7$
6. $2x \leq 6$
7. $-3x > 9$
8. $x/2 > 3$
9. $x/-4 < 3$
10. $2x - 3 > 4$
11. $3x - 2 \leq 6x + 4$
12. $2x + 4 < 2x + 7$
13. Rework Example 5.30 by first subtracting $5x$ from each side of the inequality.

14. In the "Application" subsection, under the assumptions of part a, what does Ann need to score on test 4 to earn an A?
15. In the "Application" subsection, under the assumption of part c, can Brad now earn a C after grades of 57, 48, and 72 on the first three tests? If so, what does he need to score on the fourth test?
16. Carol has earned 30 credits in college with a grade point average (GPA) of 1.90. She plans to take 6 credits in summer school. What does she need to average for those 6 credits in order to pull her GPA up to a 2.00?
17. A tank contains 10 gallons of water and 1 gallon of antifreeze. How much antifreeze needs to be added so that the ratio of antifreeze to water is at least 1 to 4?
18. A high school miler wants to run a 5-minute mile in his next race. That's an average of 12 miles per hour. He figures he'll start out slowly and finish with a strong kick. Accordingly, he runs the first half-mile at 10 miles per hour. How fast must he run the last half-mile so that his time for the race is 5 minutes or less?
19. The first few terms of the Fibonacci sequence that we listed are 1, 1, 2, 3, 5, 8, 21, 34. What is the next term?
20. In solving Fibonacci's Rabbit Problem, the last two terms of interest are 233 and 377. What is the next term? In other words, how many pairs of rabbits would there be after 13 months?
21. Give an example of a linear inequality in two variables.
22. Give an example of an inequality in two variables that is not a linear inequality.

1. In our world we deal with inequality more than equality; for example, you likely are taller than or shorter than your teacher rather than the same height, and you likely are taking more credits or fewer credits than your neighbor rather than the same number. Because of this, should the mathematics taught in high school spend more time working with inequalities than it currently does? Why or why not?

2. (*Library research*) Write a two- to five-page paper on either al-Khwārizmī, Omar Khayyam, Fibonacci, Recorde, or Harriot. Consult at least two references. Books listed in the bibliography are good starting points.

3. Recorde first used our symbol for "=" and Harriot introduced our symbols for "<" and ">." Invent an alternate symbol for use in place of =, <, or >. Describe the merits of your symbol.

4. The *Liber abbaci* was written to introduce Hindu-Arabic numerals, a place-value system, to people accustomed to Roman numerals, which is not a place-system.
   a. How does this compare with a child learning to count and do arithmetic? Is replacing the old system harder because old techniques have to be put aside, or is it easier because familiarity with one system of numeration is already known? Support your answer.
   b. Another analogy is that it is like replacing our foot-pound-quart system of measurement with the metric system. Either amplify this analogy or state another analogy and amplify it.

5. Bring to class a pineapple, a pine cone, or some other object occuring in nature that is related to the Fibonacci sequence. Show the class the relationship.

6. (*Library research*) Write a one- or two-page paper describing a place where the Fibonacci sequence occurs, other than the places discussed in this section.

7. (*Library research*) Look through one or more issues of the *Fibonacci Quarterly* and find something related to the Fibonacci sequence that you can understand. Write a one- to three-page description of what you found. Give a specific bibliographic reference. Your description should be understandable to people in your class.

8. Table 5.6 lists six properties for solving inequalities. Consolidate this to three properties.

9. This exercise is for a written debate. Some people will do part a, others will do part b.
   a. Argue why you think solving linear inequalities is more difficult than solving linear equations.
   b. Take someone's answer to part a and refute it.

10. Suppose x denotes a person's height, in inches, at age 2 and y denotes the person's adult height in inches. Translate the inequality $y \le 2x + 3$ into words.

11. What is so big about linear things? Your algebra course has material on solving linear equations, the application of linear equations, and graphing linear equations; part of this section talks about solving linear inequalities; and another part of this section discusses linear inequalities in two variables. Why has so much space and time been spent on *linear* equations and inequalities?

12. Outline (in outline form) the history of efforts to solve polynomial equations.

13. Construct a time line showing efforts to solve polynomial equations.

14. (*Library research*) Report on the history of developments in linear programming.

15. (*Library research*) Describe two real-world applications of linear programming. (Airlines and AT&T are users of linear programming.)

16. Algebra is very old, while one of its subfields, linear programming, is relatively new. Draw a time line listing some major events in the development of algebra.

CHAPTER 5 REVIEW EXERCISES

1. Write a paragraph describing what algebra is.

2. a. Distinguish between rhetorical algebra and symbolic algebra.
   b. Which of these two types was used first?
   c. How far back in time do we have evidence of the first type? How recent is the second type?

3. Find a prime number larger than 80.

4. Describe how algebra can be regarded as generalized arithmetic.

5. Add: $\dfrac{2x - 3}{x + 2} + \dfrac{x + 4}{5x - 1}$

6. Simplify: $\dfrac{(x^2 - x - 6)(x + 4)}{x^2 + x - 12}$

7. Carry out the division and simplify: $\dfrac{x^2 - 4}{x^2 + 2x - 3} \div \dfrac{x^2 - 6x + 8}{x^2 - x - 12}$

8. a. Give an example of a linear equation in one variable.
   b. Give an example of an equation in one variable that is not a linear equation.

9. A store advertised 20% off fashion watches. One watch had a sale price of $36.76. What was the regular price?

10. A utility company charges $.36059 per hundred cubic feet of gas plus a $7.50 monthly service charge. If a person's monthly bill is for $45.00, how many hundred cubic feet of gas did the person use in that month?

11. **a.** Which of the following equations define y as a function of x?
   i. $3x + 2y = 6$
   ii. $x^2 - y = 0$
   iii. $y^2 - x = 0$

   b. Pick any one of the equations in part a that does *not* define y as a function of x, and explain why it does not.

12. If $f(x) = 3x^2 - x + 4$, find
   a. $f(2)$ **b.** $f(0)$

13. State an example of a function f such that $f(1) = 3$ and $f(2) = 6$.

14. Give an example of a function of more than one variable.

15. For a baseball pitcher, the earned-run average (ERA) is the number of earned runs per nine innings pitched. It is a function of earned runs allowed and innings pitched. If we let R denote earned runs allowed and IP denote innings pitched, then ERA = $R/(\text{IP}/9)$ or ERA = $9R/\text{IP}$.

   a. Suppose a pitcher has allowed 32 earned runs in 89 innings. What is his earned-run average? (It's typical to round the ERA to two decimal places.)

   b. If this pitcher pitches one more inning without allowing a run, what will his ERA be?

16. **a.** Calculators have a function key $\boxed{\sqrt{\ }}$ for the square root function.
   i. Is this a linear function?
   ii. Is this an exponential function?

   b. i. Is the function in Review Exercise 12 a linear function?
   ii. Is it an exponential function?

   c. i. Give an example of a linear function. Don't use the functions from part a or b.
   ii. Give an example of an exponential function. Don't use the functions from part a or b.

   d. Give an example of a linear inequality in two variables.

   e. Give an example of an inequality in two variables that is *not* a linear inequality.

17. Consider the function given by $y = 7x^2 - 6x - 1$.

   a. The graph of this function is a parabola. Does it open upward or downward?

   b. If you answered upward to part a, write an equation of a parabola that opens downward. If you answered downward to part a, write an equation of a parabola that opens upward.

   c. Find the vertex of this parabola.

   d. Sketch the graph of this parabola.

   e. Find, if they exist, the real zeros of this function.

   f. i. What is the value of the discriminant?
   ii. What does this value tell you?

18. A swim club has 70 family memberships, each of which pays $400 a year. The club is considering what to charge for next year. One of the officers believes that for each additional $5 charged, the club will lose one family.

   a. What should the club charge to get maximum revenue?

   b. What is this maximum revenue?

19. Let $f(x) = 3.27^x$ and $g(x) = (0.327)^x$.

   a. Which of the following graphs is the graph of f?

   b. Which of the following graphs is the graph of g?

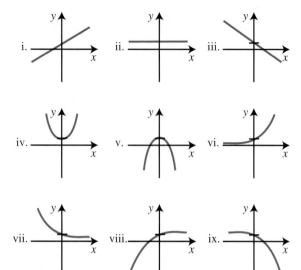

i.   ii.   iii.

iv.   v.   vi.

vii.   viii.   ix.

20. Calculate the following values.
   a. 3^4 **b.** 4^3 **c.** $(-3)^4$ **d.** $(-4)^3$
   e. -2^5 **f.** 2^{-5} **g.** $(1/2)^3$ **h.** $(1/2)^{-3}$

21. Plot points and sketch the graph of the function $y = 3^{-x}$.

22. In developing the formula for the amount present when a principal is invested for a period of time at compound interest, we encountered the following expression: $[P(1 + i/m)^2] + [P(1 + i/m)^2] \cdot i/m$. Write this expression in a simplified form.

23. **a.** If $2000 is invested for 4 years at 6% compounded monthly, how much will be available?

   b. How much should be invested at 6% compounded monthly in order to have $3000 at the end of 4 years?

   c. If an amount P is invested at an interest rate i compounded continuously for n years, the amount available is $A = Pe^{in}$. If $2000 is invested at 6% compounded continuously for 4 years, how much will be available?

24. If a car costs $20,000 today and inflation is 3% per year, what will a comparable car cost in 25 years?

25. A report issued by the World Bank in the mid-1990s said that the world's urban populations totaled 1.4 billion people in 1990 and are growing by 3.8% a year. If that growth rate continues, what will be the size of the world's urban populations in 2010?

26. What is the value of e to the nearest tenth?

27. Describe a real-world situation that uses the number e.

28. For carbon-11 the formula for radioactive decay is $Q = Q_0 e^{-.035t}$, where Q is the amount present after t minutes and Q_0 is the original amount present.

   a. If 3.00 g are present originally, how much will be present after 2 minutes?

   b. Which is the best estimate of the half-life of carbon-11?
   $\frac{1}{2}$ minute,   2 minutes,   11 minutes,   20 minutes

29. In the formula $Q = Q_0 e^{-.035t}$ (see Review Exercise 28), _____ is an exponential function of _____. (Fill in the blanks with Q, Q_0, e, or t.)

30. On June 8, 1994, a magnitude 8.2 earthquake occurred in Bolivia. This was how many times as powerful as the North-

ridge, California, earthquake of January 17, 1994, which measured 6.7 on the Richter scale?

31. Briefly describe the history of algebra from the ninth century through the sixteenth century. Include a description of the cultural conditions that influenced the development, the nature of the problems to which algebra was applied, the geographical locations of the development, and at least two people and their contributions.

32. Sketch the graph of $x < -2.5$.

33. Write an inequality to describe the graph.

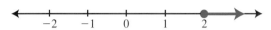

34. For each of the following values of x, state whether the inequality $2x + 3 \leq 6$ is true or false.
    a. -2 b. 0 c. $2/3$
    d. $3/2$ e. 2

35. In each of the following, place the appropriate symbol: $<$, $\leq$, $>$, $\geq$. If the symbol depends on the value of c, explain.
    a. $c + 2$ _____ $c + 3$
    b. $c - 2$ _____ $c - 3$
    c. $2c$ _____ $3c$
    d. $c/2$ _____ $c/3$

36. Solve the following inequalities and graph the solution.
    a. $3x - 2 > 10$
    b. $-3x + 4 \geq 10$

37. David has earned 44 credits in college with a grade point average of 1.860. He is taking 16 credits this term and needs an overall GPA (for 60 credits) of 2.000 at the end of the term.
    a. What does his GPA this term need to be to pull his overall GPA up to at least a 2.000?
    b. In 13 credits this term he has a 2.000 GPA. What does he need in the other 3-credit class to make his overall GPA at the end of 60 credits at least 2.000?

38. Use the following notation. In each case the equation has one variable.
    i. Solve a general first-degree (linear) equation.
    ii. Solve a general second-degree (quadratic) equation.
    iii. Solve a general third-degree equation with a formula like the quadratic formula.
    iv. Solve a general fourth-degree equation with a formula like the quadratic formula.
    v. Solve a general fifth-degree equation with a formula like the quadratic formula.

    Which ones of these five things could be done at the following times?
    a. Time of Jesus Christ
    b. Europe in the "Dark Ages" (ca. 950)
    c. George Washington president of the United States (1789–1797)
    d. The Cake Walk becomes the most fashionable dance (1900)
    e. Today

BIBLIOGRAPHY

Bolt, B. A. *Earthquakes.* New York: W. H. Freeman, 1988.

Boyer, Carl B. (revised by Uta C. Merzbach). *A History of Mathematics,* 2nd ed. New York: Wiley, 1991.

Calinger, Ronald (ed.) *Classics of Mathematics.* Englewood Cliffs, NJ: Prentice Hall, 1995.

Dunham, William. *Journey Through Genius: The Great Theorems of Mathematics.* New York: Wiley, 1990.

Eves, Howard. *An Introduction to the History of Mathematics,* 6th ed. Chicago: Saunders College, 1990.

Katz, Victor J. *A History of Mathematics.* New York: HarperCollins College, 1993.

National Council of Teachers of Mathematics (NCTM). *Historical Topics for the Mathematics Classroom.* Washington, DC: NCTM, 1969.

Oberhofer, E. S. "Different Magnitude Differences," *The Physics Teacher,* May 1991, pp. 273–274.

Stillwell, John. *Mathematics and Its History.* New York: Springer-Verlag, 1989.

van der Waerden, B. L. *A History of Algebra.* New York: Springer-Verlag, 1985.

CHAPTER 6

FUNDAMENTALS OF GEOMETRY

PROLOGUE

The poet Edna St. Vincent Millay wrote in a letter dated September 11, 1920:

> Ten years I have been forgetting all I learned so lovingly about music, and just because I am a boob. All that remains is Bach. I find that I never lose Bach. I don't know why I have always loved him so. Except that he is so pure, so relentless and incorruptible, like a principal [sic] of geometry.

Do you think of geometry as being pure, relentless, and incorruptible? In what ways does it have these properties? It's appropriate for a poet—more than, say, a scientist or an engineer—to think of it this way. However, geometry is more often associated with art than with poetry. Art by its very nature is geometric, and artists, in varying degrees, have embraced ideas from geometry. Millay admires the composer Bach because he is pure, relentless, and incorruptible. But those properties are there for

215

poets, artists, musicians, composers, and everyone in between.

Millay goes on in her letter:

Did you know I had written a sonnet to Euclid? Does it strike you as funny? It isn't funny, really. Unless, perhaps, I am funny,—which is just possible.

Her sonnet to Euclid first appeared in May 1920.

Euclid alone has looked on Beauty bare.
Let all who prate of Beauty hold their peace,
And lay them prone upon the earth and cease
To ponder on themselves, the while they stare
At nothing, intricately drawn nowhere
In shapes of shifting lineage; let geese
Gabble and hiss, but heroes seek release
From dusty bondage into luminous air.
O blinding hour, O holy, terrible day,
When first the shaft into his vision shone
Of light anatomized! Euclid alone
Has looked on Beauty bare. Fortunate they
Who, though once only and then but far away,
Have heard her massive sandal set on stone.

Are you one of the fortunate ones who, even from far away, has heard Beauty's massive sandal set on stone? Listen for it in this chapter.

Chapter opening photo: A snow crystal, pure and beautiful.

6.1 OVERVIEW AND HISTORY

GOALS
1. To outline briefly the origins of geometry.
2. To distinguish between the ancient Egyptian and ancient Greek approaches to geometry.
3. To name some ancient Greek contributors to geometry and to describe something each did.
4. To associate Euclid and the *Elements* and describe their influence.

The beginning of informal geometry cannot be dated. Elements of distance, size, shape, and volume have probably always been a part of human observation and reasoning. The shortest distance between two places is along a straight line. A string drawn taut forms part of a line. Circles occur as the outlines of the sun and moon. The heads of many flowers are circular. The rings created by a pebble tossed into water are circular. Many fruits and pebbles are spherical. Eggs and other pebbles have regular shapes. All of these geometric notions must have been observed informally before ever being approached formally.

The beginning of a formal approach to geometry is credited to the Egyptians in the Nile Valley, dating back to around 3000 B.C.E. Their geometry can be thought of as applied geometry because they developed what was necessary to fill their needs. In particular, they developed the geometry necessary for measuring and building.

Geometry is derived from the root words *geo-*, meaning "earth," and *metron*, meaning "measure." As early as 1300 B.C.E., Egyptians were taxed according to how much land they owned. Thus, if a parcel of land was shaped and subdivided as shown in ▶ Figure 6.1, it was important to be able to determine exactly how much

THE FAR SIDE By GARY LARSON

"Now watch your step, Osborne. ... The Squiggly Line people have an inherent distrust for all smoothliners."

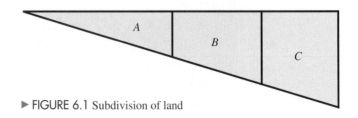

▶ FIGURE 6.1 Subdivision of land

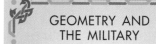

ETYMOLOGY

Geometry From Greek *geo-*, "earth," of unknown prior origin, and *metron*, "a measure." The Indo-European root is *me-*, "to measure." Thus, geometry must originally have dealt with measuring land. Although geometry gradually grew more abstract, people assumed until the beginning of the nineteenth century that the axioms and postulates of geometry naturally corresponded to the physical world as they knew it on Earth. In modern terms, however, geometry need have no physical referent at all.

GEOMETRY AND THE MILITARY

Although the ancient Greeks are thought of more for their deductive approach to geometry and the ancient Egyptians more for their applied approach to geometry, Plato, in the *Republic*, Book 7, speaks of military advantages of knowing geometry:

> It is plain that that part of geometry which bears upon strategy does concern us. For in pitching camps, or in occupying positions, or in closing or extending the lines of an army, and in all the other maneuvers of an army whether in battle or on the march, it will make a great difference to a general, whether he is a geometrician or not.

land was owned by each person. The presence of pyramids, dating back to approximately 2600 B.C.E., is evidence of Egyptian application of three-dimensional geometry. Reasonably sophisticated use of geometric concepts must have been employed during their construction.

The Greek contribution to geometry came a bit later and was significantly different in nature. The Greeks attempted to develop geometry as a deductive system. This was consistent with their knowledge of logic at that time. An early major contributor was Thales of Miletus (ca. 600 B.C.E.). His work was continued by Pythagoras (ca. 550 B.C.E.), after whom the well-known Pythagorean theorem is named. Plato (ca. 400 B.C.E.) was the next great Greek contributor. His famous school in Athens heavily emphasized knowledge of geometry. The motto over the entrance to his Academy boldly stated, "Let no one unversed in geometry enter here."

Many consider the work of Euclid (ca. 300 B.C.E.) to be the capstone of Greek contributions to geometry. He created a 13-volume set of books, the *Elements*, that provided the basis for study of plane geometry as an axiomatic system. His work is

Pyramids in Egypt, evidence of geometry applied over 4,000 years ago.

MORE ON EUCLID'S *ELEMENTS*

Euclid's *Elements* is not devoted exclusively to geometry, and its results are not the original thinking of Euclid. It contains quite a bit of number theory and elementary algebra (of a geometric nature). Most of the results were derived by earlier writers. Euclid may have created some of the proofs of these results and he may have been the first to state some of the results. However, his main contribution was the careful organization of results (theorems) in geometry that flow from one to another in a tightly knit, logical sequence.

the forerunner and basis of plane Euclidean geometry, which is currently taught in most high schools (over 2000 years after Euclid wrote the *Elements*). No other single work, except the *Bible,* has been as widely translated, published, and studied. Perhaps no other work has had as much influence on scientific thinking.

When we think of the origins of a deductive approach to geometry, we think of the ancient Greeks. Conversely, when we think of the ancient Greeks, we often think of their—especially Euclid's—development of geometry. The classical Greek period began around 600 B.C.E. and included the Pythagoreans (see Section 2.4) as progenitors. Euclid lived at about the same time as the famous philosophers Plato and Aristotle. These, along with others, emphasized the power of human reason. In doing so, they created several parts of current intellectual thought. Morris Kline puts it strongly (Kline, 1962, p. 15):

> The Greeks not only made finished products out of the raw materials imported from Egypt and Babylonia, but they created totally new branches of culture. Philosophy, pure and applied sciences, political thought and institutions, historical writings, almost all our literary forms (except fictional prose), and new ideals such as the freedom of the individual are wholly Greek contributions.

About 330 B.C.E. Alexander the Great conquered Greece and surrounding areas. He admired Greek traditions and built several Greek cities. One of these was Alexandria, Egypt, which became the center of mathematical activity in the West for the next seven centuries, a period of time known as the Hellenistic Age. During this time Alexandria came under control of the Roman Empire.

Just as the Pythagoreans played a significant role in the beginning of Greek civilization, a mathematician was part of an event that some describe as marking the end of learning in the ancient world. Hypatia was an Egyptian philosopher, mathematician, and astronomer who was known as a respected teacher. She is the first woman mathematician about whom much is known.

Hypatia was born in about 370. Her father was the mathematician Theon. She authored at least three books, one dealing with geometry, one with number theory, and one with astronomy.

The event by which she is best known in popular histories is her death. Charles Kingsley in 1881 wrote a historical novel about her entitled *Hypatia or New Foes with an Old Face.* In the Preface he warned his readers: "A picture of life in the fifth century must needs contain much which will be painful to any reader, and which the young and innocent will do well to leave altogether unread." Although modern scholarship disputes the accounts of her death put forward by Kingsley and some other older sources, all agree that she was brutally murdered in about 415 by a Christian lynch mob. Because of religious-cultural conflicts, so ended the life of the most significant woman mathematician up to that time.

WRITTEN ASSIGNMENTS 6.1

1. Write more about the origins or early history of geometry. You might focus on one of the following areas.
   a. Subconscious geometry (See the article by Eves in National Council of Teachers of Mathematics, 1969.)
   b. The ancient Babylonians
   c. The ancient Egyptians
   d. The Pythagoreans
   e. The development of geometric concepts in Africa (See Zaslavsky, 1973.)
   f. The development of geometric ideas in the Orient (See Capsule 62 in National Council of Teachers of Mathematics, 1969.)
   g. The development of geometric ideas in Mesoamerica (See Closs, 1986, Ch. 13.)

2. Write a two- or three-page paper on one of the following Greek contributors to geometry: Thales of Miletus, Pythagoras, Plato, Euclid.

3. Write more about the interaction of culture and mathematics in the death of Hypatia. (See Boyer, 1991; Calinger, 1995; Dzielska, 1995; Gittleman, 1975; and Katz, 1993.)

TERMINOLOGY AND NOTATION

GOALS
1. To work with lines, rays, and line segments.
2. To describe an angle as the union of two rays.
3. To know conditions that describe a plane.
4. To describe half-planes, parallel planes, and intersections of planes.

> The great book of Nature lies ever open before our eyes and the true philosophy is written in it. . . . But we cannot read it unless we have first learned the language and the characters in which it is written. . . . It is written in mathematical language and the characters are triangles, circles, and other geometrical figures. —Galileo

Geometry utilizes three basic undefined terms: *point*, *line*, and *plane*. We will attempt to give the reader an intuitive feel for what each of these terms represents. However, it is important to note that none of them will be defined by reference to more basic elements. (Why we start with *undefined* terms will be addressed in Section 7.1.)

One way to look at geometry is as a formally constructed discipline. Starting with the undefined terms *point, line,* and *plane* and a few accepted statements, it is possible to create a very complex system. In this chapter, we give some sense of the formalism involved in a careful development of geometry. However, because we believe it is more important for those not specializing in mathematics to have a general, intuitive feel for geometry, we will minimize formalism. In either case—formal or intuitive—certain terms and notation are essential for a discussion of the discipline. We now look at these.

Points

A *point,* geometry's first undefined term, is often represented by a dot. A dot is a two-dimensional representation of a point, which has no dimensions (no length, width, or height). It is important to recognize that dots are merely *representations* of a concept. Points (actually dots) are typically labeled with capital letters, as illustrated in ▶ Figure 6.2.

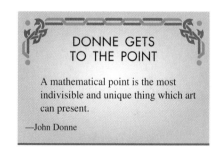

DONNE GETS TO THE POINT

A mathematical point is the most indivisible and unique thing which art can present.

—John Donne

•
A

•
B

•
C

▶ FIGURE 6.2 Representation of points

Lines

A *line,* the second of geometry's undefined terms, is a set of points formed by the intersection of two flat surfaces, as illustrated in ▶ Figure 6.3. For example, this sheet of paper and the following sheet intersect in a line; two flat walls of a room intersect in a vertical line in the corner of the room.

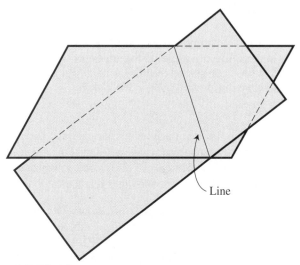

Line

▶ FIGURE 6.3 A line as the intersection of two flat surfaces

A line is one-dimensional. Any line can be uniquely represented by any two points on the line. Conversely, any two distinct points uniquely determine a line. If points A and B are on a line l, the line can be denoted by l, by $\overleftrightarrow{AB}$, or by $\overleftrightarrow{BA}$, as in ▶ Figure 6.4. Variations of lines include half-lines, rays, and line segments, which may or may not include endpoints. ● Table 6.1 summarizes these terms and their respective notations.

TABLE 6.1 Points and Lines: Terms and Notation		
Term	Illustration	Notation
Point	•	A
Line	←•————————•→ A $\quad\quad$ B	$\overleftrightarrow{AB}$
Half-line	○————————•→ A $\quad\quad$ B	$\overset{\circ\rightarrow}{AB}$
Ray	•————————•→ A $\quad\quad$ B	$\overset{\bullet\rightarrow}{AB}$
Line segment	•————————• A $\quad\quad$ B	$\overset{\bullet\bullet}{AB}$
Open line segment	○————————○ A $\quad\quad$ B	$\overset{\circ\text{—}\circ}{AB}$

It is possible to combine the set theory operations of union and intersection with parts of a line. For example, if points A and C are on line l, as shown in ▶ Figure 6.5, the intersection of rays $\overrightarrow{CA}$ and $\overrightarrow{AC}$ is the line segment $\overline{AC}$, as seen in ▶ Figure 6.6. This intersection can be written as $\overrightarrow{CA} \cap \overrightarrow{AC}$. The following example provides practice in writing the union and intersection of various parts of a line.

line l line $\overleftrightarrow{AB}$ line $\overleftrightarrow{BA}$

▶ FIGURE 6.4 Ways to denote a line

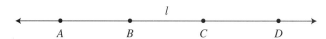

▶ FIGURE 6.5 Line l with points A, C

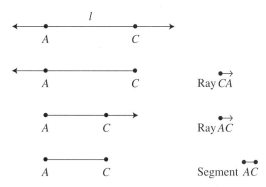

▶ FIGURE 6.6 Intersection of rays

EXAMPLE 6.1

Given the line l with points A, B, C, and D as shown in ▶ Figure 6.7, find each of the following.

a. $\overrightarrow{AC} \cap \overrightarrow{BC}$ b. $\overrightarrow{DB} \cup \overrightarrow{AD}$

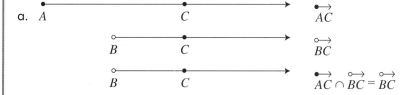

▶ FIGURE 6.7 Line l with points A, B, C, D

SOLUTION Finding the answers here may be aided by sketching first.

a.

In this case, the half-line $\overset{\circ}{\overrightarrow{BC}}$ is a subset of ray $\overrightarrow{AC}$. Therefore, the intersection is the subset $\overset{\circ}{\overrightarrow{BC}}$.

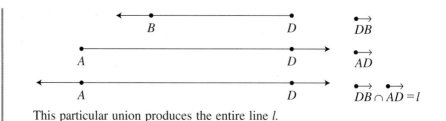

This particular union produces the entire line *l*.

Additional practice with problems of this type is provided in the exercises at the end of this section.

Angles

An *angle* is the union of two rays with a common endpoint. To illustrate, rays $\overleftrightarrow{BA}$ and $\overleftrightarrow{BC}$ form the angle $\angle ABC$, as shown in ▶ Figure 6.8.

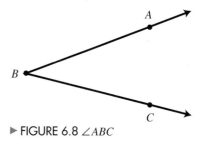

▶ FIGURE 6.8 $\angle ABC$

The common endpoint is the *vertex* of the angle, and each ray is a *side* of the angle. Note that an angle consists only of points on the two rays that form the angle; it does not contain any points "inside" or "outside" the angle. A more detailed treatment of angles follows in the next section.

Planes

Typically, we characterize a *plane,* geometry's third undefined term, as a flat surface. For example, the floor is part of a plane and the chalkboard is part of a plane. Just as lines can be extended infinitely far in either direction, planes have no boundaries. Whereas, a point has no dimensions and a line is one-dimensional, a plane is two-dimensional. A plane can be represented by any three *noncollinear* points (not on a line) in that plane. Conversely, any three noncollinear points uniquely determine a plane. A typical representation is shown in ▶ Figure 6.9.

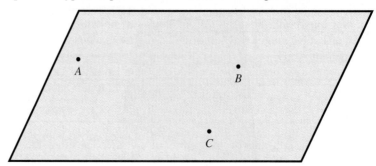

▶ FIGURE 6.9 Three noncollinear points determine a plane

YOUR FORMULATION

A point is determined by two intersecting lines. A line is determined by two points or by two intersecting planes. A plane is determined by three noncollinear points.

a. What are two other ways to determine a plane?
b. A camera tripod or a three-legged stool illustrates a plane determined by three noncollinear points. For each way to determine a plane that you found in part a, give an example to illustrate it.

The three-point determination of a plane is very useful. For example, camera tripods are always stable because the three contact points between the legs and the resting surface uniquely determine a plane. Contrast this with a four-legged table or chair, which may be unstable. The bottoms of three of the legs determine a plane. The table or chair will be stable when the bottom of the fourth leg is exactly in the plane determined by the bottom of the other three legs.

Two planes that do not intersect are called *parallel planes.* Two planes that do intersect (but are not the same plane) form a line of intersection, as shown previously in Figure 6.3.

Just as a point on a line can create two half-lines, a line in a plane creates two half-planes, as shown in ▶ Figure 6.10.

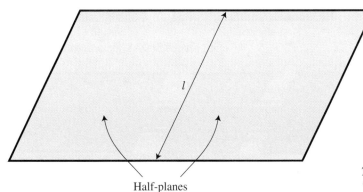

Half-planes

The Dutch artist M. C. Escher called this 1954 woodcut "Three Intersecting Planes." Can you find them?

▶ FIGURE 6.10 Half-planes

Two different lines, l_1 and l_2, in a plane may intersect in a common point A. If they do not intersect, the lines are parallel. These cases are shown in ▶ Figure 6.11.

The study of figures in a plane is called Euclidean (plane) geometry and is the basis of the typical high school course in geometry.

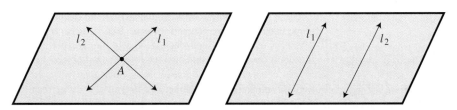

▶ FIGURE 6.11 Lines in a plane

1. Must ray $\overrightarrow{AB}$ go from point A to the right? If so, why? If not, give an example to illustrate.

2. Given line l with points A, B, C, D, as shown, find each of the following.

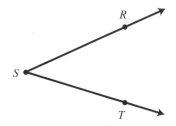

   a. $\overset{\circ\rightarrow}{AC} \cap \overset{\bullet\rightarrow}{BD}$

   b. $\overset{\bullet\rightarrow}{DC} \cup \overset{\circ\rightarrow}{AB}$

   c. $\overset{\circ\rightarrow}{BA} \cup \overset{\bullet\rightarrow}{BD}$

   d. $\overset{\circ\rightarrow}{BA} \cup \overset{\circ\rightarrow}{BD}$

   e. $\overset{\bullet\rightarrow}{AB} \cap \overset{\bullet\rightarrow}{BC}$

   f. $\overset{\circ\rightarrow}{DA} \cap \overset{\bullet\bullet}{AB}$

3. $\angle RST$ is the union of what two rays?

4. The caption for Figure 6.9 says that three noncollinear points determine a plane. Is the word *noncollinear* necessary? Why or why not?

5. Consider the plane containing your paper. Draw a line on this plane so that one half-plane contains the upper left-hand corner of your paper and the other half-plane contains the lower right-hand corner.

6. Describe portions of a building that are contained in each of the following:

   a. Two parallel planes

   b. Two intersecting planes

WRITTEN ASSIGNMENTS 6.2

1. (This exercise is for those who took geometry in high school.) In this section, points, lines, and planes have been described. What is different in the way they are described here from the way they were described to you in high school? (We're still talking about the same objects.)

2. (This exercise is for those who did *not* take geometry in high school.) Why did you choose not to take geometry in high school? Was it avoidance of all mathematics or avoidance of geometry in particular? In retrospect, was it a good choice? Why or why not?

3. Compare and contrast your own experience and what you have heard others say comparing algebra and geometry. Is algebra hard and geometry easy, or vice versa?

4. Do you think there is a gender difference in ability to do algebra versus ability to do geometry? Do you believe that females do better in algebra while males do better in geometry, or do you believe the reverse of this, or do you believe there is no gender difference? Offer rationale and evidence to support your position.

5. This section was developed with point, line, and plane as undefined terms. Can you do better than that? That is, can you get by with only two of these as undefined and the third defined in terms of the other two? Try to do this, then criticize what you have written.

6. The fact that three noncollinear points determine a plane explains why three-legged objects like camera tripods are always stable, why four-legged objects like chairs and tables sometimes are not stable, and why two-legged objects like banners connected to two poles are very unstable. Why, then, do birds and animals have either two or four legs rather than three legs?

6.3 ● ANGLES

GOALS
1. To understand the following terms associated with angles: *vertex, sides, interior, exterior, vertical, adjacent, acute, obtuse, right, complement, supplement, transversal, alternate interior, alternate exterior,* and *corresponding.*
2. Given an angle, to estimate its measure and, given a measure, to construct an angle with that measure.
3. To know the relations between measures of complementary angles, supplementary angles, vertical angles, alternate interior angles, alternate exterior angles, and corresponding angles.

Terminology and Notation

In the last section, we learned that an angle is the union of two rays with a common endpoint. If rays $\overrightarrow{AB}$ and $\overrightarrow{AC}$ form an angle as shown in ▶ Figure 6.12, the angle is denoted ∠*BAC* or ∠*CAB*. This notation requires only that the common endpoint be listed in the middle. In Figure 6.12, the common endpoint, *A,* is the *vertex* of the angle. Rays $\overrightarrow{AB}$ and $\overrightarrow{AC}$ are the *sides* of the angle. Points between the sides (extended) of an angle are on the *interior* of the angle. Those outside the extended sides are on the *exterior* of the angle.

Much of the study of angles in plane geometry is based on angles formed by intersecting lines. When two lines intersect as shown in ▶ Figure 6.13, angles on the same side of one line are called *adjacent angles.* Thus, ∠*AOB* and ∠*BOC* are

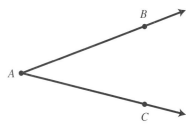

▶ FIGURE 6.12 Rays form an angle

A Tinker Toy construction has many angles.

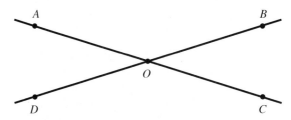

▶ FIGURE 6.13 Angles formed by intersecting lines

adjacent angles but ∠AOB and ∠COD are not adjacent angles. Angles formed by two intersecting lines with only the vertex in common are called *vertical angles.* In Figure 6.13, ∠AOB and ∠COD are vertical angles, as are ∠AOD and ∠BOC. Adjacent and vertical angles are not unique. An interesting exercise is to name as many pairs of such angles as possible from two intersecting lines.

Measure of an Angle

Measuring angles is different from measuring distances on a line. The "distance" between sides of an angle has no meaning because you could get different distances depending on which points on the sides were chosen for your measurement. Thus, a different method of measurement is necessary for angles.

A standard method for measuring angles is to think of one side as the *initial side* and then to determine how much rotation would be necessary to reach the other side, referred to as the *terminal side.* The amount of rotation can be measured in degrees, radians, or gradients. We restrict our measurements to degrees. A complete revolution is 360 degrees. If you envision a circle divided into 360 identical "wedges," each wedge represents 1 degree. An angle whose measure is approximately 30 degrees is shown in ▶ Figure 6.14. The word *degrees* is frequently represented by a small circle placed above and to the right of the number. Thus, 30 degrees could be written as 30°.

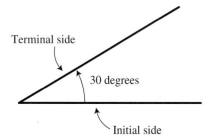

▶ FIGURE 6.14 An angle of approximately 30 degrees

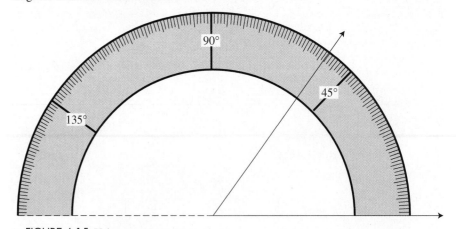

▶ FIGURE 6.15 Using a protractor

Angles can be measured with protractors, as shown in ▶ Figure 6.15. The protractor shows that this angle has a measure of approximately 53 degrees.

The measure of an angle is denoted m ∠. For example, if ∠BAC had a measure of 63°, we would write m ∠BAC = 63°. It is fairly standard to use a shorthand symbol to denote the measure of an angle. Thus, a letter variable could replace the notation m∠BAC. For instance, if θ (theta) were the letter variable, we would write θ = 63°.

Subdivisions of degrees do exist. In particular, 1 degree can be divided into 60 minutes. Each minute can be further subdivided into 60 seconds. Notations for these are ′ and ″, respectively. Thus, you could encounter an angle measurement such as 26°36′17″, which is read as "26 degrees, 36 minutes, and 17 seconds." Calculators may report angle measure in tenths or hundredths of a degree, such as 45.8°. We will restrict our discussion simply to working with whole numbers of degrees.

Angles may be classified according to their measure. Angles whose measure is between 0° and 90° are *acute angles.* Angles of measure 90° are *right angles.* Angles with measure between 90° and 180° are *obtuse angles.* Finally, angles of measure 180° are *straight angles.* This information is summarized in ● Table 6.2.

Right angles, which played a significant role in the development of geometry and, later, trigonometry, are formed when two lines are perpendicular. Right angles are typically denoted as shown in ▶ Figure 6.16.

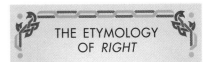
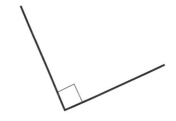

▶ FIGURE 6.16 Right angles

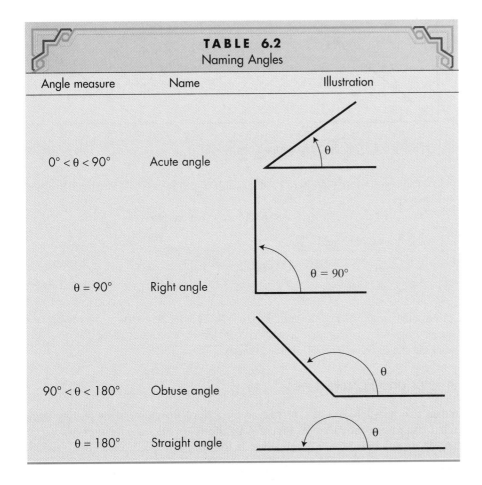

TABLE 6.2
Naming Angles

Angle measure	Name	Illustration
$0° < \theta < 90°$	Acute angle	θ
$\theta = 90°$	Right angle	$\theta = 90°$
$90° < \theta < 180°$	Obtuse angle	θ
$\theta = 180°$	Straight angle	θ

Special Angles

There are two special angle relationships worthy of mention. Two angles are called *complements* if the sum of their measures is 90°. They may also be referred to as *complementary angles*. Two angles are called *supplements (supplementary)* if the sum of their measures is 180°. When the measure of one angle is known, finding its complement or supplement is just a matter of subtraction. This is shown in the following example.

EXAMPLE 6.2

Find both the complement and supplement of each of the following two angles.
a. $m\angle BAC = 37°$
b. $\theta = 45°$

SOLUTION

a. We need to subtract to find complements and supplements. The complement of 37° is
$$90° - 37° = 53° \qquad \text{(complement of 37°)}$$

Similarly, the supplement of 37° is found to be 143°. It is helpful to envision complements and supplements as shown in ▶ Figure 6.17.

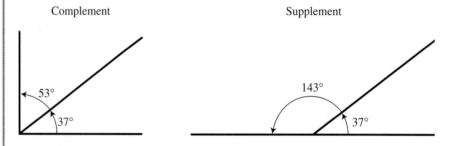

▶ FIGURE 6.17 Complement and supplement of 37°

b. The solution method here is identical to that for part a: The complement of 45° is
$$90° - 45° = 45° \qquad \text{(complement)}$$

and the supplement is
$$180° - 45° = 135° \qquad \text{(supplement)}$$

Although finding specific complements and supplements is fairly routine, the simple properties of complementary and supplementary angles are relied on extensively in the development of plane geometry.

Angles and Parallel Lines

When two parallel lines are intersected by a third line, several interesting angle relationships are produced. ▶ Figure 6.18 illustrates such a situation, where the angles are numbered for easy reference. In this diagram, l_1 is parallel to l_2, and l is

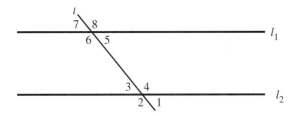

▶ FIGURE 6.18 Parallel lines intersected

the third line. The line that intersects the parallel lines is called a *transversal. Alternate interior angles* are angles formed by the intersections of line l with lines l_1 and l_2 such that they are between lines l_1 and l_2 and on opposite sides of l. In Figure 6.18, angles 6 and 4 are alternate interior angles, as are angles 5 and 3. Alternate interior angles occur in pairs. *Alternate exterior angles* are analogously defined. In Figure 6.18, one such pair would be angles 8 and 2, while the other such pair would be angles 7 and 1.

It can be shown that when l_1 is parallel to l_2, alternate interior angles have equal measure and alternate exterior angles have equal measure. This result, when combined with what we know about complementary and supplementary angles, means that we only have to know the measure of one of the eight labeled angles of Figure 6.18 to determine the measure of all of the other seven angles. To illustrate, if m∠6 = 120°, then we can conclude the following:

m∠8 = 120°	because ∠6 and ∠8 are vertical angles
m∠4 = 120°	because ∠6 and ∠4 are alternate interior angles
m∠2 = 120°	because ∠4 and ∠2 are vertical angles (also notice that ∠8 and ∠2 are alternate exterior angles)
m∠5 = 60°	because ∠6 and ∠5 are supplementary angles
m∠7 = 60°	because ∠5 and ∠7 are vertical angles
m∠3 = 60°	because ∠5 and ∠3 are alternate interior angles
m∠1 = 60°	because ∠3 and ∠1 are vertical angles (also, ∠7 and ∠1 are alternate exterior angles)

One last relationship should be noted. Angles on the same side of a transversal in the same relative positions are *corresponding angles*. In Figure 6.18, ∠5 and ∠1 are corresponding angles. Other pairs of corresponding angles are:

$$∠4 \text{ and } ∠8, \qquad ∠7 \text{ and } ∠3, \qquad ∠6 \text{ and } ∠2$$

 YOUR FORMULATION

1. Consider ∠4 and ∠8 in Figure 6.18.
   a. m∠4 = m∠6 because _____.
   b. m∠6 = m∠8 because _____.
   Therefore, m∠4 = m∠8 because they both equal m∠6.
2. Consider ∠7 and ∠3 in Figure 6.18. Give an argument like that in part 1 to show that m∠7 = m∠3.
3. In a similar manner we can show that m ∠6 = m∠2. What do these three parts show?

1. Which of the the following are alternate notations for ∠RST in the diagram?

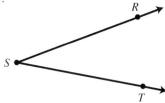

   i. ∠RTS   ii. ∠SRT   iii. ∠TSR

2. Consider ∠RST in Exercise 1.
   a. What is its vertex?
   b. What are its sides?
   c. Is R in the interior of the angle?
   d. Is R in the exterior of the angle?
   e. Draw and label E, a point in the interior of the angle.
   f. Draw and label F, a point in the exterior of the angle.
   g. Consider the point represented by the dot on the "i" in "following" in Exercise 1. Is it in the interior of ∠RST, on ∠RST, or in the exterior of ∠RST?

3. Refer to Figure 6.13. For each of the following pairs of angles, describe them as (i) adjacent, (ii) vertical, (iii) neither adjacent nor vertical, (iv) both adjacent and vertical.
   a. ∠AOB, ∠BOC
   b. ∠AOB, ∠COD
   c. ∠AOB, ∠DOA
   d. ∠BOC, ∠COD

4. Which of the following terms apply to ∠RST in Exercise 1?
   i. Acute angle
   ii. Obtuse angle
   iii. Right angle
   iv. Straight angle

5. Given line l containing points O and X as shown, draw each of the following lines.

   a. $\overleftrightarrow{OA}$ so that ∠XOA is acute
   b. $\overleftrightarrow{OB}$ so that ∠XOB is obtuse
   c. $\overleftrightarrow{OC}$ so that ∠XOC is a right angle

6. Estimate the measure of each of the following angles in the diagram.

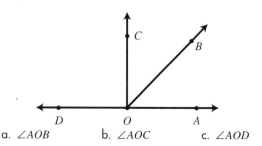

   a. ∠AOB   b. ∠AOC   c. ∠AOD

7. In each part, sketch an angle with the given measure.
   a. 30°   b. 60°   c. 180°

8. In Example 6.2a, ∠BAC and its complement are both acute. Must two complementary angles both be acute? Explain.

9. For each of the following angle measures, find both the measure of the complement and the measure of the supplement.
   a. m∠AOB = 10°   b. θ = 26°   c. m∠RST = 89°

10. In the following figure, l_1 is parallel to l_2 and eight angles are numbered.

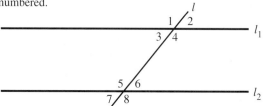

   a. What is line l called?
   b. Write the pairs of alternate interior angles.
   c. Write the pairs of alternate exterior angles.
   d. Write the pairs of corresponding angles.
   e. Let m∠2 = 45°. Find the measure of each of the other seven angles.

11. Assume two parallel lines are cut by a transversal to form eight angles, and assume the measure of one of the angles is $22\frac{1}{2}°$ (half of 45°). Sketch a picture of the three lines and the eight angles.

12. Explain how to use a protractor to find whether or not the following two lines are parallel. Do not use the protractor to extend the lines. (*Hint:* Draw a third line intersecting both of these lines.)

13. On a pool or billiard table, the measure of the angle formed when a ball hits a rail (a side of the table) equals the measure of the angle formed when it bounces off the table. Thus, on the following table ABCD, if the ball starts at P and hits a rail at Q and another rail at R, then m∠PQA = ∠BQR.
   a. If m∠PQA = 30°, what is m∠QRB? (Recall that the sum of the measures of the angles of a triangle is 180°.)
   b. What is m∠RSC?

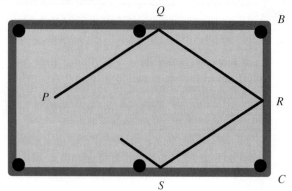

1. The terms *acute* and *obtuse* apply in other ways than in reference to angles. For example, a person might have acute appendicitis or a person's comments might be obtuse.
   a. Comment on the relationships between these uses of *acute* and *obtuse* and their geometric meanings.
   b. Are there other ways the words *acute* and *obtuse* are employed? Relate these to their geometric uses. (That same author's college professor who admonished students to sit right prefaced that comment by telling those who leaned over their desks or slouched down in their seats not to sit acute or obtuse.)
2. Degree measure of angles is based on a full revolution being 360°. Why 360? [Three references are the article "Angular Measure" in National Council of Teachers of Mathematics (1969), Boyer (1991), p. 162, and Eves, 1990.]
3. In addition to being measured in degrees, angles might also be measured in radians or gradients. Explain what each of these terms means. Illustrate by converting from each of these (degrees, radians, and gradients) to the other two.
4. In medical practice, angles can serve to describe the range of motion of elbow joints, knee joints, and hips. What are normal ranges of motion for these joints? Are there other medical uses of angles?
5. In Exercise 3, all answers were either adjacent or vertical. Must this always happen to the four angles formed by two intersecting lines? Why or why not?

6.4 POLYGONS

GOALS
1. To use terms applying to polygons in general and to triangles and quadrilaterals in particular.
2. To know and use the facts that the sum of the measures of the interior angles of a triangle is 180° and of a quadrilateral is 360° and to know how to find the sum of the measures of the interior angles of any polygon.
3. To know and use the Pythagorean theorem and to work with Pythagorean triples.
4. To work with similar triangles.

A significant part of geometry involves working with figures in a plane. You are probably aware of squares, rectangles, triangles and some other geometric figures. In this section, we define a general category of figures in a plane and then develop properties for some of the better-known ones.

Terminology

We will consider only figures formed by line segments, called polygons. A *polygon* is a figure formed when line segments enclose a region of a plane without crossing. Several polygons are shown in ▶ Figure 6.19. Line segments forming the polygon are *edges*. Their points of intersection are *vertices*. (The singular is *vertex*. It is inappropriate to say or write "vertexes" or to refer to a "vertice.")

We will further restrict our brief look at polygons to those that are convex. A *convex polygon* is one in which any two points on different edges can be joined by a line segment that does not intersect any other point of the polygon. Thus, in Figure 6.19, the polygons in parts a, b, and c are convex while those in parts d and e are not convex.

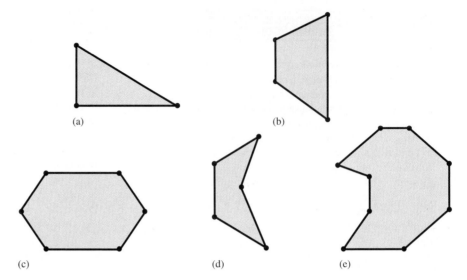

▶ FIGURE 6.19 Polygons

Angles in the enclosed region of a polygon are *interior angles*. If an edge of a polygon is extended, it forms an *exterior angle,* as shown in ▶ Figure 6.20.

A polygon is *regular* if all sides are of equal length and all interior angles have equal measure. At first glance, you may think stating both conditions is unnecessary. However, ▶ Figure 6.21 shows two different nonregular polygons: the one on the left has equal-length sides, the other one has equal-measure angles.

Polygons are classified according to the number of sides they have. For example, polygons with three sides are *triangles,* those with four sides are *quadrilaterals,* and those with five sides are *pentagons.* The names of other polygons are listed later in Table 6.6.

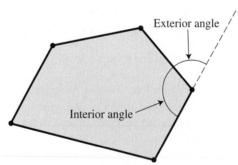

▶ FIGURE 6.20 Interior and exterior angles

▶ FIGURE 6.21 Nonregular polygons

Triangles

Let's now take a more detailed look at triangles. Recall, a triangle is a polygon with three sides. *The sum of the interior angles of any triangle is 180°.* This fact is often the basis for subsequent work with other polygons.

Triangles can be classified in either of two ways: by their edges or by their angles. ● Table 6.3 lists the three types of triangles that can be described by their edges. These are illustrated in ▶ Figure 6.22.

TABLE 6.3
Triangles as Classified by Their Edges (Sides)

Triangle Name	Description
Scalene	No two sides have equal length.
Isosceles	At least two sides have equal length.
Equilateral	All three sides have equal length.

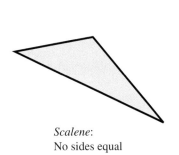
Scalene:
No sides equal

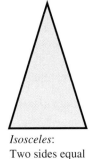
Isosceles:
Two sides equal

Equilateral:
Three sides equal

▶ FIGURE 6.22 Triangles classified by length of sides

Equal parts of a triangle are usually denoted by placing an equal number of small marks on the equal parts. For example, an isosceles triangle would have the equal sides denoted as shown in ▶ Figure 6.23a. Equal angle measures are similarly noted. Figure 6.23b shows a triangle with all three angles equal in measure.

Triangles as classified according to their interior angles are listed in ● Table 6.4 and illustrated in ▶ Figure 6.24. Note that the right angle in Figure 6.24b is indicated by a small square and that the side opposite the right angle is called the *hypotenuse.*

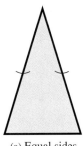
(a) Equal sides

TABLE 6.4
Triangles as Classified by Their Angles

Triangle Name	Description
Acute	All angle measures are less than 90°.
Right	One angle measure equals 90°.
Obtuse	One angle measure exceeds 90°.

(b) Equal angle measures

▶ FIGURE 6.23 Denoting equal triangle parts

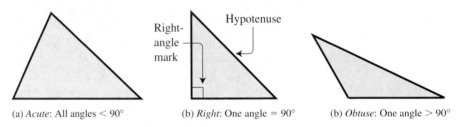

(a) *Acute*: All angles < 90°   (b) *Right*: One angle = 90°   (b) *Obtuse*: One angle > 90°

▶ FIGURE 6.24 Triangles classified by angle measure

The following example will give you some practice in identifying triangles.

EXAMPLE 6.3

Identify the type of each of the following triangles. Use both side and angle classification.

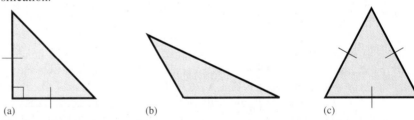

(a)   (b)   (c)

SOLUTION The triangle of part a is a right triangle, as indicated by the small square placed at the right angle. It is also an isosceles triangle, since two sides are equal in length.

The triangle of part b is an obtuse triangle. No two sides have equal measure, so it is also a scalene triangle.

The triangle of part c has all angle measures less than 90°, so it is an acute triangle. Since all three sides are equal in length, it is an equilateral triangle. Note that it is also isosceles, since two sides have equal length. However, it is customary to use the most descriptive term, which here is equilateral.

Perhaps the most widely known result concerning triangles is the Pythagorean theorem.

Geometry involves shapes. John Robinson's Genesis *uses interlocking triangles (two-dimensional objects) to produce a three-dimensional object.*

PYTHAGOREAN THEOREM

Let a, b, and c be the lengths of sides of a right triangle, with c representing the side opposite the right angle. Then

$$a^2 + b^2 = c^2$$

A right triangle with sides of length 3, 4, and 5 is shown in ▶ Figure 6.25. Note that, for this triangle, $3^2 + 4^2 = 5^2$.

The converse of the Pythagorean Theorem is also true.

CONVERSE OF PYTHAGOREAN THEOREM

Let a, b, and c be the lengths of three sides of a triangle, with c the longest. If $a^2 + b^2 = c^2$, then the triangle is a right triangle

---

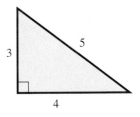

▶ FIGURE 6.25 A right triangle

The Pythagorean theorem and its converse have many applications. For example, if you have a known right triangle, it is only necessary to know the lengths of two sides in order to compute the length of the third side. A second standard application is to determine whether or not you have a right triangle. If all side lengths are known, you merely compute to see whether or not $a^2 + b^2 = c^2$. Carpenters occasionally employ this method to "square a corner." They measure 3 feet on one side, 4 feet on the other side, and then make sure the diagonal is 5 feet.

Integers that satisfy the relationship $a^2 + b^2 = c^2$ are called *pythagorean triples.* Some of them are 5, 12, 13; 7, 24, 25; 8, 15, 17; and 10, 24, 26.

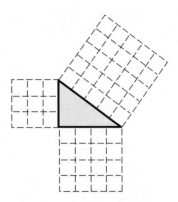

▶ FIGURE 6.26 Early form of Pythagorean theorem

One final comment about the Pythagorean theorem: Greek mathematics at the time of Pythagoras did not include the study of algebra as we know it. Thus, the Pythagorean theorem we study today $(a^2 + b^2 = c^2)$ would have been stated differently then. It undoubtedly would have discussed the area of the squares on the legs of the triangle. Such an interpretation is shown in ▶ Figure 6.26. Note that the sum of the areas for the two shorter legs equals the area for the hypotenuse. This is a geometric description of the Pythagorean theorem, but it is not a proof of it.

YOUR FORMULATION

It is generally believed that a proof Pythagoras might have given to his theorem was similar to the following argument: Use the right triangle with legs of length a and b and hypotenuse of length c to construct the accompanying square of side length $a + b$.

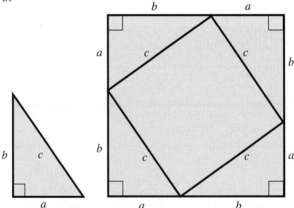

1. What is the shape of the middle figure with c on each side?
2. Why?
3. What is the area of the middle figure?
4. The area of the total figure is thus $c^2 + 4$ (area of triangle).

Now rearrange the pieces of the big square as follows:

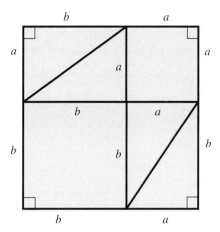

5. Now the area of the big square = _____ + 4 (area of triangle)
6. Equate the expressions for the area of the big square from parts 4 and 5 to get

$$c^2 + 4(\text{area of triangle}) = \underline{\hspace{1cm}} + 4 \text{ (area of triangle)}$$

Show that this gives $a^2 + b^2 = c^2$.

Two triangles are *similar* if and only if they have the same angle measures. This means they have the same shape but perhaps not the same size. Similar triangles are shown in ▶ Figure 6.27.

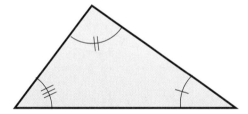

▶ FIGURE 6.27 Similar triangles

Although similar triangles do not necessarily have equal side lengths, it can be shown that *ratios* of corresponding sides are equal. To illustrate, consider ▶ Figure 6.28, which contains similar triangles with side lengths as indicated. The lengths of sides e and f can be found via ratios. Since the triangles are similar, we have $2/4 = 5/e$ as the equal ratios for corresponding sides. We solve this equation to find $e = 10$. In a similar manner, we find the length of side f: $2/4 = 6/f$, so $f = 12$.

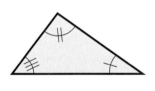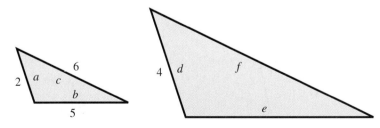

▶ FIGURE 6.28 Similar triangles with unequal side lengths

We have seen only the tip of an iceberg as far as the study of triangles is concerned. However, we will make a titanic (rather than a Titanic) effort to bypass the rest of the iceberg. We next consider some other polygons.

Quadrilaterals

A *quadrilateral* is a polygon with four sides. Several quadrilaterals are shown in ▶ Figure 6.29.

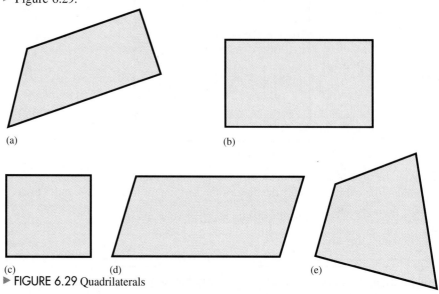

(a)

(b)

(c)　(d)　(e)

▶ FIGURE 6.29 Quadrilaterals

For any quadrilateral, the sum of the angle measures for interior angles is 360°. This can be demonstrated by means of a reasonably short geometric argument. Suppose a quadrilateral is labeled as follows:

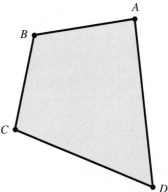

If you insert one diagonal—$\overleftrightarrow{BD}$, for example—you get the following:

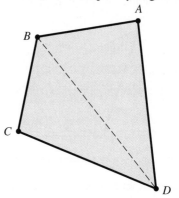

Thus, the sum of angle measures of the original quadrilateral is seen to be the sum of angle measures for $\triangle ABD$ and $\triangle BCD$. Since each of these is 180°, the total is 360°. This interesting geometric argument is extended a bit later to other polygons.

Descriptions of several special quadrilaterals are given in ● Table 6.5. Examples of each are shown in ▶ Figure 6.30.

TABLE 6.5
Special Quadrilaterals

Quadrilateral Name	Description
Square	All sides are of equal length and all angles are of equal measure (regular).
Rectangle	All angles are of equal measure (all right angles).
Parallelogram	Opposite sides are parallel.
Rhombus	All sides are of equal length.
Trapezoid	Only two sides are parallel.

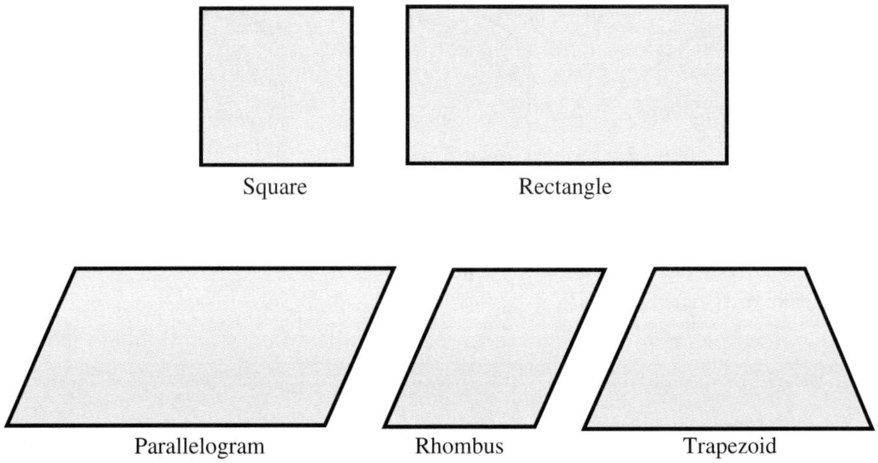

▶ FIGURE 6.30 Special quadrilaterals

Many quadrilaterals can be described by more than one name. For example, a square is also a rectangle, a parallelogram, and a rhombus. This ambiguity is partially avoided by using the most descriptive name available. Choosing the proper description is illustrated in the next example.

EXAMPLE 6.4

What is the most descriptive name for the quadrilateral shown?

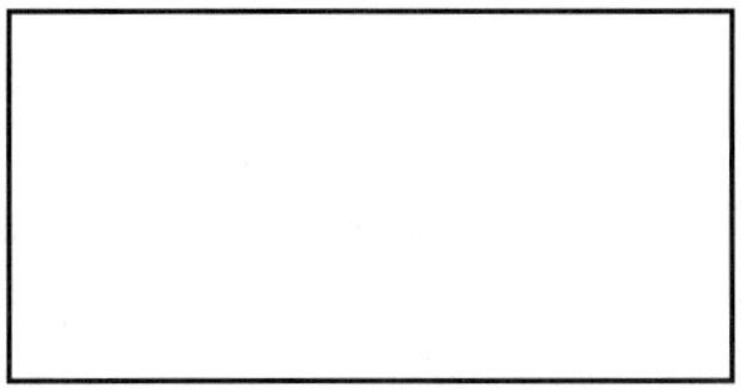

TABLE 6.6
Polygons

Number of Sides	Name
3	Triangle
4	Quadrilateral
5	Pentagon
6	Hexagon
7	Heptagon
8	Octagon
9	Nonagon
10	Decagon
11	Undecagon
12	Dodecagon
20	Icosagon

SOLUTION It is technically correct to refer to this quadrilateral as either a parallelogram or a rectangle. However, a rectangle has more conditions imposed on it than does a parallelogram. That is, opposite sides are parallel *and* angle measures are equal. Thus, the proper name is rectangle.

General Polygons

We conclude this section with a look at polygons in general. First, we note that polygons are named according to the number of sides. ● Table 6.6 lists several of these names, and ▶ Figure 6.31 contains drawings of some of them. Some of these polygons will be more familiar to you than others. For example, a regular pentagon is the shape of the Pentagon in Washington, D.C., (headquarters for the U.S. Department of Defense). A hexagon is a fairly standard shape for the exterior of nuts that can be threaded onto bolts. Open-end wrenches are made to accommodate hexagonal nuts. Other common appearances of polygon terms are requested in Written Assignment 1.

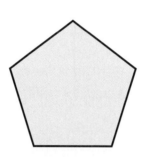
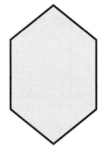
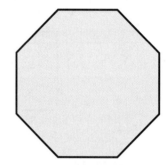

▶ FIGURE 6.31 Some polygons

The Pentagon

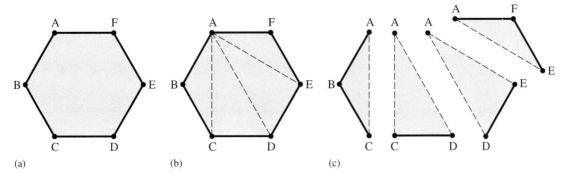

(a)   (b)   (c)

▶ FIGURE 6.32 Summing degrees for interior angles

The sum of the measures of interior angles for any polygon is not too difficult to find. To illustrate, suppose we have the hexagon in ▶ Figure 6.32a. Insert diagonals $\overleftrightarrow{AC}$, $\overleftrightarrow{AD}$, and $\overleftrightarrow{AE}$ as shown in Figure 6.32b. The original hexagon can now be visualized as four separate triangles, as in Figure 6.32c. Since each triangle has angle measures that total 180°, the angle measure sum for the hexagon is

$$180° + 180° + 180° + 180° = 4 \cdot 180° = 720°$$

This process of inserting diagonals in a polygon to form triangles is called *triangulating* the polygon. This method enables you to see that, if a polygon has n sides, it can be divided into $n - 2$ triangles. Since each of these contains 180°, the sum of angle measures of the original polygon is $(n - 2)\,180°$. Thus, we have the following general result.

The sum of angle measures for a polygon with n sides is $(n - 2)\,180°$.

We close this section with a brief look at two important properties of polygons: similarity and congruence.

Our earlier definition of similar triangles can be extended to polygons in general. Two polygons are *similar* if their corresponding angles have the same measure and their corresponding sides are in proportion. Two polygons are *congruent* if their corresponding angles have the same measure and their corresponding sides have the same length. This just means the polygons are identical in size and shape. The study of similar and congruent polygons currently constitutes much of a typical high school's first course in plane geometry.

EXERCISES 6.4

1. Refer to Figure 6.19.
   a. Which polygons are convex?
   b. Which polygons are regular?
   c. Which polygons are triangular?
   d. Which polygons are quadrilaterals?
   e. Which polygons are pentagons?

2. Draw a picture, if possible, of each of the following kinds of triangles.
   a. Scalene and acute
   b. Scalene and right
   c. Scalene and obtuse
   d. Isosceles and acute

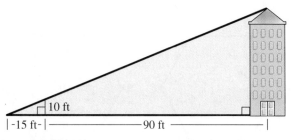

e. Isosceles and right
f. Isosceles and obtuse
g. Equilateral and acute
h. Equilateral and right
i. Equilateral and obtuse

3. For each of the following triangles, identify the type. Use both side and angle classification.

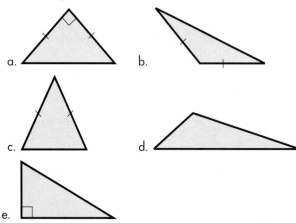

a.
b.
c.
d.
e.

4. Draw a Venn diagram relating scalene, isosceles, and equilatral triangles.

5. In the following right triangles, find the measure of the third side and identify the type of triangle, using both side and angle classification.

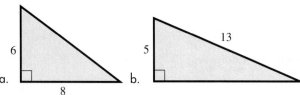

a.
b.

6. Which of the following are pythagorean triples?
a. 2, 3, 4   b. 3, 4, 5   c. 5, 12, 13

7. Assume the following triangles are similar, with side lengths as indicated. Find *a* and *b*.

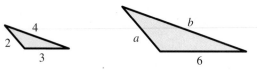

8. Assume the following triangles are similar, with side lengths as indicated. Find *e* and *f*.

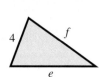 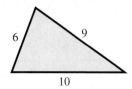

9. Given the information in the following diagram, estimate the height of the building.

10. A logo for an organization is framed by the following 3-4-5 right triangle. On a copy machine we enlarge this to an $8\frac{1}{2}'' \times 11''$ piece of paper, with the 3 side on the bottom and the 4 side on the left edge of the paper. What are the dimensions of the largest such triangle?

11. Find the most descriptive name for each of the following quadrilaterals.

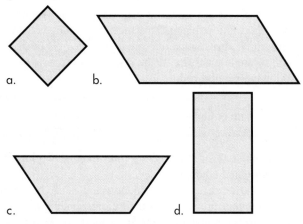

a.
b.
c.
d.

12. Draw a Venn diagram relating the set of squares, the set of rectangles, and the set of parallelograms.

13. Answer each of the following with always (A), sometimes but not always (S), or never (N).
a. When is a parallelogram a quadrilateral?
b. When is a parallelogram a rhombus?
c. When is a rhombus a parallelogram?
d. When is a rhombus a square?

14. What is the shape of a stop sign?

15. What is the shape of the Chrysler logo?

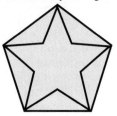

16. What is the sum of angle measures for each of the following?
    a. A pentagon
    b. An octagon
    c. A decagon
    d. A dodecagon
17. *Two polygons are similar if their corresponding angles have the same measure and their corresponding sides are in proportion.* Show that both parts of this definition are necessary by sketching each of the following.
    a. Two polygons in which corresponding angles have the same measure but in which corresponding sides are not in proportion

    b. Two polygons in which corresponding sides are in proportion but in which corresponding angles do not have the same measure
18. In the earlier examples of pythagorean triples (3, 4, 5; 5, 12, 13; 7, 24, 25; 8, 15, 17; and 10, 24, 26) either one or all three of the numbers are even.
    a. Could a pythagorean triple have no numbers even? That is, could all three of the numbers be odd? Either give an example or argue why not.
    b. Could a pythagorean triple have exactly two numbers even? Either give an example or argue why not.

WRITTEN ASSIGNMENTS 6.4

1. In this section two common occurrences of polygon terms are mentioned, the Pentagon in Washington and hexagonal nuts. State some others, and give, if necessary, a context for their use.
2. Recall or find some logos of companies or organizations that have polygonal shapes. Include a picture of each logo and a description of the polygonal shape.
3. This section introduces at least 30 terms describing polygons.
    a. Why are there duplicate terms, such as *equilateral triangle* for *regular triangle* and *square* for *regular quadrilateral?*
    b. Would you favor eliminating some terms? For example, we might say "five-sided polygon" instead of *pentagon,* or six-sided polygon" instead of hexagon. Why or why not?
4. The book *Flatland* by Edwin Abbott describes "people" who live in a two-dimensional world. Females are shaped like line segments and males are shaped like polygons.
    a. Write a report on the book.

    b. Some of the concepts in the book can be regarded as quite sexist. Do you regard them this way, or do you take them as satirical? Comment on your opinion.
5. The Pythagorean theorem is probably the most famous theorem in all of mathematics. Some would argue that it is also the most important theorem.
    a. Ask some other students to name some mathematical theorems they know. Do the responses support the claim that the Pythagorean theorem is the most famous theorem in mathematics?
    b. Ask some mathematicians if they believe the Pythagorean theorem to be the most important theorem in mathematics. Summarize all their responses, then add your own opinion based on their responses.
6. (*Library research*) In a "Your Formulation" is a proof of the Pythagorean theorem. Write another proof of the Pythagorean theorem.

6.5 PERIMETER AND AREA

GOALS
1. To use the term *perimeter* and to find the perimeter of a given polygon.
2. To know and use formulas for the perimeter and area of a triangle, of various quadrilaterals, and of a circle.
3. To use the terms *center, radius, diameter,* and *circumference of a circle* and to know the definition of π and its approximate value.
4. To compute the area of a figure made up of polygons and circles.

Early developments in geometry were based on the need to measure areas and volumes. For example, as early as 1300 B.C.E. Egyptians were taxed according to how much land they owned. The fifth century B.C.E. Greek historian Herodotus, whom the Roman orator Cicero called the Father of History, attributes the origins of geometry to this taxation.

They said also that this king divided the land among all Egyptians so as to give each one a quadrangle of equal size and to draw from each his revenues, by imposing a tax to be levied yearly. But every one from whose part the river tore away anything, had to go before him and notify what had happened. He then sent the overseers, who had to measure out by how much the land had become smaller, in order that the owner might pay on what was left, in proportion to the entire tax imposed. In this way, it appears to me, geometry originated, which passed thence to [Greece].

As you can imagine, this taxation based on land size required fairly accurate measurement and subsequent computation. Since measurement is still one of the most practical applications of geometry, we will consider how figures in a plane are "measured."

Perimeters of Polygons

The *perimeter* of a polygon is the sum of lengths of the sides of the polygon. More simply, the perimeter of a plane figure is the distance around the figure. When side lengths are known for a polygon, the perimeter is found by adding those lengths. For example, for the hexagon in ▶ Figure 6.33, the perimeter P is

$$P = 3 + 3 + 5 + 2 + 2 + 5 = 20$$

Measurement of the length of a side is in some basic unit, such as inches, feet, yards, miles, centimeters, or meters. Taking centimeters for the units in Figure 6.33, the perimeter would correctly be written as

$$P = 3 \text{ cm} + 3 \text{ cm} + 5 \text{ cm} + 2 \text{ cm} + 2 \text{ cm} + 5 \text{ cm} = 20 \text{ cm}$$

Because including the unit of measure during computation becomes tedious, solutions frequently express only the numeric values and then add the unit in the final answer. We will take this approach throughout this section.

Finding perimeters of some polygons can be shortened via formulas. Although use of these formulas is recommended, you should always understand why they are appropriate.

The first formula we recall is for the perimeter of a square. For the square in ▶ Figure 6.34 with side length s, the perimeter is

$$P = s + s + s + s = 4s$$

Thus, if you know the figure is a square, you need to know only the length of one side to compute its perimeter.

The second formula is for the perimeter of a rectangle. Suppose a rectangle has side lengths as shown in ▶ Figure 6.35, where l represents the length and w represents the width. The perimeter is

$$P = l + w + l + w = 2l + 2w$$

This same formula can be applied to any parallelogram, since opposite sides have equal length.

Other formulas could be developed, but too much formalism could hamper understanding at this point. Rather than memorize formulas, just remember that the perimeter is the distance around a figure and then add accordingly. The next example provides some practice in finding perimeters.

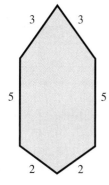

▶ FIGURE 6.33 A hexagon

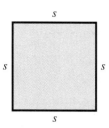

▶ FIGURE 6.34 A square

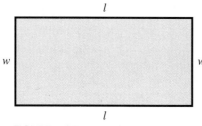

▶ FIGURE 6.35 A rectangle

EXAMPLE 6.5

Find the perimeter for each of the following figures.

a. Square

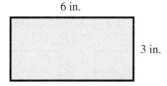

5 in.

b. Rectangle

6 in.

3 in.

c. Parallelogram + square

6 cm

Square

3 cm

Parallelogram

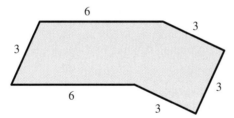

SOLUTION Figure a is a square. Since $P = 4s$, we get

$$P = 4(5) = 20 \text{ in.}$$

Figure b is a rectangle. Thus, $P = 2l + 2w$. Since $l = 6$ in. and $w = 3$ in.,

$$P = 2(6) + 2(3) = 12 + 6 = 18 \text{ in.}$$

Figure c is a parallelogram with a square adjoined to one end. This figure can be redrawn and labeled as follows:

6

3

3

6

3

3

We then add the side lengths to obtain

$$P = 3 + 6 + 3 + 3 + 3 + 6 = 24 \text{ cm}$$

2 ft

3 ft

▶ FIGURE 6.36 A rectangle

Areas of Polygons

The area of a polygon is a measure of how many unit squares could be placed in the polygon. For example, for the rectangle of ▶ Figure 6.36, whose side lengths are 3 ft and 2 ft, a unit square would be a square 1 ft × 1 ft. ▶ Figure 6.37 illustrates how six of these squares could be placed into the rectangle so that the entire rectangle is filled.

Unit square

2 ft

3 ft

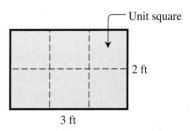

▶ FIGURE 6.37 Unit squares in a rectangle

TABLE 6.7	
Formulas for Finding the Area of Selected Polygons	
Polygon	Area Formula
Square	s^2
Rectangle	$l \cdot w$
Triangle	$\frac{1}{2}b \cdot h$
Parallelogram	$b \cdot h$
Trapezoid	$\frac{1}{2}(b_1 + b_2) \cdot h$

Dimensions for units of area are square feet, square inches, square centimeters, and so on. These are written with a small 2 above and to the right of the unit. Thus, six square feet is 6 ft^2 and 17 square centimeters is 17 cm^2. For the rectangle of Figure 6.37, we would write

$$A = 3 \times 2 = 6 \text{ ft}^2$$

There are formulas for finding the areas of some polygons. A list of these is given in ● Table 6.7. We now see how some of these formulas are developed.

Let's first consider a square whose side length is 5, as shown in ▶ Figure 6.38a. Figure 6.38b shows how 25 unit squares could be placed into the original square.

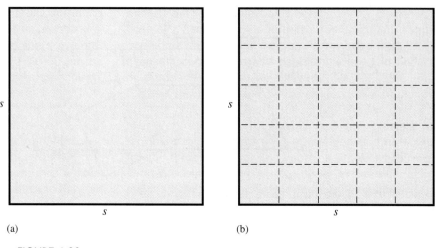

(a)   (b)

▶ FIGURE 6.38 A square

Thus, we have
$$s = 5$$
$$A = 5 \times 5 = 25$$

The area is 25 square units.

This process of finding the area of a square generalizes. That is, for a square of side length s, the area is

$$A = s^2$$

The formula for the area of a rectangle is developed similarly. For the rectangle of Figure 6.37 we merely multiplied the length by the width to get the area. In general, for a rectangle of length l and width w we have

$$A = l \times w$$

Practical examples of finding area usually are a bit more involved than just finding the product of two numbers. A typical problem is shown in the following example.

●EXAMPLE 6.6

You decide to carpet a room whose dimensions are 18 ft × 24 ft. The local carpet store is advertising the carpet you want for $14.95 a yard plus $1.95 a yard for pad and installation. [Carpeting is sold at price (such as $14.95) "a yard," but this is actually price per *square* yard.] How much will it cost you to have the room carpeted?

SOLUTION First you must be able to compute area in square yards. Since 1 yard = 3 feet, you may think of the room as shown in ▶ Figure 6.39. The area of the room, in square yards, is

$$A = 6 \times 8 = 48 \text{ yd}^2$$

We now compute the cost of the carpet. The carpet plus labor plus pad is $16.90 per square yard. Thus, the total cost c is

$$c = 48 \times \$16.90 = \$811.20$$

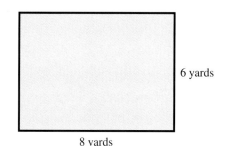

6 yards

8 yards

▶ FIGURE 6.39 Room dimensions, in yards

Formulas for the area of other polygons are not as obvious as those for squares and rectangles.

Let's next consider the triangle shown in ▶ Figure 6.40. The base has been labeled b and the altitude h. We now place this triangle in a rectangle with length b and width h, as shown in ▶ Figure 6.41.

Next, envision the rectangle separated into two rectangles, as shown in ▶ Figure 6.42. Note that the area of the portion of the triangle contained in each smaller rectangle is exactly one-half the total area. This means the total area of the triangle is one-half the area of the large rectangle. Since the area of the rectangle is $A = bh$, the area of the triangle is half that.

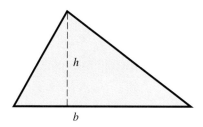

▶ FIGURE 6.40 A triangle

> ### AREA OF TRIANGLE
>
> $$A = \frac{1}{2}bh$$

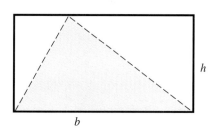

▶ FIGURE 6.41 A triangle in a rectangle

The area of a parallelogram can also be found via geometric argument. Take the parallelogram shown in ▶ Figure 6.43, with base b and altitude h. If a vertical cut were made at the corner, a triangle would be formed, as shown in ▶ Figure 6.44a.

> ### AREA OF PARALLELOGRAM
>
> $$A = bh$$

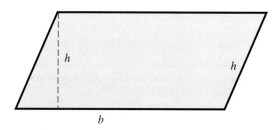

▶ FIGURE 6.43 A parallelogram

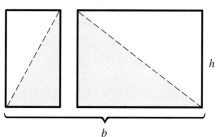

▶ FIGURE 6.42 Separated rectangles

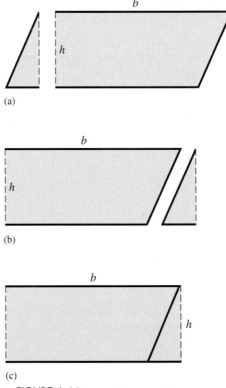

(a)

(b)

(c)

▶ FIGURE 6.44 Separating a parallelogram

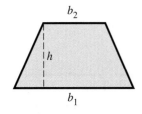

▶ FIGURE 6.45 A trapezoid

Moving this triangle to the other end of the original parallelogram produces the sketch shown in Figure 6.44b. Finally, adjoining this triangle produces the rectangle shown in Figure 6.44c. The area of this is then $A = bh$. Thus, the area of a parallelogram is the product of its base and height.

A *trapezoid* is a quadrilateral with a pair of opposite sides parallel, as shown in ▶ Figure 6.45. The method for finding the area of a trapezoid is similar to that for finding the area of a parallelogram. A regular trapezoid is a trapezoid such as pictured in Figure 6.45 in which the two base angles (formed with b_1) have the same measure and the two top angles (formed with b_2) have the same measure. By cutting a triangle off one end of a regular trapezoid and putting it on the other end, the following can be seen.

AREA OF A REGULAR TRAPEZOID

$$A = \frac{1}{2}(b_1 + b_2) \cdot h$$

This same formula holds for trapezoids that are not regular.

The following example provides practice in finding the areas of several polygons.

EXAMPLE 6.7

Find the area of each of the following:

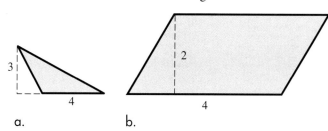
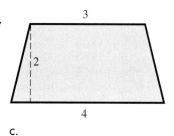

a.   b.   c.

SOLUTION We will use the formulas just developed.

a. The polygon is a triangle, so the area is

$$A = \frac{1}{2}(bh) = \frac{1}{2}(5)(3) = \frac{15}{2} = 7\frac{1}{2}$$

The area is $7\frac{1}{2}$ square units.

b. The polygon is a parallelogram with a base of 4 units and an altitude of 2 units. Using the formula $A = bh$ we get

$$A = 4 \cdot 2 = 8$$

The area is 8 square units.

c. The polygon is a trapezoid with bases $b_1 = 4$, $b_2 = 3$, and an altitude $h = 2$. We use the formula $A = \frac{1}{2}(b_1 + b_2) \cdot h$ to obtain an area of

$$A = \frac{1}{2}(4 + 3) \cdot 2$$

$$A = \frac{1}{2}(7) \cdot 2 = 7$$

Circles

The previous two subsections were concerned with polygons, that is, plane figures formed by line segments. We now look at the perimeter and the area for circles, beginning with some terminology.

▶ Figure 6.46 contains a circle with various parts labeled for reference. A *circle* consists of the set of points equidistant from a fixed point. The fixed point is the *center* of the circle. The distance from the center to points on the circle is the *radius*. The *diameter* is the length of a line segment between two points on the circle drawn through the center. More generally, a *chord* is any line segment between points on a circle. Notice that a diameter is the longest chord.

The perimeter of a circle is called the *circumference* of the circle. Since a circle does not consist of line segments, finding the circumference is different from what was done with polygons. As you might imagine, the circumference of a circle varies with the diameter of the circle. Early work with circles established the fact that the ratio of the circumference to the diameter was a constant. That is

$$\frac{\text{Circumference}}{\text{Diameter}} = \text{Constant}$$

This constant has been given the name pi (π).

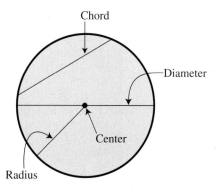

▶ FIGURE 6.46 A circle

6.5 PERIMETER AND AREA

249

ON A TANGENT

π Through the Ages

Since π is an irrational number, it is an infinite, nonrepeating decimal. Through the years, people have approximated it with increasing accuracy. Here is a chronology.

ca. 1650 B.C.E.	The Rhind mathematical papyrus from Egypt has π = 256/81 ≈ 3.16.
ca. 240 B.C.E.	First scientific attempt to calculate π was made by Archimedes, using a circle with inscribed and circumscribed polygons: 223/71 < π < 22/7. To two decimal places this gives π = 3.14.
ca. 150 C.E.	Claudius Ptolemy of Alexandria in his *Almagest,* the greatest ancient Greek work on astronomy, gives π as 377/120, which to three decimal places is 3.142.
ca. 480	Tsu Ch'ung-chih of China approximated π as 355/113, which to seven decimal places is 3.1415929. Notice that this approximation is accurate to six decimal places.
ca. 530	Aryabheta, a Hindu mathematician, approximated π as 62,832/20,000, which as a decimal is exactly 3.1416. (Although this is not as good as Tsu Ch'ung-chih's approximation, the two calculations come out of different and relatively independent civilizations.)
1579	French mathematician Francois Viete found π correct to nine decimal places by using polygons with $6 \cdot 2^{16} = 393,216$ sides.
1593	Adriaen von Roomen of the Netherlands found π correct to 15 decimal places by using polygons having 2^{30} sides.
1610	Ludolph van Ceulen of Germany found π correct to 35 decimal places by using polygons having 2^{62} sides.
1853	William Rutherford of England computed π correct to 400 decimal places.
1873	William Shanks, also of England, calculated π to 707 decimal places, but errors were later found.
1949	ENIAC, one of the first computers, calculated π to 2037 decimal places.
1959	In Paris, Francois Genuys found π to 16,167 decimal places on an IBM 704 computer.
1961	J. W. Wrench, Jr., and Daniel Shanks of the United States calculated π to 100,265 decimal places on an IBM 7090 computer.

It has been shown that π is an irrational number (cannot be expressed as a rational number of the form *a/b,* where *a* and *b* are integers). Its approximate value is

$$\pi \approx 3.1415926536$$

For most computational purposes, the approximation 3.14 will do.

The formula for circumference is usually written as

$$C = \pi d$$

or
$$C = 2\pi r$$

1967	M. Jean Guilloud and his colleagues at the Commissariat à l'Energie Atomique in Paris computed π to 500,000 decimal places on a CDC 6600 computer.
1981	Yasumasa Kanada and his colleagues at Tokyo University calculated π to 2 million digits on an NEC supercomputer.
1984	Kanada and his team expanded their calculation to 16 million digits.
1986	David Bailey at the National Aeronautics and Space Administration calculated 29 million digits of π via a Cray 2 supercomputer and a formula discovered by Jonathan and Peter Borwein.
1987, 1988	Kanada and his team first computed 134 million digits of π, then over 200 million digits, on an NEC SC-2 supercomputer.
1989 (spring)	Gregory and David Chudnovsky, Russian-born mathematicians who have lived in the United States since 1978, found 480 million digits of π.
1989 (summer)	Kanada and his team calculated 1,073,740,000 digits of π.
1989 (fall)	The Chudnovsky brothers computed π to 1,130,160,664 decimal places.
1991	The Chudnovsky brothers calculation of π reached 2.16 billion digits.
1995 (April)	Kanada and his co-workers at the University of Tokyo announced that they had calculated π to 3.22 billion decimal places.
1995	After Kanada's announcement, the Chudnovsky brothers stated that their calculation had reached more than 4 billion digits in 1994.

This chronology of π shows the longstanding interest in the famous constant. It also dramatically illustrates the immense improvement in computational devices. Mathematical knowledge follows a growth curve with time similar to the curve for the number of digits of π.

Since π is irrational, it is an infinite, nonrepeating decimal, and calculation of π to billions of digits far exceeds the needs of accuracy. Why bother? There are at least three reasons: (1) Because it's there. Calculating π to a record number of decimal places is a little like climbing Mt. Everest. (2) The calculation of π to many decimal places tests a supercomputer's speed and accuracy. (3) There are some questions to be explored regarding the distribution of digits in the decimal expansion of π.

The next example provides some practice finding the circumference of a circle.

EXAMPLE 6.8

Find the circumference of the following circle.

10 in.

SOLUTION This circle shows the diameter. Therefore, we use the formula $C = \pi d$. Taking $\pi \approx 3.14$, we get

$$C \approx (3.14)10 = 31.4 \text{ in.}$$

At this point, we can write

$$C \approx 31.4 \text{ in.}$$

Although it is incorrect to write $C = 31.4$ in. (equals sign instead of approximately equals), this is done frequently. We suggest you check with your instructor as to which form he or she prefers.

 YOUR FORMULATION

One way to approximate π is by measuring both the circumference and the diameter of several circular objects and filling in a chart such as that nearby. With careful measurement and several trials, you should see that the ratios C/d of the last column are reasonably close to 3.14.

Object	Circumference	Diameter	C/d

The area of a circle is also related to the constant π. The area of a circle with radius r is given by the formula $A = \pi r^2$. Thus, the area of a circle of radius 5 would be

$$A = \pi \cdot 5^2 = \pi \cdot 25$$

which would typically be written as

$$A = 25\pi$$

When units of measure are included, remember that area is being measured, so your answer should be given in "square" units. If the radius of the preceding example is 5 inches, the area is

$$A = 25\,\pi \text{ in.}^2$$

How is the formula $A = \pi r^2$ justified? Why does the same constant π that appeared in the formula for the circumference of a circle enter the formula for the area of a circle? Consider the circle with radius r shown in ▶ Figure 6.47a. Break

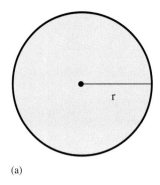
(a)

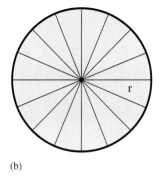
(b)

▶ FIGURE 6.47

the circle up into a large number of skinny pieces, called *sectors,* shown in Figure 6.47b. Each of these sectors is approximately a triangle with height r.

$$\text{Area of circle} = \text{sum of areas of triangles}$$
$$= \text{sum of } [\tfrac{1}{2}(\text{base})r]$$
$$= \tfrac{1}{2}r(\text{sum of bases})$$
$$= \tfrac{1}{2}r(\text{circumference of circle})$$
$$= \tfrac{1}{2}r(2\pi r)$$
$$= \pi r^2$$

Although this is not a "proof," we hope it convinces you that the result is reasonable.

Finding the area of a circle has some interesting applications. Perhaps one fairly well known to college students is finding the best price for pizza. The next example illustrates such a problem.

EXAMPLE 6.9

Domino's Pizza advertises pizzas as follows:

2 10″ pizzas $6.98
2 12″ pizzas $8.99
2 14″ pizzas $10.96

Which is the best purchase?

SOLUTION The answer is found by first finding the total area of each choice and then dividing this figure into the price, yielding the price per square inch. For a 10-in. pizza, we get

$$A = \pi r^2$$
$$= \pi(5)^2$$
$$A \approx (3.14)(25)$$

Thus $A \approx 78.5$ in.^{2} for one 10-in. pizza. For two 10-in. pizzas, the area is about

$$2(78.5) = 157 \text{ in.}^2$$

We get the unit price.

$$\text{Unit price} = \frac{\text{Price}}{A} \approx \frac{\$6.98}{157} \approx \$0.0445/\text{in.}^2.$$

Results for two 12-in. pizzas are

$$A = 2 \cdot \pi(6)^2 \approx 2(3.14)(36) = 226.08 \text{ in.}^2$$

$$\text{Unit price} = \frac{\text{Price}}{226.08} = \frac{\$8.99}{226.08} \approx \$0.0398/\text{in.}^2$$

Results for two 14-in. pizzas are

$$A = 2 \cdot \pi(7)^2 \approx 2(3.14)(49) = 307.72 \text{ in.}^2$$

$$\text{Unit price} = \frac{\text{Price}}{307.72} = \frac{\$10.96}{307.72} \approx \$0.0356/\text{in.}^2$$

If the decision to purchase is based solely on unit price, our decision would be to purchase two 14-in. pizzas.

Our concluding example allows you to combine work with polygons and circles.

●**EXAMPLE 6.10**

Find the perimeter and area of the following figure.

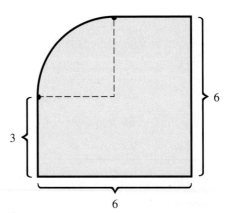

SOLUTION The solution may be aided by separating the figure into a quarter-circle, a square, and a rectangle, as illustrated in ▶ Figure 6.48. The perimeter of the original figure is indicated by solid lines or curves. The rectangle contributes $3 + 6 + 3 = 12$ to the perimeter, the square contributes $3 + 3 = 6$, and the circle contributes one-fourth of a circumference:

$$(.25)\pi(2r) \approx (.25)(3.14) \cdot 6 = 4.71$$

Thus, the total perimeter is

$$P \approx 12 + 6 + 4.71 = 22.71$$

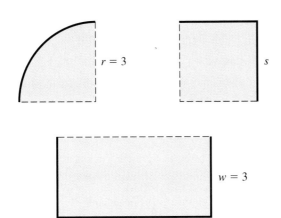

▶ FIGURE 6.48

The area is found similarly. Contributions to the total area are:

Rectangle:	$3 \times 6 = 18$
Square:	$3 \times 3 = 9$
Quarter-circle:	$(.25)\pi(3^2) \approx 7.065$

Summing these, we get

$$A \approx 34.065$$

The area is approximately 34.065 square units.

EXERCISES 6.5

1. Find the perimeter of each of the following polygons.

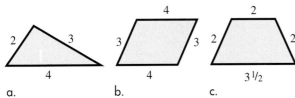

a.   b.   c.

2. A standard sheet of notebook paper is $8\frac{1}{2}'' \times 11''$.
   a. What is its perimeter?
   b. What is its area?
3. A baseball diamond is in the shape of a square, with 90 feet from base to base (including home plate).
   a. What is its perimeter?
   b. What is its area?
4. A swimming pool has the following shape.
   a. What is its perimeter?
   b. The amount of heat loss depends on the surface area. What is the surface area?

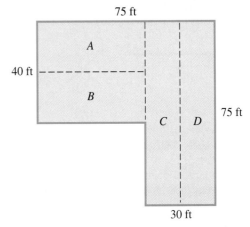

c. The pool is covered by four pool covers, as shown by the letters A, B, C, and D. Find the dimensions and the area of each of these. (A and B have the same shape, and C and D have the same shape.)

5. Find the area of each of the following polygons.

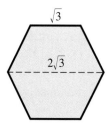

   a.   b.   c.

6. a. Suppose a regular hexagon has side length $\sqrt{3}$ as follows:

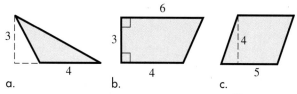

   Use the fact that the distance between opposite vertices is $2\sqrt{3}$ to find the area of the hexagon.

   b. Suppose a regular hexagon has side length s. Use the fact that the distance between opposite vertices is $2s$ to find the area of the hexagon.

7. The approximate shapes of some states follow. Distances are in miles.

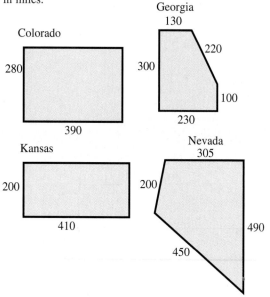

   For each state, find the perimeter, and find or estimate the area.

8. A certain kind of paint is sold in gallon or quart cans. One quart covers about 100 sq ft. How many cans of each size should be purchased to paint each of the following surfaces?

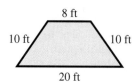

   a. The four walls of a room that is 9 ft by 12 ft with an 8-ft ceiling.

   b. The two ends of a loft, each of which is shaped as in the preceding drawing.

9. Find the circumference and area of a circle with the following dimensions.

   a. A radius of 2 feet

   b. A radius of 4 centimeters

   c. A diameter of 12 inches

10. The biggest possible circle that can be drawn on an $8\frac{1}{2} \times 11$ inch piece of paper has a diameter of $8\frac{1}{2}$ inches.

   a. What is its area?

   b. What is its circumference?

   c. What is the area of the paper that is wasted (not part of the circle)?

11. Dominos (see Example 6.9) also sells a single 14″ pizza for $5.99. How does the price per square inch of this compare with the three possibilities in Example 6.9?

12. One stereo system has a 10-inch (diameter) woofer while another has an 8-inch woofer, so the woofer in the first system has $\frac{10}{8} = 1\frac{1}{4}$ times the diameter of the woofer in the second system. Find the relation between the areas of the two woofers.

13. A car has tires with a 24-inch diameter. How many revolutions does a wheel on this car make in going a mile (5280 ft)?

14. This exercise explores the effect on area of doubling both linear dimensions.

   a. What is the effect on the area of a square if the length of each side is doubled?

   b. What is the effect on the area of a triangle if the base and height are both doubled?

   c. What is the effect on the area of a circle if the radius is doubled?

   d. What is the effect on the area of a circle if the diameter is doubled?

15. A Norman window has a rectangular base topped by a semi-circle, as shown. Suppose a Norman window has a rectangular part that is 2 feet wide and 3 feet high. What is the surface area of the window? What is its perimeter?

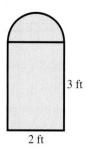

16. A round table has a diameter of 41 inches without a leaf. It has one 12-inch leaf that, when inserted, makes the overall shape an oval 41 inches wide by 53 inches long.

   a. What is the surface area of the round table (without leaf)?

   b. What is the surface area of the oval table?

1. Find the roots words for *perimeter* and *area*. Discuss the origin of these terms.

2. This exercise deals with the chronology of π. Information can be found by looking in your library or periodicals database under *pi*.

   a. Describe in more detail the technique used by at least one person to calculate π.

   b. Write a paper of two or more pages on one of the people listed in the chronology of π. Include information on that person's calculation of π.

   c. When π was first calculated to several decimal places, the calculation was important for accurate computation of perimeters and areas of circles. Current calculations of π to billions of digits far exceed needs of accuracy. Give more detail on the three reasons given for calculating π to more and more decimal places. Perhaps add more reasons.

   d. Describe the background on the use of the symbol π.

   e. This section has used π in connection with the area or perimeter of a circle. Describe another place where the constant π appears.

3. One of the three problems of antiquity is squaring the circle, or the quadrature of the circle. Describe precisely what this problem is, trace its history, and report on its current status.

4. A wedding ring has a simple circular shape. As a circle can be thought of as unending, so the wedding ring can be thought to symbolize the couple's unending love for each other.

   a. Write more about the symbolism from geometry in a wedding ring.

   b. Describe other places where a circle serves symbolic purposes. What is the imagery intended by the circle's use?

5. a. Investigate the problem of enclosing the maximum area with a fixed amount of fencing. In particular, if you have 60 feet of fencing, in what shape should you arrange it so that it encloses the largest possible area? Present an analysis to other members of the class to support your answer.

   b. What implications does your answer in part a have for house construction? Explain.

6. In basketball, the phrase "shooting from the perimeter" refers to shooting the basketball from relatively far away from the basket. Is this use of *perimeter* related to the perimeter of a polygon? Explain.

1. Outline in one or two paragraphs the *origins* of geometry.

2. What are the root words for *geometry,* and what do they mean?

3. Describe the difference between the ancient Egyptian and the ancient Greek approaches to geometry.

4. Name two ancient Greek contributors to geometry, and describe something that each one did.

5. a. Who wrote the *Elements?*

   b. Describe the influence of this work.

6. Given line *l* with points *W, X, Y, Z* as shown, find each of the following if possible. If not possible, so state.

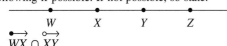

   a. $\overrightarrow{WX} \cap \overrightarrow{XY}$

   b. $\overrightarrow{ZY} \cup \overrightarrow{WY}$

   c. Two rays whose intersection is $\overline{XY}$

   d. Two rays whose union is $\overleftrightarrow{XY}$

7. ∠*ABC* is the union of what two rays?

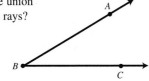

8. The statement "Three points determine a plane" is not correct. Insert a word to make it correct.

9. In addition to the condition stated in Review Exercise 8, two intersecting lines determine a plane. State another combination of points and lines that determines a plane.

10. Describe two physical objects that lie in parallel planes.

11. Draw a line on this paper so that one half-plane contains the statement of Review Exercise 11 and the other half-plane contains the page number.

12. Given two planes, π_1 and π_2, describe all possibilities for their intersection.

13. Consider angle ∠*ABC* in Review Exercise 7.

   a. Which of these are alternate notations for it?

   ∠*ACB*, ∠*BAC*, ∠*BCA*, ∠*CAB*, ∠*CBA*

   b. What is its vertex?

   c. What are its sides?

   d. Is *A* in the interior of the angle?

   e. Is *A* in the exterior of the angle?

   f. Draw and label *I,* a point in the interior of the angle.

   g. Draw and label *E,* a point in the exterior of the angle.

   h. Is it an acute angle?

   i. Is it an obtuse angle?

   j. Is it a right angle?

k. Is it a straight angle?

l. Estimate, to within 10 degrees, its measure.

m. Draw an angle adjacent to it. Describe the adjacent angle.

14. Draw a pair of vertical angles. Describe which angles they are.

15. a. Construct an angle of measure 135°.

   b. Does this angle have a complement? If so, draw one and tell its measure.

   c. Does this angle have a supplement? If so, draw one and tell its measure.

16. In the following figure l_1 is parallel to l_2 and eight angles are numbered.

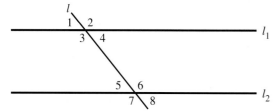

   a. What is line l called?

   b. Write the pairs of alternate interior angles.

   c. Write the pairs of alternate exterior angles.

   d. Write the pairs of corresponding angles.

   e. Suppose m $\angle 4 = 40°$. Find the measure of each of the other seven angles.

17. Draw a Venn diagram showing the relationship among pentagons, quadrilaterals, and triangles.

18. a. Draw a picture of a quadrilateral, if one exists, with four equal angles but that is not regular.

   b. Draw a picture of a quadrilateral, if one exists, with four equal sides but that is not regular.

19. Draw a picture of a polygon, if one exists, that is *not* convex.

20. a. Draw a picture of a regular pentagon.

   b. What is the measure of each interior angle?

   c. Sketch an exterior angle and label it α.

21. Draw a picture, if possible, of each of the following kinds of triangles. If no such triangle exists, so state.

   a. Scalene but not acute

   b. Isosceles but not equilateral

   c. Equilateral but not acute

   d. Right and isosceles

22. In the following right triangles, find the measure of the third side.

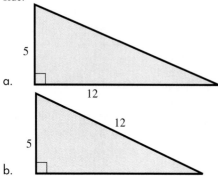

a.

b.

23. Give an example of a pythagorean triple.

24. Sketch a triangle that is similar to the triangle in Review Exercise 22a but that has a different size.

25. Use the information in the following diagram to estimate the height of the building.

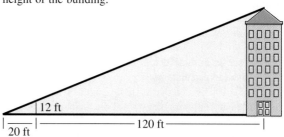

26. Draw a picture of each of the following, if one exists. If no such polygon exists, so state.

   a. A parallelogram that is not a rectangle

   b. A rhombus that is not a square

   c. A trapezoid that is not a parallelogram

27. a. How many sides does a hexagon have?

   b. Name a polygon with more than six sides, and tell how many sides it has.

28. What is the sum of the measures of the interior angles of a hexagon?

29. Several newsmagazines are $8'' \times 10\frac{1}{4}''$.

   a. What is the perimeter of one of these?

   b. What is its area?

30. Consider a standard $8\frac{1}{2}'' \times 11''$ piece of paper. Suppose a triangular piece is cut out from the midpoint of the top to the midpoint of the right side. What is the area of the (larger) piece that remains?

31. A small tire is 205 millimeters in diameter.

   a. What is the circumference of the tire?

   b. How many revolutions does the wheel holding such a tire make in going a mile (1 mile = 1609.344 meters)?

32. A circular 12-inch skillet has a diameter at the top of 12 inches. What is the area of the top of the skillet?

33. Draw a picture of a circle and a chord that is not a diameter.

34. (*Multiple choice*) In any circle, let A denote its area, c denote its circumference, d denotes its diameter, and r denote its radius. Then π is which of the following ratios? A/c, A/d, A/r, c/A, c/d, c/r, d/A, d/c, d/r, r/A, r/c, r/d

35. Suppose a 400-meter track is to be built with 100-meter straightaways on both sides and semicircular ends. How wide between the straightaways should the track be, to the nearest centimeter? The 400-meter distance is measured on the inside of the track.

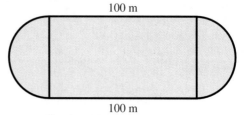

Abbot, Edwin. *Flatland: A Romance of Many Dimensions.* Boston: Little, Brown, 1929; repr., Emerson, Verglanck, NY, © 1982.

Boyer, Carl B. (revised by Uta C. Merzbach). *A History of Mathematics,* 2nd ed. New York: Wiley, 1991.

Calinger, Ronald (ed.). *Classics of Mathematics.* Englewood Cliffs, NJ: Prentice-Hall, 1995.

Closs, Michael P. (ed.). *Native American Mathematics.* Austin: University of Texas Press, 1986.

Deakin, Michael A. B. "Hypatia and Her Mathematics," *The American Mathematical Monthly,* March 1994, pp. 234–243.

Dzielska, Maria. *Hypatia of Alexandria.* Cambridge, MA: Harvard University Press, 1995.

Eves, Howard. *An Introduction to the History of Mathematics,* 6th ed. Chicago: Saunders College, 1990.

Gittleman, Arthur. *History of Mathematics.* Columbus, OH: Charles E. Merrill, 1975.

Katz, Victor J. *A History of Mathematics: An Introduction.* New York: HarperCollins College Publishers, 1993.

Kingsley, Charles, *Hypatia, or New Foes with an Old Face.* London: Macmillan, 1881.

Kline, Morris. *Mathematics: A Cultural Approach.* Reading, MA: Addison-Wesley, 1962.

"The Mountains of Pi," *The New Yorker,* March 2, 1992, v. 68, n. 2, p. 36.

National Council of Teachers of Mathematics. *Historical Topics for the Mathematics Classroom.* Washington, DC: NCTM, 1969.

Zaslavsky, Claudia, *Africa Counts.* Boston: Prindle, Weber & Schmidt, 1973.

CHAPTER 7

TOPICS IN GEOMETRY

CHAPTER OUTLINE

PROLOGUE

The end of the twentieth century is an appropriate time to look back at some topics in the development of geometry. So climb in the time machine (Is that using the fourth dimension?) and travel back to the time the only book to have appeared in more editions than the *Bible*—Euclid's *Elements*—was written.

Greece of 2300 years ago was the dominant intellectual civilization of its time. It prided itself on its ability to reason. The geometry of the Babylonians and Egyptians, the prevailing prior civilizations, was based on carefully recorded observations and on procedures that seemed to work. Much was known, but not in the carefully reasoned way the Greeks approached knowledge. Aristotle suggested that a scientific work needed to begin with definitions and axioms. That's exactly what the *Elements* did! Euclid defined some terms, stated 10 axioms and postulates, and then stated and proved a number of theorems. In this ap-

261

proach to geometry the prototype of the axiomatic method was born.

Let's move ahead in time roughly a millennium to what in Europe is often called the Dark Ages. Euclid's book has come into conflict with the one book published in more editions than it, the *Bible*. Staunch Christians of the time are arguing that truth comes from God through revelation in the church rather than from an axiomatic system. Geometry, like most intellectual activity, is put on hold.

As the time machine moves us through the Renaissance, we see a rebirth of learning generally, including learning in geometry. The ability of mathematics to predict events in nature makes it more and more revered. Other disciplines are compared with it. What can be more precise than numbers or geometric descriptions? Where other disciplines may come close to the truth about nature, mathematics is absolutely true.

Our time machine moves us to the early 1800s. Just prior, two countries—the United States and France—have built their governments on axiomatic foundations. However, it took revolutions for them to do this. A revolution is about to take place in mathematics, in the part we've been following, geometry.

Three mathematicians, in different countries, each find that one of Euclid's postulates can be assumed to be false instead of true, and a completely consistent axiomatic system results. Mathematics does not have absolute truth. It is only as true as the axioms and postulates that are assumed to be true. Non-Euclidean geometry is born. Instead of one monolithic geometry—Euclidean geometry—there are now several geometries, each based on its own axiomatic system.

It's about time to return to the present. On our trip, we've seen Euclidean geometry rise as an axiomatic system, be heralded as a source of absolute truth, then have its status as bearing absolute truth taken away and replaced by multiple geometries. Through all of this turmoil the importance of the axiomatic method is maintained. But let's remember what geometry deals with: basically, shapes. Some figures have congruent shapes and some have similar shapes. As we head through the twentieth century, we find people interested in patterns or shapes that exhibit self-similarity.

Their search is aided by the development of computers that can repeat those self-similar shapes over and over very quickly. The figures that are created are regarded to have a fractional dimension and are called *fractals*. Fractal geometry is born. It produces beautiful images, some of which are used in science fiction movies to produce landscapes from outer space. Picture such an ethereal landscape as you reflect on our journey: a foundation in the axiomatic method, an eruption with non-Euclidean geometry, and images of self-similar figures, all topics in geometry.

Chapter opening photo: In Square Limit, *M.C. Escher used ideas prompted by non-Eudidean geometry and anticipated ideas of fractal geometry.*

 GEOMETRY AS AN AXIOMATIC SYSTEM

GOALS
1. To describe what an axiomatic system is and to relate the development of Greek geometry as an axiomatic system.
2. To work with any given finite geometry axiomatic system.
3. To describe projective geometry and to follow some of its reasoning.
4. To describe analytic geometry and to give some analytic proofs.

In this section, we get away from the specifics of geometric terms and concepts. Instead, we look at geometry as a logical, deductive system. Historically, one of the great achievements of the early Greek mathematicians was the creation of the postulational process. In order to establish a statement in this process, we must show that it is a logical consequence of previously established statements. This postulational process distinguished Greek geometry from that of the Babylonians and Egyptians before them. It forms a mode of thinking in all of mathematics today.

Axiomatic Via Latin, from Greek *axioma*, "that which is thought fitting; decision; self-evident principle." The Indo-European root is *ag-*, "to drive, to lead." A subsidiary Greek meaning, "to weigh," led to *axioma*, literally "something weighty." In mathematical terms, axioms are concepts felt "weighty" or worthy enough that you can base a logical system on them.

Fractal coined by the mathematician Benoit Mandelbrot (who was born in 1924 and whose last name means "almostbread" in German). The word is a variant of *fractional,* since a fractal is a type of fractional dimension found in many naturally occurring physical phenomena. The Indo-European root is *bhreg-*, which is more recognizable in the native English *break* than in the Latin cognate *fractus*, "broken," which is the source of *fractal.*

Why was the Greek creation of the postulational process a great achievement? Think about how we establish truth in an academic discipline, in a political system, or in a religious system. At least one way to establish something as "true" is to argue that it follows from a set of basic principles. In an academic discipline, those principles are what is taught in the fundamentals course. In a political system, the basic principles are embodied in a constitution and bylaws. In a religious system, the basic principles form its beliefs. The axiomatic method, or postulational method, is a way to establish truth by basing what is accepted in the system on a set of fundamental principles.

Most high school geometry courses are based on the process developed by the early Greeks. If you took such a course, perhaps you remember proving results via a sequence of "given" statements, such as the following problem and solution (some of the reasons, though correct, are not statements that have been established in this chapter).

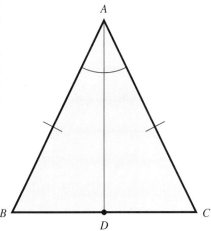

Problem:

Given: Triangle ABC with $AB = AC$ and AD bisects $\angle BAC$
Prove: $BD = DC$

Proof:

Statement	Reason
1. $AB = AC$	1. Given
2. $m\angle B = m\angle C$	2. The measures of angles opposite equal sides in a triangle are equal.
3. AD bisects $\angle BAC$	3. Given
4. $m\angle BAD = m\angle DAC$	4. Definition of angle bisector
5. $\triangle ABD$ is congruent to $\triangle ACD$	5. ASA: If two triangles have two pairs of angles with equal measures and the included sides are equal, then the triangles are congruent.
6. $BD = DC$	6. Corresponding parts of congruent triangles are equal.

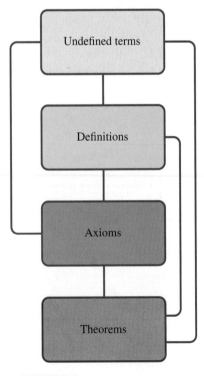

In this section we are not concerned with any specific result or statement. Rather, we are looking at how to achieve results using previously established statements and results. This process is referred to as *axiomatics*. A system that utilizes this process is an *axiomatic system*. In more recent mathematics, axiomatics has become a formal field of study.

Axiomatic Systems

There must be a starting point in any axiomatic system. Geometry as we've presented it has three undefined terms: *point, line,* and *plane.* (See Section 6.2.) These undefined terms, in turn, help to produce definitions. For example, *point* and *line* plus a tacit understanding of *betweenness* helped us define a *line segment.*

Why does an axiomatic system have undefined terms? Trying to define everything is fraught with problems. One problem is comprehension. Consider this definition of a *chufa:* "a sedge whose tuberow roots are eaten in southern Europe." Until we know what a sedge is, this definition only partly describes a chufa. Consequently, we push definitions back to common understandings, which become our undefined terms.

A second problem with trying to define everything is circularity. Suppose that prior to defining *line segment,* we had tried to define *point* and *line.* We might have said that a *point* is the intersection of two lines and a *line* is a line segment extended. Note that these definitions are circular: A line is a line segment extended, and a line segment is part of a line.

Think of visiting a civilization whose language you do not know and without a dictionary to translate from and to English. To communicate with people there you would have to point and gesticulate until you had some common understanding of the meanings of some words. From that point on, other terms could be described to you based on those you already knew. In the same way, an axiomatic system begins with undefined terms.

The undefined terms serve to create definitions. Our earlier definition of a line segment uses the undefined terms *point, line,* and *betweenness.* Much of the material in Chapter 6 dealt with the definition of terms.

An axiomatic system must also have some initial statements from which to build. These initially assumed statements are called *axioms* or *postulates* (Although early Greek mathematicians made a distinction between postulates and axioms, most mathematicians today use these terms interchangeably.) Statements that follow logically from the undefined terms, definitions, and axioms or postulates are called *theorems.* Mathematicians seek to find these results, then argue that they follow logically from other results that are known.

An axiomatic system consists of undefined terms, definitions, axioms or postulates, and theorems. ▶ Figure 7.1 diagrams an axiomatic system.

Euclid's Elements

Euclid (330?–275? B.C.E.) is perhaps the most famous of the Greek mathematicians. His work, the *Elements* (ca. 300 B.C.E.), has been the basis for teaching geometry for over 2000 years.

In the *Elements,* after 23 definitions (he tried to define everything), Euclid then listed the following 10 statements (5 axioms and 5 postulates):

A1   Things that are equal to the same thing are also equal to one another.
A2   If equals are added to equals, the wholes are equal.

A3	If equals are subtracted from equals, the remainders are equal.
A4	Things that coincide with one another are equal to one another.
A5	The whole is greater than any one part.
P1	It is possible to draw a straight line from any point to any other point.
P2	It is possible to produce a finite straight line indefinitely in that straight line.
P3	It is possible to describe a circle with any point as center and with a radius equal to any finite straight line drawn from the center.
P4	All right angles are equal to one another.
P5	If a straight line intersects two straight lines so as to make the interior angles on one side of it together less than two right angles, these straight lines will intersect, if independently produced, on the side on which are the angles that together are less than two right angles.

Euclid then proceeded to derive 465 propositions (theorems) from these 10 statements.

Euclidean Geometry

Euclid's *Elements* form the basis of Euclidean geometry. Today's high schools actually teach mostly Euclidean geometry, which usually consists of two parts: formulating proofs and performing constructions with only a straightedge and compass.

All proofs and constructions in Euclidean geometry can be logically derived from the 10 statements contained in the *Elements*. Thus, Euclidean geometry is an axiomatic system. In fact, it is the archetypal axiomatic system.

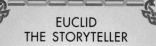

YOUR FORMULATION

(*For people who have taken geometry*)

1. Give an example of a theorem that was proved in your geometry class.
2. Give an example of a construction with only a straightedge and compass that was done in your geometry class.

A Finite Geometry Axiomatic System

Another example of an axiomatic system, one that is much less familiar and hence emphasizes the axiomatic structure, is a finite geometry. We will explore such an axiomatic system. Again, we are not concerned with any specific result or statement but are looking at the process of achieving results.

With permission of Bob Thaves.

The undefined terms are *point* and *line*. We need no definitions to understand the following postulates.

P1   Any two distinct lines have exactly one point in common.
P2   Every point is on exactly one pair of lines.
P3   There are exactly four lines.

And we develop several theorems—some of them very modest—about this system. Results are named T1, T2, and so forth (T stands for theorem). We reason both from a picture and from a table.

By P3, there are exactly four lines, call them l_1, l_2, l_3, and l_4. There are exactly six pairs of lines: l_1 and l_2, l_1 and l_3, l_1 and l_4, l_2 and l_3, l_2 and l_4, and l_3 and l_4. Therefore, we have established the following:

T1   There are exactly six points.

Geometry is often visual. In the axiomatic system with which we are now working, we will look at the lines, the points, and their relationships both in a table and in a picture. Our picture won't be that of ordinary Euclidean geometry, since P2 says that every point is on exactly one pair of lines and in ordinary Euclidean geometry every point is on an infinite number of pairs of lines. In our picture, points will be the "dots" that you usually associate with points. Lines are sets of points. Since T1 states that there are exactly six points, a line does not have an infinite number of points on it. Lines, in our pictures, may also not be "straight." ▶ Figure 7.2 summarizes the information and names the points. Notice that P2 is satisfied by checking in either the table or the picture that each point is on exactly one pair of lines.

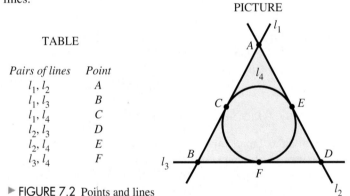

TABLE

Pairs of lines	Point
l_1, l_2	A
l_1, l_3	B
l_1, l_4	C
l_2, l_3	D
l_2, l_4	E
l_3, l_4	F

▶ FIGURE 7.2 Points and lines

Here's another theorem:

T2   There are exactly three points on each line.

One proof of this theorem follows.

Proof: The three postulates and T1 show that the only possible configuration for this geometry is shown in the table except for the names of points and lines in Figure 7.2. The picture presents the information in the table in diagramatic form. Consider the following list of all four lines and the points on them.

Line:	contains	Points:
l_1		A, B, C
l_2		A, D, E
l_3		B, D, F
l_4		C, E, F

Therefore, each line contains exactly three points.

Define two points to be *parallel points* if and only if they have no line in common. For example, by the notation of Figure 7.2, *A* and *F* are parallel points. Here are two theorems about parallel points that can be proved (see the Exercises).

T3   Each point has exactly one point parallel to it.
T4   On a line not containing a given point there is one and only one point that is parallel to the given point.

In this finite geometry axiomatic system, two undefined terms and three postulates lead to at least four theorems. Our focus is on the process of achieving results, which is a key part of the axiomatic method. The axiomatic method is the basic method of establishing truth in geometry and in all of mathematics.

Variations in Geometry

Euclidean geometry is but one of several forms of geometry. Although it is beyond the scope of this text to look in depth at other forms of geometry, we will briefly mention and describe three of them: projective geometry, analytic geometry, and transformational geometry. It is important to be aware that each variation is an *axiomatic system*. That is, the system consists of:

Undefined terms
Definitions
Basic assumptions
Statements that can be logically deduced from the basic assumptions

Projective Geometry.   *Projective geometry* is a branch of geometry that deals with the qualitative and descriptive properties of geometrical figures. It does not use measures of line segments and angles. Rather, it analyzes properties of angles and intersections after what is called a projective transformation.

Projective geometry originated out of the interests of some Renaissance artists of the fourteenth and fifteenth centuries in dealing with perspective. They tried to represent objects the way we see them by giving a two-dimensional painting the depth of our three-dimensional world. This they accomplished by having lines of sight meet at an imagined, distant point. (See ▶ Figure 7.3.) This idea of perspective

▶ FIGURE 7.3 Illustration of perspective is seen in Leonardo da Vinci's *Last Supper.*

Railroad tracks appear to meet in the distance.

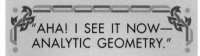
becomes clear when we look at receding railroad tracks that appear to meet in the distance. Thus we are interested in the descriptive property that the lines appear to meet rather than the property that the distance between the two railroad tracks remains constant.

The axiomatic system for projective geometry incorporates (in axiom P3) the idea that any two lines meet in a point (as railroad tracks or lines of sight in the photo appear to do). We take as undefined the terms point and line and postulate the following:

P1   There exists at least one line.
P2   Given any two distinct points, there is exactly one line on both of them.
P3   Given any two distinct lines, there is exactly one point on both of them.
P4   There are at least three points on any line.
P5   Not all points are on the same line.

The Exercises provide more work with this axiomatic system for projective geometry.

Analytic Geometry.   *Analytic geometry* is a method of geometry in which a correspondence is made between points in a plane and ordered pairs of real numbers. This makes possible a correspondence between curves in the plane and equations in two variables. This is the correspondence we used in Chapter 5 to plot points and to graph linear, quadratic, exponential, and other functions. The development of geometric principles from Euclidean geometry can then be realized by analyzing equations of the form

$$f(x, y) = 0$$

where f is a function. Thus, the task of proving a theorem in geometry can be shifted to that of proving a theorem in algebra.

The essential axiom of this axiomatic system is that points on the plane can be matched to pairs of real numbers by a one-to-one mapping that preserves geometric properties. Thus, axioms and definitions in geometry are translated into algebra, results are proved algebraically, and the results are then translated back to geometry.

Rene Descartes and Pierre de Fermat are usually regarded as the developers of analytic geometry, dating from a publication by Descartes in 1637. It is in his honor that the Cartesian coordinate system is named.

Let's illustrate an analytic proof of a theorem in geometry. Assume that we have already proved the theorem that if P is the point with coordinates (x, y) and Q is the point with coordinates (x', y'), then the midpoint of $\overset{\bullet\ \bullet}{PQ}$ has coordinates

$$\left(\frac{x + x'}{2}, \frac{y + y'}{2} \right)$$

We use this fact to prove that the line through the midpoint of two sides of a triangle is parallel to the third side. Consider the triangle in ▶ Figure 7.4, with coordinates $P(x_1, y_1)$, $Q(x_2, y_2)$, and $R(x_3, y_3)$. Let M be the midpoint of $\overset{\bullet\ \bullet}{PR}$ and N be the midpoint of $\overset{\bullet\ \bullet}{QR}$. We wish to show that $\overleftrightarrow{MN}$, the line through the two midpoints, is parallel to the third side, $\overset{\bullet\ \bullet}{PQ}$. The coordinates of M are

$$\left(\frac{x_1 + x_3}{2}, \frac{y_1 + y_3}{2} \right)$$

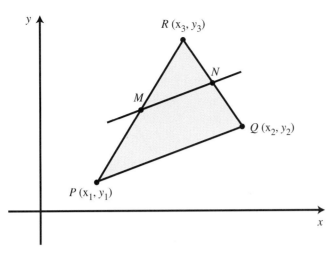

R (x_3, y_3)

N

M

Q (x_2, y_2)

P (x_1, y_1)

▶ FIGURE 7.4 A triangle

and of N are $\left(\dfrac{x_2 + x_3}{2}, \dfrac{y_2 + y_3}{2} \right)$

The slope of $\overleftrightarrow{MN}$ is

$$\frac{(y_2 + y_3)/2 - (y_1 + y_3)/2}{(x_2 + x_3)/2 - (x_1 + x_3)/2} = \frac{(y_2 - y_1)/2}{(x_2 - x_1)/2} = \frac{y_2 - y_1}{x_2 - x_1}$$

The slope of $\overleftrightarrow{PQ}$ is $(y_2 - y_1)/(x_2 - x_1)$. Since the slope of $\overleftrightarrow{MN}$ equals the slope of $\overleftrightarrow{PQ}$, $\overleftrightarrow{MN}$ is parallel to $\overleftrightarrow{PQ}$. In other words, the line through the midpoints of two sides of a triangle is parallel to the third side.

Notice how this analytic proof uses the algebraic description of the slope of a line rather than appealing to postulates and previously proved theorems. The work was done algebraically, then the conclusion was translated back to geometry.

Transformational Geometry *Transformational geometry* is a more sophisticated method of analyzing and developing geometric properties. Essentially, equations are written that transform (move) points in the plane. With transformations, a certain basic set of properties is assumed. These are then utilized to develop an axiomatic system. This method requires more preparation than ordinary Euclidean geometry because students need to be competent in both algebra and properties of functions. For example, students will work with problems like the following:

If $T_1(x, y) = (x + 1, y)$ and $T_2(x, y) = (x, y + 1)$, show that the composition $(T_1 \circ T_2)(x, y)$ can be written as the single transformation $T(x, y) = (x + 1, y + 1)$.

Projective, analytic, and transformational geometry are but a few of the known forms and methods of geometry. You need not be aware of specific results in any of these variations. However, you should recognize that each of them represents an axiomatic system. That is, each starts with a few undefined terms, definitions, and some basic assumptions and proceeds to establish principles by means of a logical sequence of previous results.

Rene Descartes, developer of analytic geometry.

Exercises 1 and 2 refer to the finite geometry axiomatic system.

1. Use the notation of Figure 7.2 to establish T3 by writing for each point the point that is parallel to it.
2. Complete the following table using the notation of Figure 7.2. Observe that it verifies T4.

Point	Line not on point	Point on line parallel to given point
A	l_3	F
A	l_4	F
B	l_2	E
B		
C		
C		
D		
D		
E		
E		
F		
F		

Exercise 3 refers to projective geometry.

3. We develop a model for projective geometry. By P1, there is at least one line, call it *l*. By P4, there are at least three points on *l*, call them *A, B, C*. By P5, not all points are on the same line, so there is a point, call it *D*, not on *l*.

• *D*

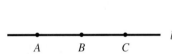

By P2, there is exactly one line on *A* and *D*, call it *i;* exactly one line on *D* and *B*, call it *j;* and exactly one line on *D* and *C*, call it *k*.
   a. Complete the model with as few points and lines as possible. (*Hint:* You should have seven lines.)
   b. Verify that each of the five postulates is satisfied in this model.

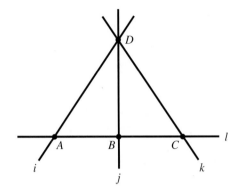

4. Use the following method from analytic geometry to show that the diagonals of a rectangle bisect each other. In particular, consider any rectangle, and position it as shown. Then *A* has coordinates of the form $(a, 0)$ and *C* has coordinates of the form $(0, c)$.

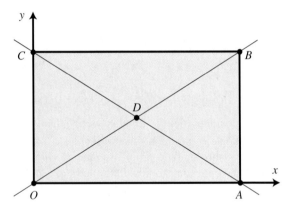

   a. What are the coordinates of *B?*
   b. Find an equation for $\overleftrightarrow{OB}$.
   c. Find an equation for $\overleftrightarrow{AC}$.
   d. Find the intersection of the lines found in parts b and c. This gives the coordinates of *D*.
   e. Verify that *D* is the midpoint of $\overleftrightarrow{OB}$ and of $\overleftrightarrow{AC}$. Therefore, the diagonals of a rectangle bisect each other.

1. Discuss how each of the following can be viewed as an axiomatic system.
   a. Your favorite sport
   b. Card games
   c. Our legal system

2. Axiomatic systems were integral to the Greek way of thinking. They were emphasized less during the Middle Ages but appeared again after the Renaissance. Political theorists embraced ideas of axiomatics. Both the American and French revolutions culminated in constitutions crafted by political

thinkers steeped in axiomatics. It is no coincidence that our Declaration of Independence resolves, "We hold these truths to be self-evident . . ." These self-evident truths are the axioms on which the United States is founded. Write more about the political thinkers around the time of the U.S. and French revolutions. How did they incorporate the axiomatic method in their political thinking?

3. We stated that Euclid is perhaps the most famous of the Greek mathematicians. However, he is not usually regarded as the greatest. That honor is reserved for Archimedes. Write a biography of Archimedes. Why is he regarded as greater than Euclid? Why is Euclid perhaps more famous?

4. Write more about the origins of projective geometry. Some of the principal figures were architects Brunelleschi and Alberti, artists da Vinci, Dürer, and Franccochi, and mathematicians Desargues and Pascal. Later mathematical developments were made by Poncelet, Steiner, Möbius, Charles, and Plücker, and still later by others.

5. Analytic geometry was developed by Rene Descartes. Write more about him and his development of analytic geometry.

6. Dealing with axiomatic systems in mathematics has the danger of making mathematics seem overly formal and contrived. Write your personal reactions to dealing with geometry as an axiomatic system.

● 7.2 ● NON-EUCLIDEAN GEOMETRY

GOALS
1. To do some work with spherical geometry.
2. To describe the work of several people with Euclid's fifth postulate.
3. To describe what non-Euclidean geometry is and to distinguish between hyperbolic and elliptic geometry.
4. To work in non-Euclidean geometry, with a model, and to reflect on the ramifications of non-Euclidean geometry.

The material in this section may challenge some of your conventional ways of thinking geometrically. If you took a geometry course, it probably was Euclidean geometry. Thus, you have been trained to think in Euclidean terms. Furthermore, Euclidean geometry is a geometry of the plane. For example, the sum of the degrees of interior angles of a triangle is 180°, as shown in ▶ Figure 7.5. However, if you consider the sum of measures of the angles of a triangle on a sphere, you get different results. Since "lines" on a sphere are really arcs, the triangle in ▶ Figure 7.6 is seen to have more than 180° as the sum of the measures of interior angles.

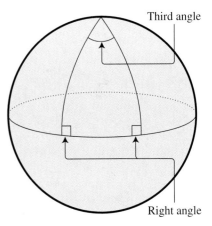

▶ FIGURE 7.6 A triangle on a sphere

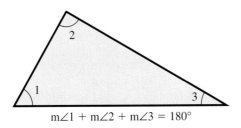

m∠1 + m∠2 + m∠3 = 180°

▶ FIGURE 7.5 Sum of the degrees of interior angles of a triangle in a plane

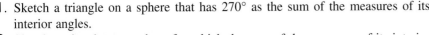

YOUR FORMULATION

The triangle in Figure 7.6 has an angle sum of more than 180°.

1. Sketch a triangle on a sphere that has 270° as the sum of the measures of its interior angles.
2. Sketch a triangle on a sphere for which the sum of the measures of its interior angles is greater than 270°.

The previous illustration and the Your Formulation should help you to realize that Euclidean geometry results do not necessarily hold for surfaces that are not planes (i.e., on a plane).

Dominance of Euclidean Geometry

Euclid's work in geometry dominated geometric thought among mathematicians from 300 B.C.E. until the early 1800s. It was the geometry of the world in which people lived; thus, assumptions and conclusions were readily accepted.

The Parallel Postulate

Recall that the axioms and postulates assumed by Euclid included the following:

P5   If a straight line intersects two straight lines so as to make the interior angles on one side of it together less than two right angles, these straight lines will intersect, if independently produced, on the side on which are the angles that together are less than two right angles.

This statement, known as "the parallel postulate," is not as self-evident as the first nine. In effect, it says: Given a line and a point not on the line, one and only one line can be drawn through the given point parallel to the given line. (See ▶ Figure 7.7.)

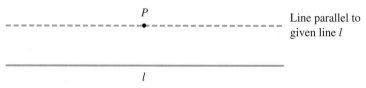

▶ FIGURE 7.7 Euclid's parallel postulate

Euclid apparently realized that this axiom differed from the rest. He avoided using it in the first theorems he proved, which involved the other nine axioms and postulates. Finally, after proving 28 propositions (theorems), he used the parallel axiom for Proposition 29.

Two Millennia of Work

For over 2000 years after the appearance of Euclid's *Elements,* mathematicians were very concerned about the parallel postulate. Their work can be broadly classified into two categories: trying to prove that the fifth postulate could be derived from the others, and trying to replace the fifth postulate by a more self-evident statement.

Several tried to prove that postulates could be derived from the other nine. If it could have been shown that the last postulate was dependent on the others, then only a list of nine axioms would be needed to build the Euclidean system of geometry.

Early Efforts   The first-known attempt to prove the dependence of the parallel postulate was by Ptolemy (ca. 150 C.E.). He failed because he inadvertently assumed something that was equivalent to Euclid's fifth postulate. In other words, he assumed what he was trying to prove, so his reasoning was invalid.

Mathematicians continued working on the dependence of the parallel postulate for approximately 2000 years. This led to the published results of the Italian Jesuit priest Girolamo Saccheri (1667–1733) and Swiss German mathematician Johann Heinrich Lambert (1728–1777). We will look at a broad outline of Saccheri's arguments in order to see how non-Euclidean geometry evolved.

Saccheri's Work Saccheri was trying to prove Euclid's fifth postulate from Euclid's other axioms and postulates. He accepted the first nine axioms and the first 28 propositions. He then considered the quadrilateral *ABCD* as shown in ▶ Figure 7.8.

▶ FIGURE 7.8 Quadrilateral *ABCD* studied by Saccheri

Saccheri assumed *A* and *B* were right angles and that the measures of sides $\overrightarrow{AD}$ and $\overrightarrow{BC}$ were equal. Thus, he assumed what is shown in ▶ Figure 7.9.

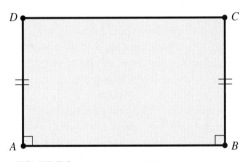

▶ FIGURE 7.9 Assumptions of Saccheri

He then drew some supplementary lines and used congruence theorems from Euclid's first 28 propositions to show that angles *D* and *C* were equal. Next he considered three possibilities for equal angles *C* and *D*:

1. *C* and *D* are equal as right angles.
2. *C* and *D* are equal as acute angles.
3. *C* and *D* are equal as obtuse angles.

These were referred to as the *hypotheses of the right angle, of the acute angle,* and *of the obtuse angle,* respectively.

Saccheri's plan was to show that possibilities for the acute-angle and obtuse-angle assumptions would lead to contradictions, making the remaining possibility the right-angle assumption. He had great difficulty in arriving at contradictions. In the course of his work, however, Saccheri actually proved several classical results of what is today called non-Euclidean geometry. However, he was so intent on arriving at a contradiction that he failed to realize the underlying principle of non-Euclidean geometry:

Each assumption for angles C and D generates a logically consistent system!

More will be said about this later.

Subsequent Efforts   The second major effort of several mathematicians from the time of Euclid until 1800 was to replace the parallel postulate with a more self-evident statement. One of these was given by Joseph Fenn in 1769. He was able to show that the following statement was equivalent to Euclid's fifth postulate:

> Two intersecting lines cannot both be parallel to a third straight line.

Fenn's statement is equivalent to the axiom proposed by John Playfair (1748–1819) in 1795:

> Through a given point *P* not on a line *l*, there is one and only one line in the plane of *P* and *l* that does not meet *l*.

This is the form of the parallel postulate that is usually found in most high school textbooks and that was illustrated in Figure 7.7.

The equivalent axioms of Fenn, Playfair, and others all had a common property: They could not be shown to be provable from the first nine axioms of Euclid.

Results of the 1800s

After more than 2000 years of work by geometers, two types of remarkable results were developed in the nineteenth century. One of these was that the parallel postulate was indeed independent of the first nine axioms. In other words, the parallel postulate cannot be proved from the first nine axioms. Contributors to this result included mathematicians Eugenio Beltrami, Arthur Cayley, Felix Klein, and Henry Poincaré. Mathematical developments related to this work became the basis for much of the mathematical development of the twentieth century.

The second type of result developed in the nineteenth century showed that the hypothesis of the acute angle stated by Saccheri led to a consistent geometric system. That is, the nine axioms of Euclid and the acute-angle hypothesis (see Figure 7.9) produced a system as logically correct as the nine axioms and the right-angle hypothesis. The three mathematicians generally credited with independently developing the major results in this were Carl Friedrich Gauss (1777–1855) of Germany, Janos Bolyai (1802–1860) of Hungary, and Nicolai Lobachevsky (1793–1856) of Russia. Their work became known in the 1820s.

In order to get a somewhat oversimplified idea of how the acute-angle hypothesis can generate a logically consistent geometry, consider the three-dimensional surface shown in ▶ Figure 7.10, which resembles two infinitely long horns placed bell to bell. This surface is referred to as a *pseudosphere*. Clearly, lines on a pseudosphere do not resemble lines on a plane. Furthermore, if a quadrilateral is embedded on a pseudosphere, as shown in ▶ Figure 7.11, results involving angles' measures and sums will not be the same as for quadrilaterals embedded in a plane.

The geometry applicable to a pseudosphere is called *hyperbolic geometry*. However, Lobachevsky was so instrumental in the development of this geometry that

Janos Bolyai

Nicolai Lobachevsky

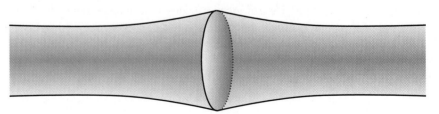

▶ FIGURE 7.10 A non-Euclidean surface

One of the Three Greatest Mathematicians of All Time: Carl Friedrich Gauss

Among the developers of non-Euclidean geometry—Gauss, Bolyai, and Lobachevsky—the best known is Gauss. Gauss was a child prodigy. At the age of 3 he found an error in his father's bookkeeping. At the age of 18 he solved a problem that had been around for more than 2000 years, since the time of Euclid, by showing how to construct a regular 17-sided polygon with straightedge and compasses. In his doctoral dissertation, written at the age of 20, he gave the first entirely satisfactory proof of what is now known as the fundamental theorem of algebra. In addition to work in geometry and algebra, Gauss contributed in a variety of other areas of mathematics, physics, and astronomy.

From 1807 on he worked as director of the observatory and professor of mathematics at the University of Göttingen. Since he believed that most of his students were not well prepared, he avoided classroom teaching as much as possible. He believed

the returns were not worth the investment of his time (Boyer, 1991, p. 508). His assessment of relative worth was justified. Gauss is regarded as the greatest mathematician of the nineteenth century and, along with Archimedes and Newton, as one of the three greatest mathematicians of all time.

Carl Friedrich Gauss makes his mark. He made contributions in statistics and other areas as well as in geometry.

most mathematicians today refer to this as *Lobachevskian geometry*. One of its features is that the sum of the angles of a triangle is less than 180°.

A second type of non-Euclidean geometry was developed from the obtuse-angle assumption by Georg Friedrich Bernhard Riemann (1826–1866). Riemann's geometry was closely related to the geometry of a sphere. On that surface (see Figure 7.6), the sum of the angles of a triangle can exceed 180°. The geometry developed by Riemann is referred to as *Riemannian, spherical,* or *elliptic geometry*. In elliptic geometry the sum of the angles of a triangle is always greater than 180°.

Ramifications of Non-Euclidean Geometry

The development of non-Euclidean geometries in the 1800s shook the foundations of mathematics and was notable in intellectual history. Morris Kline (1962, p. 553)

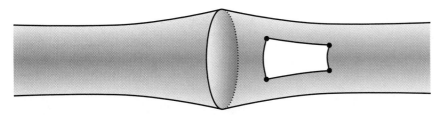

▶ FIGURE 7.11 Quadrilateral on a pseudosphere

says, "The two concepts which have most profoundly revolutionized our intellectual development since the nineteenth century are evolution and non-Euclidean geometry." The development of non-Euclidean geometries forced mathematicians to alter their approach to the discipline. According to Eves (1990, p. 501):

> [A] much more far-reaching consequence . . . was the liberation of geometry from its traditional mold. A deep-rooted and centuries-old conviction that there could be only the one possible geometry was shattered, and the way was opened for the creation of many different systems of geometry. With the possibility of creating such purely "artificial" geometries, it became apparent that geometry is not necessarily tied to actual physical space. The postulates of geometry became, for the mathematician, mere hypotheses whose physical truth or falsity need not concern him. The mathematician may take his postulates to suit his pleasure, just so long as they are consistent with one another.

The concept of mathematics as "absolute truth" was shattered. Mathematics had a centuries-old reputation for being the most objective of disciplines. Statements in mathematics are not subject to personal tastes as in the arts or humanities, to changes in social behavior as in the social sciences, or to deviant patterns as in the natural and physical sciences. Truth in mathematics was regarded as absolute. It must necessarily be so. This belief was exploded with the demonstration that a logically consistent non-Euclidean geometry could be developed. Truth is not absolute; it is relative to the axioms that make up the axiomatic system.

With Euclid's 10 axioms and postulates, we get the geometry we know. Change just one postulate—Euclid's fifth—and we get a non-Euclidean geometry that is as consistent (free of contradictions) as is Euclidean geometry.

Several new branches of mathematics arose from the work of nineteenth century geometers. They realized they were no longer tied to the physical world in their thinking. Mathematicians had to revise their understanding of mathematics and its relation to the physical world. Thinkers in other disciplines had to realize that a logical system of thought modeled reality only as well as did the axioms on which the system was based.

<hr>

EXERCISES 7.2

1. Sketch a triangle on a sphere with three right angles so the sum of the degrees of the interior angles is 270°.

2. In this exercise we establish a result about the sum of the measures of the interior angles of a triangle in Euclidean, hyperbolic, or elliptic geometry. In all three geometries, we accept Euclid's five axioms and first four postulates. The side-angle-side congruence theorem can be proved from these nine axioms and postulates, so it holds in all three geometries.

   In triangle ABC, let D be the midpoint of $\overset{\bullet\!-\!\bullet}{AC}$ and E be the midpoint of $\overset{\bullet\!-\!\bullet}{BC}$. Construct $\overset{\bullet\!-\!\to}{AF}$ perpendicular to $\overset{\longleftrightarrow}{DE}$ as shown. Construct $\overset{\bullet\!-\!\bullet}{DG} = \overset{\bullet\!-\!\bullet}{FD}$ as shown. Construct $\overset{\bullet\!-\!\bullet}{EH} = \overset{\bullet\!-\!\bullet}{GE}$

   as shown. Draw $\overset{\bullet\!-\!\bullet}{BH}$. In triangles ADF and DGC, $\overset{\bullet\!-\!\to}{AD} = \overset{\bullet\!-\!\to}{DC}$, since D is the midpoint of $\overset{\bullet\!-\!\to}{AC}$. $\overset{\bullet\!-\!\bullet}{FD} = \overset{\bullet\!-\!\bullet}{DG}$, by construction. m$\angle FDA = $ m$\angle CDG$, since they are vertical angles. Therefore, triangle ADF is congruent to $\triangle DGC$, by side-angle-side.

   a. Argue why triangle CGE is congruent to triangle EHB.

   b. Argue why m$\angle EHB = 90°$.

   c. Argue why $\overset{\bullet\!-\!\bullet}{AF} = \overset{\bullet\!-\!\bullet}{BH}$.

   Therefore, figure $ABHF$ is a Saccheri quadrilateral.

   d. Argue why m$\angle FAD = $ m$\angle DCG$.

   e. Argue why m$\angle GCE = $ m$\angle EBH$.

The sum of the measures of the interior angles of triangle ABC

$= m\angle BAC + m\angle ACB + m\angle CBA$

$= m\angle BAC + m\angle DCG + m\angle GCE + m\angle CBA$

$= m\angle BAC + m\angle FAD + m\angle EBH + m\angle CBA$ (parts d and e)

$= m\angle FAB + m\angle HBA$

f. Argue that in hyperbolic geometry this sum is less than 180°, in Euclidean geometry this sum equals 180°, and in elliptic geometry this sum is greater than 180°.

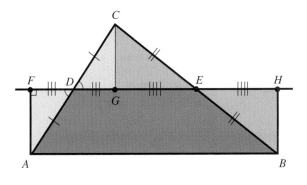

3. Demonstrate what the sum of the measures of the interior angles of a quadrilateral is in each of the following. (*Hint:* Divide the quadrilateral into two triangles.)
   a. Hyperbolic geometry
   b. Euclidean geometry
   c. Elliptic geometry

4. Consider spherical geometry (see Figure 7.6). Consider lines to be great circles on the sphere; i.e., a line is the intersection of the sphere and a plane through the center of the sphere. In this geometry fill in the appropriate modification to Playfair's axiom.

   Through a given point *P* not on a line *l*, there is _____ line that does not meet *l*.

5. Consider geometry on a pseudosphere (see Figure 7.10). In this geometry fill in the appropriate modification to Playfair's axiom.

   Through a given point *P* not on a line *l*, there is _____ line that does not meet *l*.

WRITTEN ASSIGNMENTS 7.2

1. Two statements equivalent to Euclid's fifth postulate, one due to Fenn and the other to Playfair, are stated in this section. Look up and describe other statements equivalent to Euclid's fifth postulate. (You might try a book on non-Euclidean geometry or a section on non-Euclidean geometry in a college geometry book.)

2. Gauss, Bolyai, and Lobachevsky are credited with discovering non-Euclidean geometry. Did they collaborate, did one plagiarize from the others, or was their work independent? If two or more of them made their discoveries independently, why wasn't it done before? In other words, isn't it very unusual that after 2000 years, two or more people discover non-Euclidean geometry at the same time? How can it be explained?

3. Write a one- or two-page description of non-Euclidean geometry. Imagine that you are writing it for your high school English teacher.

4. Since the world might really be non-Euclidean, do you think non-Euclidean geometry should be taught in addition to or instead of Euclidean geometry in high school? Why or why not? State your position in a way that would influence a high school board of education.

5. Describe your personal reactions to knowing there are non-Euclidean geometries.

6. By replacing Euclid's fifth postulate with an alternate postulate, two different geometries emerged. Describe another subject of study in which a traditionally accepted basic assumption has been challenged. What caused the challenge? What has been the result?

7. Visual perception has been described in terms of non-Euclidean geometry. Write a report on this. Some references are A. A. Blank, "Curvature of Binocular Space," *Journal of the Optical Society of America,* Vol. 51 (1961), pp. 335–339; A. A. Blank, "The Luneberg Theory of Binocular Vision," *Journal of the Optical Society of America,* Vol. 43 (1953), pp. 717–727; R. K. Luneberg, "The Metric of Binocular Vision Space," *Journal of the Optical Society of America,* Vol. 40 (1950); and R. K. Luneberg, *The Mathematical Analysis of Binocular Vision,* Princeton, Princeton University Press, 1947.

A fractal

GOALS
1. To describe what a fractal is.
2. To give at least two examples of fractals.
3. To outline the historical development of fractals.
4. Given a description of a fractal and the first few steps in its construction, to continue its construction.

Fractals are a relatively new and fascinating branch of geometry. Simply stated, *fractals* are designs that exhibit *self-similarity*. To illustrate, consider an "H fractal," formed as follows. First draw an "H" where the sides are two-thirds the length of the line segment connecting the sides. This is illustrated in ▶ Figure 7.12.

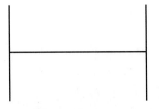

▶ FIGURE 7.12 Base for H fractal

In the second stage, let each of the original sides be a connecting line segment for a new "H," where the sides are two-thirds the length of the connecting segment, as shown in ▶ Figure 7.13.

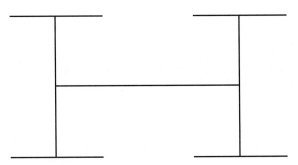

▶ FIGURE 7.13 Second stage for H fractal with some enlarging

The third stage consists of using each new "side" as the connecting segment for a new "H," where each new side is two-thirds the length of the connecting segment. This third stage is illustrated in ▶ Figure 7.14.

The continuation of this process produces various stages of the H fractal. This particular H fractal after 10 stages is illustrated in ▶ Figure 7.15.

YOUR FORMULATION

1. Draw the fourth stage for an H fractal.
2. Pick out the first four stages for an H fractal in the H fractal after 10 stages that is shown in Figure 7.15.

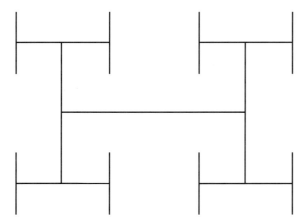

▶ FIGURE 7.14 Third stage for an H fractal

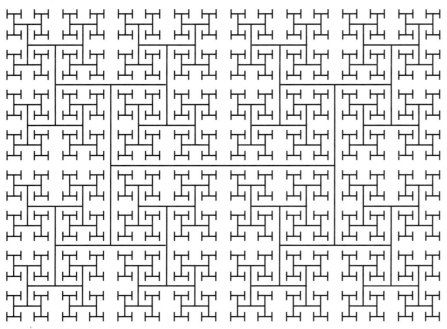

▶ FIGURE 7.15 An H fractal after 10 stages with some enlarging

Several things should be noted about this first example of a fractal. First, reconsider the term *self-similarity*. Each stage consists of recreating the base figure several times on a reduced scale. Thus, each addition is similar to the original figure. Second, the process can be continued indefinitely. However, practical considerations cause us to terminate the process. Third, using a reduction of two-thirds was an arbitrary choice. There are infinitely many choices available for reduction scales.

Using this initial example of a fractal, we consider a more formal definition.

DEFINITION

A *fractal* is a geometrical figure in which an identical motif repeats itself on an ever-diminishing scale.

Historical Development

The formal study of fractals dates from the mid-1970s. This coincides with the widespread availability of computer graphics. Computer programs that produce fractal images are relatively easy to write. The current emphasis on fractal geometry as a field of study is generally attributed to Benoit B. Mandelbrot. His famous book *The Fractal Geometry of Nature,* published in 1983, is considered to be the most significant work in fractal geometry.

Although the formal study of fractals is very recent, some rudimentary work with these figures was done in the late 1800s and early 1900s. George Cantor discussed the mathematical properties of the Cantor-point-set around 1870. As you will see, this creates a fractal by the current definition. Many consider this to be the oldest fractal. Henri Poincaré's works circa 1890 include sketches that can be thought of as fractals-to-be.

ON A TANGENT

Benoit Mandelbrot: The Father of Fractals

Benoit Mandelbrot (1924–   ), the inventor of fractals, was born in Poland. He moved to France when he was 12. He attended schools in France, with a brief stay in the United States at the California Institute of Technology. His doctoral thesis at the University of Paris, completed in 1952, involved mathematical linguistics and statistical thermodynamics. Its title, *Games of Communication,* reflected the influence of John von Neumann, whose coauthored book *Theory of Games and Economic Behavior* marked the start of the subject of game theory. A year later von Neumann sponsored Mandelbrot at the Institute for Advanced Study in Princeton, New Jersey. Mandelbrot's thesis and *Theory of Games and Economic Behavior* were both attempts to put together and develop a set of mathematical ideas from several different disciplines.

He joined IBM in 1958 and soon became an IBM Fellow. At IBM he was given freedom to choose and carry out his work in areas related to his specialization, in order to promote creative achievement.

His diversity of interests is reflected in the variety of visiting professorships he has held: Visiting Professor of Economics, Applied Mathematics, and Mathematics at Harvard, of Engineering at Yale, and of Physiology at the Albert Einstein College of Medicine. The unifying thread in this diversity has been fractal geometry.

His development of fractal geometry took place over time. By 1961–62, Mandelbrot was convinced that he had identified a phenomenon present in many aspects of nature. He continued to refine his ideas through the 60s using his geometric thinking. He comments on thinking in shapes (Albers and Alexanderson, 1985, p. 217):

It may have become true that people who think best in shapes tend to go into the arts, and that people who go into science or mathematics are those who think in formulas. On these grounds, one might argue that I was misplaced by going into science, but I do not think so. Anyhow, I was lucky to be able—eventually—to devise a private way of combining mathematics, science, philosophy, and the arts.

One stage of a fractal image of Mandelbrot's famous book.

In 1904, Helge von Koch presented the Koch curve, often referred to as the Koch snowflake. This curve and its properties will be developed later. In 1915, the Polish mathematician Vaclav Sierpinski created the Sierpinski triangle, also called Sierpinski's sieve. The work of Pierre Fatou and Gaston Julia circa 1918 led to many fractals-to-be. However, lack of computer-generating capabilities limited their work to hand-drawn images with very few stages.

Examples of Fractals

We now develop several fractals and examine some of their properties. Further information about these fractals can be found in the references at the end of this chapter. We first look at three classical fractals, Sierpinski's triangle, the Koch curve (and Koch snowflake), and the Cantor-point-set.

◆ EXAMPLE 7.1

Sierpinski's triangle was developed by Vaclav Sierpinski in 1915. We begin this fractal by starting with a triangle such as shown in ▶ Figure 7.16. Then we draw lines connecting the midpoints of the sides and cut out the center triangle, as shown in ▶ Figure 7.17.

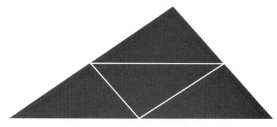

▶ FIGURE 7.16 Base design for Sierpinski's triangle

Draw lines connecting the midpoints
of the sides:

Cut out the center triangle:

▶ FIGURE 7.17 The first stage of Sierpinski's triangle

To create the next stage, we repeat this process with each of the three triangles formed by the previous process. This is illustrated in ▶ Figure 7.18. To produce successive stages, continue the process of deleting center triangles from each of the remaining triangles. Three additional stages of this triangle are illustrated in ▶ Figure 7.19.

▶ FIGURE 7.18 The second stage of Sierpinski's triangle

Do it yet again:

and again:

and again:

▶ FIGURE 7.19 Successive stages of Sierpinski's triangle

An illustration of the self-similarity of a fractal is contained in ▶ Figure 7.20. Note that the basic motif is apparent for any scale of the triangle.

A fractal

looks the same

over all ranges

of scale.

This is called *self similarity*.

▶ FIGURE 7.20 Self-similarity in Sierpinski's triangle

●**EXAMPLE 7.2**

In 1904, Helge von Koch developed a fractal called the *Koch curve* or *Koch snowflake* with interesting mathematical properties. Here's von Koch's idea.

Level 0   Begin with a line segment of length L.

Level 1   Delete the middle third and replace it with a "peak" whose sides represent two sides of an equilateral triangle.

The length of this curve is now $(4/3)L$.

Level 2   Repeat this process on each of the four line segments to obtain the following, with total length $(4/3)(4/3)L = (4/3)^2 L$.

Level 3   Repeat this process yet again to obtain the following, whose length is $(4/3)^3 L$.

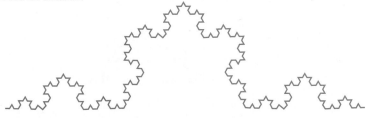

Successive iterations produce a design that looks like the curve in ▶ Figure 7.21. After infinite iteration, the distance between the endpoints along this fractal would be infinite.

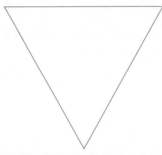

▶ FIGURE 7.21 The Koch curve

Perhaps more well known than the Koch curve is the Koch snowflake. This is produced in a manner very similar to that used to produce the Koch curve. The difference is that we start with an equilateral triangle.

We now apply the previous process to each side of the triangle to obtain the following:

Again we apply the replacement rule to this result (iterate) to get the following:

After another iteration, we have the following:

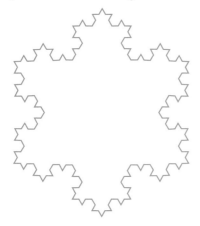

After infinitely more iterations, we get the Koch snowflake.

Some interesting mathematical properties of the Koch snowflake exist. First, each segment, however small, is infinitely long. Second, this curve does not have a tangent line at any point. Third, the region bounded by the fractal is finite. These properties led mathematicians of von Koch's era to refer to the curve as a "pathological curve." Later, Koch's snowflake was called Koch Island by B. B. Mandelbrot.

EXAMPLE 7.3

The *Cantor-point-set,* considered to be the oldest fractal (dating from circa 1870) is developed as follows.

Level 0   Start with a line segment of length L.

———————————————————————

Level 1   Delete the middle third to obtain the following, leaving the total length of the remaining segments as $(2/3)L$.

———————   ———————

Level 2   Repeat this process with the two remaining segments to get the following, meaning the length of the remaining segments is now $(2/3)(2/3)L = (2/3)^2 L$.

——   ——   ——   ——

Level 3   Repeat again to obtain eight segments, whose total length is $(2/3)^3 L$.

— —   — —   — —   — —

As you can see, the sum of the lengths approaches zero while the number of points remaining stays infinite. Mathematicians indicate the Cantor-point-set by assuming the original line segment is the interval [0, 1] (the set of all points between 0 and 1, including 0 and 1) on a number line. The points remaining then are shown by the set

{0, 1, 1/3, 2/3, 1/9, 2/9, 7/9, 8/9, 1/27, 2/27, 7/27, 8/27, 19/27, 20/27, 25/27, 26/27, . . .}

These three classical examples set the stage for extensive subsequent work. We will continue by looking at some more examples and see how computers have been used for illustrating fractals.

The next example was created by Gaston Julia and mentioned in a publication dated 1919. Unfortunately, depicting stages of his fractals was far too complex for his era. Although the mathematics was understandable, the beauty of his work was not appreciated until computer graphics were able to illustrate his fractals. The basic idea for creating Julia fractals is to associate (map) points in the plane with other points in the plane by means of mathematical functions.

EXAMPLE 7.4

We now look at two Julia fractals. The first one is produced by associating points on a checkerboard with other points in the plane by using the mapping

$$x' = x^2 - y^2, \qquad y' = 2xy$$

where the checkerboard region is given by $1 \le x \le 2$, $1 \le y \le 2$. This fractal is shown in ▶ Figure 7.22.

Before transformation

After transformation

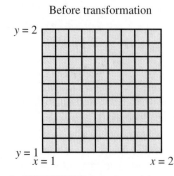
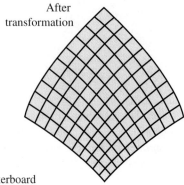

$y = 2$
$y = 1$
$x = 1$ $x = 2$

▶ FIGURE 7.22 Julia fractal from a checkerboard

The second Julia fractal is derived from a more complex mapping. We omit the mathematical development, but show it, in ▶ Figure 7.23, for its beauty.

▶ FIGURE 7.23 Julia fractal from a simple function

Perhaps the most famous of all fractals is that referred to as the Mandelbrot set. It was created by B. B. Mandelbrot and popularized by A. K. Dewdney (1985). The existence of high-speed computers and color graphics allowed the beauty of this fractal to be observed and its properties to be examined in detail.

EXAMPLE 7.5

We begin by showing the famous image of the Mandelbrot set in ▶ Figure 7.24.

▶ FIGURE 7.24 The Mandelbrot set

Blowups of portions of the boundary of the Mandelbrot set reveal amazingly beautiful patterns. ▶ Figure 7.25 shows a sequence of blowups from a portion of the boundary.

▶ FIGURE 7.25 Blowups of the boundary of the Mandelbrot set

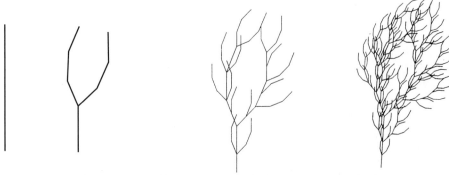

(a) Generate a bush fractal with this replacement rule

(b) The second and third iterations

▶ FIGURE 7.26 Creating a bush fractal

(c) After two more iterations and some enlarging

Fractals in Nature

No work on fractals would be complete without a discussion of fractals that occur in nature. Trees are good examples of fractals. The tree trunk can be considered the base unit. As a branch grows from the main trunk, it too has smaller branches growing from it. This process continues as the tree ages. Along these lines, a bush fractal can be generated by replacing a line segment with a relatively simple "branch." The growth of a bush fractal is illustrated in ▶ Figure 7.26.

Another commonly occurring natural phenomenon that exhibits fractal-type behavior is a river and its tributaries. The main river is the base unit, the major tributaries are secondary stages, streams into the tributaries are tertiary stages, etc.

Many spirals are considered to be types of fractals. Again, we will omit the mathematical development and just illustrate the spirals. ▶ Figure 7.27 shows an ammonite, and ▶ Figure 7.28 shows an *Architectonica maxima* (center) and *perspectiva,* which are excellent examples of growth spirals.

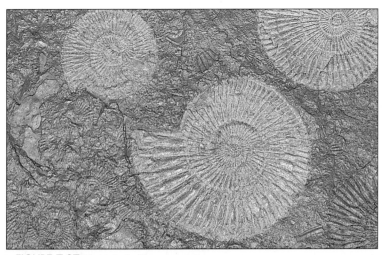

▶ FIGURE 7.27 The spiral of an ammonite

▶ FIGURE 7.28 The spiral of *Architectonica maxima* (center) and *perspectiva.*

Perhaps the most simple spiral is the spiral of Archimedes. This spiral is generated in such a way that, if you draw a line from the center to the outer ring, all the successive intersections of the line with the spiral are the same distance apart. Grooves of a record and the windings of a coiled ribbon both form this kind of spiral.

For those interested in fractals in nature, there are two excellent works to examine; see Mandelbrot (1983) and Michael McGuire (1991).

EXERCISES 7.3

1. One variation of the Koch snowflake is to start with an equilateral triangle and proceed as in the chapter except to have the peaks on the interior of the triangle. Level 0 and the first step are shown here. Using a full sheet of paper, create the next level.

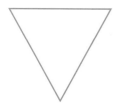 

2. a. Construct level 4 of the Cantor-point-set.
   b. The following points remain in the Cantor-point-set. Write those with denominators of 81.

   {0, 1, 1/3, 2/3, 1/9, 2/9, 7/9, 8/9, 1/27, 2/27, 7/27, 8/27, 19/27, 20/27, 25/27, 26/27, . . .}

3. The Box Fractal, attributed to Dick Oliver (1992), is formed as follows:
   a. Draw a square.
   b. Divide the square into nine smaller squares.
   c. Remove the middle square along each edge.
   This process is repeated with each remaining square to get successive levels of development. The base figure and level 1 are shown here. Create level 2.

4. Dragon Fractals can be created by starting with a long, thin strip of paper, folding the strip, repeating the folding, then, unfolding the paper and looking at the edge of the unfolded paper. For example, if you fold in half each time with the right half folded on top of the left half, a third-level development is as shown here. Create as many more levels of this fractal as you can.

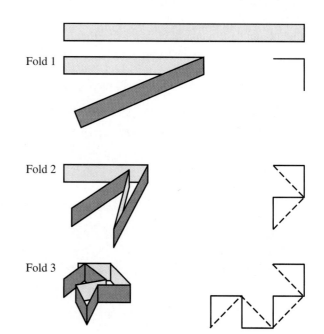

Fold 1

Fold 2

Fold 3

5. Minkowski's Sausage is a fractal formed as follows:

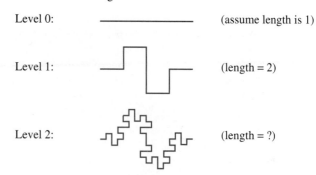

Level 0: ———————— (assume length is 1)

Level 1: (length = 2)

Level 2: (length = ?)

a. What is the length of the level 2 curve?
b. The level 0 curve has 1 segment and the level 1 curve has 8 segments. How many segments does the level 3 curve have?
c. What is the length of the level 3 curve?

1. Get from the library a copy of either Mandelbrot's *The Fractal Geometry of Nature* (1983) or McGuire's *An Eye for Fractals* (1991). Peruse the book, paying particular attention to the illustrations. Using ideas from the book, take a camera and create your own collection of fractals in nature. Prepare a display for your class.

2. Construct a time line showing the development of fractals.

3. Write a letter (you need not send it) to your high school geometry teacher (or any high school math teacher) describing what fractals are and offering advice on whether or not they should be included in a high school geometry course.

4. The formal study of fractals emerged with the development of the computer. Write one to three paragraphs describing why the development of computers was important to the formal study of fractals.

CHAPTER 7 REVIEW EXERCISES

1. Describe what an axiomatic system is. Draw a diagram to show the relationship of its parts.

2. State one of the weaknesses of Euclid's *Elements*.

3. Consider the finite geometry axiomatic system with undefined terms *point* and *line* and the following postulates:

   P1   Any two distinct lines have exactly one point in common.
   P2   Every point is on exactly one pair of lines.
   P3   There are exactly four lines.

   Here are two theorems:

   T1   There are exactly six points.
   T2   There are exactly three points on each line.

   a. Suppose the four lines (P3) are called *a, b, c,* and *d.* By P1, any two distinct lines have exactly one point in common. Therefore, lines *a* and *b* have a point in common, call it *ab.* Let *ac* denote the point in common on lines *a* and *c.* By T1 there are exactly six points. Two of them are *ab* and *ac.* What are the other four?

   b. T2 says there are exactly three points on each line. What are the three points on line *b?*

4. Describe projective geometry and its origins.

5. Projective geometry uses as undefined terms *point* and *line* and has the following five postulates:

   P1   There exists at least one line.
   P2   Given any two distinct points, there is exactly one line on both of them.
   P3   Given any two distinct lines, there is exactly one point on both of them.
   P4   There are at least three points on any line.
   P5   Not all points are on the same line.

   We develop a model for projective geometry. By P1, there is at least one line, call it *l.* By P4, there are at least three points on *l,* call them *P, Q, R.* By P5, not all points are on the same line, so there is a point, call it *S,* not on *l.*

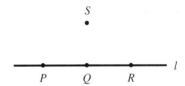

By P2, there is exactly one line on *P* and *S,* call it *m;* exactly one line on *Q* and *S,* call it *n;* and exactly one line on *R* and *S,* call it *o.*

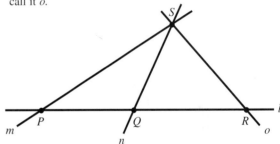

Complete the model with as few points and lines as possible. (*Hint:* You should have seven lines.)

6. Describe analytic geometry and its origins.

7. Verify via analytic methods that the diagonals of a rectangle are equal, as follows. Consider rectangle *OABC,* placed in a

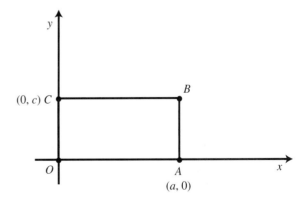

coordinate system with O at the origin, A on the x-axis, and C on the y-axis, as shown. Then O has coordinates $(0, 0)$, A has coordinates $(a, 0)$, and C has coordinates $(0, c)$.

  a. What are the coordinates of B?

  b. Recall that the distance between points (x_1, y_1) and (x_2, y_2) is

$$\sqrt{(x_2 - x_1)^2 + (y_2 - y_1)^2}$$

  Use this formula to find the distance AC.

  c. Find the distance OB.

  d. Show that $AC = OB$.

8. Briefly describe transformational geometry.

9. Describe a way that spherical geometry is different from plane geometry by showing a result that is true in one but not in the other.

10. Sketch a triangle on a sphere for which the sum of the degrees of the interior angles is between 180° and 270°.

11. Describe Euclid's approach to his fifth postulate.

12. Pick one of the following statements about Euclid's fifth postulate that is false, and explain why it is false.

  a. It is true for both Euclidean and non-Euclidean geometries.

  b. It is the only one that has not been proved.

13. Describe the work of Saccheri and Lambert on the fifth postulate and approximately when it was done.

14. a. State Playfair's axiom.

  b. How is it related to Euclid's fifth postulate?

15. a. Who are usually regarded as the developers of non-Euclidean geometry?

  b. About when?

16. Describe what non-Euclidean geometry is.

17. Distinguish between hyperbolic and elliptic geometry.

18. The awareness of non-Euclidean geometry meant there was another kind of geometry known, but it had a much more profound meaning. Describe that meaning and its ramifications.

19. Discuss which is correct—Euclidean or non-Euclidean geometry.

20. Use the pseudosphere model of elliptic geometry to illustrate a result that is true in elliptic geometry but not true in Euclidean geometry.

21. Describe what a fractal is.

22. A Binary Tree fractal is generated by the following process:

  i. Start with a "T" such that the horizontal length is twice the vertical length.

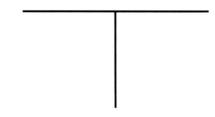

  ii. At each end of the horizontal segment, repeat the design where the new vertical length is one-half the previous vertical length.

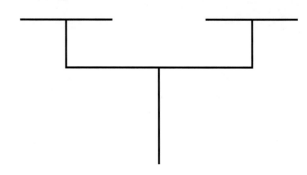

  a. Construct this fractal for as many levels as you can.

  b. What is the sum of the total vertical lengths for each stage of construction?

  c. What is the total height of this fractal for each stage of construction? Indicate this by showing the sum of fractions.

  d. What number does the total height get close to but never reach?

23. Give an example of a fractal other than the one in Review Exercise 22.

24. Outline the historical development of fractals.

BIBLIOGRAPHY

Albers, Donald J., and Alexanderson, G. L., eds. *Mathematical People.* Boston: Birkhauser, 1985.

Boyer, Carl B. (revised by Uta C. Merzbach). *A History of Mathematics,* 2nd ed. New York: Wiley, 1991.

Devaney, R. L. *A First Course in Chaotic Dynamical Systems.* Reading, MA: Addison-Wesley, 1992.

Dewdney, A. K. "Computer Recreation: A Computer Microscope Zooms In for a Look at the Most Complex Object in Mathematics," *Scientific American,* August 1985, pp. 16–24.

Eves, Howard. *An Introduction to the History of Mathematics,* 6th ed. Chicago: Saunders College, 1990.

Goldberger, A. L., Rigney, D. R., and West, B. J. "Chaos and Fractals in Human Physiology," *Scientific American,* February 1990, pp. 42–49.

Grebogi, C., Ott, E., and Yorke, J. "Chaos, Strange Attractors, and Fractal Basin Boundaries in Nonlinear Dynamics," *Science,* October 1992, pp. 632–638.

Horgan, J. "Fractal Shorthand," *Scientific American,* February 1988, p. 28.

Kline, Morris. *Mathematics: A Cultural Approach.* Reading, MA: Addison-Wesley, 1962.

Lamb, J. "Where Math Really Figures," *Management Today,* December 1987, p. 129.

Lauwerier, Hendrik A. *Fractals: Endlessly Repeated Geometrical Figures.* Princeton, NJ: Princeton University Press, 1991.

Mandelbrot, Benoit B. *The Fractal Geometry of Nature.* San Francisco: W. H. Freeman, 1983.

McGuire, Michael. *An Eye for Fractals: A Graphic & Photographic Essay.* Redwood City, CA: Addison-Wesley, 1991.

National Council of Teachers of Mathematics. *Historical Topics for the Mathematics Classroom.* Washington, DC: NCTM, 1969.

Oliver, Dick. *Fractal Vision: Put Fractals to Work for You.* Carmel, IN: Sams, 1992.

Studt, Tim. "Finding Order in Chaotic Mixing," *R & D Magazine,* February 1994, pp. 82–84.

Trudeau, Richard. *The Non-Euclidean Revolution.* Boston: Birkhauser, 1987.

CHAPTER 8

PROBABILITY

PROLOGUE

At least 33 states in the United States have state-run lotteries that together collect a total of over $20 billion annually. In 1992 an Australian syndicate, the International Lotto Fund, won a $27 million Virginia lottery by purchasing up to $5 million worth of $1 lottery tickets. This angered some lottery players. Subsequently, Virginia and some other states have taken steps to discourage

a single group from buying a huge number of tickets. In particular, Virginia lottery outlets must sell to all waiting customers before selling a large number of tickets to just one customer.

Is the strategy of buying all the tickets in a lottery a good money-making scheme? For example, if each student at the University of Virginia paid an extra $300 tuition a year, then over

$5 million would be generated. If the university then won $27 million, it could substantially reduce tuition.

If you are a resident of a state with a state-run lottery, should you support legislation prohibiting a single group's buying all the tickets or making it difficult for a single group to do so? Is it unfair to permit such a group purchase?

There are several problems with the scheme of buying all the possible ticket choices. First, to buy all the possible tickets in the Virginia lottery would require purchasing about 7.1 million tickets, rather than just 5 million. Second, there might be multiple winners. This would reduce the individual payoff from $27 million to a lesser amount. Third, it would take a huge expenditure of time to buy all the possible ticket choices. If each transaction took 5 seconds, say, it would take over 400 24-hour days to buy 7.1 million tickets.

Different lotteries have different games for the players to play. How does the preceding analysis change from game to game? Is it always the case that a "sure thing" is a risky investment?

Chapter opening photo: What's the probability there's a winner here?

 8.1 HISTORICAL BACKGROUND

GOALS
1. To describe the development of probability, including some of its significant historical periods, some of the people involved and their contributions, and what prompted the developments.
2. To state Fermat's Last Theorem and describe its current status.

The weather report predicts a 30% chance of rain. What's the probability you will pass your next test? What's the chance of winning the lottery? The probability your flight will be late is .15. These are all sentences dealing with probability. Whenever there is uncertainty about the occurrence of an event, probability enters the picture.

All disciplines make use of probability. The art historian wonders about the likelihood that a painting was done by a famous artist rather than by a forger. The biologist considers the probability that two brown-eyed parents will have a blue-eyed offspring. The sociologist observes that the probability that a criminal will return to jail is .8.

YOUR FORMULATION

Give an example of how the discipline of your major (or an area you are considering for a major) involves probability.

We use understandings of probabilities in our daily lives. We don't stay home from work even though the chance of being injured at work today is 1 in 25,000. You might take one instructor rather than another because you have a higher probability of getting a good grade with the first instructor. Health and safety risks are frequently described. Death and taxes notwithstanding, very little is certain. We live constantly with uncertainty, and uncertainty is described by probability.

The Birth of Probability Theory

Questions of uncertainty or chance have likely been considered by people throughout history. However, there is no record of the mathematical treatment of probability until the late 1400s and the early 1500s. At that time Italian mathematicians considered some probability questions arising from games of chance involving dice. The famous scientist Galileo was one of those to write about such questions. However, the birth of probability theory is usually dated at 1654. A gambler found that his theoretical reasoning on a game of chance did not agree with what he observed. He posed the problem to Blaise Pascal, a French mathematician. Pascal corresponded with Pierre de Fermat, another French mathematician. From their correspondence the theory of probability emerged.

Why Probability Theory Emerged in the Seventeenth Century

Why, when people had been gambling for thousands of years (Maistrov, 1974, p. 14), did probability theory finally emerge in the seventeenth century? M. G. Kendall (1956) puts forward four reasons why probability theory did not emerge earlier:

1. The absence of ideas of combinations and ways of working with them
2. The superstition of gamblers
3. The absence of a notion of chance events (the Greeks and medieval scholars looked for regularity only in mathematics and in heavenly things)
4. Religious barriers to the development of concepts surrounding gambling

At the same time, a number of features of the intellectual and social climate of the sixteenth and early seventeenth centuries set the stage for a mathematical theory of probability.

- A number of developments in mathematics in Western Europe:
  - The acceptance of 0 (zero) as a number
  - The use of negative and irrational numbers
  - The use of decimal fractions
  - Methods in algebra to solve third- and fourth-degree equations

ON A TANGENT

History's Most Celebrated Math Problem

The solution to what has been called history's most celebrated math problem was announced on June 23, 1993. *Fermat's Last Theorem* states that if a natural number n is greater than 2, then there do not exist natural numbers x, y, and z such that $x^n + y^n = z^n$. Fermat had jotted the "theorem" in the margin of a book, along with the comment "I have assuredly found an admirable proof of this, but the margin is too narrow to contain it." Fermat's proof was never published or circulated. For over 350 years nobody had been able to prove the result.

The announced proof was devised by Andrew Wiles, an English mathematician now at Princeton University. His "proof" takes 200 pages. Announcement of Wiles "proof" made the front page of the *New York Times* on June 24, 1993. The subtitle of a July 5, 1993, *Time* article on the "proof" said: "History's most celebrated math problem is solved at last." Many major publications reported the news. Wiles' result was announced at the end of three 1-hour lectures in Cambridge, England. The full proof awaited finalizing for publication. By late summer 1993 it became clear that there was a gap in Wiles' proof. During the next year Wiles revised part of his proof and by the fall of 1994 copies of his proof were being circulated. It has withstood scrutiny. In addition to proving Fermet's Last Theorem, his techniques have made major advances in an area of mathematics called number theory.

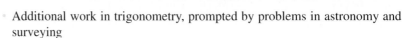

The front page of the New York Times *announces the proof of Fermat's Last Theorem.*

Andrew Wiles

Andrew Wiles is the son of a theologian at Oxford University in England. He first saw Fermat's Last Theorem at the age of 10 in a book from his local public library. It made him want to be a mathematician and to solve the problem. He spent a considerable amount of time as a teenager trying to solve it. When he completed his doctorate and became a professional mathematician, he realized that much more than teenage enthusiasm was needed to solve the problem. His work on it was put on hold. In 1986 work by Gerhard Frey of the University of the Saarlands in Germany and Kenneth Ribet of the University of California at Berkeley culminated in the result that Fermat's Last Theorem would be true if a key part of something called the Taniyama conjecture could be proved. This reignited Wiles. He worked on the theorem in the attic of his Princeton, New Jersey, home until his announcement of an almost-proof on June 23, 1993. From the fall of 1993 until the fall of 1994 he worked on filling the gap in the proof.

- Additional work in trigonometry, prompted by problems in astronomy and surveying
- The use of symbols for mathematical concepts, such as + (plus), − (minus), and $\sqrt{}$ (square root)
- The development of economic attitudes that made probability useful
- Pressure from society to obtain solutions to practical problems
- Academic competition among mathematicians

No matter which of these reasons may be true, it is clear that social and intellectual attitudes and developments discouraged the development of probability theory before the mid-1600s and made the time ripe for its development at that time.

The Bernoulli Family

The Bernoulli family of Switzerland is to mathematics what the Bach family of Germany is to music. About 60 to 70 Bachs were musicians or composers, most of them in the 1700s. Perhaps a dozen members of the Bermoulli family were recognized mathematicians. This is more than any other family in history. Like the Bachs, much of their work was done in the 1700s. The Bernoulli mathematical family tree crosses five generations, into the mid-1800s.

The Bernoulli mathematics legacy began with brothers Jakob (1654–1705) and Johann (1667–1748), who also qualify as the most famous pair of brothers in mathematics. However, their mathematics was more noteworthy than their brotherly love. One problem they both worked on was that of the *catenary,* the curve formed by a heavy cord hanging between two fixed points. Galileo had considered the problem about a century earlier. Jakob, the older brother, first considered the problem. William Dunham (1990, p. 192) describes what happened:

Unfortunately, he got nowhere on this vexing problem. After a year's unsuccessful effort, Jakob was chagrined to see the correct solution published by his young brother Johann. For his part, the upstart Johann could hardly be considered a gracious winner, as seen in his subsequent recollection of the incident:

The efforts of my brother were without success; for my part, I was more fortunate, for I found the skill (I say it without boasting, why should I conceal the truth?) to solve it in full . . . It is true that it cost me study that robbed me of

rest for an entire night . . . [B]ut the next morning, filled with joy, I ran to my brother, who was still struggling miserably with this Gordian knot without getting anywhere, always thinking, like Galileo, that the catenary was a parabola. Stop! Stop! I say to him, don't torture yourself any more to try to prove the identity of the catenary with the parabola, since it is entirely false.

It is amusing to note the time required for Johann's succesful solution. To sacrifice the rest "of an entire night" on a problem that Jakob had wrestled with for a year qualifies as a first-class insult if there was one.

Johann's inability to get along with Jakob carried over to his relationship with his son Daniel, who was also a mathematician. It is reported that Daniel's success, including winning prizes for which both he and his father were competing, so irritated Johann that Johann expelled Daniel from the family home.

Although the Bernoullis competed against one another, they won notable success in mathematics. Jakob Bernoulli made substantial advances in the early stages of calculus. His *Ars Conjectandi* ("The Art of Conjecturing") was the next major accomplishment in probability after the Fermat–Pascal correspondence. Johann also made advances in calculus and probability. He was the first to verify correctly that a cycloid is the curve that satisfies what is called the brachistochrone problem: Find the path along which an object falls most rapidly between one point and a lower point not directly underneath it (find the fastest playground slide). Johann was also a teacher of Leonhard Euler (see Chapters 3 through 6).

Jakob Bernoulli

Johann Bernoulli

Included among the early developers of probability theory are Christian Huygens, several members of the Bernoulli family (Jakob, Johann, and Johann's sons Nicolaus and Daniel) and Abraham de Moivre. Ian Hacking writes (Maistrov, 1974, p. vii).

Quantum theory tells us that everything is governed by laws that are irreducibly probabilistic. Many philosophers in recent years have claimed that all our learning from experience must be understood in terms of probability theory. Perhaps no other mathematical concept of recent times has so completely permeated both our practical and our theoretical lives.

WRITTEN ASSIGNMENTS 8.1

1. (*Library research*) Find out more about Pascal. Write a two- or three-page paper describing the relation of mathematics and religion in his life.

2. (*Library research*) Find out more about Fermat. Write a two- or three-page paper focusing on some aspect of his life, such as his stating a number of results without proof (see Dunham, 1990, pp. 158–159) or his career as an attorney.

3. a. Fermat published little. Why do you think this was so?

b. Do you consider it important for a scholar to publish? Why or why not?

4. (*Library research*) What is the current status of a proof of Fermat's Last Theorem? Write a paragraph to a page describing what has developed since April 1996, when the On a Tangent box on "History's Most Celebrated Math Problem" was written. Include a photocopy of at least one source or a complete bibliographic reference to it.

8.2 BASIC CONCEPTS OF PROBABILITY

GOALS
1. To describe the sample space of an experiment.
2. To know and use the two basic probability rules.
3. To know, use, identify, and give examples of three approaches to assigning probabilities.

The probability of an event is the chance that the event occurs. According to classical probability theory, the probability of an event is the ratio of the number of times an event happens to the total number of outcomes. In this section we will mention three procedures to assign a probability to an event, all based on the classical concept. We start by looking at some probabilities that we understand already.

If you flip a coin, the probability of its landing heads up is 1/2. If you roll a die (singular of "dice"), the probability of its having a 4 on the top face is 1/6 and the probability of its being an even number is 3/6 = 1/2. If you pick a card from an ordinary deck of 52 cards, the probability of its being red is 1/2, the probability of its being a club is 13/52 = 1/4, the probability of its being an ace is 4/52 = 1/13, and the probability of its being the king of diamonds is 1/52.

The following probabilities are less apparent than those in the preceding paragraph but have been reported. The probability that an American will graduate from college is a little less than 1/5. In Cleveland or Newark, the probability that a resident has either completed college or attended for four years or more is 6/100; in Austin, Texas, it is 31/100. The probability that a college freshman will graduate is about 1/2. If a person earns a bachelor's degree, the probability is 1/7 that the person will pursue a master's degree and about 3/1000 that the person will complete a Ph.D. If a person goes to work, the probability is about 1/70 that in the next year the person will be injured on the job. The probability that a randomly selected American will die this year is about 1/119, and the probability that a randomly selected American will live past age 65 is about .8.

Experiments, Outcomes, and Sample Spaces

The actions *flip a coin, roll a die,* and *pick a card* are examples of what are called *experiments.* By an experiment we mean an action whose possible outcomes can be described. The possible *outcomes* of the experiment of flipping a coin and observing the top face are *heads* and *tails.* The possible outcomes of the experiment of rolling a die and observing the number on the top face are *1, 2, 3, 4, 5, 6* (see ●Table 8.1). The possible outcomes of the experiment of drawing a card and recording its suit are *club, diamond, heart, spade.* The collection of all possible outcomes of an experiment is called the *sample space* of the experiment. The sample space of the experiment of flipping a coin and observing the top face is {heads, tails}. The sample space of the experiment of rolling a die and observing the number on the top face is {1, 2, 3, 4, 5, 6}. The sample space of the experiment of draw-

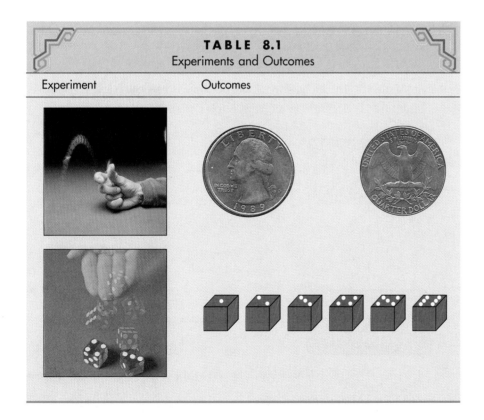

TABLE 8.1
Experiments and Outcomes

Experiment	Outcomes

TABLE 8.2		
Experiments, Outcomes, and Sample Spaces		
Experiment	Outcomes	Sample Space
Flip a coin and observe the top face.	Head, tail	{H, T}
Roll a die and observe the number on the top face.	1, 2, 3, 4, 5, 6	{1, 2, 3, 4, 5, 6}
Draw a card and record its face value.	2, 3, . . . , 10, J, Q, K, A	{2, 3, . . . , 10, J, Q, K, A}

ing a card and recording its face value is {2, 3, 4, 5, 6, 7, 8, 9. 10, J, Q, K, A}. This is summarized in ⬤ Table 8.2.

Mathematical Modeling and Assigning Probabilities

The real-world problem is to describe how likely some combination of specific outcomes is. We produce a mathematical model that is a sample space with a probability assigned to each outcome in the sample space. To do this we make assumptions about the probabilities that are associated with each outcome. The person creating the model (see ▶ Figure 8.1) makes these assumptions and assigns probabilities. To do this we make assumptions about the probabilities that are associated with each outcome. The modeler makes these assumptions and assigns probabilities.

In assigning probabilities to the outcomes of an experiment, there are two basic rules followed.

Probability Rule 1.   The probability of any outcome is between 0 and 1, inclusive.

Probability Rule 2.   The sum of the probabilities of all the outcomes is 1.

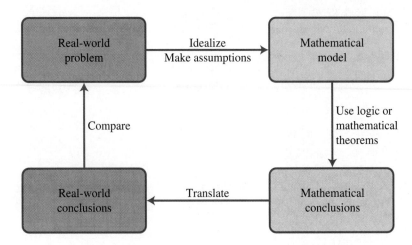

If necessary, refine the model

▶ FIGURE 8.1 Mathematical modeling

In the experiment of rolling a die, the assignment of probabilities is

Outcome	2	not 2
Probability	$\frac{1}{6}$	$\frac{5}{6}$

Notice that this satisfies both probability rules.

A probability is never negative (Probability Rule 1). The smallest value that a probability can take is 0. An example of an outcome of an experiment that has probability 0 is getting an English common word (not a proper name) starting with q and followed by a letter other than u.

The largest a probability can be is 1 (Probability Rule 1).

 YOUR FORMULATION

Give an example of an outcome of an experiment that has probability 1.

Probability Distributions

The previous tables that *list* the *outcomes* of the experiment and the *probabilities* associated with each outcome are called *probability distributions*. If the event "e" is denoted e, we write $P(e)$ to denote the probability that "e" occurs; i.e., $P(e)$ is the probability of the outcome "e."

The following probability assignment is *not* appropriate:

Outcome	e	not e
Probability	.1	.8

This is because $.1 + .8 \neq 1$. The following probability assignment is also inappropriate, since 2.5 is not between 0 and 1:

Outcome	e	not e
Probability	.1	2.5

EXAMPLE 8.1

In each of the following probability distributions fill in probabilities, if possible, to satisfy the two probability rules. For those where it is not possible, explain why not.

a.

Outcome	A	B
Probability	.2	

b.

Outcome	A	B	C
Probability	.6	.2	

c.

Outcome	A	B	C
Probability	.7	.5	

d.

Outcome	A	B	C
Probability		−0.3	0.7

e.

Outcome	A	B	C
Probability			.1

SOLUTION

a. The probability of outcome B must be .8, written $P(B) = .8$, so that $P(A) + P(B) = 1$.

b. $P(C) = .2$ so that $P(A) + P(B) + P(C) = 1$.

c. It is not possible to find such a probability distribution, since to make $P(A) + P(B) + P(C) = 1$, $P(C)$ must be $-.2$, which violates Rule 1.

d. It is not possible to find such a probability distribution, since Rule 1 is already violated with $P(B) = -0.3$.

e. There is an infinite number of ways to complete this probability distribution. Here are two of them:

   i. $P(A) = .5$ and $P(B) = .4$
   ii. $P(A) = 1/3$ and $P(B) = 17/30$

We can assign probabilities to the outcomes of an experiment by means of three types of procedures:

- Axiomatic
- Long-run relative frequency
- Subjective

Axiomatic Approach   If a coin is flipped, an axiomatic assignment of probabilities would assign $P(\text{head}) = 1/2$ and $P(\text{tail}) = 1/2$. The two outcomes should be equally likely and the probabilities should sum to 1, so each probability must be 1/2. More generally, if an experiment has n outcomes and each outcome is expected to occur equally often, then each outcome would have probability $1/n$. The assignment $P(\text{head}) = 1/2$ is done axiomatically.

Long-Run Relative Frequency Approach   If we take the long-run relative frequency approach to find $P(\text{heads})$, we flip the coin many times. The fraction of times the coin turns up heads would be $P(\text{heads})$ and the fraction of times the coin turns up tails would be *P*(tails). The Frenchman Comte de Buffon (1707–1788) flipped a coin 4040 times and got 2048 heads (Moore, 1979, p. 240), so his long-run relative frequency assignment was *P*(heads) = 2048/4040 ≈ .5069. Noted statistician Karl Pearson (1857–1936), founder of the journal *Biometrika,* flipped a coin 24,000 times and got 12,012 heads (Moore, 1979, p. 240), so his long-run relative frequency assignment was *P*(heads) = 12,012/24,000 = .5005.

Let's consider some cases where the long-run relative frequency approach arises naturally. If you pick a letter at random from a randomly selected book, the probability of picking an *e* is .13. This is based on looking at many pieces of English writing and counting the number of *e*'s and the number of letters, then dividing to get about .13. The National Household Survey on Drug Abuse found that in the early 1990s 54.7% of 18–25-year-olds reported ever having used an illicit drug. If an 18–25-year-old person were picked at random, the probability that that person would report ever having used an illicit drug is .547. In the cases cited, we have assigned a probability to an outcome based on the relative likelihood of that outcome. In general, the long-run relative frequency approach says the probability of an outcome is the ratio of the number of times that outcome happens to the total number of outcomes observed.

Notice that "long-run" in "long-run relative frequency" refers to the number of attempts rather than the length of time over which an experiment takes place. In flipping the coin, many coin flips were performed. In picking a letter, many letters were chosen. In the drug abuse survey, many people were surveyed.

Subjective Approach   We take the subjective approach to assigning probabilities whenever neither the axiomatic approach nor the long-run relative frequency approach is readily available. For example, what is the probability that the Chicago Cubs will win the World Series next year? Any assignment of probability here is just someone's opinion. Although the person may use data and analyze players, ultimately the assignment is that person's subjective belief. In this book we will not talk about how to make good subjective probability assignments. However, keep in mind that it is an approach to assigning probabilities.

SOME PROBABILITIES CALCULATED BY MEANS OF THE LONG-RUN RELATIVE FREQUENCY APPROACH	
Probability	Event
.45	A person uses mouthwash at least once a day
.35	A person twists an Oreo apart before eating it
.26	An 18–24-year-old says a favorite spring ritual is gardening
.25	An ice cream serving is vanilla
.20	A first-time college freshman is from out of state
.14	An adult American is a veteran
.05	A U.S. household is phoneless

What is the probability that the Chicago Cubs will win the next World Series?

EXAMPLE 8.2

You roll a die and record whether or not the number on the top face is a square ($1 = 1^2$ or $4 = 2^2$; 2 is not a square). Find the probability distribution for this experiment.

SOLUTION Since there are six equally likely faces that might appear on the top of the die and a square happens on exactly two of those ($1 = 1^2$ and $4 = 2^2$), the probability of a square is $2/6 = 1/3$. A nonsquare occurs on four faces, so the probability of a nonsquare is $4/6 = 2/3$. The probability distribution is

Outcome	Square	Nonsquare
Probability	$\frac{1}{3}$	$\frac{2}{3}$

Notice that the two probability rules are satisfied and we have made the probability assignments via the axiomatic method.

EXAMPLE 8.3

In the 1968 U.S. presidential election, electoral votes were cast as follows:

Candidate	Electoral Vote
Richard Nixon	301
Hubert Humphrey	191
George Wallace	46

If an elector is chosen at random, what is the probability the elector voted for Richard Nixon?

SOLUTION Altogether there were 538 electoral votes cast, of which 301 went to Nixon, so the probability an elector voted for Nixon is 301/538.

EXERCISES 8.2

In each of Exercises 1–3, describe the sample space for the given experiment. In other words, describe the set of all possible outcomes.

1. Draw a card from a deck of 52 cards and record its color.
2. Pick a person in North America and record the gender of the person.
3. Pick a vowel from the alphabet.

In Exercises 4–12, fill in probabilities, if possible, to satisfy the two probability rules. For those where it is not possible, explain why not.

4.

Outcome	A	B
Probability	.3	

5.

Outcome	A	B
Probability		1

6.

Outcome	A	B
Probability	1.1	

7.

Outcome	A	B
Probability		−0.1

8.

Outcome	A	B	C
Probability	.1	.2	

9.

Outcome	A	B	C
Probability	.5	.6	

10.

Outcome	A	B	C
Probability	.9	.1	

11.

Outcome	A	B	C
Probability	.3		

12.

Outcome	A	B	C	D
Probability	$\frac{1}{2}$	$\frac{1}{6}$	$\frac{1}{4}$	

13. For each part, indicate whether or not the probability assignment satisfies the two probability rules.

a.

Outcome	A	B
Probability	1	0

b.

Outcome	A	B
Probability	$\frac{1}{3}$	.7

c.

Outcome	A	B	C
Probability	.3	.3	.3

d.

Outcome	A	B	C
Probability	0.8	0.7	−0.5

14. Find the probability distribution for the following experiment. Use the axiomatic method.

Roll a die and observe whether the top number is odd or even.

15. Use the axiomatic method to determine, before a card is selected at random from a standard 52-card deck, the probability that the card selected is each of the following.
   a. Black
   b. A diamond
   c. A queen
   d. A red jack
   e. The king of hearts

16. The 1990 census showed the following racial composition of the U.S. population.

White	199,686,070
Black	29,986,060
American Indian, Eskimo, or Aleut	1,959,234
Asian-Pacific Islander	7,273,662
Other race	9,804,847
Total	248,709,873

A person is picked at random from the 1990 U.S. population. What is the probability of each of the following?
   a. The person is an Asian-Pacific Islander.
   b. The person is not white.

17. Describe a situation in which you would make a subjective probability assignment.

18. Two dice are to be rolled. Before they are rolled, find the probability of the following.
   a. Their sum will be 2.
   b. Their sum will be 3.
   c. Their sum will be 4.
   d. Both dice will have the same number.
   e. Their sum will be 10 or higher.

19. In 1995 there were about 5.66 billion people in the world. The principal languages spoken were:

Language	Number of Speakers
Mandarin	975 million
English	478 million

If a person is chosen at random from the world population in 1995, what is the probability of each of the following?
   a. The person speaks Mandarin.
   b. The person does not speak English.

20. The infant mortality rate (the rate of deaths before age 1) in the United States is 8.3 per 1000. If an infant is picked at random in the United States, what is the probability of each of the following?
   a. The infant will die before age 1.
   b. The infant will survive until age 1.

1. Some people believe that textbooks, when talking about probability, spend too much space talking about coins, dice, and cards. Would you suggest not including these illustrations? Why or why not? If they were not included, what would you put in their place?

2. Give an example, other than one in this section, of an assignment of probabilities using long-run relative frequencies.

3. Estimate the probabilities of three outcomes in your life. For example, estimate the probability that at a random time in the day you will be sleeping or the probability that you will get an A on the next test in this class.

8.3 SOME COUNTING RULES

GOALS
1. To know and use the multiplication rule.
2. To work with permutations of a set.
3. To calculate the number of permutations and the number of combinations of n things taken r at a time.
4. To solve counting problems.

There are times when we would like to know how many ways there are to do something without listing all the possibilities. For example, if you are looking at a new car that comes in a choice of five colors, with a choice of four interiors and a choice of three wheel covers, how many different possibilities are there? We will develop some general principles on counting rules, then answer this question.

Choices, choices, choices.

The Multiplication Rule

Suppose a family has two cars and three drivers. How many car-driver pairs are possible? Let's call the cars R and Y (for red and yellow) and the drivers A, B, and C. The possible pairs are RA, RB, RC, YA, YB, and YC. We could also have calculated that there are six pairs by the following argument.

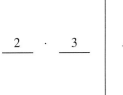

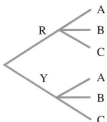

Draw two blanks, one for the car and one for the driver.

We have two choices for the car that goes in the first blank—either R or Y.

For each choice of car, there are 3 choices of driver.

Altogether there are 2 · 3 = 6 car-driver pairs.

This argument can be generalized to a situation in which we do one thing in *m* ways and a second thing in *n* ways. Together the two things can be done in $m \cdot n$ ways. This is called the *multiplication rule*. It can be extended from two things to many things, as the next example illustrates.

EXAMPLE 8.4

Answer the question posed in the first paragraph of this section.

SOLUTION

$$\underline{5} \cdot \underline{4} \cdot \underline{3} = 60$$

— Number of paint colors
— Number of interiors
— Number of wheelcovers

EXAMPLE 8.5

A designator for a building consists of two letters. For example, Pearce Hall is designated PE. How many two-letter designators are possible?

SOLUTION

$$\underline{26} \cdot \underline{26} = 676$$

— Number of choices for the first letter
— Number of choices for the second letter

Permutations

New we apply the multiplication rule to a situation in which we wish to count the number of possible arrangements of a finite set. For example, consider Hagar's suggestion in the cartoon above. How many arrangements could Helga have? There are

cook, clean, wash
cook, wash, clean
clean, cook, wash
clean, wash, cook
wash, clean, cook
wash, cook, clean

Would Hagar dare suggest doing one sequence a day and taking the seventh day off? In any case, there are six possible arrangements. A rearrangement of a set of objects is called a *permutation* of the set of objects. Thus there are six permutations of the three words *cook, clean,* and *wash*.

Let's extend this idea. How many permutations are there of a four-element set? We'll argue as we did with the multiplication rule.

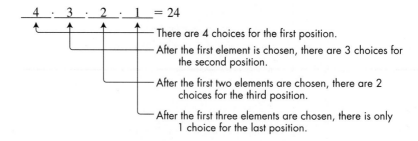

$$\underline{4} \cdot \underline{3} \cdot \underline{2} \cdot \underline{1} = 24$$

There are 4 choices for the first position.

After the first element is chosen, there are 3 choices for the second position.

After the first two elements are chosen, there are 2 choices for the third position.

After the first three elements are chosen, there is only 1 choice for the last position.

There are 24 permutations of a four-element set. If the set is {A, B, C, D}, four of the permutations are ABCD, ACBD, CDAB, and DCBA.

For a positive integer n, the expression

$$n \cdot (n-1) \cdot (n-2) \cdots 2 \cdot 1$$

is denoted $n!$ and is called n *factorial*. In particular,

$$3! = 3 \cdot 2 \cdot 1 = 6, \qquad 4! = 4 \cdot 3 \cdot 2 \cdot 1 = 24, \qquad \text{and} \qquad 5! = 5 \cdot 4 \cdot 3 \cdot 2 \cdot 1 = 120$$

YOUR FORMULATION

Here we investigate the number of permutations of an *n*-element set.

1. A three-element set has six permutations. A four-element set has 24 permutations. Find the number of permutations of a five-element set:

   _____ _____ _____ _____ _____ .

2. Find how many permutations an *n*-element set has by completing the blanks in the following argument.

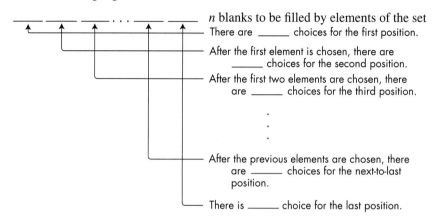

_____ _____ _____ . . . _____ _____   *n* blanks to be filled by elements of the set

There are _____ choices for the first position.

After the first element is chosen, there are _____ choices for the second position.

After the first two elements are chosen, there are _____ choices for the third position.

.
.
.

After the previous elements are chosen, there are _____ choices for the next-to-last position.

There is _____ choice for the last position.

Summarize what you have found about the number of permutations of an *n*-element set.

If you have a calculator with a factorial key on it, you can easily calculate *n*! for whole number values of *n* by entering *n* and then pressing the factorial key. For example, to calculate *5*! press *5,* then the factorial key. Our argument says that a set with *n* elements has *n*! permutations. In particular, a set with 5 elements has $5! = 120$ permutations.

> If a set has *n* elements, there are *n*! permutations of it.

A related class of problems is illustrated by the next two examples.

⊸EXAMPLE 8.6

There are 30 students in a class. In how many ways can a seating chart be constructed for the students who will sit in the first three seats.

SOLUTION

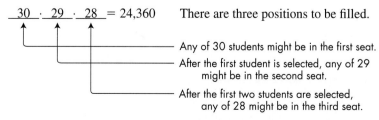

$\underline{30} \cdot \underline{29} \cdot \underline{28} = 24{,}360$ There are three positions to be filled.

Any of 30 students might be in the first seat.

After the first student is selected, any of 29 might be in the second seat.

After the first two students are selected, any of 28 might be in the third seat.

Notice that the solution to Example 8.6 uses an extension of the multiplication rule: The first student can be chosen in any one of 30 ways, the second student in any one of 29 ways, and the third student in any one of 28 ways, so together the three things can be done in $30 \cdot 29 \cdot 28$ ways.

The general problem that Example 8.6 illustrated was to find the number of arrangements of some (perhaps all) objects of a set, or what is called the number of permutations of n objects taken r at a time. In Example 8.6, the number of permutations of 30 things taken 3 at a time was found to be 24,360.

To calculate the number of permutations of n objects taken r at a time, denoted $P(n, r)$, we suggest the method used in Examples 8.6 for smaller values of r. For large values of r, a formula for $P(n\ r)$ is helpful. Consider a set of n objects from which we select an arrangement of r of those objects. How many such arrangements are there? We use the procedure of Example 8.6.

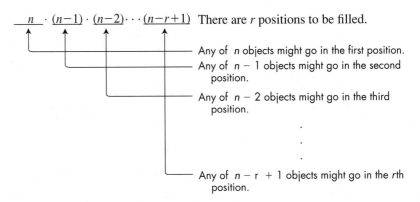

$$\underline{n} \cdot \underline{(n-1)} \cdot \underline{(n-2)} \cdots \underline{(n-r+1)} \quad \text{There are } r \text{ positions to be filled.}$$

Any of n objects might go in the first position.
Any of $n - 1$ objects might go in the second position.
Any of $n - 2$ objects might go in the third position.
.
.
.
Any of $n - r + 1$ objects might go in the rth position.

Therefore,

$$P(n, r) = n \cdot (n - 1) \cdot (n - 2) \cdots (n - r + 1)$$

$$= n \cdot (n - 1) \cdot (n - 2) \cdots (n - r + 1) \cdot \frac{(n - r)(n - r - 1) \cdots 2 \cdot 1}{(n - r)(n - r - 1) \cdots 2 \cdot 1}$$

$$= \frac{n!}{(n - r)!}$$

$$(1) \quad P(n, r) = \frac{n!}{(n - r)!}$$

Previously we calculated $P(30, 3) = 24,360$. The same result can be calculated by applying equation (1) with a calculator that has a factorial key or a factorial menu option:

$$P(30, 3) = \frac{30!}{(30 - 3)!} = \frac{30!}{27!} = 24,360$$

●EXAMPLE 8.7

An athletic conference has 10 teams. The conference tournament involves the top eight teams. In how many ways can these be selected and ranked?

SOLUTION We will answer the question using two different methods.

Method 1: $\quad \underline{10} \cdot \underline{9} \cdot \underline{8} \cdot \underline{7} \cdot \underline{6} \cdot \underline{5} \cdot \underline{4} \cdot \underline{3} = 1,814,400$

Method 2: $\quad P(10, 8) = 10!/2! = 1,814,400 \quad$ (This method lends itself to a calculator with a factorial key.)

Combinations

When we deal with permutations, the order of the elements makes a difference. The permutations ABC and ACB are different. For instance, in Example 8.6 the seating arrangement Nick, Ginny, Juan would be different from the seating arrangement Nick, Juan, Ginny. Sometimes we are interested in the number of ways r objects can be selected from a set of n objects, without regard to the order of the elements. For example, in how many ways can a committee of three students be selected from a class of 30 students.

Let's start our analysis with a smaller example. From the set $S = \{a, b, c, d\}$ how many two element subsets can be chosen? If they are chosen *and ordered*, there are $\underline{4} \cdot \underline{3} = 12$ permutations:

$$
\begin{array}{cccccc}
ab & ac & ad & bc & bd & cd \\
ba & ca & da & cb & db & dc
\end{array}
$$

If order doesn't matter, there are only half as many subsets since, for example, ab and ba form the same subset, $\{a, b\}$. Therefore, S has 6 two-element subsets. How many three-element subsets does S have? If they are chosen *and ordered,* there are $\underline{}4\underline{} \cdot \underline{}3\underline{} \cdot \underline{}2\underline{} = 24$ permutations. If order doesn't matter, whenever we pick a three-element subset—say, $\{a, b, c\}$—it gets counted each way these three elements can be rearranged. There are $\underline{}3\underline{} \cdot \underline{}2\underline{} \cdot \underline{}1\underline{} = 6$ rearrangements of the three elements. Therefore, each three-element subset gets counted six times. Consequently, there are $24/6 = 4$ distinct three-element subsets: $\{a, b, c\}$, $\{a, b, d\}$, $\{a, c, d\}$, and $\{b, c, d\}$.

An r-element subset of an n-element set is called a *combination* of n things taken r at a time. The number of combinations of n things taken r at a time is denoted $C(n, r)$. In the preceding paragraph we found that $C(4, 2) = 6$ and $C(4, 3) = 4$.

To find $C(4, 0)$ we find the number of subsets of a four-element set that have 0 elements. There is only one such subset, namely, the empty set, so $C(4, 0) = 1$.

To find $C(4, 1)$ we find the number of subsets of a four-element set that have 1 element. They are $\{a\}$, $\{b\}$, $\{c\}$, $\{d\}$, so $C(4, 1) = 4$.

We could also have calculated $C(4, 3)$ by noting that a three-element subset leaves out exactly one element of S. Consequently, we could leave out a, b, c, or d, leaving the three-element subsets $\{b, c, d\}$, $\{a, c, d\}$, $\{a, b, d\}$ and $\{a, b, c\}$; so $C(4, 3) = 4$.

To find $C(4, 4)$ we find the number of subsets of a four-element set S that have four elements. There is only one, namely, S itself, so $C(4, 4) = 1$.

By extending the analysis for $C(4, 2)$ and $C(4, 3)$ to $C(n, r)$, the number of r-element subsets of an n-element set, we get the following equation.

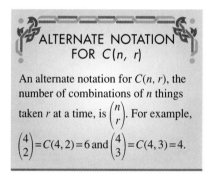

ALTERNATE NOTATION FOR $C(n, r)$

An alternate notation for $C(n, r)$, the number of combinations of n things taken r at a time, is $\binom{n}{r}$. For example, $\binom{4}{2} = C(4, 2) = 6$ and $\binom{4}{3} = C(4, 3) = 4$.

$$
(2)\quad C(n, r) = \frac{P(n, r)}{r!} = \frac{n(n-1)(n-2) \cdots (n - r + 1)}{r!}
$$

This can be written as follows:

$$
(3)\quad C(n, r) = \frac{n!}{r!\,(n - r)!}
$$

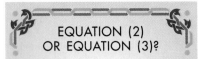

EQUATION (2)
OR EQUATION (3)?

Which equation should you use, (2) or (3)? If your calculator has a factorial key, equation (3) is quicker. If you are doing calculations by hand or if your calculator lacks a factorial key, equation (2) is quicker. Example 8.8 uses both formulas.

For example,

$$C(4, 2) = \frac{4!}{2!(4-2)!} = \frac{4!}{2!2!} = 6$$

$$C(4, 3) = \frac{4!}{3!(4-3)!} = \frac{4!}{3!1!} = 4$$

EXAMPLE 8.8

In how many ways can a committee of three students be selected from a class of 30 students?

SOLUTION Since order doesn't matter, we can apply equation (2) to get

$$C(30, 3) = \frac{30 \cdot 29 \cdot 28}{3 \cdot 2 \cdot 1} = 4060$$

Alternatively, formula (3) gives

$$C(30, 3) = \frac{30!}{3!27!} = 4060$$

Which Counting Method Do I Use?

We have talked about three counting methods—the multiplication rule, the number of permutations of n things taken r at a time, and the number of combinations of n things taken r at a time. When we encounter a situation in which a counting method is useful, there are three suggestions to keep in mind.

1. If not all the objects are from the same set (a paint color, an interior, and a wheel cover; a car and a driver) or if objects may be repeated (use the same number or letter again), use the multiplication rule.
2. Permutations involve the order of the objects. Also, with either the multiplication rule or counting the number of permutations, we can start by putting blanks to represent choices.
3. Combinations do not involve the order of objects.

Let's try these suggestions in several examples.

EXAMPLE 8.9

An instructor tells a class that a quiz will consist of two questions from a list of six questions. How many quizzes are possible?

SOLUTION The order of questions does not matter here, and the objects are from the same set of six questions. Therefore, we're dealing with combinations. In particular, we want to know the number of combinations of 6 things taken 2 at a time.

$$C(6, 2) = \frac{6 \cdot 5}{2 \cdot 1} = 15$$

There are 15 possible quizzes.

EXAMPLE 8.10

A university course abbreviation consists of three letters followed by three digits, e.g., MAT 203, ENG 101. How many course abbreviations are possible?

SOLUTION

$$\underbrace{26 \cdot 26 \cdot 26}_{\text{Letters}} \cdot \underbrace{10 \cdot 10 \cdot 10}_{\text{Digits}} = 17{,}576{,}000$$

There are 17,567,000 such abbreviations. Notice that we are counting neither permutations nor combinations, since the elements are chosen from different sets (letters or numbers).

EXAMPLE 8.11

From the applicants for president of a college there are 10 semifinalists.

a. If members of the screening committee are asked to rank the 10 semifinalists, how many rankings are possible?
b. If only the top five are to be selected and ranked, how many lists are possible?
c. If the top five are to be selected but not ranked, how many lists are possible?

SOLUTION

a. $\underline{10} \cdot \underline{9} \cdot \underline{8} \cdot \underline{7} \cdot \underline{6} \cdot \underline{5} \cdot \underline{4} \cdot \underline{3} \cdot \underline{2} \cdot \underline{1} = 3{,}628{,}800$
b. $\underline{10} \cdot \underline{9} \cdot \underline{8} \cdot \underline{7} \cdot \underline{6} = 30{,}240$
c. Since order doesn't matter, we are looking for combinations. In particular,

$$C(10, 5) = \frac{10 \cdot 9 \cdot 8 \cdot 7 \cdot 6}{5 \cdot 4 \cdot 3 \cdot 2 \cdot 1} = 252$$

The next example requires more than one counting method.

EXAMPLE 8.12

Students at a high school are going to select a homecoming court consisting of three boys and three girls. The three boys will be selected from a list of nine boys and the three girls will be selected from a list of ten girls. How many courts are possible?

SOLUTION

$$\underbrace{C(9, 3) \cdot C(10, 3)}_{} = \frac{9 \cdot 8 \cdot 7}{3 \cdot 2 \cdot 1} \cdot \frac{10 \cdot 9 \cdot 8}{3 \cdot 2 \cdot 1} = 10{,}080$$

Number of ways to pick 3 boys from 9
Number of ways to pick 3 girls from 10

EXERCISES 8.3

1. If a department designator consists of three letters, such as MAT for mathematics or ELE for elementary education, how many designators are possible?

2. In selecting a menu for a banquet there are two choices of salad, six choices of entree, five choices of vegetable, and four choices of dessert. How many menus are possible.

3. List all the permutations of {A, B, C, D}.

4. Calculate 2! and 6!.

5. In Example 8.6 the solution was found by calculating
$$P(30, 3).$$
State a situation related to your major that deals with $P(n, r)$, the number of permutations of n things taken r at a time.

6. An art gallery has nine paintings from which five will be selected and hung on a wall. How many different arrangements are possible? In other words, in how many ways can five paintings be selected and then arranged?

In Exercises 7–11, calculate the indicated number.

7. $P(10, 2)$
8. $P(10, 10)$
9. $C(10, 2)$
10. $C(10, 8)$
11. $C(10, 10)$
12. A basketball team has 14 players on it. If a coach does not take into consideration what position a player plays (for example, for a drill), how many teams of five players can be formed?
13. A monogram on a sweater consists of two initials, such as DN or MM. How many monograms are possible?
14. There are 13 chapters in this book. In how many possible orders could they have been arranged?

15. A group of seven people enters a restaurant. Instead of waiting for a table, they decide to split up into a group of four and a group of three. In how many different ways can they do this?
16. In a family with two children there are four family patterns with respect to gender of the children: BB, BG, GB, and GG. How many patterns are there for a family with the following?
   a. Three children
   b. Four children
17. Suppose you go out to eat with 11 other people, and altogether there are six females and six males. Two tables will be used, each with three females and three males. How many different sets of people could be at the table with you?
18. Sweet Honey in the Rock is a quintet of vocalists. They sit or stand on the stage in a row. After nearly every song they switch some positions where they are sitting or standing. In how many different orders could the five of them appear? Do not differentiate between sitting and standing.

WRITTEN ASSIGNMENTS 8.3

1. Some people say that having four kids in a family is more than twice as much work as having two. One of the authors believes that this can be supported by noting that conflict usually arises between two children. In a family with two children there is only one pair, while in a family with four children there are $C(4, 2) = 6$ pairs of children, so conflict is six times as likely. Let's call these numbers the conflict index of a family.

Number of children	1	2	3	4	5	6
Conflict index	0	1		6		

   a. Complete the table for the conflict index.
   b. Do you agree that the amount of work in raising a family is related to the conflict index? Do the conflict index numbers shown match your sense of the relative conflict in large versus small families? In other words, does the conflict index provide a good mathematical model for conflict within a family? Write one to three paragraphs describing why or why not.

2. Recall your experience with combination locks. Should they be called permutation locks rather than combination locks? Explain.
3. Construct a flowchart to answer the following question: "Which counting method do I use?"
4. The following is one example of what is called in the literature "the traveling salesman problem."

   Suppose you want to visit the six largest cities in the country by going from one to the other without going through any city twice and to return to the city where you started. You'd like to do this by the least expensive route, so you plan to write down all the possible routes and calculate the cost of each. Since a route, say, NLCHPS (this represents New York to Los Angeles to Chicago to Houston to Philadelphia to San Diego), and its reverse, SPHCLN, have the same cost, call them the same route. How many routes are possible?

   Find out what is known about a solution to the traveling salesman problem, and write one or two pages describing what you find.

8.4 ● DETERMINING PROBABILITIES USING COUNTING RULES

GOAL   To determine the probability of an event using counting rules.

In determining the probability of an event it is often convenient to use the counting rules discussed in the previous section. We proceed by looking at three examples.

EXAMPLE 8.13

A coin is flipped four times. What is the probability of each of the following?

a. All four flips come up heads.
b. Two flips are heads and two flips are tails.

SOLUTION How many sequences of heads and tails are possible?

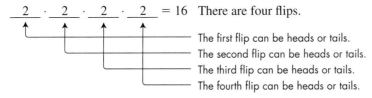

$$\underline{2} \cdot \underline{2} \cdot \underline{2} \cdot \underline{2} = 16 \quad \text{There are four flips.}$$

The first flip can be heads or tails.
The second flip can be heads or tails.
The third flip can be heads or tails.
The fourth flip can be heads or tails.

This can also be seen from the following tree diagram.

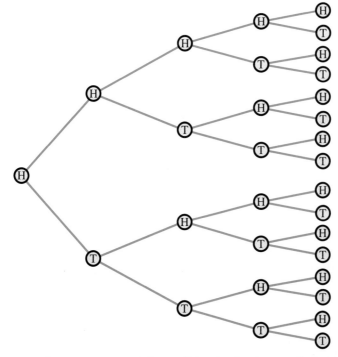

a. Only one of the sequences produces all heads. (In the tree diagram this is the sequence found by taking the top branch all the way through.) Let's assume that any two sequences are equally likely. Then the probability that all four flips will come up heads is 1/16.

b. To have two flips be heads and two be tails, exactly two of the four positions in the sequence are filled by heads, for example, HHTT and HTHT. In how many ways can two of the four positions be filled by heads? Since order doesn't matter, we want the number of combinations of 4 things taken 2 at a time. This is

$$C(4, 2) = \frac{4 \cdot 3}{2 \cdot 1} = 6$$

Therefore, the probability that two flips are heads and two are tails is 6/16 = 3/8. Does this surprise you? The most common incorrect answer is 1/2. In the tree diagram, pick out the six paths that have two heads and two tails.

 YOUR FORMULATION

Example 8.13 showed that if a coin is flipped four times, the probability that exactly half the flips are heads is 3/8. What happens for a different number of flips?

1. Suppose two coins are flipped. What is the probability that one flip is heads and one is tails?
2. As the number of flips increases (from 2 to 4 to 6, etc.), what happens to the probability that exactly half the flips are heads?
   a. It goes up.   b. It stays the same.   c. It goes down.
3. Verify that your answer to part 2 is correct in the case when six coins are flipped by calculating the probability that exactly three flips are heads.

EXAMPLE 8.14

A dog has a litter of six puppies. What is the probability of each of the following?

a. All six are female.

b. Three are female and three are male.

SOLUTION How many birth orders are there in terms of gender?

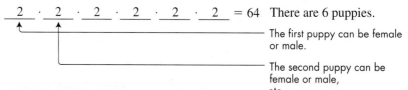

a. Only one of the birth orders produces six female puppies, namely, FFFFFF. We assume that any two birth orders are equally likely, so the probability that all six puppies are female is

$$\frac{\text{Number of orders in which all six are female}}{\text{Number of birth orders}} = \frac{1}{64}$$

b. To have three female and three male puppies, exactly three of the six positions in the birth order (three of the six blanks) are filled by females. Since the order among the three female positions doesn't matter, we want the number of combinations of 6 things taken 3 at a time. This is

$$C(6, 3) = \frac{6 \cdot 5 \cdot 4}{3 \cdot 2 \cdot 1} = 20$$

Therefore, the probability that exactly three puppies are female and three are male is 20/64 = 5/16.

EXAMPLE 8.15

Michigan has a lottery game called Lotto 47 in which the player chooses six distinct numbers between 1 and 47 inclusive. To win the grand prize, the player must pick all the numbers correctly. What is the probability of winning the grand prize?

SOLUTION

$$\text{Probability of winning} = \frac{\text{Number of ways to win}}{\text{Number of ways to pick 6 numbers from 47}}$$

$$= \frac{1}{C(47, 6)} = \frac{1}{10{,}737{,}573} \approx .000000093$$

This probability of about 1 in 11 million means that about every 11 million times you play Lotto 47 you could expect to win once.

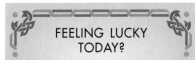

FEELING LUCKY TODAY?

To get 10,737,573 Lotto 47 tickets, you could buy 7000 tickets each day of four years in college (365 days a year for three years and 366 days for a leap year) and receive 510,573 Lotto 47 tickets as a graduation present.

EXERCISES 8.4

1. A coin is flipped 10 times. What is the probability of each of the following?
   a. All 10 flips come up heads.
   b. Five flips are heads and five flips are tails.
2. Refer to Example 8.14. Complete the following table.

Number of female puppies	0	1	2	3	4	5	6
Probability				$\frac{20}{64}$			$\frac{1}{64}$

Although this is not in reduced form it is convenient to leave it this way to check that the sum of the probabilities is 1.

3. On an eight-question multiple-choice test, each question has five possible answers. One test taker guesses at each answer. What is the probability of each of the following?
   a. The test taker gets all eight questions correct.
   b. The test taker gets four questions right and four questions wrong.
4. A bridge hand consists of 13 cards from a 52-card deck. What is the probability that a bridge hand contains the following?
   a. All four aces
   b. No aces

One of 635, 013, 559, 600 possible bridge hands.

5. Consider the Lotto 47 game referred to in Example 8.15. Suppose people complain that too few people win the grand prize. Two ways to modify it are to pick six numbers from a smaller set, say, 46 numbers, or to pick fewer numbers, say, five, from 47.
   a. If a prize is awarded for choosing six numbers from 46 and having all six be correct, what is the probability of winning a prize?
   b. If a prize is awarded for choosing five numbers from 47 and having all five be correct, what is the probability of winning a prize?
6. A jury pool consists of 14 men and 16 women. Twelve jurors are to be selected at random from the jury pool.
   a. What is the probability that the jury is all male?
   b. What is the probability that the jury is exactly half male?
7. In pocket billiards (pool) there are 15 balls, numbered 1 through 15. Those numbered 1 through 7 are called solids and those numbered 9 through 15 are called stripes, and the number-8 ball is special. Suppose on the break (the shot that starts the game) two balls go into pockets. If it is equally likely that each ball will go in a pocket, what is the probability of each of the following?
   a. Both are solids.
   b. Both are the same type, i.e., both solids or both stripes.
8. In a clothing store's promotional sale, for each item you buy you can draw a capsule that tells whether you receive a 25% discount or a 10% discount. There are 100 capsules, 40 of which say 25% discount and 60 of which say 10% discount. If you buy three items and consequently draw three capsules, what is the probability of each of the following?
   a. All three are for a 25% discount.
   b. Exactly two are for a 25% discount.
9. Let p be the probability of getting exactly two heads in the flip of four coins and p' be the probability of getting exactly three heads in six flips of a coin. How do p and p' compare?
   i. $p < p'$
   ii. $p = p'$
   iii. $p > p'$

WRITTEN ASSIGNMENTS 8.4

1. Refer to the puppy example of Example 8.14 and Exercise 2. Write a paragraph or two to describe intuitively to a fellow student why 1/2 is too large an answer to Example 8.14b.
2. Consider Example 8.15 about Michigan's Lotto 47 game. Suppose you could talk other people into going together with you and buying up all the possible lottery tickets, i.e., buy every possible combination of six of the 47 numbers.
   a. Discuss the merits of this investment. Beware of tying with someone else.
   b. How long would it take to buy all the tickets? Make some assumption about the time needed to purchase a ticket.
3. There are 20 applicants for a certain position. Five are women and 15 are men. Four applicants will be selected for an interview. Suppose the four applicants selected are all men. If any two applicants are equally likely to have an interview, having all four interviewees be men would happen by chance only 28% of the time. Does this lead you to conclude that the selection process is biased against women? Why or why not.

PROBABILITIES OF COMPOUND EVENTS

GOALS
1. To calculate probabilities involving the union of two events.
2. To calculate conditional probabilities.
3. To understand the concepts of mutually exclusive events and independent events and the relation between them.
4. To calculate probabilities involving the intersection of two events.

In this section we examine the probabilities of compound-events—events whose outcomes are described in terms of two or more other events. For example, the 1990 census showed that 11.9% of the U.S. population age 5 years and over lived in California, 13.8% of the population age 5 years and over spoke a language at home other than English, and 3.7% of the population age 5 years and over lived in California and spoke a language at home other than English. If a person age 5 or over is picked at random from the U.S. population, what is the probability the person is from California or speaks a language at home other than English (or both)?

In Section 8.2 a sample space was described as the set of all possible outcomes of an experiment. An *event* is a collection of outcomes. For example, if a card is drawn from a standard deck and the card recorded, the event "get an ace" occurs when any of the four aces is drawn. The event is a subset of the sample space. Set notation can describe the union and intersection of two events. If A and B are events, $A \cup B$ is the set of all outcomes in A or B or both and $A \cap B$ is the set of all outcomes in both A and B. These compound events are described by the Venn diagrams in ▶ Figures 8.2 and 8.3.

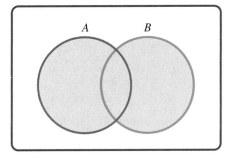
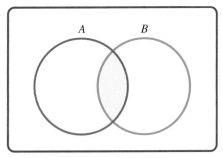

▶ FIGURE 8.2 $A \cup B$ ▶ FIGURE 8.3 $A \cap B$

For example, if a student is selected from the college population and the person's classification and gender are recorded, let

S denote the outcome "sophomore"
J denote the outcome "junior"
F denote the outcome "female"
M denote the outcome "male"

Then $S \cup F$ denotes the outcome "sophomore or female" and $S \cap M$ denotes the outcome "sophomore and male." Can we find probabilities such as $P(S \cup F)$ and $P(S \cap M)$ without knowing explicitly all the outcomes in the sample space?

EXAMPLE 8.16

A student is selected at random from a certain college. Suppose the probability of selecting a sophomore is .26, the probability of selecting a female is .53, and the probability of selecting a female sophomore is .14. What is the probability the student is a sophomore or a female?

SOLUTION We could add the percentage of sophomores and the percentage of females to get $26\% + 53\% = 79\%$. However, this counts sophomore females twice—once as sophomores and once as females. To compensate for this we subtract out the 14% who have been double-counted. Therefore, the percentage of students who are sophomores or females is $26\% + 53\% - 14\% = 65\%$, so the probability a student is a sophomore or a female is .65.

The conclusion of Example 8.16 can be written as follows:

$$P(S \cup F) = P(S) + P(F) - P(S \cap F)$$

In general, if A and B are events, then, the following holds.

> (4) $P(A \cup B) = P(A) + P(B) - P(A \cap B)$

Notice that $P(A \cap B)$ is subtracted to eliminate double-counting.

EXAMPLE 8.17

Answer the question in the first paragraph of this section.

SOLUTION Let's refer to the event of selecting a person age 5 or over who lives in California as CA and the event of selecting a person age 5 or over who speaks a language at home other than English as NE. The probability that a randomly selected person age 5 or over is a Californian, or a non-English speaker is

$$P(CA \cup NE) = P(CA) + P(NE) - P(CA \cap NE) = .119 + .138 - .037 = .220$$

It sometimes happens that $A \cap B = \varnothing$. In that case, events are called *mutually exclusive*. For example, in picking a voter, events *Y:* the voter is under 25 (young), and *S:* the voter is over 65 (senior), are mutually exclusive. When events are mutually exclusive, $P(A \cup B) = P(A) + P(B)$, as shown in Example 8.18.

EXAMPLE 8.18

The probability that a state's general sales and use tax is under 4% is .06, and the probability it is over 6% is .10. What is the probability that a state's general sales and use tax is either under 4% or over 6%?

SOLUTION Let S represent the event that a state's general sales and use tax is under 4% (think S = small) and B represent the event that it is over 6% (think B = big). Then

$$P(S \cup B) = P(S) + P(B) - P(S \cap B) = .06 + .10 - 0 = .16$$

There are no outcomes in $S \cap B$: The tax cannot be both under 4% and over 6%.

Notice that S and B are mutually exclusive events.

Conditional Probability

If it is known that one event has happened, that knowledge might alter the calculation of the probability of another event. For example, suppose you have randomly selected one card from a deck of 52 cards. If you are told that it is a red card, what is the probability it is a heart? Since hearts and diamonds are equally likely, the probability is 1/2.

This is a *conditional probability* and it is denoted $P(A \mid B)$, read as "the probability of A given B," where B represents what is known before computing the probability of A. In general, if A and B are events, to find the probability of A given B, i.e., to find $P(A \mid B)$, consider that B has happened ("given B"), and focus on the

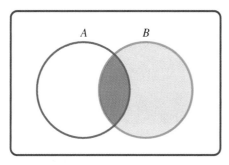

▶ FIGURE 8.4 *A* given *B*

(light and dark) shaded areas in ▶ Figure 8.4. Given that the outcome is also in *A*, we focus on the part of *A* that is also in *B* (dark blue), so the following holds.

(5) $P(A \mid B) = \dfrac{P(A \cap B)}{P(B)}$

This makes sense only if $P(B) \neq 0$.

In the deck of cards example, if we let *A* denote the outcome of selecting a heart and *B* the outcome of selecting a red card, we have $P(B) = 1/2$ and $P(A \cap B) = 1/4$. Thus

$$P(A \mid B) = \frac{1/4}{1/2} = 1/2$$

●EXAMPLE 8.19

Refer to Example 8.16, where $P(S) = .26$, $P(F) = .53$, and $P(S \cap F) = .14$.

a. If a female student is selected at random from this college, what is the probability she is a sophomore?

b. If a sophomore is selected, what is the probability the person is female?

SOLUTION

a. $P(S \mid F) = \dfrac{P(S \cap F)}{P(F)} = \dfrac{.14}{.53} \approx .264$

b. $P(F \mid S) = \dfrac{P(F \cap S)}{P(S)} = \dfrac{.14}{.26} \approx .538$

Dependent and Independent Events

If we refer to the census data at the beginning of this section, it can be shown that $P(NE \mid CA) = .311$ and $P(NE) = .138$. Knowing that a person was from California more than doubled the probability the person was a non-English speaker. It is said that events NE and CA are *dependent,* since knowing that one has happened changes the likelihood that the other one happened.

In general, two events A and B are *independent* provided $P(A \mid B) = P(A)$ or, equivalently, $P(B \mid A) = P(B)$. In other words, the occurrence of B has not affected the probability of the occurrence of *A,* or, equivalently, the occurrence of *A* has not affected the probability of the occurence of *B*. Two events that are not independent are called *dependent*. In other words, A and B are dependent provided $P(A \mid B) \neq P(A)$ or, equivalently, $P(B \mid A) \neq P(B)$.

●EXAMPLE 8.20

In drawing a card from a standard 52-card deck, are the events "get a club" and "get an ace" independent?

SOLUTION Denote the events by *C* and *A,* respectively. $P(A \mid C) = 1/13$ and $P(A) = 4/52 = 1/13$. Since $P(A \mid C) = P(A)$, *C* and *A* are independent.

●EXAMPLE 8.21

Refer to Example 8.19 and determine whether the events "a female is selected" (*F*) and "a sophomore is selected" (*S*) are dependent or independent.

SOLUTION If we reexamine Example 8.19, we see that $P(S \mid F) \approx .264$ and $P(S) = .26$. Although these numbers are close in value, they are not equal. Therefore, the events are dependent.

Relation Between Mutually Exclusive and Independent Events

Let's now look at the relationship between mutually exclusive events and independent events. Suppose *A* and *B* are mutually exclusive events, with one of them having some positive probability of occurring (see ▶ Figure 8.5). For example, let *A* be the event "select a red card" and *B* be the event "select a black card" when one card is randomly selected from a standard deck of cards. In this case, $P(A) = .5$ and $P(A \mid B) = 0$. Thus, the events are dependent, since $P(A \mid B) \neq P(A)$. This reasoning leads to the following result:

If *A* and *B* are mutually exclusive events and either $P(A) > 0$ or $P(B) > 0$, then *A* and *B* are dependent.

This conditional sentence leads us directly to a result for independent events. If *A* and *B* are independent, with $P(A) > 0$ or $P(B) > 0$, then *A* and *B* are not mutually exclusive. [Recall that $(p \Rightarrow q) \Leftrightarrow (\sim q \Rightarrow \sim p)$.] To illustrate, let *R* be the event you "select a red card" and *A* be the event you "select an ace" when one card is randomly selected from a standard deck. These events are independent [$P(R \mid A) = 1/2$ and $P(R) = 1/2$]. Therefore, they cannot be mutually exclusive.

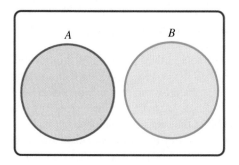

▶ FIGURE 8.5 Mutually exclusive events

YOUR FORMULATION

Suppose events A and B are dependent. Must they be mutually exclusive? Choose one of the following answers, and support it.

1. A and B are mutually exclusive.
2. A and B are not mutually exclusive.
3. A and B might be either mutually exclusive or not mutually exclusive.

The Probability of A and B

If A and B are events with known probabilities $P(A)$ and $P(B)$, what is the probability that A and B both occur? In other words, what is $P(A \cap B)$? Equation (5) provides an answer.

$$P(A \mid B) = \frac{P(A \cap B)}{P(B)} \qquad \text{This is equation (5).}$$

Solve for $P(A \cap B)$ to obtain $\qquad P(A \cap B) = P(A \mid B)P(B)$.

A similar result holds if you start with $P(B \mid A)$.

In summary, the probability of A and B, $P(A \cap B)$, can be found by the following.

> (6) a. $P(A \cap B) = P(A \mid B)P(B)$
> b. $P(A \cap B) = P(B \mid A)P(A)$

LIGHTEN YOUR LOAD

Either part of equation (6) can be derived from the other by interchanging A and B. Either part can be used to find $P(A \cap B)$. We showed how equation (6) can be derived from equation (5). By reversing the steps, equation (5) can be derived from equation (6). Therefore, it is unnecessary to memorize three separate equations [equation (5) and both parts of equation (6)]. Instead, you can remember any one of them and derive the other two from it.

EXAMPLE 8.22

The probability that a 45-year-old male dies before his next birthday is .00218. If two 45-year-old males are selected at random, what is the probability of each of the following?

a. Both die before their next birthdays.
b. Neither of them dies before his next birthday.
c. At least one dies before his next birthday.

SOLUTION

a. P(both die before their next birthdays)
= P(first dies before his next birthday)
    × P(second dies before his next birthday | first dies before his next birthday)
= (.00218)(.00218)   We treat the two events as independent: The first male's death before his next birthday should not affect the probability that the second male will die before his next birthday.
≈ .00000475

b. P(neither dies before his next birthday)
= P(both live until their next birthdays)
= P(second lives until his next birthday | first lives until his next birthday)
    × P(first man lives until his next birthday)
= (.99782)(.99782)
≈ .99564

c. P(at least one dies before his next birthday) $= 1 - P$(neither dies before his next birthday)

$\approx 1 - .99564 \qquad$ (from part b)

$= .00436$

Notice that problems that were solved in Section 8.4 using counting rules can also be solved—usually more easily—by the product rule of equation (6). We revisit Example 8.14 in the following example.

EXAMPLE 8.23

A dog has a litter of six puppies. What are the following probabilities?

a. All six are female.
b. Three are female and three are male.

SOLUTION Assume that the probability that a puppy is female is 1/2; i.e., $P(F) = 1/2$.

a. P (all six are female) $= \dfrac{1}{2} \cdot \dfrac{1}{2} \cdot \dfrac{1}{2} \cdot \dfrac{1}{2} \cdot \dfrac{1}{2} \cdot \dfrac{1}{2} = \dfrac{1}{64}$

P (first puppy is female)
P (second is F | first is F)
P (third is F | first and second are F)
etc.

b. One way to get three female and three male is MMMFFF. This probability is

$$\frac{1}{2} \cdot \frac{1}{2} \cdot \frac{1}{2} \cdot \frac{1}{2} \cdot \frac{1}{2} \cdot \frac{1}{2} = \frac{1}{64}$$

Another way is MMFFMF, with probability

$$\frac{1}{2} \cdot \frac{1}{2} \cdot \frac{1}{2} \cdot \frac{1}{2} \cdot \frac{1}{2} \cdot \frac{1}{2} = \frac{1}{64}$$

How many ways are there? We want to fill exactly three of the positions in the birth order with females. The number of ways to do this is

$$C(6, 3) = \frac{6 \cdot 5 \cdot 4}{3 \cdot 2 \cdot 1} = 20$$

Thus $\qquad P(\text{3F and 3M}) = 20 \cdot \dfrac{1}{64} = \dfrac{5}{16}$

This is the same answer we arrived at in Example 8.14.

Recall that when events A and B are independent, $P(B \mid A) = P(B)$. In this case, equation (6) becomes the following.

(7) When A and B are independent, $P(A \cap B) = P(A)P(B)$.

Notice that the events of interest in Examples 8.22 and 8.23 were independent.

The Birthday Problem

We conclude this section with a problem commonly known as the birthday problem. Here's a particular form of it: If 30 people are in a room, what is the probability that two or more of them have a birthday on the same day (not necessarily the same year). The problem can be varied by replacing 30 by another number. In modeling the problem we make the simplifying assumption that the same number of people are born on each day of the year, so that if a person is picked at random, the probability the person was born today is 1/365.

The process for solving the birthday problem can be illustrated by considering the situation for only four people. In so doing, the calculations will be simplified. (Reducing a problem to a simpler case is one of the techniques emphasized in the problem-solving sections of Chapter 1.) Since for any event B, $P(B$ happens$)$ + $P(B$ doesn't happen$)$ = 1, $P(B$ happens$)$ = $1 - P(B$ doesn't happen$)$, the solution for four people is:

What is the probability that two or more have the same birthday?

$$P(\text{two or more have the same birthday}) = 1 - P(\text{no two have the same birthday})$$
$$= 1 - 1 \cdot \frac{364}{365} \cdot \frac{363}{365} \cdot \frac{362}{365}$$

The first person has a birthday with probability 1.
The second person has a birthday different than the first person, with probability $\frac{364}{365}$. We assume that any two birthdays are equally likely. People born on February 29 are counted as if they were born on March 1 or February 28.

$$\approx 1 - .984 = .016$$

We next apply this process to other numbers to obtain the data shown in ● Table 8.3 for the probability that two or more have the same birthday.

TABLE 8.3	
Probability that Two People Have the Same Birthday	
Number of People in a Room	Probability Two or More Have the Same Birthday
4	.016
10	.117
22	.476
23	.507*
30	.706
40	.891
50	.970
60	.994

*Notice that in a set of 23 people the probability is over 1/2 that two or more have the same birthday while in a set of 22 people the probability is under 1/2.

1. Suppose A and B are events with $P(A) = .2$, $P(B) = .3$, and $P(A \cap B) = .1$. Find $P(A \cup B)$.

2. Suppose E and F are events with $P(E) = .45$, $P(F) = .57$, and $P(E \cup F) = .85$. Find $P(E \cap F)$.

3. Suppose I and J are events with $P(I) = .72$, $P(I \cup J) = .91$, and $P(I \cap J) = .42$. Find $P(J)$.

4. In 1995 about 5.1% of the world population lived in North America, about 33.6% were Christian, and about 4.4% were North American Christians. If a person were picked at random in 1995, what is the probability the person was a North American or a Christian?

5. In 1991, 18% of the states had over 5 million automobiles registered, 26% of the states had over 100,000 motorcycles registered, and 16% of the states had both over 5 million automobiles registered and over 100,000 motorcycles registered. If a state is selected at random, what is the probability that it had over 5 million automobiles or over 100,000 motorcycles registered?

6. Which statement is true for any events A and B?
   a. $P(A \cup B) < P(A \cap B)$
   b. $P(A \cup B) \leq P(A \cap B)$
   c. $P(A \cup B) = P(A \cap B)$
   d. $P(A \cup B) \geq P(A \cap B)$
   e. $P(A \cup B) > P(A \cap B)$

7. Give an example of two events A and B so that $P(A \cup B) = 0$.

8. Give an example of two events A and B so that $P(A \cup B) = 1$.

9. Give an example of two events that are mutually exclusive.

10. Suppose a card is drawn from a standard 52-card deck. Give an example of an event that is mutually exclusive from the event "get a club."

11. The probability that a state's record high temperature is 120°F or over is .18, and the probability that a state's record high temperature is under 110°F is .22. What is the probability that a state's record high temperature is the following?
    a. Either under 110°F or 120°F or over
    b. At least 110°F but less than 120°F
    c. At least 110°F
    d. Under 120°F

12. Suppose A and B are events with $P(A) = .3$, $P(B) = .4$, and $P(A \cap B) = .1$. Find the following.
    a. $P(A \mid B)$
    b. $P(B \mid A)$
    c. Are A and B mutually exclusive?

13. Refer to events S and B in Example 8.18. Find the following.
    a. $P(S \mid B)$
    b. $P(B \mid S)$

14. If events A and B are mutually exclusive and neither event has zero probability, find the following.
    a. $P(A \mid B)$
    b. $P(B \mid A)$

15. Consider the following data on tropical storms and hurricanes, 1886–1991.

Month	Number of Tropical Storms (Including Hurricanes)	Number of Tropical Storms that Reached Hurricane Intensity
Jan–Apr	3	1
May	14	3
June	56	23
July	69	36
Aug	216	150
Sept	302	189
Oct	188	96
Nov	43	22
Dec	6	3
Total	897	523

a. If a tropical storm occurred in October, what is the probability it reached hurricane intensity?

b. Given that a tropical storm reached hurricane intensity, what is the probability it occurred in October?

16. The distribution of medals for the 1994 Olympic Winter Games in Lillehammer, Norway, is shown. Assume that each athlete who won a medal won exactly one medal.

	Gold	Silver	Bronze	Total
Norway	10	11	5	26
Germany	9	7	8	24
Russia	11	8	4	23
Italy	7	5	8	20
United States	6	5	2	13
Canada	3	6	4	13
Switzerland	3	4	2	9
Austria	2	3	4	9
S. Korea	4	1	1	6
Finland	0	1	5	6
Japan	1	2	2	5
France	0	1	4	5
Netherlands	0	1	3	4
Sweden	2	1	0	3
Kazakhstan	1	2	0	3
China	0	1	2	3
Slovenia	0	0	3	3
Ukraine	1	0	1	2
Belarus	0	2	0	2
Great Britain	0	0	2	2
Uzbekistan	1	0	0	1
Australia	0	0	1	1

a. If a randomly selected athlete won a medal at these games, what is the probability the athlete was from Norway?

b. If a randomly selected medalist was from Canada, what is the probability the medal was gold?

17. Use the data of Exercise 15. Is a tropical storm's reaching hurricane intensity independent of its happening in September?

18. Refer back to the data of Exercise 16. Is a medal going to the United States independent of its being gold?

19. In drawing a card from a standard 52-card deck, consider the following events.

    R: draw a red card
    D: draw a diamond
    K: draw a king

    a. Are R and D independent?
    b. Are R and K independent?
    c. Are D and K in dependent?

20. If events A and B are mutually exclusive and both can happen, when are they independent—always, sometimes but not always, or never?

21. The probability that an 80-year-old male will die before his next birthday is 0.07407. If two 80-year-old males are selected at random, what are the following probabilities?

    a. Both live until their next birthday.
    b. At least one lives until his next birthday.
    c. Both die before their next birthdays.

22. Suppose five cards are drawn from a standard 52-card deck. What is the probability that all five are spades?

23. In the game of pool, there are 15 numbered balls: 1–7 are called solids, there is the eight-ball, and 9–15 are called stripes. Suppose when the balls are broken two balls go in pockets. Regard this as a random event. What is the probability of each of the following?

    a. Both are solids.
    b. Both are the same type (either both solids or both stripes).
    c. One is the eight ball.

24. In the mid 1990s, Mark Price of the Cleveland Cavaliers had the leading career free throw percentage in the National Basketball Association, at 90.6%. Suppose in a game he had two free throw attempts.

    a. What is the probability he made both of them?
    b. What assumption(s) do you make in answering part a?
    c. What is the probability he missed both of them?
    d. What is the probability he made at least one of them?

25. A sock drawer contains six blue socks and eight gray ones, and two socks are chosen at random. What are the following probabilities?

    a. Both are blue.
    b. Both are gray.
    c. They are the same color.

26. In a raffle for a single prize, 300 tickets are sold.

    a. What is the probability that a person who buys one ticket will win?
    b. What is the probability that a person who buys four tickets will win?

27. Three people are in a room. Calculate the probability that two or more have the same birthday.

28. To calculate the probability that, in a room with four people, two or more have the same birthday, we assumed there are 365 birthdays, each equally likely. Suppose there are 366 birthdays (count February 29), each equally likely.

    a. To three decimal places, by how much do the probabilities differ?
    b. To four decimal places, by how much do the probabilities differ?

29. Consider the census data from the first paragraph of this section: 11.9% are Californians, 13.8% are non-English speakers, and 3.7% are both Californians and non-English speakers. If a Californian is picked at random, what is the probability that that person is a non-English speaker? (*Hint:* Let CA denote the event: the person is Californian, and let NE denote the event: The person is a non-English speaker. Percentages of the U.S. population in each of three places are indicated in the figure.)

---

1. Equation (4) says that for events A and B, $P(A \cup B) = P(A) + P(B) - P(A \cap B)$. Write a paragraph describing in your own words why subtracting $P(A \cap B)$ is necessary.

2. A TV weather forecaster reportedly said that the probability of rain on Saturday was 50% and the probability of rain on Sunday was 50%, so there was a 100% chance of rain on the weekend. Is this correct? Why or why not?

3. Explain why the term *mutually exclusive* makes sense. That is, two events A and B are mutually exclusive provided $A \cap B = \emptyset$ or $P(A \cap B) = 0$. Why does this fit with the ordinary meaning of the words mutually exclusive?

4. In the subsection headed "Relation Between Mutually Exclusive and Independent Events," you are asked to recall that $(p \Rightarrow q) \Leftrightarrow (\sim q \Rightarrow \sim p)$. What is p and what is q? What do you do with the hypothesis that $P(A) > 0$ or $P(B) > 0$?

5. Describe in your own words the relation between the concepts *mutually exclusive* and *independent*.

6. In reference to the birthday problem, Les Krantz, in his 1992 book *What the Odds Are,* says, "If the group contains 366 or more random individuals, it is almost certain at least two will have matching birthdays." State "almost certain" as a probability, and support your answer.

THE ODDS ARE . . .

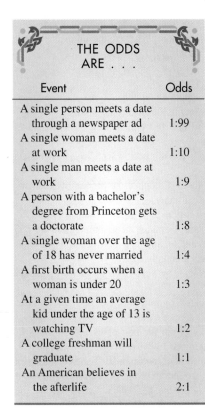

Event	Odds
A single person meets a date through a newspaper ad	1:99
A single woman meets a date at work	1:10
A single man meets a date at work	1:9
A person with a bachelor's degree from Princeton gets a doctorate	1:8
A single woman over the age of 18 has never married	1:4
A first birth occurs when a woman is under 20	1:3
At a given time an average kid under the age of 13 is watching TV	1:2
A college freshman will graduate	1:1
An American believes in the afterlife	2:1

8.6 ODDS AND EXPECTED VALUE

GOALS
1. To convert odds to probability and probability to odds.
2. To calculate expected values.

Odds

Often the likelihood of an event is stated by giving odds rather than a probability. For example, the odds against the Chicago Cubs' winning the World Series next year might be 100 to 1, denoted 100:1, or the odds that a particular boxer will win a fight are 3 to 2, or the odds against your winning a particular prize in a contest are 10,000 to 1. What do these statements mean? We translate them in terms of probabilities.

Saying the odds against the Chicago Cubs' winning the World Series next year are 100 to 1 means that it won't happen 100 times for every 1 time it will happen. Consequently, the probability they will not win it is 100/101 or, equivalently, the probability they will win it is 1/101. Saying the odds that a particular boxer will win a fight are 3 to 2 means that it will happen three times for every two times it does not happen. Therefore, the probability this boxer will win is 3/5 or, equivalently, the probability this boxer will lose is 2/5. In general, saying the odds are p to $q(p:q)$ in favor of an event means that it will happen p times for every q times it does not happen, so the probability is $p/(p + q)$ that the event will happen. Saying that the odds are p to $q(p:q)$ against an event means that the event does not happen p times for every q times it does happen, so the probability the event doesn't happen is $p/(p + q)$.

To translate from a probability to odds, suppose the probability of an event is a/b. Then the event happens a times for every $(b - a)$ times that it does not happen. Consequently, the odds in favor of the event are $a:(b - a)$.

 YOUR FORMULATION

Suppose the probability of an event is a/b. What are the odds against the event?

What are the odds for the boxer on the left?

EXAMPLE 8.24

a. What does it mean to say the odds against your winning a particular prize in a contest are 10,000 to 1?

b. What is the probability you will win this particular prize?

SOLUTION

a. The probability that you will not win the prize is 10,000/10,001.

b. 1/10,001.

EXAMPLE 8.25

In Example 8.3 it is concluded that the probability an elector after the 1968 U.S. presidential election voted for Richard Nixon is 301/538. What are the odds that an elector voted for Nixon?

SOLUTION 301 electors voted for him and 538 − 301 = 237 did not, so the odds that an elector voted for Nixon are 301 to 237.

FRANK AND ERNEST © by Bob Thaves

With permission of Bob Thaves.

Expected Value

Suppose a student guesses on every question of a 20-question multiple-choice test. What is the expected number right? Suppose a coin is flipped 10 times. How many heads are expected? These questions all are about expected value.

The *expected value* of the outcome of an experiment is the average (technically, the mean) value of the outcome if the experiment were to be repeated many times. In order to develop this concept, suppose the number of cars arriving at a fast-food restaurant follows the pattern shown in ● Table 8.4. The expected arrival pattern over a 10-minute period is shown in ● Table 8.5. Now, the average (mean) number of arrivals per 1-minute interval is

$$\frac{\text{Total number of arrivals}}{10} = \frac{0(3) + 1(4) + 2(2) + 3(1)}{10}$$

$$= 0\left(\frac{3}{10}\right) + 1\left(\frac{4}{10}\right) + 2\left(\frac{2}{10}\right) + 3\left(\frac{1}{10}\right)$$

$$= 0 + .4 + .4 + .3 = 1.1$$

Thus, on the average, 1.1 arrivals per minute are expected.

TABLE 8.4	
Arrivals per minute	Probability
0	0.3
1	0.4
2	0.2
3	0.1

TABLE 8.5	
Arrivals per minute	Number of 1-min intervals
0	3
1	4
2	2
3	1

7. The pattern of cars arriving per hour at a car wash during the past week is as follows. What is the expected number of cars arriving per hour?

Cars Arriving per Hour	Probability
3 or fewer	0
4	.10
5	.25
6	.40
7	.25
8 or more	0

8. If a coin is flipped six times, what is the expected number of heads?

9. In the game Monopoly, if a player's token lands on the space marked Community Chest, then the player draws a card from a stack of 12 cards (there should be more, but some of the cards have been lost). Two of the cards instruct the player to collect $200; three of them instruct the player to collect $100; two of them instruct the player to pay $50; and, for the remaining five cards, no money is exchanged. What is the expected value for the player who lands on Community Chest?

10. In a 50–50 drawing, tickets are sold for a fixed price. The holder of the winning ticket splits the pot of money with the sponsoring organization. For example, if 104 tickets are sold at $1 each, the holder of the winning ticket gets $52 and the sponsoring organization gets $52.

a. If 104 tickets are sold at $1 each, what is the expected value of the purchase of a ticket?

b. If 104 tickets are sold at $1 each, what is the expected value of the purchase of two (2) tickets?

c. If 20 tickets are sold at $1 each, what is the expected value of the purchase of a ticket?

d. Can you predict, in general, what the expected value of the purchase of a 50–50 ticket is?

11. One way of scoring a multiple-choice exam to correct for guessing is to subtract points for an incorrect answer. Suppose a multiple-choice question has five possible answers and that a correct answer gets 1 point, an incorrect answer costs the taker 1/5 of a point, and a blank answer is worth 0 points.

a. If a person guesses on a question, what is the expected score?

b. If a person is certain that one answer is incorrect and then guesses among the remaining four answers, what is the expected score?

12. Convert to odds the probability "one in nine hundred kazillion" in the following Luann cartoon.

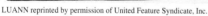

LUANN reprinted by permission of United Feature Syndicate, Inc.

WRITTEN ASSIGNMENTS 8.6

1. Explain why when odds are stated as p:q, where p and q are positive integers, p and q have no factors larger than 1 in common. In other words, explain why p and q are relatively prime. In particular, why are odds of 4:2 not stated?

2. Refer to Exercise 10. Sometimes in a 50–50 drawing, a person gets a bargain rate for buying multiple tickets. For example, tickets might cost $1 each or six for $5. Describe what such a bargain-price arrangement does to the expected value of the purchase of a ticket.

4. Describe a lottery scheme in use in your locale, and find its expected value.

5. Pascal argued for the value of believing in the religious concept of life after death by saying that its worth if it is attained is infinite, so even if it has only a small probability of being true, its expected value is positive. How do you react to his argument?

1. Name three people who were involved with the development of probability, state a contribution each of them made, and give the approximate date of the person's work.

2. Describe three developments that prompted advances in probability.

3. State Fermat's Last Theorem, and describe its current status.

4. Describe the sample space for the experiment of selecting a time zone in the continental United States.

5. The probability that a person's bachelor's degree is in business/management is 1/10. What is the probability that a person's bachelor's degree is not in business/management?

6. In each of the following, tell whether or not the probability assignment is legitimate, i.e., satisfies the two probability rules. If an assignment is not legitimate, change, if possible, one (1) number in order to make the probability assignment legitimate. If this is not possible, so state.

   a.
Outcome	A	B	C
Probability	.1	.2	.3

   b.
Outcome	A	B	C
Probability	.4	.5	.6

   c.
Outcome	A	B	C	D
Probability	.7	.8	.9	.10

7. In the board game Payday there are 31 days in an imaginary month. A player rolls a die and moves the number of spaces shown on the top face to a day on the calendar. On each day there is a description of what happens. On days 1, 3, 5, 11, 16, 19, 22, and 24 the player collects mail, which in turn may give further instructions. On days 2, 8, and 31 the player collects money. On days 4, 12, 15, and 25 the player gets a "Deal" card, which permits the purchase of a "deal." On days 6, 13, 18, and 23 the player pays money. Days 7, 14, 21, and 28 are "Sweet Sundays," which require no further action. Days 9, 17, 20, and 30 are labeled "Buyer" and permit the player to receive compensation for a "Deal" card. Days 10, 26, 27, and 29 have special instructions.

   a. If the player is at the start (day 0), what are the following probabilities for the first roll?
      i. The player collects mail.
      ii. The player gets a "Deal" card.
   b. At a random point in the game, what is the probability that the player is on the following?
      i. "Sweet Sunday"
      ii. A day to collect mail

8. The assignment of probabilities in Review Exercise 7 is by the axiomatic method. Give an example of a problem or situation in which the axiomatic method would *not* be an appropriate method to assign probabilities. You need not assign the probabilities; just state the problem or situation.

9. On a three-question multiple-choice quiz each question has five possible answers. How many sequences of answers are possible? (A, D, C and A, A, E are two possible sequences.)

10. List all the permutations of the letters C, A, R.

11. Calculate 5!.

12. Six drawers were taken out of a chest of drawers for moving it into an apartment. If any drawer will fit into any opening, in how many ways can the six drawers be replaced?

13. Calculate $P(9, 3)$, the number of permutations of 9 things taken 3 at a time.

14. Calculate $C(9, 3)$, the number of combinations of 9 things taken 3 at a time.

15. A resident assistant in a residence hall has 21 women she must assign to 10 rooms. Nine of the rooms will have two people each, and one room will have three people in it.
   a. In how many ways can she assign the three people to the one three-person room?
   b. Amy and Felicia are two of the 21 women. If the room assignment is random, what is the probability that Amy and Felicia are both in the three-person room?

16. A publication lists 50 notable movies of the year. In how many ways could the first-, second-, and third-best movies of the year be selected from this list if any two are equally likely to be selected?

17. A coin is flipped eight times. What are the following probabilities?
   a. All eight flips come up tails.
   b. Four flips are heads and four flips are tails.

18. As of January 1, 1995, eight of the 51 "states" (50 states and the District of Columbia) have a minimum age of over 16 for a regular driver's license. If four of these 51 "states" were picked at random, what are the following probabilities?
   a. All four have a minimum age of 16 or less for a regular driver's license.
   b. Two have a minimum age of 16 or less for a regular driver's license and two have a minimum age of over 16.

19. Give an example of an experiment and two events A and B such that:
   a. $P(A \cup B) = P(A) + P(B)$
   b. $P(A \cup B) \neq P(A) + P(B)$

20. The following figures refer to percentages of students enrolled in grades K through 12 in a private school in 1990–91: 17.6% were in a school with enrollment of less than 150, 54.7% were in a Catholic school, and 3.8% were in a Catholic school with enrollment of less than 150. What percentage were in either a Catholic school or a school with enrollment of less than 150 (or both)?

21. Suppose one student is selected from among the students at your school. Describe an event that is mutually exclusive from the event "get a sophomore."
22. The U.S. Department of Interior's list of endangered animal species as of June 3, 1993, is shown.

Group	U.S. Only	Foreign Only
Mammals	37	249
Birds	57	153
Reptiles	8	64
Amphibians	6	8
Fishes	55	11
Snails	12	1
Clams	50	2
Crustaceans	10	0
Insects	13	4
Arachnids	3	0
TOTAL	251	492

a. If an endangered species of animal is from the U.S. only, what is the probability it is a fish?

b. If an endangered species is a crustacean, what is the probability it is from the U.S. only?

c. Suppose an endangered species is chosen. Are the events "mammal" and "from U.S. only" independent? Why or why not? Be specific.

23. (*Multiple choice*)
   a. If two events are mutually exclusive and both can happen, then they
      i. Must be independent
      ii. Might be independent or might be dependent
      iii. Must be dependent
   b. If two events are independent and both can happen, then they
      i. Must be mutually exclusive
      ii. Might be mutually exclusive or might not be mutually exclusive
      iii. Cannot be mutually exclusive

24. The probability that a college freshman will graduate is about 0.5. If two college freshmen are picked at random, what is the probability they both will graduate?

25. The all-time top-50 American movies in terms of rental income include 11 movies made before 1980. If two all-time top-50 American movies are selected at random, what is the probability that both were made before 1980?

26. Six (6) people are in a room. Calculate the probability that two or more have the same birthday.

27. Suppose a weather forecast says that the probability of rain is 40%.
   a. What are the odds for rain?
   b. What are the odds against rain?

28. Suppose the odds against a particular horse's winning a race are 4:1. What is the probability that the horse will win the race?

29. In the World Series the first team to win four games is the world champion. If the two teams are evenly matched and each team has an equal chance of winning each game, the probabilities for varying-length series are as shown. What is the expected length (to the nearest tenth of a game) of a World Series with evenly matched teams in which each team has an equal chance of winning each game?

Length of Series	Probability
4	.1250
5	.2500
6	.3125
7	.3125

BIBLIOGRAPHY

Calinger, Ronald, ed. *Classics of Mathematics.* Englewood Cliffs, NJ: Prentice Hall, 1995.

Dunham, William. *Journey Through Genius: The Great Theorems of Mathematics.* New York: Wiley, 1990.

Eves, Howard. *An Introduction to the History of Mathematics,* 6th ed. Chicago: Saunders College, 1990.

Gittleman, Arthur. *History of Mathematics.* Columbus: Merrill, 1975.

Hald, Anders. *A History of Probability and Statistics and Their Application Before 1750.* New York: Wiley, 1990.

Kendall, M. G. "The beginnings of a probability calculus," *Biometrica 43,* pp. 1–14, 1956.

Krantz, Les. *What the Odds Are.* New York: Harper Perennial, 1992.

Maistrov, L. E. *Probability Theory—A Historical Sketch.* New York: Academic Press, 1974.

Moore, David S. *Statistics: Concepts and Controversies,* San Francisco: W. H. Freeman, 1979.

STATISTICS

PROLOGUE

Statistics is the science that deals with data—their collection, analysis, presentation, and inferences about the population based on information contained in a sample. Each of the aspects of statistics has probably affected you. You may have been interviewed by someone collecting data in a survey. You analyzed data when you decided which col-lege to attend, which food to buy, or when you considered who was the smartest person in your high school class. Data are constantly presented to you in classes and in the media. One of the distinguishing characteristics of the newspaper *USA Today* is its presentation of statistical data. Those data presented to you are often used to make infer-ences: If 54% of

the people surveyed favor the incumbent over the challenger in a political race, we might infer that the incumbent will win. In medical studies we usually infer something about a larger group of people based on what happened to the group studied. Results of a survey are often reported with a margin of error. This assists in making an inference.

The goal of this chapter is to make you a more intelligent user of statistics. We will try to reinforce what you already know, then add some new ideas.

Chapter opening photo: Statistics is the science that deals with data.

GRAPHICAL PRESENTATION OF DATA AND SAMPLING

GOALS
1. To understand the difference between descriptive and inferential statistics.
2. To work with histograms and stem-and-leaf displays.
3. To identify a population and a sample.
4. To distinguish between a random and a nonrandom sample.

The part of statistics dealing with the presentation of data is called *descriptive statistics;* the part dealing with making inferences from data is called *inferential statistics.* In descriptive statistics, data can be described either graphically (pictorially) or numerically. In this section we will discuss some graphical ways of presenting data. The subsequent two sections will describe numerical ways to present data.

ON A TANGENT

When Did Statistics Begin?

Since statistics is the science that deals with data, we might ask when data collecting began. It's likely that the first systematic data were collected in connection with censuses. The rulers of ancient Egypt, Greece, and Rome took censuses of people and quantities of grain. In ancient Rome, the population and property of every family were counted every 5 years and, beginning with the rule of Emperor Augustus (27 B.C.E.–17 C.E.) this census was extended to all the territory in the Roman empire. With the fall of the Roman empire regular national censuses disappeared until the beginning of the eighteenth century, although during the Middle Ages extensive periodic censuses of people or economic data were taken in England, Russia, and Venice.

In London of the 1500s and 1600s "Bills of Mortality" were published, prompted in part by epidemics of the plague. John Graunt (1620–1674), a self-described

"Citizen of London," examined the number of burials and the causes of death (originally, either disease or accident), put the information in a table, then commented on his findings in a 1662 publication, *Natural and Political Observations Mentioned in a Following Index and Made Upon the Bills of Mortality.* Graunt's book had great influence in leading others to analyze data. His work has been described as one of the first efforts in sociology (Tankard, 1984, p. 9). Others employed his methods, including extending them to analysis of economic data, and called them *Political Arithmetic.* About 1800 these investigations came to be called *statistics.*

Thus, accounts of deaths gave life to a new discipline in the mathematical sciences. Whereas geometry and parts of algebra are over 2000 years old, statistics is relatively new.

Histograms

A *histogram* is a form of a bar graph.

EXAMPLE 9.1

Nearby is a histogram describing the final percentage grades of students in a particular class. Percentages are rounded to the nearest whole percent.

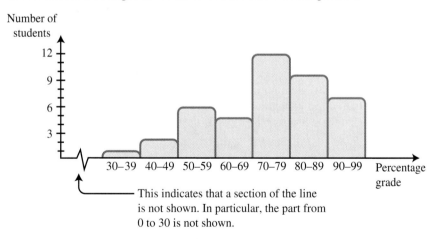

This indicates that a section of the line is not shown. In particular, the part from 0 to 30 is not shown.

a. What is the lowest percentage that anyone in class has?
b. How many students have percentages in the 80s?
c. How many students have percentages under 60?
d. How many students are there in this class?

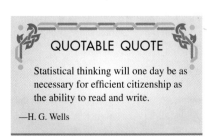

a. The bar representing the one person with the lowest percentage includes percentages from 30 through 39. Therefore, the lowest percentage of anyone in the class is between 30 and 39, inclusive.

b. 9. This is the number of students associated with the bar representing 80–89.

c. $1 + 2 + 6 = 9$. Add the numbers of students in the three left-most bars.

d. $1 + 2 + 6 + 5 + 12 + 9 + 7 = 42$. Add all the numbers of students.

EXAMPLE 9.2

Here is a histogram describing the percentage change in U.S. population, by state and the District of Columbia, from 1980 to 1990. Percentages are given to the nearest tenth of a percent. Use the histogram to answer the following questions.

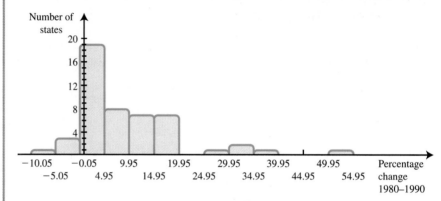

a. What was the largest percentage decline in population from 1980 to 1990?

b. What was the largest percentage increase?

c. How many states lost population from 1980 to 1990?

d. How many states had population increases of 10% or more?

SOLUTION

First we comment on the boundaries between bars at 4.95, 9.95, 14.95, etc. These numbers ending in .95 or .05 avoid ambiguity. Since percentages recorded in the data are given to the nearest tenth, no piece of data ends in .95 or .05. Consequently, given a piece of data, it is clear in which bar it is. Had the boundaries between bars been, for example, 5.0, 10.0, 15.0, etc., we would not have been able tell from the histogram in which bar the percentage 5.0 would fall. To avoid this ambiguity, the histogram is drawn so the boundaries between classes are numbers that are not possible pieces of data.

a. The most negative possible number that could appear here is −10.0, so the largest decline was at most 10.0%. (Actually, it was 8.0%, for West Virginia.)

b. The largest possible number here is 54.9, so the largest increase is at most 54.9%. (Actually, it was 50.1%, for Nevada.)

c. The states that lost population are in the two bars at the left. There are four of them.

d. We add the numbers in the bars beginning at 9.95: $7 + 7 + 1 + 2 + 1 + 1 = 19$.

We will now construct a histogram from the sales tax data in the *Information Please Almanac,* 1996 edition, shown in ● Table 9.1.

TABLE 9.1
State General Sales and Use Taxes*

State	Percentage Rate	State	Percentage Rate	State	Percentage Rate
Alabama	4	Kentucky	6	North Dakota	5
Alaska	0	Louisiana	4	Ohio	5
Arizona	5	Maine	6	Oklahoma	4.5
Arkansas	4.5	Maryland	5	Oregon	0
California	6	Massachusetts	5	Pennsylvania	6
Colorado	3	Michigan	6	Rhode Island	7
Connecticut	6	Minnesota	6.5	South Carolina	5
Delaware	0	Mississippi	7	South Dakota	4
D.C.	5.75	Missouri	4.225	Tennessee	6
Florida	6	Montana	0	Texas	6.25
Georgia	4	Nebraska	5	Utah	4.875
Hawaii	4	Nevada	6.5	Vermont	5
Idaho	5	New Hampshire	0	Virginia	3.5
Illinois	6.25	New Jersey	6	Washington	6.5
Indiana	5	New Mexico	5	West Virginia	6
Iowa	5	New York†	4	Wisconsin	5
Kansas	4.9	North Carolina	4	Wyoming	4

*Local and county taxes, if any, are additional.
†New York City, 8.25%.

One question that always arises in constructing a histogram is how wide a bar should be or, equivalently, how many bars there should be. In general, if there is a small number of measurements, there should be only a few bars, and if there is a large number of measurements, there should be more bars. In this case, there is a natural width of 1 percentage point to a bar, with bars going from one whole number to, but not including, the next whole number. For example, one interval might be from 4% to, but not including, 5%. Using these intervals we tabulate the data as shown in ● Table 9.2. From this table we construct the histogram shown in ▶ Figure 9.1. Here the histogram is interpreted in part in conjunction with Table 9.2, which describes the intervals more clearly. Without the

TABLE 9.2
Tabulated Sales Tax Data

Interval	Tally	Frequency
0.0–0.9	ЦН	5
1.0–1.9		0
2.0–2.9		0
3.0–3.9	||	2
4.0–4.9	ЦН ЦН |||	13
5.0–5.9	ЦН ЦН ||||	14
6.0–6.9	ЦН ЦН ЦН	15
7.0–7.9	||	2

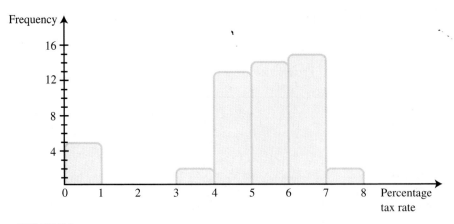

FIGURE 9.1 Histogram of sales tax data

table we wouldn't know whether a state with a tax rate of 7% is included in the tallest bar or the one to the right of it.

Notice that the histogram shows clearly that the most common tax rates are in the 4%–5%–6% range; i.e., they are between 4.0% and 6.9%, inclusive. It also shows that among the states with sales taxes, no tax rate was lower than 3.0% and none was higher than 7.9%.

Here are some principles to keep in mind when constructing a histogram.

CONSTRUCTING A HISTOGRAM

1. Make all the intervals the same width.

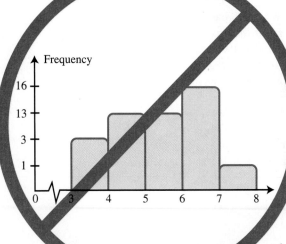

2. Choose the intervals so there is no ambiguity about which of two adjacent intervals contains a given number. If the measurements are whole numbers, the boundaries between intervals would typically end in .5. For example, one interval might be 2.5–4.5 and the next interval 4.5–6.5. Since 4.5 is not a possible measurement, no ambiguity arises. With the data in Table 9.1 the boundaries between intervals could have ended in .9. The first interval could had been from −0.1 to 0.9, the second from 0.9 to 1.9, the third from 1.9 to 2.9, etc. In the histogram of Example 9.2 the points of division (−5.05, −0.05, 4.95, etc.) are chosen for this reason: Since the percentages are given to the nearest tenth, one of the data points will not fall on a boundary.

3. On the vertical axis, allow the same space for each unit of frequency. **Beware!** Do not construct a histogram like the one to the left.

CALCULATOR NOTE

If you have access to a graphing calculator or a computer, you may be able to have the machine help construct a histogram. Some graphing calculators have built-in routines that, with some prompting by you, will construct a histogram. A variety of computer software packages, including statistical software and spreadsheets, will do the same thing.

Stem-and-Leaf Display

A method of presenting data pictorially that has become popular in the last few years is the stem-and-leaf display. We illustrate it in ► Figure 9.2, using the data of Table 9.1. The stem will be the left-most digit, in this case the whole-number digit. For each sales tax percent, we record in the "leaf" column, the next digit, i.e., the first digit to the right of the decimal point. Thus, if there is more than one digit to the right of the decimal point, we truncate the remaining positions. For example, 4.25 becomes 4.2 purposes of our plot.

The table in Figure 9.2 was constructed much as the tallying was done for a histogram. As a number appears in the table, the appropriate entry under "leaf" is recorded. The leaf entries for stem 4 are, in order, 0 for the 4.0% tax in Alabama, 5 for the 4.5% tax in Arkansas, 0 for the 4.0% tax in Georgia, etc. The leaves can be arranged in order or left the way they first appear. For example, the leaves with stem 4 could be written as 0000000025589.

Notice that because of the way the intervals were chosen in Table 9.2, the histogram of Figure 9.1 presents vertically what the stem-and-leaf display of Figure 9.2 presents horizontally. Notice, also, that in this and all stem-and-leaf displays, all of the original data are preserved, at least in a truncated form. In particular, we know that the tax rates in the 3% range are 3.0% and 3.5%. This specificity is lost in the histogram.

A disadvantage of the stem-and-leaf display is that it will not work well with larger data sets. In such cases the leaves will run off the edge of the paper.

State General Sales and Use Tax

Stem	Leaf
0	0 0 0 0 0
1	
2	
3	0 5
4	0 5 0 0 9 0 2 0 0 5 0 8 0
5	0 7 0 0 0 0 0 0 0 0 0 0 0
6	0 0 0 2 0 0 0 5 5 0 0 0 2 5 0
7	0 0

► FIGURE 9.2 Stem-and-leaf display

●EXAMPLE 9.3

A stem-and-leaf display for the closing prices of 20 selected stocks listed on the New York Stock Exchange on a given day is shown. The closing price is in dollars per share. For example, the entry with stem 1 and leaf 6 denotes a stock that closed at $16 per share.

Stem	Leaf
0	1 3 6 7
1	6
2	3 4 6
3	3 6 9
4	1 5
5	2 4 5 8
6	3
7	5 7

a. What was the lowest closing price among the 20 stocks?
b. What was the highest closing price among these 20 stocks?
c. How many of the 20 stocks closed under $25 per share?

SOLUTION

a. The lowest closing price was $1 per share (01).
b. The highest closing price was $77 per share.
c. Seven of the stocks—those in the first two stems plus those closing at 23 and 24—closed under $25 per share.

The histogram and the stem-and-leaf display are only two of many ways of describing data pictorially. However, the principles that apply to constructing and interpreting them also apply to many other ways of representing data pictorially.

The Best Visual Presentation of Data Ever

Information can be presented pictorially by methods other than histograms, stem-and-leaf displays, pie charts, or other familiar methods. The pictorial presentation of data is, in part, a matter of design and art. Perhaps the best visual presentation of data ever done was by the French engineer Charles Joseph Minard (1781–1870) in his depiction of Napoleon's ill-fated invasion of Russia in 1812.

Napoleon's army of 422,000 marched into Russia in June 1812 near the Niemen River. Czar Alexander of Russia pulled back his troops rather than be caught in a mismatched battle. As the Russian troops retreated, they burned fields of grain and slaughtered livestock rather than leave them for the French. This weakened Napoleon's army, and some men deserted to search for food. On Minard's map, the width of the band indicates the size of the French army at each location. On September 14, Napoleon's forces, reduced to less than one-fourth their original size, entered Moscow. Alexander had set fire to the city rather than surrender it to the French. Napoleon stayed in Moscow for 5 weeks before deciding in the middle of October to turn back. In November, his starving army ran into snow and cold (see the temperature graph along the bottom of the map), as well as Russian raiders. In the middle of December, the 10,000 remaining French soldiers, about 2.4% of the original invading force, struggled out of Russia.

The map-graph displays five different measurements: the size of the army, its location on a two-dimensional coordinate system (distance east and distance north), the direction of the army's movement, and the temperature on various dates during the withdrawal.

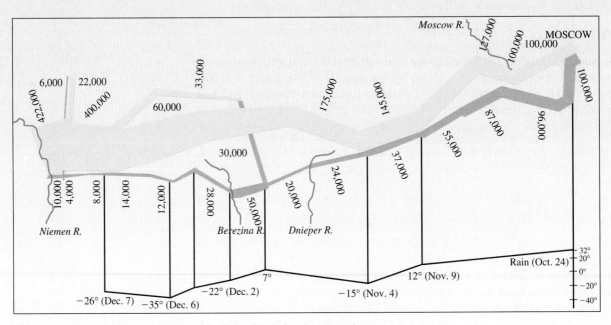

Redrawing of Charles Minard's 1861 graph of Napoleon's Russian campaign.

Sampling

We make a distinction between a population and a sample. A *population* is the set of all measurements that characterize some phenomenon. A *sample* is a subset of a population. For example, the grade point averages of everybody at your school would be a population from which the grade point averages of people in your class would be a sample. We will take a more relaxed attitude toward describing a population and a sample by saying in this example that the population is the set of all students in your school and that a sample is the set of people in your class.

EXAMPLE 9.4

In order to measure the public's perception of the U.S. president's job performance, 800 people from around the country are selected and asked to rate his job performance on a scale of 0 through 10.

a. Describe the population.
b. Describe the sample.

SOLUTION

a. The population is the set of all people in the United States. More precisely, it is the set of performance ratings by all the people in the country, i.e., it is a set with over 200 million elements, each of which is a number from 0 through 10.
b. The sample is the set of 800 people surveyed. More precisely, it is the set of 800 job ratings of the president. It is a set with 800 elements, each a number from 0 through 10.

A sample is *random* provided it is chosen by means of a procedure that gives any member of the population the same probability of being chosen as any other member. If we were to pick five baseball players by listing all the major league players by salary, picking the first five would not produce a random sample. Nor would a random sample be produced by listing them alphabetically and picking the first five. A random sample could be chosen by writing each name on a piece of paper, mixing the pieces of paper thoroughly, and choosing five. Often, tables of random numbers or a random-number generator on a computer are used to select a random sample.

EXAMPLE 9.5

Explain why the following sampling method is not random: A local fine arts series wants to determine what kind of music its potential audience likes the most, so it asks the 150 attendees at a jazz concert it has sponsored what kind of music they like best.

SOLUTION The people who attend the jazz concert probably like jazz and people who choose not to attend are less likely to like jazz. Consequently, the 150 attendees at the concert are more likely to say their favorite music is jazz than are most other groups of 150 people.

 YOUR FORMULATION

Describe a sampling method (other than the methods in Example 9.5 and the paragraph preceding it) that is not random. Explain why your method is not random.

1. Describe the science of statistics.
2. Describe the difference between descriptive and inferential statistics.
3. The ages at inauguration of 42 U.S. presidents (Cleveland is counted twice) are shown in the histogram. Use it to answer these questions.

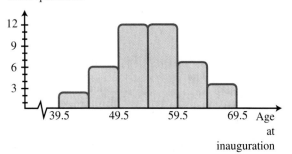

a. What is the youngest age at which a U.S. president has been inaugurated?
b. What is the oldest age at which a U.S. president has been inaugurated?
c. How many presidents were in their 50s when they were inaugurated?
d. How many presidents were under 60 when they were inaugurated?
4. Construct a histogram for the data in the following table.

Interval	Frequency
9.5–11.5	3
11.5–13.5	1
13.5–15.5	6
15.5–17.5	9
17.5–19.5	7
19.5–21.5	5

5. The land area of the states, to the nearest square mile, is given in the table.

State	Land Area	State	Land Area
Alabama	50,750	Indiana	35,870
Alaska	570,373	Iowa	55,875
Arizona	113,642	Kansas	81,823
Arkansas	52,075	Kentucky	39,732
California	155,973	Louisiana	43,566
Colorado	103,730	Maine	30,865
Connecticut	4,845	Maryland	9,775
Delaware	1,955	Massachusetts	7,838
Florida	53,997	Michigan	56,809
Georgia	57,919	Minnesota	79,617
Hawaii	6,423	Mississippi	46,914
Idaho	82,751	Missouri	68,898
Illinois	55,593	Montana	145,556

State	Land Area	State	Land Area
Nebraska	76,878	Rhode Island	1,045
Nevada	109,806	South Carolina	30,111
New Hampshire	8,969	South Dakota	75,898
New Jersey	7,419	Tennessee	41,220
New Mexico	121,364	Texas	261,914
New York	47,224	Utah	82,168
North Carolina	48,718	Vermont	9,249
North Dakota	68.994	Virginia	39,598
Ohio	40,953	Washington	66,582
Oklahoma	68,679	West Virginia	24,087
Oregon	96,003	Wisconsin	54,314
Pennsylvania	44,820	Wyoming	97,105
		U.S. Total	**3,536,280**

a. Construct a histogram for these data.
b. Construct a stem-and-leaf plot for these data.
6. The prices of dishwashers reviewed by *Consumer Reports* in October 1995 are shown in the display. The stem is in hundreds of dollars.

Stem	Leaf
2	9
3	6 9 3 8 5 7 7
4	7 0 6 0 7 0 5
5	6 6 5 1
6	
7	5
8	4

a. How much does the least expensive dishwasher cost?
b. How much does the most expensive one cost?
c. How many dishwashers were reviewed?
d. How many dishwashers cost under $400?

In each of Exercises 7–9, describe the population and the sample. If the population is not specifically mentioned in the information given, describe it in a reasonable way.

7. A newspaper reports that 62% of its readers favor a change in Social Security benefits. The result is based on responses of 210 people.
8. A survey is done to determine what radio station is the favorite in your listening area. Assume 250 people are surveyed and 50 choose your favorite station.
9. The Physicians' Health Study followed 21,996 physicians to check, among other things, the effect of aspirin on heart attacks.

In each of Exercises 10–11, describe why the sampling method is not random. Also describe how this method of sampling might change the results from those of a random sampling method.

10. A state representative wants to judge public reaction to a proposed bill to cut back on heating assistance to the poor. To do this 200 people in the district are polled by a telephone survey using random-digit dialing.

11. A teacher wanted to know if students understood the topic that she had just discussed, so she asked each of the five students sitting in the front row a question about it.

12. A sample is a subset of a population. Two subsets are the empty set, which provides no information, and the entire population, which defeats the purpose of sampling. Call other subsets nontrivial. How many nontrivial samples are there if the population has the following?
    a. 3 elements   b. 6 elements   c. 12 elements

WRITTEN ASSIGNMENTS 9.1

1. At the start of the chapter is a quote from H. G. Wells: "Statistical thinking will one day be as necessary for efficient citizenship as the ability to read and write." Do you believe that day is now, in the future, or never? Explain your position.

2. State an advantage and a disadvantage of a stem-and-leaf plot compared with a histogram.

3. In a newspaper or magazine find a pictorial presentation of some data.
   a. Are the data presented in a histogram, a stem-and-leaf plot, or some other form?
   b. Write some summary comments about the data.

4. An old adage says that a picture is worth a thousand words. State why you believe this is true for pictorial presentation of data.

5. In a newspaper or magazine, find the results of a survey or a statistical study.

   a. Describe both the population and the sample.
   b. Do you believe the sample is random? Why or why not?

6. Sometimes we are interested in a representative sample. Roughly speaking, a *representative sample* represents the population from which it is drawn. For example, a representative sample of sales tax rates might consist of a low rate, a medium rate, and a high rate, such as {0, 5.0, 6.0}. A representative sample of four sales tax rates might be {0, 4.0, 5.0, 6.0}.
   a. A test in a course is a sample of the material that has been covered. On a 10-question test on this chapter, would you expect it to be a random sample or a representative sample? Explain.
   b. Is every random sample a representative sample? Why or why not?
   c. Is a representative sample a random sample? Why or why not?

9.2   MEASURES OF CENTRAL TENDENCY AND LOCATION

GOALS
1. To calculate as precisely as possible the mean and the median of a set of data given numerically or pictorially.
2. To work with the mean and the median.
3. To use the terms *skewed to the left, skewed to the right,* and *symmetric.*
4. To work with quartiles and percentiles.

In the previous section we looked at ways of presenting data pictorially. In this section we begin to explore numerical ways of doing the same thing. In particular, we look at ways to describe where the center of a set of data is and ways to describe the position of a measurement in a set of data.

Measures of Central Tendency

The measure of central tendency that you've used for many years is the *mean,* which is the sum of all the measurements divided by the number of measurements. In a sample, we denote the mean by $\bar{x}$, read "x bar," and in a population we denote the mean by the Greek letter μ (mu).

EXAMPLE 9.6

The three most highly paid baseball players in 1993 were Ryne Sandberg at $6,475,000, Bobby Bonilla at $6,200,000, and Dwight Gooden at $5,916,667. Find their mean salary.

SOLUTION Consider these three as a sample of all baseball players and denote the mean by

$$\bar{x} = \frac{\$6,475,000 + \$6,200,000 + \$5,916,667}{3} = \frac{\$18,591,667}{3} \approx \$6,197,222$$

The mean "averages out" high and low measurements and gives a number in the center of a set of measurements. Notice that it need not be one of the measurements. $6,197,222 is not the salary of any of the three players.

A second measure of central tendency is the *median*. After the data are arranged in order, it is the middle measurement, if there is one; otherwise, it is the mean of the two middle measurements.

EXAMPLE 9.7

Find the median of the numbers 4, 0, 8, 1, 9, 9, 3.

SOLUTION First put the numbers in order: 0, 1, 3, 4, 8, 9, 9. Notice that 4 is the middle measurement: Three measurements are below it and three are above it. Thus the median is 4.

EXAMPLE 9.8

Find the median of the numbers 5, 2, 3, 2, 8, 8.

SOLUTION First put the numbers in order:

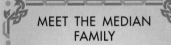

Notice that there are two middle measurements, 3 and 5. The median is the mean of these, which is 4.

Notice that when there is an odd number of measurements, as in Example 9.7, the median is one of the measurements, namely, the middle one. When there is an even number of measurements, as in Example 9.8, the median is the mean of the two middle measurements.

When calculating the mean, every measurement matters. When calculating the median, some measurements can be changed without affecting the median. For example, in Example 9.7 the mean is

$$\frac{4 + 0 + 8 + 1 + 9 + 9 + 3}{7} = 4\frac{6}{7}$$

If the 0 is changed to a 1, the mean changes to 5 but the median remains 4.

BEWARE!

Keep in mind that to find the median, the numbers must be written in order. In Example 9.7 if we had not written the numbers in order first, we might have said the median was 1, which is the middle of the symbols but not the middle value of the numbers.

MEET THE MEDIAN FAMILY

This is a story about Carol and Paul Median, she 32 and he 34, married for 1 year and living just south of Bloomington, Indiana. They are as average as can be. Last year Paul earned $28,449, and Carol made $23,479, precisely the midpoint paychecks for men and women their age. It also happens that their part of Indiana is the population center of the United States. Since no real-life couple is truly typical, however, I have created the fictional Medians, going by Census figures, to give a sense of the exact middle of American life.

Andrew Hacker, "Meet the Median Family," *Time,* January 29, 1996

The mean is affected by extreme measurements, while the median is not. For example, for the numbers 1, 2, 4, 5, 8, the mean is 4 and the median is 4; but if the 8 is changed to 88 to give the set of numbers 1, 2, 4, 5, 88, the new mean is 20 while the new median is the same as the old median, 4.

Because the mean is affected by extreme measurements, the median is frequently used as a typical measurement from a set of data. For example, if the yearly incomes of people in a car pool are $17,000, $19,000, $20,000, $24,000, and $80,000, the median income of $20,000 is much more typical of the incomes of people in the car pool than is the mean of $32,000. Because of this, median incomes are often reported rather than mean incomes. For example, the Census Bureau reports that the median household income in the United States in 1993 was $31,241.

When data are presented pictorially, what can be said about the mean and the median? If the data are presented in a stem-and-leaf display, both the mean and the median can be calculated, since all the individual measurements are present (at least with two significant figures). (To be precise we should say "approximated" rather than "calculated" because of the possible truncation of data in a stem-and-leaf display. For example, if the data set has two measurements, 1.99 and 2.99, the mean and median are both 2.49. In a stem-and-leaf display these might be represented as 1.9 and 2.9. From these data, we would calculate both the mean and the median to be 2.4, which is an approximation to the true value of 2.49.) If the data are presented in a histogram, we can locate in which interval (bar) the median is; but it is difficult, at best, to say much about the mean.

EXAMPLE 9.9

For the stem-and-leaf display of Figure 9.2 in the preceding section, and repeated here, calculate the mean sales and use tax and the median sales and use tax.

State General Sales and Use Tax

Stem	Leaf
0	0 0 0 0 0
1	
2	
3	0 5
4	0 5 0 0 9 0 2 0 0 5 0 8 0
5	0 7 0 0 0 0 0 0 0 0 0 0 0
6	0 0 0 2 0 0 0 5 5 0 0 0 2 5 0
7	0 0

SOLUTION The mean is

$$\frac{3.0 + 3.5 + \cdots + 7.0}{51} = \frac{238.0}{51} \approx 4.67$$

Thus, the mean sales and use tax is about 4.67%.

To find the median, we notice that there are 51 measurements, so we seek a number such that 25 measurements are above it and 25 are below it. Stems 0, 1, 2, 3, and 4 contain a total of 20 measurements, and the last two stems (6 and 7) contain a total of 17 measurements. Therefore, the median is in stem 5. Since all but one entry there ends in 0, the median is 5.0%.

EXAMPLE 9.10

Calculate the mean and median tax rates from the histogram of Figure 9.1.

SOLUTION For the mean about all we can say is that it's somewhere in the middle, maybe around 5% to 6%. From Example 9.9 we see that this estimate is not very accurate. To estimate the median, notice that there are 20 measurements in the bars from 0 to 5 combined and 17 measurements in the last two bars combined. We are seeking a number such that 25 of the tax rates are below it and 25 are above. That number is in the interval 5.0–5.9, so the median tax rate is between 5.0% and 5.9%, inclusive.

If a set of data has the property that when its frequency distribution is graphed there is a long tail to the right, then the set of data is said to be *skewed to the right*. This is equivalent to saying that big numbers spread farther from the middle than small ones. For example, ▶ Figure 9.3a presents a set of data that is skewed to the right. When a set of data has the property that when its frequency distribution is graphed there is a long tail to the left, then the set of data is said to be *skewed to the left*. This is equivalent to saying that small numbers spread farther from the middle than large ones. Figure 9.3b shows a set of data that is skewed to the left. If a set of data is neither skewed to the left nor skewed to the right—i.e., if it is not skewed at all—then it is said to be *symmetric*. Figure 9.3c shows a set of data that is symmetric.

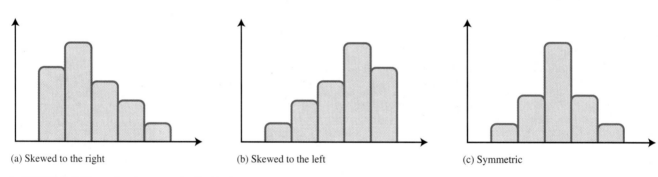

(a) Skewed to the right

(b) Skewed to the left

(c) Symmetric

▶ FIGURE 9.3 Skewed and symmetric distributions

When a set of data is skewed to the right, the mean is larger than the median because the large measurements pull the mean up above the middle. Similarly, when a set of data is skewed to the left, the mean is less than the median. And when a set of data is symmetric, the mean equals the median. The converses of these statements follow: If the mean of a set of data is larger than the median, the data are skewed to the right; if the mean is less than the median, the data are skewed to the left; and if the mean equals the median, the data are symmetric.

 YOUR FORMULATION

Why do the converses of the preceding statements follow? After all, if a statement is true, its converse is not necessarily true.

EXAMPLE 9.11

Are the data displayed in Figure 9.2 in the preceding section and repeated here skewed to the left, skewed to the right, or symmetric?

State General Sales and Use Tax

Stem	Leaf
0	0 0 0 0 0
1	
2	
3	0 5
4	0 5 0 0 9 0 2 0 0 5 0 8 0
5	0 7 0 0 0 0 0 0 0 0 0 0 0 0
6	0 0 0 2 0 0 0 5 5 0 0 0 2 5 0
7	0 0

SOLUTION Since the small measurements (the 0.0's) are further removed from the middle (the median) than are the larger measurements, the data are skewed to the left. An alternate analysis is that the mean is about 4.67 and the median is 5.0 (as calculated in Example 9.9). Since the mean is less than the median, the data are skewed to the left.

Sometimes the way a set of data is skewed affects the measure of central tendency a person wants to use. For example, in negotiations between major league baseball owners and the players' association, which measure of central tendency will the owners want to use in order to argue that the average player's salary is quite high? Which measure will the players association mention in order to argue for more money? (Try to answer those two questions before reading farther.)

Major league players have a minimum salary, with a number of players at or near the minimum and a few very highly paid players (in 1995 the top 100 players out of 824 received 54% of the total payroll). Those high salaries cause the mean to be greater than the median. (In 1995 the median salary was $275,000 and the mean salary was $1,089,621.) Consequently, the owners will point to the mean to argue that the salaries are quite high and the players' association will point to the median in arguing for more money.

Measures of Location

The median of a set of data is a number located so that half the measurements are above that number and half are below it. This idea can be extended to other measures of location. If data are put in numerical order, then divided into four equal parts, the three points of division are called *quartiles,* denoted Q_1, Q_2 and Q_3. (See ▶ Figure 9.4.) The lower quartile, denoted Q_1, is a number such that one-quarter of the measurements are less than Q_1 and three-quarters of the measurements are greater than Q_1. More precisely, we define the *lower quartile, Q_1,* as the median of the measurements below the median.

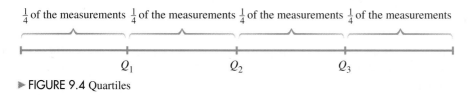

$\frac{1}{4}$ of the measurements $\frac{1}{4}$ of the measurements $\frac{1}{4}$ of the measurements $\frac{1}{4}$ of the measurements

▶ FIGURE 9.4 Quartiles

The second quartile, Q_2, is a number such that two-quarters (one-half) of the measurements are greater than Q_2. More precisely, Q_2 is defined to be the median. The upper quartile, Q_3, is the median of the upper half of the measurements.

EXAMPLE 9.12

Find the upper and lower quartiles of the state general sales and use taxes listed in Table 9.1 of the previous section and displayed in Example 9.11 in this section.

SOLUTION The stem-and-leaf display in Figure 9.2 has the data partially in order. The median is 5.0. The data "below" the median are

$$0, 0, 0, 0, 0, 3, 3.5, 4, 4, 4, 4, 4, 4, 4, 4, 4.2, 4.5, 4.5, 4.8, 4.9, 5, 5, 5, 5, 5$$

↑

Median of data in lower half

The median of these is 4.0, so the lower quartile is 4.0. The data "above" the median are

$$5, 5, 5, 5, 5, 5, 5, 5.7, 6, 6, 6, 6, 6, 6, 6, 6, 6, 6.2, 6.2, 6.5, 6.5, 6.5, 7, 7$$

↑

Median of data in upper half

The median of these is 6.0, so the upper quartile is 6.0.

Another measure of location is a *percentile*. The pth percentile of a set of data is a number such that $p\%$ of the data are less than the number and $(100 - p)\%$ of the data are above number. For example, the 39th percentile is a number such that 39% of the data are less than the number and 61% of the data are greater than the number.

Percentiles are used with larger sets of data. If you took the ACT or SAT exam, your scores were reported in percentiles. Scores on other standardized tests are often reported in percentiles. Weights and heights of children are sometimes stated in percentiles (see Exercises 17 and 18). Percentiles and quartiles are both measures of location. They tell where a particular measurement is located in a set of data.

Percentiles, quartiles, and the median are related as follows:

$$25^{th} \text{ percentile} = \text{lower quartile}$$
$$50^{th} \text{ percentile} = \text{middle quartile} = \text{median}$$
$$75^{th} \text{ percentile} = \text{upper quartile}$$

For each of these lines you should be able to convert any one of the measurements on a line (such as lower quartile) to the other measurement(s) on that line (25th percentile).

In summary, a measure of central tendency is a single number that tells where the center of a set of data is, and other measures of location give benchmark positions in a set of data. With these measures we can describe an entire set of data by means of one or a relatively few numbers.

1. a. Give an example of a set of measurements in which the mean is one of the measurements.
   b. Give an example of a set of measurements in which the mean is not one of the measurements.

In Exercises 2–4, calculate the mean and the median.

2. 1, 7, 3, 1, 5, 2, 1, 6
3. 3, 3, 3, 1, 3, 0, 7, 3, 0
4. The lengths of the months of the year (count February as 28 days)
5. In Example 9.8 change one number so that the median does change.
6. In assessing property values, the median home value in a neighborhood is frequently examined. Suppose on one block there are seven houses and the values of the houses are $100,000, $105,000, $110,000, $115,000, $120,000, $140,000, and one more value that is higher than any of these.
   a. What is the median value?
   b. What can be said about the mean value? Be as specific as possible.
7. Consider the set of numbers 1, 2, 3, 4, x, where x is some unknown number.
   a. What can be said about the median? Be as specific as possible.
   b. What can be said about the mean value? Be as specific as possible.
8. For the data pictured in the stem-and-leaf display in Example 9.3, calculate the mean and median.

9. From the histogram in Exercise 3 of Section 9.1, estimate the mean and the median ages of U.S. presidents at inauguration.
10. Sketch a histogram for a set of data that is skewed to the left.
11. Give an example of a set of three numbers that is skewed to the left.
12. (*Multiple choice*) The median household income in 1993 was $31,241. Which is true about the mean household income in 1993?
    a. It was less than $31,241.
    b. It was $31,241.
    c. It was greater than $31,241.
    d. Not enough information is given to answer this question.
13. For 1994 high school graduates the mean composite ACT score was 20.8 and 51% of those taking it scored 20 or below. What can you say about the median?

In each of Exercises 14–16, find the lower quartile and the upper quartile of the given data.

14. 3, 4, 7, 8, 9, 12, 13, 13
15. 0, 7, 6, 9, 2, 3, 7, 4, 6
16. The data in Exercise 5 of Section 9.1 on land area of states
17. For 11-year-old boys, the 90th percentile in height is 60 inches. Explain what this means.
18. Suppose a high school has 80 girls in its graduating class. The 95th percentile of heights for 17- and 18-year-old girls is $68\frac{1}{4}$ inches. How many girls in the school's graduating class would you expect to be over $68\frac{1}{4}$ inches tall?

1. The mean that we described is more precisely called the *arithmetic mean*. Two other means are the *geometric mean* and the *harmonic mean*. Look up, possibly in a statistics book, the description of one of these two. Define that particular mean, give two examples where it is calculated, and compare it with the arithmetic mean.
2. Describe a set of data other than one described in this chapter or in class for which the following are true.
   a. The mean is less than the median.
   b. The mean is greater than the median.
3. The graph to the right shows median net worth of Americans in 1990, by age groups. Why do you think the Census Bureau reports *median* net worth rather than mean net worth? Are they trying to hide something, are they too lazy or pressed for time to calculate the mean, or is there some other reason?
4. The word *average* in mathematical usage usually means "mean" while in common usage it also means "medium" or "typical." In the latter usage, it's more like a median. Look up a dictionary definition of *average,* and write about your view of the appropriateness and clarity of the word *average* to mean "mean." For example, an AP news story in April 1993 stated, "The average major league salary on opening day was $1,120,254."

HIGH NET WORTH
Americans between the ages of 65 and 75 have the highest net worth of any citizens.

Median Net Worth

Under 35
$6,078

35–44
$33,183

45–54
$57,466

55–64
$80,032

65–69
$83,478

70–75
$82,111

75+
$61,491

9.3 ◀ MEASURES OF SPREAD

GOALS
1. To explain why a measure of central tendency is not sufficient to describe a set of data.
2. To calculate and interpret the range of a set of data.
3. To work with box plots.
4. To work with the variance and standard deviation of a set of data.

Measures of central tendency provide a quick description of a set of data. Knowing that the mean monthly income in 1993 for all persons 18 years old and over whose highest degree earned was an associate's degree was $1985 and the mean monthly income for those whose highest degree earned was a bachelor's degree was $2625 gives us an idea of salaries for people with associate's degrees and bachelor's degrees. How much do we know by knowing a mean or a median?

Consider the following example: The houses on one block have values of $80,000, $83,000, $85,000, $87,000, and $90,000, and on another block the houses have values of $25,000, $30,000, $85,000, $85,000, and $200,000. Notice that on both blocks the mean and median home values are $85,000. However, the spreads, or ranges, of values of houses on the two blocks are very different. In this section, we examine some ways to describe how much data are spread.

Range

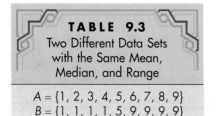

TABLE 9.3
Two Different Data Sets with the Same Mean, Median, and Range

$A = \{1, 2, 3, 4, 5, 6, 7, 8, 9\}$
$B = \{1, 1, 1, 1, 5, 9, 9, 9, 9\}$

The *range* of a set of data is the largest measurement minus the smallest. In our example, the range on the first block is $10,000, while on the second block the range is $175,000. The range describes the distance between the extreme measurements.

The range is a rather crude way of measuring the spread of a set of data. Consider, for example, the two sets of data in ● Table 9.3. Notice that both sets have mean 5, median 5, and range 8, yet the sets are quite different. Mentioning the quartiles helps to distinguish the sets: The lower quartile of A is 2.5, while the lower quartile of B is 1; the upper quartile of A is 7.5, while the upper quartile of B is 9.

A wide spread in housing costs.

Houses without much spread in prices.

Box Plots

A way to describe pictorially the spread of a set of data is the *box plot,* or *box-and-whisker diagram.* A box plot graphically shows five numbers to represent a set of data. The five numbers pictured in a box plot are the smallest measurement, the lower quartile, the median, the upper quartile, and the largest measurement. A box plot for data set *A* in Table 9.3 is shown in ▶ Figure 9.5.

The median marks the middle of the set of data. The box shows the middle half of the measurements. At a glance the location of the middle half of the measurements is quite visible. The whiskers mark the location of the lowest quarter of the measurements and the highest quarter of the measurements. The pattern of this box plot is typical. It has a box in the middle that spreads on both sides of the median, and it has whiskers on each end.

The box plot for set *B* is unusual in that the smallest measurement is also the lower quartile and the largest measurement is also the upper quartile. Consequently, the box plot for data set *B* in Table 9.3, shown in ▶ Figure 9.6, appears to have its whiskers plucked out. It's unusual to have the smallest measurement also be the lower quartile or to have the largest measurement also be the upper quartile.

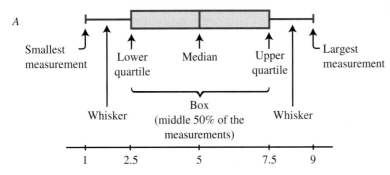

▶ FIGURE 9.5 Box plot for data set *A*

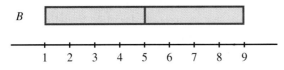

▶ FIGURE 9.6 Box plot for data set *B*

●EXAMPLE 9.13

Construct a box plot for the state general sales and use taxes listed in Table 9.1 of Section 9.1.

SOLUTION From previous work, we know that the median is 5.0, the lower quartile is 4.0, and the upper quartile is 6.0. The long whisker on the left says the data are skewed to the left.

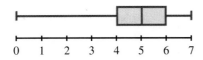

The following example illustrates how box plots can help to compare two sets of data.

EXAMPLE 9.14

Use the two box plots to compare the ages at inauguration of U.S. presidents and of presidents' wives.

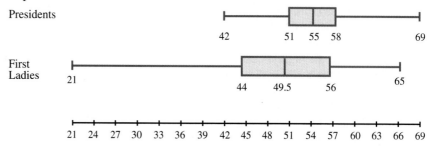

SOLUTION The distribution of the ages of presidents is skewed to the right, while the distribution of the ages of First Ladies is skewed to the left. On the average presidents have been older than First Ladies. The ages of First Ladies spread out considerably more than do the ages of the presidents (44 years vs. 27 years). The youngest First Lady was 21 years younger than the youngest president, while the oldest president was 4 years older than the oldest First Lady. The oldest quarter of presidents have ages only a few years higher than those of the First Ladies.

Standard Deviation

Another way to describe the spread of a set of data is to describe how much they spread out from the middle. Let's use the mean to describe the middle and consider the data from the start of this section on the values of houses on the first block: $80,000, $83,000, $85,000, $87,000, and $90,000. Let's look at the deviations of each value from the mean of $85,000 and average those deviations, as shown in ● Table 9.4. We conclude that the average deviation from the mean is 0. This is not a very good description of how much the values spread out: It says that on the aver-

TABLE 9.4
Deviations from the Main

Value: x (in thousands of $)	Deviation: $x - \mu = x - 85$
80	−5
83	−2
85	0
87	2
90	5

$$\text{Average deviation} = \frac{(-5) + (-2) + 0 + 2 + 5}{5} = \frac{0}{5} = 0$$

age they spread out none. The problem is that the positive and negative deviations "cancel each other out," i.e., they add to give 0. It is a useful fact—whose argument we will not give—that for any set of data, the sum of the deviations from the mean is zero. This is a way to check the accuracy of the calculations of deviations from the mean.

The idea of trying to average deviations from the mean has some merit. The problem was that the positive and negative deviations cancelled each other out, so let's change all the deviations to positive numbers. There are two ways to do this. One is to take the absolute value of each deviation and average those absolute values. This produces a parameter called the *mean absolute deviation.* However, this method is seldom employed.

A much more prevalent method is to compute the *standard deviation.* We get the standard deviation by noting that a second way to get rid of negative numbers is to square them. Here, we will square the deviations and average them to produce a parameter called the *variance, denoted* σ^2, which is the mean of the squared deviations. We use the information from Table 9.4 to calculate the variance in ⬤ Table 9.5.

What is the variance? It is the average of the squares of deviations from the mean. Our goal was to describe how much, on the average, measurements spread out from the mean. We've now calculated the average of the *squares* of deviations (in order to get rid of negative numbers). To compensate for the squaring process, we take the square root of the variance to produce the standard deviation, σ. With the data in Table 9.5, $\sigma = \sqrt{11,600,000} \approx 3406$. Since the house values appear to be given to the nearest thousand dollars, i.e., with two significant digits, we will report the standard deviation with one more significant digit and say it is about $3410.

To calculate the standard deviation of a population the following procedure can be followed.

1. Find the mean, μ.
2. Calculate the deviation of each measurement from the mean: $x - \mu$.
3. Square each deviation: $(x - \mu)^2$
4. Average the squares of deviations to get the variance:

$$\sigma^2 = \frac{\Sigma(x - \mu)^2}{n}$$

   (The symbol Σ means to take the sum of all the values of the variables coming after the symbol.)
5. Take the square root of the variance to get the standard deviation: $\sigma = \sqrt{\sigma^2}$.

TABLE 9.5
Calculation of Variance

Value: x	Deviation: $x - \mu = x - 85,000$	Deviation squared: $(x - \mu)^2 = (x - 85,000)^2$
80,000	−5,000	25,000,000
83,000	−2,000	4,000,000
85,000	0	0
87,000	2,000	4,000,000
90,000	5,000	25,000,000
	Sum 0	58,000,000

$$\text{variance} = \sigma^2 = \frac{58,000,000}{5} = 11,600,000$$

This procedure is applied again in the next example. The data for this example are the values of the houses on the second block mentioned at the start of this section: $25,000, $30,000, $85,000, $85,000, and $200,000.

 YOUR FORMULATION

Before doing the calculations in Example 9.15, explore your understanding of the standard deviation by answering the following multiple-choice question.

If σ denotes the standard deviation of the set of data in Example 9.15, how will σ compare with 3410, the standard deviation of the house values on the other block?

a. $\sigma < 3410$ b. $\sigma = 3410$ c. $\sigma > 3410$

You can check your answer by studying Example 9.15 and the reasoning for the answer right after the example.

• EXAMPLE 9.15

Calculate the variance and the standard deviation of the house prices $25,000, $30,000, $85,000, $85,000, and $200,000.

SOLUTION

x	Deviations $x - \mu$	Squares of Deviations $(x - \mu)^2$
25,000	−60,000	3,600,000,000
30,000	−55,000	3,025,000,000
85,000	0	0
85,000	0	0
200,000	115,000	13,225,000,000
Sum 4,225,000	0	19,850,000,000

$$\mu = \frac{4,225,000}{5} = 85,000 \qquad \sigma^2 = \frac{19,850,000,000}{5} = 3,970,000,000$$

$$\sigma = \sqrt{3,970,000,000} \approx 63,008$$

Notice that our calculations are on track, since the sum of the deviations (middle column) is 0.

We round both the variance and the standard deviation to one more significant digit than the original measurements (with two significant digits), so the variance is 3,970,000,000 and the standard deviation σ is about 63,000.

The answer to the multiple-choice question in the Your Formulation is answer c: $\sigma > 3410$. How should we have known this before the calculation? The house prices on this second block spread out more than those on the first block, so for the second block the deviations will be bigger, as will the squares of deviations, the average of the squares of deviations (the variance), and its square root (the standard deviation). More generally, the variance and standard deviation of a set of data measure how much the data spread out. The more the data spread out, the larger will be the variance and the standard deviation.

The variance and standard deviation just described are for a population. When working with a sample, the variance is denoted s^2 (standard deviation = s) and is obtained by dividing by $n - 1$ rather than n. This is done for technical reasons that we will not pursue. In this section we will use only the population variance and population standard deviation. If you have a calculator with a key on it to give the standard deviation, check to see whether it gives the population standard deviation, sometimes denoted $\boxed{\sigma_n}$, or the sample standard deviation, sometimes denoted $\boxed{\sigma_{n-1}}$.

Empirical Rule

A relation between the range of a set of data and its standard deviation is given by the following.

EMPIRICAL RULE

If a histogram for a set of data is approximately mound shaped, then 6 standard deviations, centered at the mean, cover almost all the measurements and 4 standard deviations, centered at the mean, cover about 95% of the measurements.

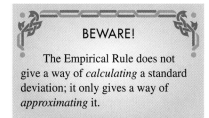

BEWARE!

The Empirical Rule does not give a way of *calculating* a standard deviation; it only gives a way of *approximating* it.

This "rule" is not a precise mathematical statement. In particular it is not proved as a theorem. Instead it can be used to relate *approximately* the range and standard deviation of a certain type of set of data.

The hypothesis that a histogram is approximately mound shaped would include histograms like those in ▶ Figure 9.7 but not like those in ▶ Figure 9.8.

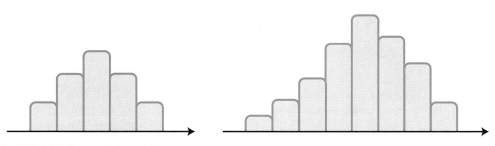

▶ FIGURE 9.7 Mound-shaped histograms

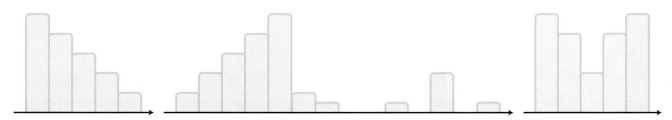

▶ FIGURE 9.8 Not-mound-shaped histograms

To illustrate a use of the Empirical Rule, consider the data on tax rates from Section 9.1 stated in Tables 9.1 and 9.2 and pictured in Figure 9.1. The range of tax rates is 7 (7−0). Let σ denote the standard deviation of these tax rates. By the Empirical Rule, $6\sigma \approx 7$, so $\sigma \approx 7/6 \approx 1.2$. This gives an approximation to the size of the standard deviation.

The part of the Empirical Rule that says, "4 standard deviations" refers to the "middle" 95% of the data. There are 51 tax rates in the set of data, and 95% of 51 is 48.45, or about 48. We toss out three extreme rates—say, two 0s and a 7.0. The remaining 95% of the tax rates stretch from 0 to 7.0. The Empirical Rule says that 4 standard deviations cover these 7 units; i.e., $4\sigma \approx 7$, so $\sigma \approx 7/4 = 1.75$. This is another approximation to the size of the standard deviation. Although this approximation of 1.75 differs from the approximation 1.2 we arrived at previously, they are both close to the actual standard deviation.

The estimate of a standard deviation as just obtained from the range of a set of data is called the *range approximation to the standard deviation.* It is usually accurate within a factor of 2. On the tax rate data, the calculation that $\sigma \approx 1.2$ suggests that the true value of σ is between 0.6 (=1/2 of 1.2) and 2.4 (=2 × 1.2).

EXAMPLE 9.16

Estimate the standard deviation of the ages of people in the United States.

SOLUTION First consider the hypothesis of a mound-shaped histogram. A histogram of ages would be skewed to the right, since there would be a long tail of ages of older people, but it is close enough to being mound shaped that the Empirical Rule can be applied. Only a relatively few people have ages up around 100, so the "4 standard deviations" part of the Empirical Rule will be used. Between what ages will about 95% of the people be? Probably between 0 and 84. If σ denotes the standard deviation of the ages of people in the United States, then $4\sigma \approx 84$, so $\sigma \approx 21$. The true value of σ is expected to be between $1/2 \times 21 = 10.5$ and $2 \times 21 = 42$. Since the population is large, the range approximation to the standard deviation is expected to be better than if the population were small, so the standard deviation is expected to be closer to 21 than to 10.5 or 42.

The range approximation to the standard deviation can be used both to estimate a standard deviation, as in Example 9.16, and to check on the calculation of a standard deviation. For example, one error that is sometimes made is to report a variance as a standard deviation. If this error had been made with the data of Table 9.5, the standard deviation would have been reported as 11,600,000. However, by the range approximation method, the true value of σ should be somewhere between 1250 and 5000. This indicates an error has been made.

Summary

If a set of data is described by both a measure of central tendency and a measure of spread, we can get a fairly accurate understanding of the data. Ways to describe the spread include the range, the variance, the standard deviation, and a box plot. The standard deviation and the range are related by the Empirical Rule.

1. Calculate the range of 1, 3, 5, 7, 9.
2. Calculate the range of 4, 1, 0, 1, 9, 9, 3.
3. For the state general sales and use taxes shown in Table 9.1 of Section 9.1, calculate the range. Include states with no statewide sales and use taxes.

In Exercises 4 and 5, give an example, if possible, of a set of data whose range is the given number. If no example exists, so state.

4. 5   5. $\frac{1}{2}$
6. Give an example of a set of data with mean 3 and range 2.
7. Give an example of a set of data with median 2, range 4, and largest measurement 5.
8. Suppose a set of data has smallest measurement 1 and range 6. Be as specific as possible in answering each of these.
   a. What is the largest measurement?
   b. What can be said about the median?
   c. What can be said about the mean?
9. Set B in Table 9.3 has the lower quartile equal to the smallest measurement and the upper quartile equal to the largest measurement. The corresponding box plot has no whiskers.
   a. Could a box plot have no box? That is, could the lower quartile, the median, and the upper quartile be the same?
   b. If so, give an example of such a set of data. If not, explain why not.

In each of Exercises 10–12, construct a box plot for the given data. The data sets are the same as Section 9.2, Exercises 14–16.

10. 3, 4, 7, 8, 9, 12, 13, 13.   11. 0, 7, 6, 9, 2, 3, 7, 4, 6.
12. The data in Exercise 5 of Section 9.1 on land area of states.
13. A summary of the scores on two tests in a class are shown in the box plots. Compare the results of the two tests. Include a statement of which test had the better results.

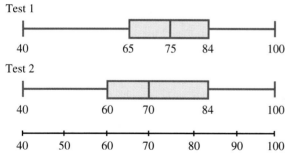

In each of Exercises 14 and 15, calculate the variance and the standard deviation.

14. 1, 2, 5   15. 2, 3, 7, 8
16. (*Multiple choice*) For the data 4, 5, 6, 7, 8, the mean is 6 and the standard deviation is $\sqrt{2}$. If the 4 is replaced by 5 and the 8 is replaced by 7 to give the data 5, 5, 6, 7, 7, the mean will still be 6. Which one of the following will be true of the standard deviation?

i. less than $\sqrt{2}$ ii. equal to $\sqrt{2}$ iii. greater than $\sqrt{2}$
17. (*Multiple choice*) Let σ_s denote the standard deviation of the ages of students at your college or university and σ_f denote the standard deviation of the ages of faculty at your school. Which one is true?
   i. $\sigma_s < \sigma_f$ ii. $\sigma_s = \sigma_f$ iii. $\sigma_s > \sigma_f$
18. The variance of a set of data is 4. What is its standard deviation?
19. The standard deviation of a set of data is 16. What is the variance?

For Exercises 20 and 21, find the range approximation to the standard deviation and compare it with the calculated standard deviation.

20. 1, 2, 5 (see Exercise 14)
21. 2, 3, 7, 8 (see Exercise 15)
22. IQ scores can be represented by a mound-shaped histogram with a mean of 100. Use your knowledge of IQ scores to estimate the standard deviation of IQ scores.
23. (*Multiple choice*) A package of pretzels is marked "net wt. 12 oz." Suppose 1000 "12 oz." packages of this manufacturer's pretzels are carefully weighed. Which is the best estimate of the standard deviation of these weights? A negative number, 0, 0.01, 0.1, 1, 12, a number larger than 12
24. (*Multiple choice*) For the data 1, 1, 5, 5, what is the standard deviation? A negative number, 0, 1, 2, 3, 4, 5, a number larger than 5
25. Can a set of data have a negative standard deviation? If so, give an example of such a set. If not, explain why not.
26. The Securities and Exchange Commission looked at the possibility of requiring investment companies to tell investors some measure of the risk of their mutual funds. The standard deviation is one measure of the risk involved (see the Margin Note "Here comes the SD" in this section). Fill in the blanks in the following excerpts from *Business Week,* May 22, 1995, p. 143.
   Standard deviation is the risk measure many academics prefer. It looks at how much a fund's returns diverge from
   a. _____. Since it doesn't relate to any index, it can be used to compare funds from different asset classes. For example, based on monthly returns, the 3-year standard deviation of the average U.S. diversified equity fund is 9.84 percentage points. Bond funds rate 5.15.
   b. Why do bond funds have a smaller standard deviation than equity funds?
   Standard deviation is useful because it can give a range of returns you can reasonably expect in a coming year. Returns for a fund with a standard deviation of 4 and an average return of 5% should fall between 1% and 9% two-thirds of the time. Some professionals argue that doesn't do you much good, since there's still a
   c. _____ [what fraction or percentage] chance the returns will be out of that range.

1. In his seventh-grade math class, Steve learned how to calculate the mean and median of a set of numbers. Write a description for Steve of why the mean and median are not sufficient to describe a set of data.

2. Suppose you have a choice of two teachers, A or B, for a course next term. Last year the mean grade in A's class was 3.0 and the mean grade in B's class was 2.5. Describe a situation in which you would rather have teacher B than teacher A. Why would you rather have B in this situation?

3. It is common for newspaper or magazine articles to describe the average of a set of data—i.e., to describe its center—but it is much less common to have articles describe how much the data spread out. Why do you think this is?

4. Describe why the standard deviation is a better measure of the spread of a set of data than a range is.

5. Since the range approximation to the standard deviation is so much quicker a way of getting a standard deviation than using the formula

$$\sqrt{\frac{\Sigma\,(x - \mu)^2}{n}}$$

why did we even bother with that formula?

6. Can two sets of data have the same range but different standard deviations? If so, give an example. If not, explain why not.

7. Can two sets of data have the same standard deviation but different ranges? If so, give an example. If not, explain why not.

8. Are the mean and the range of a set of data related? If so, explain why. If not, explain why not.

9. The standard deviation of a particular set of data is 9. What does this mean?

9.4 NORMAL DISTRIBUTIONS

GOALS
1. To distinguish between discrete and continuous random variables.
2. To calculate values of the function defining a standard normal random variable.
3. To calculate probabilities for a standard normal random variable.
4. To calculate probabilities for a normal random variable.

Continuous Random Variables

The experiments looked at in Chapter 8 all had a countable number of outcomes: Flip a coin to get a head or a tail. Count the number of heads when four coins are flipped to get possible outcome 0, 1, 2, 3, or 4. Roll a die and count the number on the top face to get possible outcome 1, 2, 3, 4, 5, or 6.

In this section, experiments whose outcomes can take on an entire interval of values are considered. The variables that take on an entire interval of values are called *continuous random variables,* whereas the variables that take on a countable number of outcomes (such as those of the preceding paragraph) are called *discrete random variables.* Some examples of continuous random variables are the speeds at which we might drive (measured on a speedometer with a needle rather than on a digital speedometer), the height of a person, temperature when measured on a mercury thermometer rather than on a digital thermometer, and the length of time a song takes (measured on a watch with a sweep second hand rather than on a digital watch).

Continuous Probability Distributions

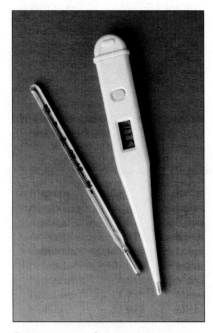

Temperature can be a continuous random variable or a discrete random variable.

The graphical way to describe probabilities for a discrete random variable is with a histogram. For example, the probability $p(x)$ of getting x heads ($x = 0$, 1, 2, 3, or 4) when four coins are tossed is shown in ▶ Figure 9.9. With a continuous random

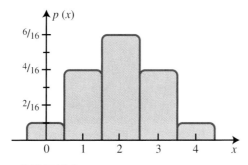

▶ FIGURE 9.9 A histogram

variable there is an infinite number of *x* values to consider, and they are in an interval rather than going from one to another (such as 0, 1, 2, 3).

The probability values for a continuous random variable are described by a curve like that shown in ▶ Figure 9.10. This can be regarded as a generalization of a histogram. The function describing the curve is called a *probability density function*. The area *A* between two *x*-values *a* and *b* is the probability that *x* is between *a* and *b*. In other words, if *P* refers to a probability density function, $P(a < x < b) = A$. Probability Rule 2 of Section 8.2 (the sum of the probabilities of all the outcomes is 1) means the area under the curve and above the *x*-axis is 1.

With a discrete random variable, Probability Rule 1 states that the probability of any outcome is between 0 and 1. For a continuous random variable, the probability of any particular outcome $x = a$, like the speed being $43\frac{1}{7}$ mph, is the area under the curve from *a* to *a* ($43\frac{1}{7}$ to $43\frac{1}{7}$), which is 0. That is, with a continuous random variable, the probability of any particular outcome is 0. Because of this,

$$P(a < x < b) = P(a \le x < b) = P(a < x \le b) = P(a \le x \le b)$$

Therefore, it doesn't matter whether endpoints are included or not.

With a continuous random variable, probabilities are calculated via calculus and are often available in tables. We will focus on one class of continuous random variables, called a *normal random variable*.

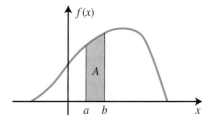

▶ FIGURE 9.10 A probability density function

The Normal Distribution

The normal distribution is considered to be the most important probability distribution in statistics. There are infinitely many normal curves, each of which has a graph that is mound shaped or bell shaped, like the curve in ▶ Figure 9.11.

The normal distribution is important for two reasons. The first is that many naturally occurring measurements, including heights, temperatures, times, IQs, and test scores, have distributions that are approximately normal. The second is that the normal distribution is often involved when making an inference about the mean of a population or about a proportion in a population. (The topic of inferential statistics is discussed in the next section.)

Although there are infinitely many normal curves, probabilities with any of them can be found once we know how to find probabilities with one particular known as the *standard normal distribution*. The standard normal variable is typically denoted *z*, and the curve describing the probability distribution is given by

$$f(z) = \frac{1}{\sqrt{2\pi}} e^{-z^2/2}$$

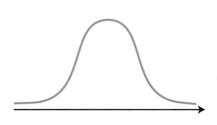

▶ FIGURE 9.11 A normal curve

Abraham DeMoivre and Pierre-Simon Laplace

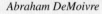

The first use of the concept of the normal distribution was by Abraham DeMoivre (1667–1754). Demoivre was born in France but spent most of his life in England, where he was a friend of Isaac Newton, one of the developers of calculus and the legendary thinker under an apple tree. DeMoivre in 1733 proved some results about a curve related to what is now called the normal curve. He used this curve to approx-

Abraham DeMoivre

imate some probabilities of a particular type of discrete random variable. His results were published in the second and third editions of his *Doctrine of Chances.*

A story is told about DeMoivre's death (Eves, 1990, p. 429) that each day he noticed that he required 15 minutes more sleep than on the preceding day. When the sequence of sleep times reached 24 hours, DeMoivre died.

An extension of DeMoivre's work with what is now called the normal curve was done by the French mathematician Pierre-Simon Laplace (1749–1827). Laplace's result was termed the normal law by H. Poincaré. Maistrov says (Maistrov, 1974, p. 148): "Only after this

work of Laplace did the widespread applications of probability theory become feasible as a scientifically justified method." The mathematical historian Carl Boyer says (Boyer, 1991, p. 492), "The theory of probability owes more to Laplace than to any other mathematician."

Not only was Laplace a mathematician, he made one attempt at government. Napoleon I, a great admirer of scientists, named Laplace Minister of the Interior. Laplace was better at mathematics than administration, however. Napoleon said that he "carried the spirit of the infinitely small into the management of affairs" (Boyer, 1991, p. 492).

Pierre-Simon Leplace

Some function values are calculated in ● Table 9.6. Exercise 5 asks you to calculate the two blank values in the table. The graph of the standard normal distribution is shown in ▶ Figure 9.12. Notice the following properties of the standard normal curve.

The areas under the standard normal curve are given in ● Table 9.7. For example, if we look up $z = 1.23$ in the table (under z, go down to 1.2, then go across from z

PROPERTIES OF THE STANDARD NORMAL CURVE

1. The area under the curve is 1.
2. The curve is symmetric about the line $z = 0$; i.e., the part to the left of $z = 0$ can be reflected in the line $z = 0$ to produce the part to the right of $z = 0$. This means that for any number z, $f(-z) = f(z)$.
3. The curve extends indefinitely in both the positive and negative z-directions.
4. Almost all the area under the curve is between $z = -3$ and $z = +3$.

TABLE 9.6
Some Values of the Standard Normal Probability Density Function

z	f(z)
0	.399
±1	.242
±2	_____
±3	_____

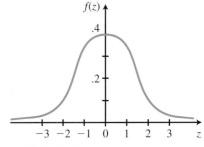

► FIGURE 9.12 Standard normal curve

TABLE 9.7
Standard Normal Probabilities

z	.00	.01	.02	.03	.04	.05	.06	.07	.08	.09
.0	.0000	.0040	.0080	.0120	.0160	.0199	.0239	.0279	.0319	.0359
.1	.0398	.0438	.0478	.0517	.0557	.0596	.0636	.0675	.0714	.0753
.2	.0793	.0832	.0871	.0910	.0948	.0987	.1026	.1064	.1103	.1141
.3	.1179	.1217	.1255	.1293	.1331	.1368	.1406	.1443	.1480	.1517
.4	.1554	.1591	.1628	.1664	.1700	.1736	.1772	.1808	.1844	.1879
.5	.1915	.1950	.1985	.2019	.2054	.2088	.2123	.2157	.2190	.2224
.6	.2257	.2291	.2324	.2357	.2389	.2422	.2454	.2486	.2517	.2549
.7	.2580	.2611	.2642	.2673	.2704	.2734	.2764	.2794	.2823	.2852
.8	.2881	.2910	.2939	.2967	.2995	.3023	.3051	.3078	.3106	.3133
.9	.3159	.3186	.3212	.3238	.3264	.3289	.3315	.3340	.3365	.3389
1.0	.3413	.3438	.3461	.3485	.3508	.3531	.3554	.3577	.3599	.3621
1.1	.3643	.3665	.3686	.3708	.3729	.3749	.3770	.3790	.3810	.3830
1.2	.3849	.3869	.3888	.3907	.3925	.3944	.3962	.3980	.3997	.4015
1.3	.4032	.4049	.4066	.4082	.4099	.4115	.4131	.4147	.4162	.4177
1.4	.4192	.4207	.4222	.4236	.4251	.4265	.4279	.4292	.4306	.4319
1.5	.4332	.4345	.4357	.4370	.4382	.4394	.4406	.4418	.4429	.4441
1.6	.4452	.4463	.4474	.4484	.4495	.4505	.4515	.4525	.4535	.4545
1.7	.4554	.4564	.4573	.4582	.4591	.4599	.4608	.4616	.4625	.4633
1.8	.4641	.4649	.4656	.4664	.4671	.4678	.4686	.4693	.4699	.4706
1.9	.4713	.4719	.4726	.4732	.4738	.4744	.4750	.4756	.4761	.4767
2.0	.4772	.4778	.4783	.4788	.4793	.4798	.4803	.4808	.4812	.4817
2.1	.4821	.4826	.4830	.4834	.4838	.4842	.4846	.4850	.4854	.4857
2.2	.4861	.4864	.4868	.4871	.4875	.4878	.4881	.4884	.4887	.4890
2.3	.4893	.4896	.4898	.4901	.4904	.4906	.4909	.4911	.4913	.4916
2.4	.4918	.4920	.4922	.4925	.4927	.4929	.4931	.4932	.4934	.4936
2.5	.4938	.4940	.4941	.4943	.4945	.4946	.4948	.4949	.4951	.4952
2.6	.4953	.4955	.4956	.4957	.4959	.4960	.4961	.4962	.4963	.4964
2.7	.4965	.4966	.4967	.4968	.4969	.4970	.4971	.4972	.4973	.4974
2.8	.4974	.4975	.4976	.4977	.4977	.4978	.4979	.4979	.4980	.4981
2.9	.4981	.4982	.4982	.4983	.4984	.4984	.4985	.4985	.4986	.4986
3.0	.4987	.4987	.4987	.4988	.4988	.4989	.4989	.4989	.4990	.4990

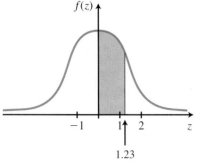

▶ FIGURE 9.13 $P(0 < z < 1.23)$

to the heading .03), we find .3907. This is the shaded area shown in the graph at the top of the table. In ▶ Figure 9.13 the shaded area is the probability that

$$0 < z < 1.23.$$

Other standard normal curve probabilities are calculated in the next example.

EXAMPLE 9.17

If z is a standard normal random variable, calculate the following probabilities.

a. $P(-1.23 < z < 0)$
b. $P(z > 0.58)$
c. $P(z < -1.91)$
d. $P(0.82 < z < 1.51)$
e. $P(-2.37 < z < -2.16)$
f. $P(-3.07 < z < 1.42)$
g. $P(z < 2.28)$
h. $P(z > -0.16)$

SOLUTION In each of these it is helpful to have a sketch of the area under the standard normal curve corresponding to the given probability.

a.

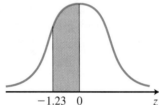

Because of the symmetry of the normal curve, this is the same as

$$P(0 < z < 1.23) = .3907$$

b.

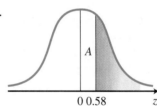

First find A = .2190. Then:

shaded area = $P(z > 0.58)$ = .5000 − .2190
= .2810

Area to the right of z = 0 ─────┐
A ─────────────────┘

c.

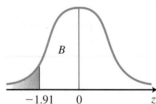

First find B = .4719. Then:

shaded area = $P(z < -1.91)$ = .5000 − .4719
= .0281

d.

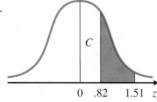

$P(0.82 < z < 1.51$ = shaded area

= (shaded area + C) − C

= $P(0 < z < 1.51)$
  − $P(0 < z < 0.82)$

= .4345 − .2939 = .1406

(This allows us to use areas in the form of those in the normal curve table)

e.

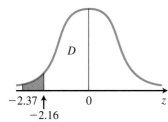

$P(-2.37 < z < -2.16)$ = shaded area

$\qquad\qquad\qquad\quad$ = shaded area $+ D) - D$

$\qquad\qquad\qquad\quad$ = $P(-2.37 < z < 0)$
$\qquad\qquad\qquad\qquad\quad - P(-2.16 < z < 0)$

$\qquad\qquad\qquad\quad$ = $.4911 - .4846 = .0065$

f.

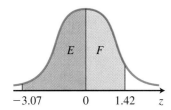

$P(-3.07 < z < 1.42)$ = shaded area

$\qquad\qquad\qquad\quad$ = $E + F$

$\qquad\qquad\qquad\quad$ = $P(-3.07 \le z < 0)$
$\qquad\qquad\qquad\qquad\quad + P(0 < z < 1.42)$

$\qquad\qquad\qquad\quad$ = $.4989 + .4222 = .9211$

g.

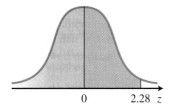

$P(z < 2.28)$ = shaded area

$\qquad\qquad\quad$ = $P(z \le 0) + P(0 < z < 2.28)$

$\qquad\qquad\quad$ = $.5000 + .4887 = .9887$

h.

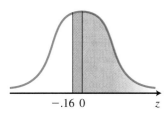

$P(z > -0.16)$ = shaded area

$\qquad\qquad\quad$ = $P(-0.16 < z < 0) + P(z \ge 0)$

$\qquad\qquad\quad$ = $.0636 + .5000$

$\qquad\qquad\quad$ = $.5636$

The standard normal distribution is used to find probabilities associated with a general normal random variable. The probability density function for a normal random variable x with mean μ and standard deviation σ is given by

$$f(x) = \frac{1}{\sigma\sqrt{2\pi}}\, e^{-(1/2)[(x-\mu)/\sigma]^2}$$

Notice that if $\mu = 0$ and $\sigma = 1$, then x has a standard normal distribution. Our purpose in stating this probability density function is to show that a normal curve is given by a specific mathematical function rather than just being a curve that is mound shaped or bell shaped. Three different normal distributions are shown in ► Fig. 9.14.

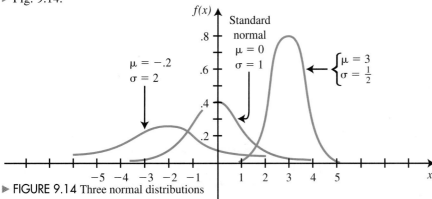

► FIGURE 9.14 Three normal distributions

Most of the following properties of the normal curve are illustrated in Figure 9.14.

The general normal random variable x can be converted to a standard normal random variable z by letting

$$z = \frac{x - \mu}{\sigma}$$

This *z-score* tells the number of standard deviations x is from the mean μ. The difference $x - \mu$ is the directed distance (it might be positive, negative, or zero—see ▶ Figure 9.15) from μ to x, and when $x - \mu$ is divided by σ, the result is the number of standard deviations x is from μ.

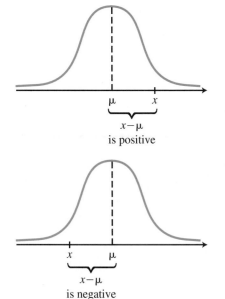

▶ FIGURE 9.15 Number of standard deviations from the main

PROPERTIES OF THE NORMAL CURVE

1. The area under the curve is 1.
2. The curve is symmetric about the line $x = \mu$; so when μ is positive, the center (or highest point) of the curve shifts to the right, and when μ is negative, the curve shifts to the left.
3. The curve extends indefinitely in both the positive and negative x-directions.
4. When σ is small (close to 0), the curve is tall and skinny; when σ is large, the curve is more spread out.
5. Almost all the area under the curve is within 3σ of μ.

To find the probability that a normal random variable x with mean 5.0 and standard deviation 2.0 is between 5.0 and 6.5, you can apply the following procedure. Sketch a graph of the normal distribution and note the mean 5.0 and the x-values of interest: 5.0 and 6.5.

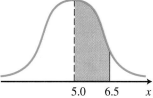

1. How many standard deviations is the x-value of 5.0 from the mean? This is the z-value corresponding to $x = 5.0$.
2. How many standard deviations is the x-value of 6.5 from the mean?
   a. How many units is it from the mean?
   b. 1 standard deviation is 2.0 units. How many standard deviations is the number you obtained in part a?
   The number you found in part b is the number of standard deviations 6.5 is from the mean.
3. (*Fill in the blank.*) $P(5.0 < x < 6.5) = P(0 < z < .75) = $ _____

EXAMPLE 9.18

Let x be a normally distributed random variable with mean 25 and standard deviation 5. Find each of the following.

a. $P(20 \le x \le 28)$ b. $P(26 \le x \le 36)$
c. $P(19 \le x \le 23)$ d. $P(x < 18)$
e. $P(x \ge 33)$

SOLUTION

a.

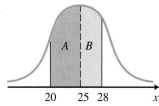

For $x = 20$, $z = \dfrac{20 - 25}{5} = \dfrac{-5}{5} = -1$

so $A = .3413$. For $x = 28$,

$$z = \frac{28 - 25}{5} = \frac{3}{5} = .6,$$

so $B = .2257$.

$$P(20 \le x \le 28) = A + B = .5670$$

b.

For $x = 26$, $z = \dfrac{26 - 25}{5} = \dfrac{1}{5} = .2$

so $C = .0793$. For $x = 36$,

$$z = \frac{36 - 25}{5} = \frac{11}{5} = 2.2$$

so $C + D = .4861$.

$$\begin{aligned} P(26 \le x \le 36) &= (C + D) - C \\ &= .4861 - .0793 = .4068 \end{aligned}$$

c.

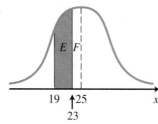

For $x = 19$,

$$z = \frac{19 - 25}{5} = \frac{-6}{5} = -1.1$$

so $E + F = .3643$. For $x = 23$,

$$z = \frac{23 - 25}{5} = \frac{-2}{5} = -.4$$

so $F = .1554$.

$$P(19 \le x \le 23) = (E + F) - F$$
$$= .3643 - .1554 = .2089$$

d.

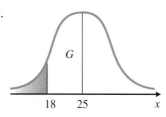

For $x = 18$,

$$z = \frac{18 - 25}{5} = \frac{-7}{5} = -1.4$$

so $G = .4192$.

$$P(x < 18) = \text{shaded area}$$
$$= .5000 - .4192 = .0808$$

e.

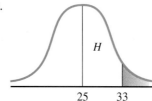

For $x = 33$,

$$z = \frac{33 - 25}{5} = \frac{8}{5} = 1.6$$

so $H = .4452$.

$$P(x \ge 33) = .5000 - .4452 = .0548$$

Applications of the Normal Distribution

Earlier we stated that one of the reasons for the importance of the normal distribution is that many naturally occurring measurements have distributions that are approximately normal. The next two examples and Exercises 18–21 illustrate uses of the normal distribution.

EXAMPLE 9.19

The grade point averages (GPAs) of students at a certain college are approximately normally distributed with a mean of 2.7 and a standard deviation of 0.5. What percentage of students have GPAs under 2.0?

SOLUTION

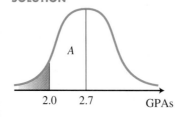

For a GPA of 2.0,

$$z = \frac{2.0 - 2.7}{0.5} = \frac{-.7}{.5} = -1.4$$

so $A = .4192$, and $P(\text{GPA} < 2.0) = .0808$. About 8.08% of students have GPAs under 2.0.

EXAMPLE 9.20

The heights of adult males are normally distributed with a mean of 69.2 inches and a standard deviation of 2.8 inches.

a. What percentage of adult males are between 5 feet and 6 feet tall?
b. What percentage of adult males are over 6 feet tall?

SOLUTION

a. A height of 5 feet is 60 inches and a height of 6 feet is 72 inches.

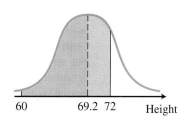

The z-value for 60 inches is
$$z = \frac{60 - 69.2}{2.5} = \frac{-9.2}{2.5} = -3.68$$

The z-value for 72 inches is
$$z = \frac{72 - 69.2}{2.5} = \frac{2.8}{2.5} = 1.12$$

When we look in the normal curve table for $z = 3.68$, we don't find it. However, note that

$P(-3.68 < z < 0) = P(0 < z < 3.68) > .4990$

$P(0 < z < 1.12) = .3686.$

$P(60 < \text{height} < 72 = P(60 < \text{height} < 69.2) + P(69.2 \le \text{height} < 72)$

$\qquad = P(-3.68 < z < 0) + P(0 \le z < 1.12) > .4990 + .3686$

$\qquad = .8676$

so $P(60 < \text{height} < 72) > .8676$. The largest the probability would be

$$5000 + .3686 = .8686.$$

Therefore, about 87% of adult males are between 5 feet and 6 feet tall.

b.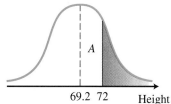

In part a the area A is calculated to be .3686.

Then

$P(\text{height} > 6 \text{ ft}) = \text{area of shaded part}$
$\qquad = .5000 - .3686$
$\qquad = .1314$

Therefore, about 13% of adult males are over 6 feet tall.

This section has introduced you to the normal distribution, the most important distribution in statistics. One of the reasons for its importance is the description of many naturally occurring measurements, as illustrated by the preceding two examples. Another reason for its importance will be described in the next section on statistical inference.

In each of Exercises 1–4, identify the random variable as discrete or continuous.

1. The number of tails on 10 flips of a coin.
2. The time it takes to run 100 meters, measured with a sweep second hand.
3. The time it takes to run 100 meters, measured on a digital stopwatch.
4. The incomes of people in the United States.
5. In Table 9.6 calculate each of the following to three decimal places.
   a. $f(\pm2)$
   b. $f(\pm3)$

In each of Exercises 6–15 assume that z has a standard normal distribution. Find the indicated probability.

6. $P(z > 0)$ (can be done without tables)
7. $P(0 < z < 2.83)$
8. $P(-2.20 < z < 0)$
9. $P(z > 0.55)$
10. $P(z < 2.27)$
11. $P(z > -0.12)$
12. $P(z < -1.07)$
13. $P(1.68 < z < 2.31)$
14. $P(-2.24 < z < -0.09)$
15. $P(-1.72 < z < 2.80)$
16. Assume x has a normal distribution with mean 80 and standard deviation 9. Find each of the following probabilities.
   a. $P(x < 70)$
   b. $P(x \geq 70)$
   c. $P(60 < x \leq 72)$
   d. $P(55 \leq x \leq 93)$
17. Assume x has a normal distribution with mean 0.21 and standard deviation 0.06. Find each of the following probabilities.
   a. $P(x \leq .10)$
   b. $P(x \geq .10)$
   c. $P(.16 \leq x \leq .31)$
   d. $P(.25 < x \leq .39)$
18. The heights of American women of ages 18 to 24 are approximately normally distributed with a mean of 64 inches and a standard deviation of 2.5 inches.
   a. What percentage of American women of ages 18 to 24 are between 5 feet and 6 feet tall?
   b. What percentage of American women of ages 18 to 24 are under 5 feet tall?
19. The Environmental Protection Agency (EPA) has stated that the gasoline mileage for a particular model of car is 25 mpg. Also, this is the mean mileage, and the mileage figures for this model are normally distributed with a standard deviation of 1.5 mpg.

a. What is the probability that a car of this model gets less than 21 mpg?
b. What is the probability that a car of this model gets 28 or more mpg?

20. The pulse rate, in beats per minute, of the adult male population between the ages of 18 and 25 in the United States has a normal distribution with a mean of 72 beats per minute and a standard deviation of 9.7 beats per minute. Pulse rates between 50 and 85 beats per minute are considered typical.
   a. What percentage of adult males between the ages of 18 and 25 in the United States have a typical pulse rate?
   b. The pulse of the average woman is a little faster—about 78 beats per minute. What percentage of men have pulse rates this high or higher?

21. The theory of "grading on the curve" is to assume that scores on a test are approximately normally distributed with mean μ and standard deviation σ. A grade of C is assigned to scores between μ − .5σ and μ + .5σ, B's and D's go an additional standard deviation each way: B's from μ + .5σ to μ + 1.5σ and D's from μ − 1.5σ to μ − .5σ. A's and E's are beyond these: A's for scores higher than μ + 1.5σ and E's for scores less than μ − 1.5σ.

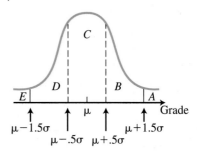

a. Find the approximate percentage of scores for each grade A, B, C, D, E.
b. A test with mean of 72.3 and standard deviation of 8.7 is to be graded on the curve. Construct the grading scale.

22. (*Multiple choice*) Which one of the following is correct?
   i. Every mound-shaped distribution is a normal distribution.
   ii. Some, but not all, mound-shaped distributions are normal distributions.
   iii. No mound-shaped distribution is a normal distribution.

23. (*Multiple choice*) Which one of the following is correct?
   i. Every normal distribution is mound-shaped.
   ii. Some, but not all, normal distributions are mound-shaped.
   iii. No normal distribution is mound-shaped.

1. Write a paragraph describing the relation between mound-shaped distributions and normal distributions.

2. In 1995, there were 824 major league players. The median salary was $275,000 and the mean salary was $1,089,621. The best-paid 12% of the players received 54% of the total payroll—$485 million. The remaining 724 players received a total of $413 million. Do you think major league baseball salaries for 1995 were normally distributed? Why or why not? Find comparable information for the current year. Discuss whether or not these salaries are normally distributed.

3. The Empirical Rule of Section 9.3 is based on the normal curve.
   a. Write a paragraph or two describing the relationship between the Empirical Rule and the normal curve.
   b. Extend the Empirical Rule to say what percentage of the measurements are covered if you go 1 standard deviation each way from the mean.

4. (*Library research*) *The Bell Curve: Intelligence and Class Structure in American Life* is a controversial book by Richard J. Herrnstein that was published in 1994.
   a. Describe a criticism of the book that contains statistical concepts (Check the key term *bell curve* in a periodical database to locate a number of articles criticizing the book.)
   b. Draw a diagram showing the sequences of terms or concepts that need to be described in order for someone at the start of this course to understand the criticism you've described in part a.
   c. For the criticism you've described in part a, provide an argument either to support Herrnstein's position or to support the criticism of his position.

9.5 STATISTICAL INFERENCE

GOALS
1. To describe what a sampling distribution is.
2. To work with the sampling distribution of $\bar{x}$.
3. To understand the relation between sample size, confidence level, and margin of error.
4. To calculate and interpret a margin of error.

Up to this point in this chapter we have discussed descriptive statistics, the part of statistics dealing with the presentation of data. Data may be presented graphically or numerically. Graphical methods include histograms, stem-and-leaf displays, and box plots. Numerical methods include both measures of central tendency and measures of spread. In this section we look at *inferential statistics,* the part of statistics dealing with drawing conclusions about a population based on information contained in a sample. As Morris Kline suggests when he describes statistics as "the mathematical theory of ignorance," we deal with a situation in which we know something about a sample, such as its mean, and we try to say something about the population mean, of which we are ignorant.

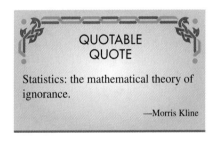

QUOTABLE QUOTE

Statistics: the mathematical theory of ignorance.

—Morris Kline

To illustrate the kinds of inference we might want to make, consider the following example: A consumers' group is concerned about whether or not a cereal manufacturer is filling its "15-oz" boxes of breakfast flakes to the 15-oz level. To test this, the consumers' group plans to sample randomly a number of "15-oz" boxes of the cereal from area stores, weigh them, and determine their mean weight. From this statistic the consumers' group will decide whether or not to put pressure on the cereal manufacturer.

There are several questions that need to be considered. How many boxes of cereal should be sampled? If the sample mean weight is less than 15 oz, how much less than 15 oz should it be before pressure is put on the cereal manufacturer? How likely is it that an error will be made?

Do the 15 ounce boxes really contain 15 ounces of cereal?

Sampling Distributions for $\bar{x}$

The concept underlying statistical inference is that of a *sampling distribution*—a probability distribution of what is called a statistic. A *statistic* is a numerical measurement in the sample that is the basis for estimating a measurement in the population. For example, the sample mean is used to estimate the population mean.

To see the concept of a sampling distribution with a computationally manageable example, consider the cereal example just presented. Suppose the manufacturer makes five boxes of cereal (the population) with weights (unknown to the consumers' group) of 14.8, 15.0, 15.0, 15.2, and 15.4 oz. The consumers' group takes a sample of size 2 and calculates the sample mean $\bar{x}$. There are

$$C(5, 2) = \frac{5 \cdot 4}{2 \cdot 1} = 10 \text{ samples of size 2}$$

They are described in ● Table 9.8. The probability distribution for $\bar{x}$ is shown in ●Table 9.9, and a relative frequency histogram for $\bar{x}$ is shown in ▶ Figure 9.16.

The mean weight of the population is 15.08 oz, while the sample means vary. There is a 20% chance that the sample mean will be less than 15.0 oz. In other words, if the conjecture that the mean weight of all the cereal boxes is less than 15.0 oz is tested on a sample of size 2, there is a 20% chance of making a mistake.

TABLE 9.8
Samples of Size 2

Sample	$\bar{x}$
14.8, 15.0	14.9
14.8, 15.0	14.9
14.8, 15.2	15.0
14.8, 15.4	15.1
15.0, 15.0	15.0
15.0, 15.2	15.1
15.0, 15.4	15.2
15.0, 15.2	15.1
15.0, 15.4	15.2
15.2, 15.4	15.3

TABLE 9.9
Probability Distribution for $\bar{x}$

$\bar{x}$	14.9	15.0	15.1	15.2	15.3
$p(\bar{x})$	$\frac{2}{10}$	$\frac{2}{10}$	$\frac{3}{10}$	$\frac{2}{10}$	$\frac{1}{10}$

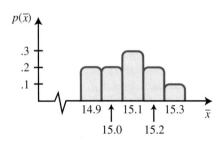

▶ FIGURE 9.16 Histogram for $\bar{x}$

Notice also that if the conjecture that the mean weight of all the cereal boxes is less than 15.0 oz is tested on a sample of size 1, there is also a 20% chance of making a mistake. Let's try samples of size 3:

$$C(5, 3) = \frac{5 \cdot 4 \cdot 3}{3 \cdot 2 \cdot 1} = 10$$

There are 10 of these, shown in ● Table 9.10. The probability distribution for $\bar{x}$ is shown in ● Table 9.11, and a relative frequency histogram for $\bar{x}$ is shown in ▶ Figure 9.17.

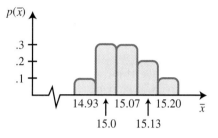

TABLE 9.10 Samples of Size 3	
Sample	$\bar{x}$
14.8, 15.0, 15.0	14.93
14.8, 15.0, 15.2	15.00
14.8, 15.0, 15.4	15.07
14.8, 15.0, 15.2	15.00
14.8, 15.0, 15.4	15.07
14.8, 15.2, 15.4	15.13
15.0, 15.0, 15.2	15.07
15.0, 15.0, 15.4	15.13
15.0, 15.2, 15.4	15.20
15.0, 15.2, 15.4	15.20

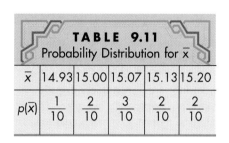

TABLE 9.11 Probability Distribution for $\bar{x}$				
$\bar{x}$	14.93	15.00	15.07	15.13	15.20
$p(\bar{x})$	$\frac{1}{10}$	$\frac{2}{10}$	$\frac{3}{10}$	$\frac{2}{10}$	$\frac{2}{10}$

▶ FIGURE 9.17 Histogram for $\bar{x}$

If the conjecture that the mean weight of all the cereal boxes is less than 15.0 oz is tested on a sample of size 3, there is a 10% chance of making a mistake. Does the chance of making a mistake decline further if the conjecture is tested on a sample of size 4?

YOUR FORMULATION

1. In our cereal example, how many samples of size 4 are there?
2. Complete ● Table 9.12 by listing the other samples of size 4 and, for each sample, its mean $\bar{x}$.
3. Complete ● Table 9.13 to give the probability distribution for $\bar{x}$.
4. Complete ▶ Figure 9.18 to give the histogram for $\bar{x}$.
5. If the conjecture that the mean weight of all the cereal boxes is less than 15.0 oz is tested on a sample of size 4, what is the probability of making a mistake?

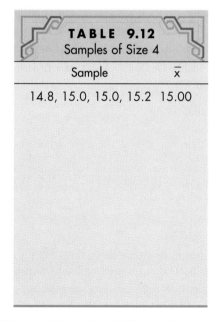

TABLE 9.12 Samples of Size 4	
Sample	$\bar{x}$
14.8, 15.0, 15.0, 15.2	15.00

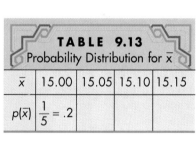

TABLE 9.13 Probability Distribution for $\bar{x}$			
$\bar{x}$	15.00	15.05	15.10	15.15
$p(\bar{x})$	$\frac{1}{5} = .2$			

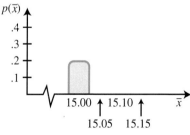

▶ FIGURE 9.18 Histogram for $\bar{x}$

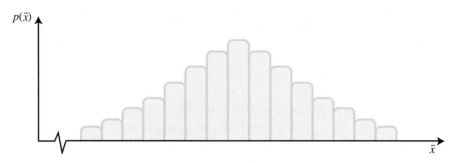

▶ FIGURE 9.19 Histogram for $\bar{x}$

Of course, if the conjecture were tested with a sample of size 5, the whole population would be the sample and the chance of making a mistake because of sampling would be eliminated.

Consider now the more realistic situation that the cereal manufacturer produces thousands of boxes of "15-oz" breakfast flakes. If samples of, say, size 100 are taken, there is a huge number of samples possible. If a mean $\bar{x}$ is calculated in each sample, a relative frequency histogram for $\bar{x}$ values can be generated that will look something like the one in ▶ Figure 9.19.

This example with cereal illustrates several points about statistical inference concerning a population mean μ based on information from a sample mean $\bar{x}$.

* The inference almost always has a chance of being wrong.
* The probability of error because of sampling depends on the distribution of $\bar{x}$.
* The probability distribution of $\bar{x}$ (called the *sampling distribution of $\bar{x}$*) changes as the sample size changes.
* The probability of error due to sampling generally decreases as the sample size increases.

Rather than calculate the sampling distribution of $\bar{x}$ for a population and sample size of interest, the following properties of the sampling distribution of $\bar{x}$ can be used. The proofs of these properties can be found in many mathematical statistics books.

PROPERTIES OF THE SAMPLING DISTRIBUTION OF $\bar{x}$

1. The mean of the sampling distribution of $\bar{x}$ equals the mean of the population.
$$\mu_{\bar{x}} = \mu$$

2. The standard deviation of the sampling distribution is
$$\frac{\text{standard deviation of population}}{\sqrt{\text{sample size}}}$$
$$\sigma_{\bar{x}} = \frac{\sigma}{\sqrt{n}}$$

3. If the sample size is large, the sampling distribution of $\bar{x}$ is approximately normal.

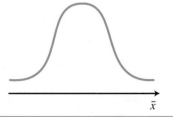

In Property 3, how many need to be in the sample for it to be "large"? For most populations that will be sampled, a sample of size 30 or larger is large enough for the normal distribution to give a good approximation to the distribution of $\bar{x}$.

EXAMPLE 9.21

Suppose our cereal manufacturer claims that the mean weight of boxes of "15-oz" breakfast flakes is 15.05 oz with a standard deviation of 0.40 oz.

a. If the manufacturer's claim is true, what is the probability that a sample of 100 boxes of "15-oz" breakfast flakes will have a mean weight of less than 15 oz?

b. What about a sample of 500 boxes?

SOLUTION First consider the distribution of $\bar{x}$. Since the sample size is large, it is approximately normal. Its mean $\mu_{\bar{x}} = \mu = 15.05$.

a.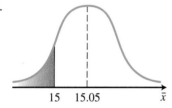

Its standard deviation

$$\sigma_{\bar{x}} = \frac{\sigma}{\sqrt{n}} = \frac{0.40}{\sqrt{100}} = \frac{0.40}{10} = .04$$

We're asked to find $P(\bar{x} < 15)$, which is the shaded area under the normal curve. For $\bar{x} = 15$, $z = -1.25$. From the normal curve table, we find

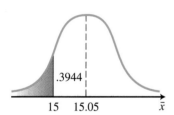

$$P(-1.25 < z < 0) = .3944$$

so

$$P(15 < \bar{x} < 15.05) = .3944$$

Then $P(\bar{x} < 15) = .5000 - .3944 = .1056$. The answer is .1056. Thus it is unlikely that a sample of 100 boxes of "15-oz" breakfast flakes will have a mean weight of less than 15 oz if the manufacturer's claim is true.

b. In this case,

$$\sigma_{\bar{x}} = \frac{\sigma}{\sqrt{n}} = \frac{0.40}{\sqrt{500}}$$

and for $\bar{x} = 15$,

$$z = \frac{15 - 15.05}{0.40/\sqrt{500}} \approx -2.80.$$

From the normal curve table, we find $P(-2.80 < z < 0) = .4974$

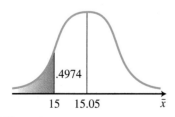

so

$$P(15 < \bar{x} < 15.05) = .4974$$

Then

$$P(\bar{x} < 15) = .5000 - .4974 = .0026.$$

The answer is .0026, or almost 0.

The conclusions of Example 9.21 can be stated differently: If the manufacturer's claim is true and if the consumers' group decides to put pressure on the cereal manufacturer if the sample produces a mean weight of less that 15.0 oz, then with a sample of 100 boxes the consumers' group will make a mistake 10.56% of the time and with a sample of 500 boxes it will make a mistake 0.26% of the time.

EXAMPLE 9.22

Suppose the consumers' group believes that sampling 500 boxes will cost more than it wants to spend. It would like to sample enough boxes so that the chance of making a mistake because of sampling is 1% or less. How many boxes should it sample?

SOLUTION We know from the preceding example that we need at least 100 boxes, so our sample is large and hence the distribution of $\bar{x}$ is approximately normal. It has a mean of 15.05 and a standard deviation

$$\sigma_{\bar{x}} = \frac{\sigma}{\sqrt{n}} = \frac{0.40}{\sqrt{n}}$$

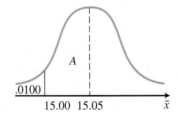

To keep the probability of sampling error at 1% or less, the area in the left tail must be .0100 or less. The area A will be .5000 − .0100 = .4900. From the normal curve table, the corresponding z-value for an area of .4900 is about −2.33. (It's negative because 15.00 is to the left of the mean 15.05.)

$$-2.33 = z = \frac{15.00 - 15.05}{\sigma_{\bar{x}}} = \frac{-0.05}{\sigma/\sqrt{n}} = -.05\frac{\sqrt{n}}{\sigma} = -.05\frac{\sqrt{n}}{0.40}$$

$$-2.33 = -.125\sqrt{n}$$

$$\sqrt{n} = \frac{-2.33}{-.125} = \frac{2.33}{.125}$$

$$n = \left(\frac{2.33}{.125}\right)^2 = 347.4496$$

To guarantee having the error no larger than 1%, we round n up to 348. The consumers' group should sample at least 348 boxes.

One practical problem with the previous analysis is that it requires knowledge of the mean and standard deviation of the population. However, it is often the mean of the population that we are trying to estimate. What do we do in these cases? It can be shown that if the sample size is large (30 or more), then a good estimate of σ is calculated by

$$s = \sqrt{\frac{\Sigma(x - \bar{x})^2}{n - 1}}$$

In other words, s is like the population standard deviation except we divide by $n - 1$ instead of by n. It may be denoted on your calculator by $\boxed{\sigma_{n-1}}$.

The following example illustrates how an inference can be made about a population mean μ if only $\bar{x}$ and s are known.

EXAMPLE 9.23

A consumers' group believes that a fast-food chain's cherry pies have an average of less than 2.0 cherries per pie. They take a random sample of 50 pies and find a mean number of cherries of $\bar{x} = 1.9$ with a sample standard deviation of $s = 0.5$. If they release to the media a statement that the fast-food chain's cherry pies have an average of less than 2.0 cherries per pie, what is the probability of making an error caused by sampling?

SOLUTION To make an error, the actual mean number of cherries per pie must be 2.0 or more. We proceed with the assumption that $\mu = 2.0$. Since the sample size is large, the distribution of $\bar{x}$ is approximately normal. It has a mean $\mu_{\bar{x}} = \mu = 2.0$ and a standard deviation

$$\sigma_{\bar{x}} = \frac{\sigma}{\sqrt{n}} \approx \frac{s}{\sqrt{n}} = \frac{0.5}{\sqrt{50}}$$

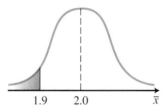

The probability of making an error caused by sampling is $P(\bar{x} \leq 1.9)$. For $\bar{x} = 1.9$,

$$z = \frac{1.9 - 2.0}{\sigma_{\bar{x}}} \approx \frac{-0.1}{0.5/\sqrt{50}} \approx -1.41$$

$$P(\bar{x} \leq 1.9) = P(z \leq -1.41)$$

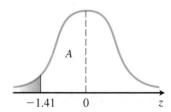

First we find the area $A = P(-1.41 \leq z \leq 0)$ by looking up $z = 1.41$ in the normal curve table and getting $A = .4207$. Then the shaded area $= .5000 - .4207 = .0793$. The probability of making an error caused by sampling is about .0793.

Sampling Distribution for $\hat{p}$

In addition to making inferences about the mean of a population, it is frequently desirable to make inferences about the proportion of the population that has a certain characteristic. Public opinion polls usually make an inference about the pro-

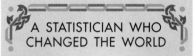
A STATISTICIAN WHO CHANGED THE WORLD

W. Edwards Deming developed the industrial process of *quality control*. As a part of that process, a sample is taken partway through production. If the data from the sample are too far away from what is expected, then the production process is corrected. The effect of his thinking is summed up in his obituary in *Time*, January 3, 1994.

W. Edwards Deming, 93, American industrial-efficiency expert and guru of the postwar Japanese economic miracle [died] in Washington. Deming was a modern illustration of the biblical truth that a prophet is without honor in his own land. Educated in mathematics and physics, he worked with Bell Labs' Walter Shewhart during the 1930s developing quality-control theories that stressed achieving uniform results during production rather than through inspection at the end of the production line. During World War II Deming successfully applied his approach to the making of airplane parts. Ignored by postwar American industry, the irascible Deming took his gospel to Japan in 1950, where it was embraced. His ideas finally took root in the U.S. in the 1980s, when the Detroit auto industry asked for his help in competing with the very Japanese firms he had inspired.

SHOE

portion or percentage of the population that believes a certain way such as the percentage that approve of what the president is doing, the percentage that plan to vote for a certain candidate in an upcoming election, or the proportion that take an afternoon nap. Academic and business studies look at proportions such as the proportion of people who are overweight, the proportion of consumers who might buy a new product, and the percentage of bolts that are defective.

Let p denote the proportion of the population that has a certain characteristic. We measure this by checking the proportion $\hat{p}$ (read this as "p-hat") in a sample of size n that has the characteristic. As with making inferences about means, different samples of size n will produce different values of $\hat{p}$. One randomly selected group of 100 people may have 54 people approving of what the president is doing ($\hat{p} = .54$), while another randomly selected group of 100 people may have 61 people approving of what the president is doing ($\hat{p} = .61$). The sample statistic $\hat{p}$ has a distribution that underlies making inferences about p. We state, without proof, properties of the sampling distribution of $\hat{p}$.

PROPERTIES OF THE SAMPLING DISTRIBUTION OF $\hat{p}$

1. The mean of the sampling distribution of $\hat{p}$ equals p.

$$\mu_{\hat{p}} = p$$

2. The standard deviation of the sampling distribution of $\hat{p}$ is $\sqrt{\dfrac{pq}{n}}$, where $q = 1 - p$.

$$\sigma_{\hat{p}} = \sqrt{\dfrac{pq}{n}}$$

3. If the sample size is large, the sampling distribution of $\hat{p}$ is approximately normal.

In Property 3, how large "large" is depends on the value of p. ● Table 9.14 gives some values of n for given p values that are large enough for the normal distribution to yield a good approximation to the distribution of $\hat{p}$.

┌●**EXAMPLE 9.24**

Suppose that 65% of the people approve of the president's performance. What is the probability that in a random sample of size 100, 60% or fewer of the people approve of the president's performance?

SOLUTION Notice that the sample size is large enough to say that the distribution of $\hat{p}$, the proportion of people in a sample of size 100 who approve of the president's performance, is approximately normal. It has mean $\mu_{\hat{p}} = p = .65$ and standard deviation

$$\sigma_{\hat{p}} = \sqrt{\frac{pq}{n}} = \sqrt{\frac{(.65)(.35)}{100}} \approx .0477$$

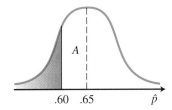

We want to find $P(\hat{p} \leq .60)$. This is the shaded area of the graph. For $\hat{p} = .60$,

$$z = \frac{.60 - .65}{\sigma_{\hat{p}}} = \frac{-.05}{\sqrt{(.65)(.35)/100}} \approx -1.05.$$

A normal curve table shows that $A = .3531$, so the shaded area is $.5000 - .3531 = .1469$. The probability in a random sample of size 100 that 60% or fewer of the people approve of the president's performance is .1469.

└●

In many questions that are asked about a population proportion, the population proportion is not known. For example, in Example 9.24 we might try to determine what percentage of all people approve of the president's performance. The only way to know this with complete accuracy is to ask everybody. In any sampling process, there is a possibility of error that can be described by means of the sampling distribution. The sampling distribution of $\hat{p}$ has mean $\mu_{\hat{p}} = p$. If we don't know p, what do we use for it? The best guess is to use $\hat{p}$. This is also the correct answer. The sampling distribution of $\hat{p}$ has standard deviation $\sigma_{\hat{p}} = \sqrt{pq/n}$. If we don't know p (and hence don't know q), how do we calculate $\sigma_{\hat{p}}$? Fortunately, the standard deviation $\sigma_{\hat{p}}$ changes very slowly with p, so a good estimate of $\sigma_{\hat{p}}$ is $\sqrt{\hat{p}\hat{q}/n}$, where $\hat{q} = 1 - \hat{p}$.

We have just described how to use the properties of the sampling distribution of $\hat{p}$ when p is unknown but $\hat{p}$ is known. How will we use them to estimate p? For example, suppose that 42% of a random sample of 812 registered voters plan to vote for candidate X in the next election. What percentage of all registered voters plan to vote for X? Our first answer is 42%.

It is common to make an estimate of something by giving a range of possible values. For instance, the expected temperature is in the low 60s; the child's height when he grows up will be between 6′0″ and 6′2″; the expected grade in the course is A or B. We will use the value of $\hat{p}$ to give an interval estimate for p. Suppose we want to construct an interval for the percentage of all registered voters who will vote for candidate X in such a way that we are 95% sure we will capture the true percentage. Since the sample size is large, the distribution of $\hat{p}$ is approximately normal. The best estimate of its mean $\mu_{\hat{p}}$ is .42 $(= \hat{p})$, and the best estimate of $\sigma_{\hat{p}}$ is

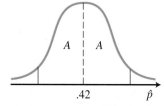

$$\sqrt{\frac{\hat{p}\hat{q}}{n}} = \sqrt{\frac{(.42)(.58)}{812}}$$

To be 95% sure the interval captures p, $A + A = .95$, so $2A = .95$ and $A = .4750$. From the normal curve table, the corresponding z-value is 1.96. The interval is

$$.42 \pm 1.96\, \sigma_{\hat{p}} = .42 \pm 1.96 \sqrt{\frac{(.42)(.58)}{812}} \approx .42 \pm .034$$

The interval is (.386, .454). Therefore, there is a 95% chance that the percentage of all voters who favor candidate X is between 38.6% and 45.4%. This is called a *95% confidence interval* for p.

EXAMPLE 9.25

A poll of 1000 registered voters in 1996 found that 45% said their state should ban casino gambling. Find a 99% confidence interval for the percentage of all registered voters who believed at that time that their state should ban casino gambling.

SOLUTION Since the sample size is large, the distribution of $\hat{p}$ is approximately normal.

$A + A = .99$, so $2A = .99$ and $A = .4950$. The corresponding z-value is 2.575. The interval is

$$.45 \pm 2.575\, \sigma_{\hat{p}} \approx .45 \pm 2.575 \sqrt{\frac{\hat{p}\hat{q}}{n}}$$

$$= .45 \pm 2.575 \sqrt{\frac{(.45)(.55)}{1000}} \approx .45 \pm .041$$

or (.409, .491).

There is a 99% chance that the true percentage of all registered voters who believed in 1996 that their state should ban casino gambling was between 40.9% and 49.1%.

Half the width of a confidence interval—i.e., the distance up or down from $\hat{p}$—is called the *margin of error*. For a 95% confidence interval, the margin of error is 1.96 $\sigma_{\hat{p}}$

EXAMPLE 9.26

Calculate the margin of error in Example 9.25.

SOLUTION It is .041, the distance up or down from $\hat{p}$. As a percentage, this is 4.1%.

Notice that the margin of error $z\sigma_{\hat{p}} = z\sqrt{\hat{p}\hat{q}/n}$ depends on both the sample size n and the confidence level (90%, 95%, 99%, etc.). Often a media report gives the margin of error without the confidence level or the confidence level without the margin of error. If the margin of error is given without the confidence level, we can usually assume a 95% confidence interval. Almost all public opinion polls state 95% confidence intervals. If the confidence level is given without the margin of error, the margin of error can be calculated as in the Example 9.26.

The sample size, the confidence level, and the margin of error are related in an intuitive way that can be made precise by the confidence interval formula

$$\hat{p} \pm z\sigma_{\hat{p}} = \hat{p} \pm z\sqrt{\frac{\hat{p}\hat{q}}{n}}$$

Statement	Intuitive Reason	Formula
If the sample size is held fixed, then as the confidence level increases, the margin of error increases.	More certainty in capturing p demands a wider net.	A higher confidence level increases z and hence the width.
If the confidence level is held fixed, then as the sample size increases, the margin of error decreases.	A larger sample gives better results, so capturing p does not require as wide a net.	Increasing n decreases $\sigma_{\hat{p}}$ and hence the width.

A statement of margin of error or sampling error can often be misinterpreted. In order to help you interpret these correctly, some incorrect interpretations will be given along with explanations of why they are incorrect. The goal is that by understanding why incorrect interpretations are wrong, you can avoid them and make a correct interpretation.

A *U.S. News/Bozell* poll of 1005 adults conducted by KRC Research & Consulting, on Feb. 8–12, 1996, and reported in *U.S. News & World Report* on April 22, 1996, p. 68, asked people if they thought incivility has worsened in the past 10 years. The yes response was 78%. "Margin of error: plus or minus 3 percentage points."

Incorrect Interpretation	Why It's Wrong
1. Between 75% and 81% of those surveyed answered yes.	Exactly 78% of those surveyed answered yes.
2. The error is that there may have been some problems in the poll.	There may have been problems in the poll, although it was taken by a reputable and experienced polling company. However, this is not what margin of error measures. It refers to the fact that the process of sampling will usually produce a sample percentage that is different than the population percentage.
3. The true percentage of all American adults who would answer yes to this question is between 75% and 81%.	There is a 95% chance that the true percentage is between 75% and 81%.
4. If repeated samples of 1005 adults were taken on Feb. 8–12, 1996, and each asked the given question, in 95% of the samples between 75% and 81% would answer yes.	This would be correct if the true percentage were 78%. However, if the true percentage were not 78% (say, 76%), then in fewer than 95% of the samples would $\hat{p}$ be .78.

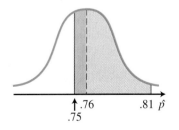

A correct interpretation is that we are 95% sure that the true percentage of all adults who believed on Feb. 8–12, 1996, that incivility has worsened in the past 10 years was between 75% and 81%. We are 95% sure, in the sense that if, on Feb. 8–12, 1996, random samples of size 1005 had been taken repeatedly and asked the given question and if "95% confidence intervals" were constructed for each sample, then 95% of these intervals would capture the true percentage of all adult Americans who would answer yes.

Summary

A major function of statistics is to make inferences about a population based on information contained in a sample. In this section we have made some introductory forays into the topic for inferences about means and proportions. The underlying concept is that of a sampling distribution. Media reports on statistical studies often use the term *sampling error* or *margin of error.* We hope that after studying this section you are a better consumer of such reports.

EXERCISES 9.5

1. Verify Property 1 in the box on "Properties of the Sampling Distribution of $\bar{x}$" for the sampling situation of Table 9.11.
2. A random sample of 30 measurements is taken from a population with mean 5 and standard deviation 2. Find $\mu_{\bar{x}}$ and $\sigma_{\bar{x}}$.
3. IQ scores have a mean of about 100 and a standard deviation of about 15. Find the probability that a random sample of 40 people has a mean IQ of 105 or more.
4. A machine is set to produce bolts with a mean length of 1.00 inch. Past experience has shown that the standard deviation of the lengths is 0.04 inches. Bolts that are too long or too short do not meet the company's quality standards and are rejected. To avoid producing too many rejects, the bolts produced by the machine are sampled from time to time to see whether the machine is still operating properly, i.e., producing bolts with a mean length of 1.00 inches. Suppose 30 bolts have been sampled and $\bar{x} = 1.02$. If the machine is operating properly, what is the probability that a sample of size 30 has $\bar{x} = 1.02$ or larger.
5. If the machine of Exercise 4 is operating properly, what is the probability that a sample of size 50 has $\bar{x} = .99$ or smaller?
6. Suppose the consumers' group of Examples 9.21 and 9.22 is willing to settle for a 5% chance of error; i.e., it would like to sample enough boxes so that the chance of making a mistake because of sampling is 5% or less. How many would it sample?
7. IQ scores have a mean of about 100 and a standard deviation of about 15. How many people must be sampled so that the probability that the mean of the sample is less than 95 is .01 or less?
8. Suppose the consumers' group of Example 9.23 decides that a probability of .0793 for a sampling error is too large. How

large should the sample be to reduce the probability of sampling error to .05?
9. Repeat Example 9.24 for a sample size of 300.
10. Suppose that for a test you know 80% of the material. Consider a 40-question exam as a sample of your knowledge.
    a. What is the probability that you will score 70% or less?
    b. What is the probability that you will score 90% or more?
11. Illustrate that $\sigma_{\hat{p}}$ changes very slowly with p by completing the following table, with answers rounded to four decimal places. Suppose p is really 0.60 and n is 50. Then $\sigma_{\hat{p}} = .0693$ but

If $\hat{p}$ is:	Then $\sigma_{\hat{p}} \approx \sqrt{\hat{p}\hat{q}/50}$ is:
.61	
.62	
.63	
.64	
.59	
.58	
.57	
.56	

12. Construct a 95% confidence interval for the poll referred to in Example 9.25.
13. A poll of approximately 1000 adults taken by Maritz Marketing Research and reported in *American Demographics,* May 1996, p. 20, found that 25% say that convenient location is the most influential factor in their fast-food restaurant choice. Construct a 99% confidence interval for the percentage of all adults who would say that convenient location is the most influential factor in their fast-food restaurant choice.

14. (*Multiple choice*) Suppose a 95% confidence interval for a proportion p is (.607, .667). With the same sample size, which of the following would be true about a 90% confidence interval for p?
   i. The interval is wider.
   ii. The interval is the same width.
   iii. The interval is narrower.
   iv. The question can't be answered unless $\hat{p}$ and n are known.

15. (*Multiple choice*) A sample of 1000 produces $\hat{p} = .637$. The 95% confidence interval for p is (.607, .667). If a larger sample also produces $\hat{p} = .637$, what will the resulting 95% confidence interval for p be?
   i. Wider
   ii. The same width
   iii. Narrower
   iv. The question can't be answered unless the sample size of the larger sample is known.

16. Candidate Y is running for office. A sample of 200 registered voters shows that 52% plan to vote for her. This gives a 95% confidence interval of 45.1% to 58.9%. She believes this interval is too wide. State two things that can be done (honestly) to reduce the width of the confidence interval.

17. A telephone poll of 800 adult Americans taken for TIME/CNN on Jan. 17–18 by Yankelovich Partners Inc. and reported in *Time,* January 29, 1996, asked: "Compared to three years ago, are you and your family better or worse off in these areas?" For the area "your overall standard of living," 46% answered "better." The margin of error, called "sampling error," is ±3%.
   a. Construct a 95% confidence interval for the percentage of all adult Americans who would have answered "better" at that time.
   b. Interpret what the sampling error of ±3% means.

18. Of the respondents in Exercise 17, 20% said "worse." The sampling error was again ±3%.
   a. Construct a 95% confidence interval for the percentage of all adult Americans who would have answered "worse" at that time.
   b. Interpret what this sampling error means.

WRITTEN ASSIGNMENTS 9.5

1. The consumers' group of Example 9.23 and Exercise 8 has come to you, as a person who has studied statistics, for advice. The group would like to do more than just produce the probability of sampling error as in Exercise 8. It would like to eliminate sampling error. Write to Mr. Con Sumer, chair of this group, telling him how to do this.

2. Find a statistical study in a newspaper or magazine. If it includes a statement of the margin of error or sampling error, show your own calculation of the margin of error. If your calculation does not match the study's calculation, try to explain why. If the article does not report a margin of error, calculate it.

CHAPTER 9 REVIEW EXERCISES

1. Describe the difference between descriptive and inferential statistics.

2. The number of U.S. representatives, by state, is shown in the histogram. Use it to answer these questions.

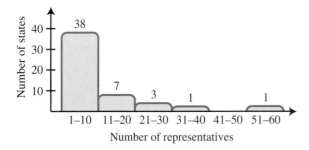

Number of representatives

   a. How many states have 10 or fewer representatives?
   b. How many states have more than 20 representatives?
   c. How many representatives does the state with the most representatives have? Answer with as much accuracy as the histogram permits, but no more.
   d. Calculate the mean number of representatives per state.
   e. (*Multiple choice*) This set of data is:
      i. Skewed to the left
      ii. Skewed to the right
      iii. Symmetric

3. Maryland has 23 counties and one independent city. We will refer to all 24 as units. Their populations, as counted by the 1990 Census, are shown here. Construct a histogram to represent this set of data.

Unit	Population
Allegany	74,946
Anne Arundel	427,239
Baltimore city	736,014
Baltimore county	692,134
Calvert	51,372
Caroline	27,035
Carroll	123,372
Cecil	71,347
Charles	101,154
Dorchester	30,236
Frederick	150,208
Garrett	28,138
Hartford	182,132
Howard	187,328
Kent	17,842
Montgomery	757,027
Prince Georges	729,288
Queen Annes	33,953
Saint Mary's	75,974
Somerset	23,440
Talbot	30,549
Washington	121,393
Wicomico	74,339
Worcester	35,028

4. Construct a stem-and-leaf plot to represent the data in Review Exercise 3.

5. The home run distances, in feet, down the left field line in major league baseball parks are shown in the following stem-and-leaf display (source: *The World Almanac 1996*).

Stem	Leaf
31	2 5 5
32	5 5 5 7
33	0 0 0 0 0 0 0 1 2 3 3 5 5 5 8
34	0 3 7 7
35	5

a. What is the shortest home run distance down the left field line in a major league park?

b. What is the longest home run distance down the left field line in a major league park?

c. How many major league parks are there?

d. In how many major league parks is the home run distance down the left field line less than 335 feet?

e. What is the mean home run distance down the left field line in major league parks?

f. What is the median home run distance down the left field line in major league parks?

g. What is the range of the home run distances down the left field line in major league parks?

h. (*Multiple choice*) Which is the best estimate for the standard deviation of the data? (This question can be answered without calculating the standard deviation.)

i. A negative number   ii. 0   iii. 2.9   iv. 8.6   v. 40.0   vi. 74.0

6. a. Describe an advantage of a stem-and-leaf plot over a histogram.
   b. Describe an advantage of a histogram over a stem-and-leaf plot.

7. Suppose the campus newspaper interviews five students to get students' reactions to a tuition increase. Describe the population and the sample.

8. The campus radio station wants to get students' reactions to whether or not there is ample parking on campus. The station takes a sample by interviewing the first 30 students to show up at a large lecture class scheduled to meet at 9:00 A.M.
   a. Describe why this sampling method is not random.
   b. How might this method of sampling change the results from those of a random sampling method?

9. Consider the set of data 9, 9, 4, 8, 9, 3.
   a. Find the mean.
   b. Find the median.
   c. Find the range.
   d. Find the standard deviation.
   e. Suppose the 3 is replaced by a smaller number. What will be the effect on each of the following? Be as specific as possible.
      i.   The mean
      ii. The median
      iii. The range
      iv. The standard deviation
   f. Is this set of data skewed to the left, skewed to the right, or symmetric? (Which one?)

10. (*Multiple choice*) An advertisement for Nordic Track says the average weight before use was 173 lb and the average weight after 12 weeks of use was 154 lb. Suppose "average" refers to the mean. Which is true of the median weight before use?
    i. median < 173 lb   ii. median = 173 lb   iii. median > 173 lb

11. Find the lower quartile of the Maryland unit populations given in Review Exercise 3.

12. The March 1996 issue of *Consumer Reports* rates pasta sauces. The costs per ½-cup serving of 39 sauces are given. The upper quartile of these costs is $0.53. Explain what this means.

13. A student's score on the SAT exam was at the 63rd percentile. Explain what this means.

14. Explain why a measure of central tendency such as the mean is not sufficient to describe a set of data.

15. Calculate the range of the set of data in Review Exercise 3 above on Maryland unit populations.

16. The range of the set of salaries of governors of states in the United States in mid-1993 was $74,498. Explain what this means.

17. Construct a box plot for the set of data in Review Exercise 3 on Maryland unit populations.

18. The prices of rooms at 54 hotel chains was reported in the July 1994 issue of *Consumer Reports*. A box plot for these prices follows.

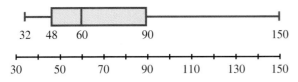

a. What was the lowest-priced room?

b. How many hotel chains had rooms under $48?

c. What is the range of prices?

d. The median price is $60. Is the mean price more than $60, exactly $60, or less than $60? (Which one?)

e. Is the set of data skewed to the left, skewed to the right, or symmetric? (Which one?)

f. Estimate the standard deviation for this set of data.

19. For the population 1, 2, 4, 4, 9, calculate the following.

a. The variance

b. The standard deviation

20. Let σ_c denote the standard deviation of temperatures in your classroom and σ_o denote the standard deviation of outdoor temperatures where your classroom is located. Are they the same size? If not, which standard deviation is larger?

21. a. The standard deviation of a set of data is 4. What is its variance?

b. The variance of a set of data is 4. What is its standard deviation?

22. Estimate the standard deviation of the set of data in Review Exercise 3 on Maryland unit populations. You need not *calculate* the standard deviation.

23. a. Is the family size of families in the United States a continuous or a discrete random variable? (Which one?)

b. If you answered "continuous," give an example of a discrete random variable. If you answered "discrete," give an example of a continuous random variable.

24. The curve describing the standard normal random variable is given by

$$f(z) = \frac{1}{\sqrt{2\pi}} e^{-z^2/2}$$

Find $f(1.5)$ to three decimal places.

25. (*Multiple choice*) Which of the following is correct?

i. Every mound-shaped distribution is a normal distribution, and every normal distribution is mound shaped.

ii. Every mound-shaped distribution is a normal distribution, but some normal distributions are not mound shaped.

iii. Some mound-shaped distributions are normal, but every normal distribution is mound shaped.

iv. No mound-shaped distribution is normal, but every normal distribution is mound shaped.

26. Assume that z has a standard normal distribution. Use a normal curve table to find the indicated probabilities.

a. $P(0 < z < 1.23)$

b. $P(z > -2.04)$

c. $P(-1.56 < z < 0.78)$

d. $P(1.90 < z < 2.94)$

27. Assume x has a normal distribution with mean 5.0 and standard deviation 2.0. Find each of the following probabilities.

a. $P(4.0 < x < 5.0)$

b. $P(x > 1.6)$

c. $P(x \leq 2.3)$

d. $P(x > 12.0)$

28. The U.S. Census Bureau found that the median price of houses in 1990 was $79,100. Assume that house prices in 1990 were normally distributed with a mean of $80,600 and a standard deviation of $37,500.

a. Why is it reasonable for the mean to be larger than the median?

b. What percentage of houses have prices over $100,000?

c. What percentage of houses are priced between $50,000 and $100,000?

d. What price on a house puts it at the 90th percentile of house prices?

29. Describe what a sampling distribution is. Illustrate with a sampling distribution of $\overline{x}$.

30. A random sample of 40 measurements is taken from a population with mean 76.2 and standard deviation 12.1. Find $\mu_{\overline{x}}$ and $\sigma_{\overline{x}}$.

31. NordicTrack reports that a 12-week study done by the National Exercise for Life Institute found that after using NordicTrack, users had an average waist size of 28.6″. Suppose "average" means "mean" and the standard deviation of waist sizes was 4.1″. In a random sample of 35 NordicTrack users after 12 weeks, what is the probability that the average waist size is each of the following?

a. Under 27.0″

b. Between 28.0″ and 30.0″

c. Over 30.0″

32. IQ scores have a mean of about 100 and a standard deviation of about 15. How many people must be sampled to be fairly sure (probability of 0.95 or more) that the mean of the sample will be between 95 and 105?

33. A poll of 800 adult Americans taken by TIME/CNN in 1994 showed that 19% of those surveyed believed you can trust the government to do what's right always or most of the time.

a. Construct a 99% confidence interval of the percentage of all adult Americans who believe this way.

b. For this poll TIME reports, "Sampling error is ±3.5%." Explain what this means.

34. (*Multiple choice*) Suppose a 90% confidence interval for a proportion p is (.405, 475). With the same sample size, what would a 95% confidence interval for p be?

i. Wider   ii. The same size   iii. narrower

iv. The question can't be answered unless more information is known.

35. (*Multiple choice*) A sample of 500 produces $\hat{p} = 0.44$. The 90% confidence interval is (.403, .477). If a larger sample also produces $\hat{p} = 0.44$, what will the resulting 90% confidence interval be?

i. Wider   ii. The same size   iii. Narrower

iv. The question can't be answered unless more information is known.

36. A survey of 50 students found that 86% believed that a particular professor was well prepared. What is the margin of error for this survey?

Annenberg/CPB Project. *For All Practical Purposes* (videotape), Programs 6–10, "Statistics."

Boyer, Carl B. (revised by Uta C. Merzbach) *A History of Mathematics,* 2nd ed. New York: Wiley, 1991.

Eves, Howard. *An Introduction to the History of Mathematics,* 6th ed. Chicago: Saunders College Publishing, 1990.

Hold, Anders. *A History of Probability and Statistics and Their Applications Before 1750.* New York: Wiley, 1990.

Maistrov, L. E. *Probability Theory—A Historical Sketch.* New York: Academic Press, 1974.

Tankard, Jr., James W. *The Statistical Pioneers.* Cambridge, MA: Schenkman, 1984.

CHAPTER 10

CONSUMER MATHEMATICS

CHAPTER OUTLINE

PROLOGUE

You may have been fortunate enough (or at least wished you were) to have parents, other relatives, or friends who set up a trust fund or annuity to help finance your college education. How much did they invest? For how long? At what interest rate? Or perhaps you want to send your own children to college and wish to set up a trust to fund their college education. And how much money needs to

be invested? For how long? At what interest rate?

As a college student you are a consumer in your local community. What can unit costs tell you? How do you compare prices of items when some or all of the items are on sale?

If you have money to invest, how much difference do different interest rates make? Is it worth the drive across town to earn an extra ¼%? Credit cards have at least one interest rate associated with them. How is the interest calculated? How much money do you pay to use your credit card?

If you buy a car and make monthly payments, how are they figured? How much of each payment is interest and how much goes toward the principal? What about house payments?

These are some of the questions that consumer mathematics helps us answer. They form the basis of this chapter.

Chapter opening photo: Consumers make a variety of decisions with money as the common thread: earning, spending, saving, and borrowing. All of these decisions involve mathematics in some way.

10.1 INTRODUCTION

GOALS   1. To describe some of the areas of consumer mathematics.
2. To describe briefly the history of interest.

Most consumer transactions involve mathematics in some way. In a straight trade—my book for your book—there is the mathematical idea of one-to-one correspondence and a consideration of value or utility, which is often measured numerically. In other types of barter, such as two pies in exchange for 3 hours of yard work, numbers enter the picture. If a monetary price is set for a product or service, then numbers again are involved.

Consumers employ mathematics in calculating taxes, comparing the prices of two products, and looking at savings from sales and the use of coupons. Some of these are discussed in Section 10.2.

Mathematics for the consumer becomes more sophisticated when dealing with interest considerations: compound interest, loans, installment purchases, annuities, and the like. All of these considerations of value in the future are tempered by the effects of inflation. These topics are covered in Sections 10.3–10.6.

Borrowing and lending

Consumer From the Latin *consumere*, from *com-*, "thoroughly," and *sumere*, "to take up, use." A consumer is one who uses a product or service, as distinguished from a producer. A consumer is one of the buying public.

The federal government sought to help consumers deal with mathematical and other questions in the Consumer Credit Protection Act of 1968, often called the Truth in Lending Act. It is intended to protect people who borrow money and who buy on credit. For loans, the lending institution is required to state the annual percentage rate (APR) to simplify the comparisons of loans whose interest rates are stated differently (e.g., $5\frac{1}{4}\%$ compounded monthly compared with $5\frac{1}{2}\%$ compounded semiannually). For people buying on credit, the business is required to include in the credit contract a statement of the total payment, the amount financed, and the finance charges.

Regulation of lending is nothing new. It began sometime after borrowing and lending began, likely before written history, probably predating even primitive forms of money. Primitive farmers may have borrowed seed until the next harvest. They may have arranged to farm someone else's land, with part of the harvest as expected payment.

About 1800 B.C.E. Hammurabi, one of the greatest kings of Babylonia, developed a system of laws known as the Code of Hammurabi, one of the first law codes in history. It covered, among other matters, tariffs, wages, loans, and debts. It permitted a legal maximum interest of $33\frac{1}{3}\%$ per year for loans of grain and 20% per year for loans of silver (Homer, 1963, p. 30). The Code of Hammurabi greatly influenced the civilizations of all Near Eastern countries.

The Israelite peoples of around 800 B.C.E. had rules governing interest (Deuteronomy 23:19–20). Interest was prohibited on loans to fellow Israelites but allowed on loans to foreigners.

Homer's *Iliad* tells of loans of cattle in ancient Greece that, if not repaid, could be claimed by legal robbery. One of the laws of Solon, a famous lawmaker known as one of the seven wise men of Greece, who lived around 600 B.C.E., was directed at Athens of his time, where most of the wealth was in the hands of a few powerful people. Farmers had been forced to mortgage their lands, using themselves and their families as collateral. Solon got a law passed that cancelled all the mortgages and freed those who had become slaves because of their inability to pay off a mortgage. In Solon's time the customary interest rate in Greece was 16–18% (Homer, 1963, p. 40).

Laws in ancient Rome also regulated credit. A law dating back to about 450 B.C.E. limited interest on loans to no more than $8\frac{1}{3}\%$ per year. Interest above the legal amount carried a penalty of fourfold damages. Personal slavery to pay off a debt was permitted, but the physical well-being of the slave was protected.

Interest has been regulated for over 4000 years and continues to be regulated today. During that time the rate of interest has varied considerably. During much of ancient times normal interest rates ranged from 8% to 12% (Homer, 1963, p. 61). In twelfth century England, people with the best credit paid $43\frac{1}{3}\%$–52% (Homer,

1963, p. 91). Loan sharks, of course, have always demanded more. In ancient Athens they charged 48% a month, for an uncompounded annual rate of 576%. In New York City today there are reports of a weekly rate of 25% (borrow $4 on Monday, repay $5 on Friday), for an uncompounded annual rate of 1300%. The ancient Greek philosopher Theophrastus, who followed Aristotle, wrote of a usurer who charged 25% a day, for an uncompounded annual rate of 9125% (in a nonleap year) (Homer, 1963, p. 6).

Although the interest rates discussed throughout this chapter may not be at usury levels, we will look at topics of interest to our pocketbooks, in the area of consumer mathematics. We'll move from the interest of usury to the use of interest.

WRITTEN ASSIGNMENTS 10.1

1. The opening of this section says, "Most consumer transactions involve mathematics in some way." Could that sentence be strengthened to read, "All consumer transactions involve mathematics in some way"? Why or why not?
2. Look up data on interest rates and construct a graphical display of them. If they are interest rates through time (some data are available in Homer, 1963; other data could be rates at a single financial institution or prime rates in the United States), then a line graph showing the interest rate as a function of time would be appropriate. If they are interest rates charged by a number of financial institutions or countries at some point in time, then a histogram, a stem-and-leaf display, or a box plot would be appropriate.

10.2 UNIT COSTS, SALES, AND COUPONS

GOALS
1. To work with unit costs.
2. To calculate tax, assessed value, or millage when the other two variables are known.
3. To work with original price, sales price, and percentage off.
4. To calculate the best buy when coupons are used.

In this section several topics in consumer mathematics will be discussed. The common feature is that all of them are described by a linear function or a linear equation.

Unit Costs

The cost of products in different-size packages can be compared by reducing items to a *unit cost:* so much per item or so much per ounce. For example, one brand of green beans costs 50¢ for a 13.5-oz can, while another brand costs 67¢ for a 16-oz can. Which is the better buy? The first brand costs $^{50}/_{13.5}$¢ per oz $\approx$ 3.7¢ per oz; the second brand costs $^{67}/_{16}$¢ per oz $\approx$ 4.2¢ per oz. Therefore, the first brand is the better buy if considerations other than cost are the same.

Many grocery stores post unit prices for products. Some states *require* stores to do this. Does yours? (See Written Assignment 1.) The unit price information is usually posted underneath the shelf where the product is displayed.

Utility companies often bill for their product on a unit basis. The following example illustrates this.

EXAMPLE 10.1

A certain gas company charges residential customers a service charge of $7.50 plus $.401120 per 100 cubic feet (CCF) of gas used, and then adds the state 4% sales tax. If during one month the customer used 168 CCF of gas, what will the bill be?

BILL DATE ▶ Apr 23.9X   RATE ▶ **Residential**

Meter Number	Last Meter Reading			Current Meter Reading			Days Billed	Units Used in 100 Cubic Feet (CCF)
	Date	Type of Reading	Reading	Date	Type of Reading	Reading		
7625608	Mar 18	Actual	7604	Apr 16	Actual	7772	29	168

SOLUTION

Customer charge	$ 7.50
168 CCF × $.401120	67.39
Subtotal	74.89
Sales tax	3.00
Total	$77.89

The bill will be $77.89.

When products are priced by adding a fixed cost, such as a customer service charge, to a unit cost to produce a total cost and any two of these three costs are known, then the third can be found. The next example provides an illustration.

EXAMPLE 10.2

Suppose a customer of the gas company in Example 10.1 has a bill for $84.56. How many CCF of gas did the customer use?

SOLUTION Let x denote the number of CCF of gas used.

Customer charge	$ 7.50
x CCF × $.401120	$.401120x$
Subtotal	$.401120x + 7.50$
Sales Tax	$.04(.401120x + 7.50)$
Total	$.401120x + 7.50 + .04(.401120x + 7.50)$

Factor out $(.401120x + 7.50)$ to get the total charge.

$$(1 + .04)(.401120x + 7.50) = 84.56$$

$$.401120x + 7.50 = \frac{84.56}{1.04}$$

$$.401120x = \frac{84.56}{1.04} - 7.50$$

$$x = \frac{1}{.401120}\left(\frac{84.56}{1.04} - 7.50\right)$$

$$x \approx 184$$

The customer used 184 CCF of gas.

In every school millage election, you can count on two things: 1. The schools need the money; 2. Nobody knows what a mill is.

Another situation that involves unit costs occurs in the levying of local taxes as millages. A *mill* is a U.S. monetary denomination of one tenth of a cent or, equivalently, one thousandth of a dollar. A millage of 40 mills taxes a property at $^{40}/_{1000}$ of a dollar for each dollar of assessed property value. This is often described by saying that for each $1000 of assessed value, the property owner is taxed $40.

EXAMPLE 10.3

Suppose a community is levying a tax of 38 mills. If a homeowner has an assessed valuation of $52,500, what is the homeowner's tax from this millage?

SOLUTION The 38 mill levy is $^{38}/_{1000}$ of a dollar for each dollar of assessed value, so the tax is

$$\frac{38}{1000}(\$52,500) = \$1995$$

If any two of the millage, the assessed value, or the tax are known, the third can be found. The next example gives an illustration.

EXAMPLE 10.4

If a house assessed at $87,000 pays a tax of $3567, what is the millage?

SOLUTION Let r denote the millage rate. Then

$$87,000r = 3567$$

$$r = \frac{3567}{87,000} = 0.041$$

The millage is 41 mills.

Sales

Stores frequently have sales in which some products are advertised at something like 20% off, 40% off, or up to 70% off. A study conducted for *Glamour* magazine and reported in *American Demographics* (Krafft, 1991) found that 75% of women buy the majority of their clothes on sale.

One way to calculate the reduced price on sale merchandise is to take the original price and subtract the amount of savings. For example, if a shirt originally priced at $33.00 is on sale at 40% off, the savings is 40% of $33.00, or .40 × $33.00, which is $13.20, so the sale price is $33.00 − $13.20 = $19.80. An alternate and slightly shorter procedure is to note that if the shirt is on sale at 40% off the original price, then the new price is 60% of the original price, or 60% of $33.00 = .60 × $33.00 = $19.80.

The original price P, the sale price S, and the discount rate d stated as a decimal or proportion (e.g., 40% = .4) are related by $S = P - dP$, or $S = (1 - d)P$. If any two of S, P, or d are known, then the third can be found. The next example illustrates such a situation.

A sale

EXAMPLE 10.5

A shoe store ad reads: "Take an additional 25% off our already-low prices." If the sale price is $42.74, what was the "already-low price"?

SOLUTION Let P denote the regular price. Then

$$P - 0.25P = \$42.74$$

$$0.75P = \$42.74$$

$$P = \frac{\$42.74}{0.75} \approx \$56.99$$

The "already-low price" was $56.99.

YOUR FORMULATION

In Example 10.5 you are given S and d and asked to find P.

1. Now we give you P and S and ask you to find d:

   A shoe store's ad reads, "20–50% off! Reg. $9.99–$49.99 SALE $6.99–$39.99." For shoes that are regularly priced $9.99 and are on sale for $6.99, what is the percentage off?

2. Here we look at the remaining situation given P and d, find S:

   A clothing store advertises 25% off. If a shirt is regularly priced at $14.99, what is the sale price?

Sometimes sales offer one or more items at regular price, along with an item at a reduced price. The next example illustrates such a sale.

EXAMPLE 10.6

A store ad says: "Second item half price when you buy the first item at regular price." What percentage off is this on the two items?

SOLUTION let P be the original price for one item. The savings is $\frac{1}{2}P$ on what would have been (before the sale) a total purchase price of $2P$, so the proportion of saving is

$$\frac{(1/2)P}{2P} = \frac{1/2}{2} = \frac{1}{4}$$

On the purchase of two items, there is 25% off. What happens is the 50% reduction on the second item gets spread across two items, yielding a reduction of 25% per item.

Coupons

The presence of coupons often makes the determination of a best economic buy difficult. Whereas a larger package of a product usually has a lower unit cost than a smaller package, with a coupon the reverse may be true.

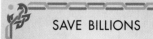
SAVE BILLIONS

Consumers "save" billions of dollars a year by using coupons. Coupon use increased from an estimated $1 billion a year in savings in 1980 to about $4 billion a year by the end of the decade. In 1995, 291.9 billion coupons were distributed, although only 5.8 billion were redeemed.

How much can we save with these?

Suppose, for example, that a certain type of pill comes in packages of three different sizes: 50 for $2.45, 100 for $4.65, and 500 for $19.40. The prices per pill are, respectively, 4.90¢, 4.65¢, and 3.88¢, so a customer's best buy is the 500-pill bottle. Now suppose a customer has a coupon for 50¢ off on one package. Which is the best buy? The prices per package now become $1.95, $4.15, and $18.90, respectively, and the costs per pill are, respectively, 3.90¢, 4.15¢, and 3.78¢, so the customer's best buy is still the 500-pill bottle. What if the store has double coupons? The prices per package now become $1.45, $3.65, and $18.40, respectively, and the costs per pill are, respectively, 2.90¢, 3.65¢, and 3.68¢, so the customer's best buy has become the 50-pill bottle.

There was nothing difficult or sophisticated in these calculations. Only perseverance was required. You can expect something like what happened in these examples to happen in general: Smaller packages might be the better buy with coupons.

Summary

In this section we have looked at consumer mathematics involving unit costs, including millages, percentages off, and the use of coupons. A common feature of all these situations is a linear equation or linear function.

EXERCISES 10.2

1. Find the unit price of each of the following: a 40-lb bag of water softener salt for $4.29 and a 50-lb bag for $5.35. Which is the better buy?

2. Find the unit price of each of the following: toilet paper at four rolls for 92¢ and six rolls for $1.49. Which is the better buy if only price is considered?

3. Pepsi is available in an eight-pack of 20-oz bottles at two for $5 or in a 2-liter bottle (67.6 oz) for 99¢. Which is the better buy?

4. The customer of the gas company of Example 10.1 uses 154 CCF of gas during one particular month. What will that month's bill be?

5. A customer of the gas company of Example 10.1 has a bill for $48.68 but misplaced the billing statement. How much gas was used?

6. An electric utility company charges a residential customer $.073357 per kilowatt-hour (kWh) plus the 5% state sales tax. What will a customer be charged for 807 kWh?

7. An electric utility company charges residential customers a flat rate per kWh, then adds on the 6% state sales tax. For 76 kWh the bill, including tax, is $62.08. What is the charge per kWh?

8. If the millage rate is 18 mills, what is the tax on a property assessed at $87,400?

9. If the millage rate is 27 mills and the tax on a property is $1009.80, what is the assessed value of the property?

10. If a property assessed at $62,200 pays a tax of $2985.60, what is the millage rate?

11. A store advertises "One-third off." If an item originally costs $42.00, what is the sale price?

12. A store advertises "25% off." If an item originally costs $31.00, what is the sale price?

13. A store has round steak on sale for $1.58 per pound. It advertises a savings of $1.71 per pound. What percentage off is this?

14. A store has items at 40% off. The sale price on one of these items is $13.95. What was the original price?

15. A store advertises, "Buy the first item at regular price and get the second item at 25% off." What is the percentage off on the two items?

16. A coupon for an automatic dishwasher powder is good for 30¢ on any size. The options are a 50-oz package for $1.88 or a 65-oz package for $2.29.

    a. With a coupon, which is the better buy?

    b. With double coupons, which is the better buy?

17. A certain dishwashing liquid has a 20¢ coupon when you buy one 22-oz size or larger *or* two 12-oz size. Suppose the 22-oz size costs $1.19, the 12-oz size costs $.69, and the 48-oz size costs $2.29.

    a. With the coupon, which size is the better buy—one 22-oz., two 12-oz, or one 48-oz?

    b. With double coupons, which is the better buy?

18. Based on this ad, answer the following.

    a. With the two 25% reductions, what is the percentage off the original price?

    b. With the two 25% reductions, what percentage of the original price does the customer pay?

25% OFF
LIST & THEN
TAKE ANOTHER
25% OFF AGAIN

EXAMPLE: $100 LIST LESS 25%
LESS 25% EQUALS $56⁰⁰
HUGE SELECTION

NEWEST STYLES, BEST
BRANDS, CITIZEN, HELBROS,
BULOVA, PULSAR, SEIKO,
LORUS, JAZ

1. Some states require grocery stores to provide unit price information. Does yours?
   a. Write a two-page persuasive argument that your state's policy on requiring (or not requiring) unit price information on grocery items is correct.
   b. Write a two-page persuasive argument that it is incorrect.
2. The gas charge in Example 10.1 gives the cost function $C(x)$, in dollars, of x CCF. The cost function is a linear function. For a linear equation, the terms *slope* and *y-intercept* apply.
   a. Write the cost function $C(x)$.
   b. What is the slope of the cost function?
   c. What is the y-intercept of the cost function?
   d. Describe a procedure that a utility company might use to charge customers that would make $C(x)$ a nonlinear function of x.
3. The first part of this section discusses unit costs on products and millage as a unit cost. Describe another (different) situation in which there is a unit cost. Give an example, like Example 10.1 or 10.3, of a calculation with the unit cost in that situation.
4. Some people question the ultimate value of coupons because of the time involved in cutting them, organizing them, and searching for them when they're needed. Make some estimates of the amount of time taken for these, and perhaps other, jobs required when dealing with coupons. Estimate the possible savings, and calculate what the payment per hour is for using coupons. Specify who the shopper is (you, your family, etc.).
5. One item that is difficult to unit price is bathroom tissue (toilet paper). It comes packaged in four-roll, eight-roll, or twelve-roll packages. Some brands have more sheets per roll than do others. Some brands have different-size sheets than do others. Investigate prices of bathroom tissue and offer advice regarding unit prices to those who might have occasion to purchase it.
6. It's not always the case that purchasing a larger quantity of a product yields a lower unit cost. Examine several products at area stores to see if you can find an instance where the customer pays a *higher* unit price when purchasing a large quantity versus a smaller quantity. Report your findings to the class. Why do you think a product would be priced in such a manner?

10.3 ▸ SIMPLE INTEREST

GOALS
1. To calculate simple interest or amount, principal, simple interest rate, or time given the other three.
2. To convert from a daily or monthly periodic rate for a credit card to an annual percentage rate, and vice versa.
3. To calculate credit card finance charges.

Interest

Interest is charged to borrow money or is paid on money invested. In short, *interest* is a charge for the use of money (by you or by the bank) or property. *Simple interest* is paid only on the amount borrowed or invested (the *principal*). It is contrasted with *compound interest,* discussed in Section 10.4, for which interest is paid on both the original principal and the previously accrued interest.

For example, the loan shark who loans you $4 on Monday with payment of $5 due on Friday is charging interest of $1, for an interest rate of $1/$4 = .25, or 25%, for the five-day period. If a bank pays 3% per year interest, compounded quarterly, then each quarter it will pay 3%/4 = 0.75% simple interest. For an investment of $100, you would earn 0.75% of $100, or 75¢ for a quarter (three months).

To state some general results involving simple interest, let

$$P = \text{principal (amount borrowed or deposited)}$$

$$r = \text{rate (stated as a decimal or proportion)}$$

$$t = \text{time}$$

$$I = \text{interest}$$

Then, interest = principal × rate × time, which symbolically is expressed as follows.

$$I = Prt$$

EXAMPLE 10.7

Find the simple interest for 6 months on $820 at 6.7% per year.

SOLUTION $\qquad I = Prt = (\$820)(.067)\left(\dfrac{1}{2}\right) = \$27.47$

↑

Since r is stated as a percentage per year, t needs to be stated in years.

The interest is $27.47.

After the calculation of interest, it is one more step to find the amount owed.

EXAMPLE 10.8

Suppose $620 is borrowed for 8 months at 1.7% per month. How much is owed at the end of 8 months?

SOLUTION The amount owed is the principal plus the interest. The interest is

$$I = Prt$$

$$= \$620(.017)(8) = \$84.32$$

Therefore, the amount owed is $620 + $84.32 = $704.32.

Amount Owed or Earned

A formula for the amount due when money is borrowed at simple interest or the amount to be received when money is invested at simple interest is amount = principal + interest, or

$$A = P + Prt = P(1 + rt)$$

$$A = P(1 + rt)$$

For example, if this formula were to be used in Example 10.8, we would have

$$A = \$620\,[1 + (.017)(8)] = \$704.32$$

Although this formula can be used to calculate the amount due, less memorization is involved with the technique of Example 10.8. The formula is especially helpful in certain other types of calculations, such as in the following example.

EXAMPLE 10.9

Suppose Alice has a check for $345 coming in 24 days. How much can she borrow today at .069% per day (a typical credit card rate) so that she will owe $345 in 24 days?

SOLUTION $A = P(1 + rt)$, so

$$P = \frac{A}{1 + rt}$$

$$= \frac{\$345}{1 + (.00069)(24)} \approx \$339.38$$

Alice can borrow \$339.38. As a check, notice that the interest she will owe is

$$Prt = \$339.38(.00069)(24) = \$5.62$$

so her repayment after 24 days will be \$339.38 + \$5.62 = \$345.00.

EXAMPLE 10.10

Lezley just got paid \$50.00. She won't need the money for a month. She can either put it in her checking account at college or mail it home to her savings account and earn an interest rate of 0.0125% per day. If she needs the money in 30 days, is it worth the 32-¢ cost of a stamp to mail a deposit home?

SOLUTION The amount of interest she will earn is

$$I = Prt$$

$$= \$50 \times .000125 \times 30 = \$.1875$$

Therefore, it is not worth the 32¢ to mail her deposit home.

YOUR FORMULATION

The formula for the amount, $A = P(1 + rt)$, involves four variables (A, P, r, and t). In this form it tells how to find A when P, r, and t are given.

1. Solve for P in terms of A, r, and t.
2. Solve for r in terms of A, P, and t.
3. Solve for t in terms of A, P, and r.

Credit Cards

Credit card companies have a variety of ways of calculating the finance charge. But most, if not all, base the monthly finance charge on a daily or monthly simple interest rate. The company will identify either a daily or a monthly periodic rate, and also an annual percentage rate. To get from the monthly periodic rate to the annual percentage rate, multiply it (in a percentage form) by 12. For example, if the monthly rate is 1.40%, then the annual percentage rate is (1.40%)(12) = 16.80%.

The finance charge for purchases and that for cash advances may be calculated slightly differently. We will look at one way of figuring the finance charge for credit purchases. The first step typically is to calculate the average daily balance. This is done by figuring the balance of the account each day (previous balance + new charges − payments or credits received), then averaging (taking the mean of) these daily balances over the number of days in the billing period (typically, 29 or 30). The finance charge is then simple interest for a month on the average daily balance.

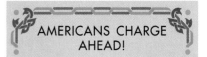
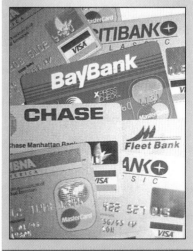

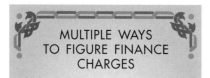

MULTIPLE WAYS TO FIGURE FINANCE CHARGES

According to *U. S. News & World Report* (p. 66) there are six ways that credit card companies calculate their finance charge. An example is given of a cardholder who pays an annual percentage rate of 19.8% and charges $1150 one month, $200 the second month, and then pays just the minimum amount due of $46. The finance charge is calculated by each of the six methods. It ranges from $18.22 to $36.84 and is different for each of the six methods.

For example, if the average daily balance were $241.58 and the monthly periodic rate were 1.40%, then the finance charge for the month would be $241.58 × .014 ≈ $3.38.

Several comments are in order regarding credit card finance charges. First, the finance charge for credit purchases usually does not apply if the bill is paid in full by the due date on the billing statement. About 34% of all bank card and retail store card users pay by the due date (Hall and Wheat, 1992, p. 29). Second, it is typical for a credit card company to have several different methods of figuring the finance charge. The method described in the preceding paragraphs, though typical, is only one method (see margin note). Third, if a credit card bill consists of both a previous balance and purchases made during the month of the billing period, the finance charge is more than just simple interest on the new end-of-the-month balance. You are invited to explore this in Written Assignment 3.

Finally, it should be emphasized that the interest rate charged for a credit card is only one factor to consider in looking at what credit card to get, if you were to get one. Other factors include the annual fee (which might be zero), the credit limits, the length of the grace period before charges begin, any rebates, and other promotional features. A typical monthly statement for a credit card is shown in ▶ Figure 10.1.

BALANCE SUMMARY			TOTALS
PREVIOUS BALANCE			398.94
AMOUNTS CREDITED	PAYMENTS	289.22	
	CREDITS	0.00	289.22
AMOUNTS DEBITED	PURCHASES	16.95	
	CASH ADVANCES	0.00	
	FEES, ADJUSTMENTS	0.00	16.95
FINANCE CHARGE	DAILY RATE CHARGE	6.89	
	TRANSACTION CHARGE	0.00	
	MIN. PERIODIC CHARGE	0.00	6.89
NEW BALANCE			133.56

CREDIT STATUS			
LIMIT	TOTAL AVAILABLE	AVAILABLE FOR CASH ADVANCE	OVERLIMIT
2500	2366	2366	

FINANCE CHARGE SUMMARY	BALANCE SUBJECT TO FINANCE CHARGE METHOD C	DAYS IN BILLING CYCLE	DAILY PERIODIC RATE	CORRESPONDING ANNUAL PERCENTAGE RATE	ANNUAL PERCENTAGE RATE
PURCHASES	369.24	31	0.060246	21.99	21.99
CASH ADVANCES	0.00	31	0.060246	21.99	

▶ FIGURE 10.1 Finance charge calculation

EXERCISES 10.3

1. a. Find the simple interest for 3 months on $380 at 8.1% per year.
   b. Find the amount to be repaid after 3 months.
2. a. Find the simple interest for 2 years on $1500 at 12% per year.
   b. Find the amount to be repaid after 2 years.
3. a. Find the simple interest for 3 weeks (21 days) at 0.083333% per day.
   b. Find the amount to be repaid after 3 weeks.
4. Aaron can invest $100 for 21 days at 0.011111% per day. How much simple interest will he earn?
5. How much money must Lezley of Example 10.10 invest for 30 days at 0.0125% per day in order to earn the price of a 32¢ stamp?

6. How much must Aaron of Exercise 4 invest for 21 days at 0.011111% per day in order to earn the price of a 32¢ stamp?
7. Bill expects an income tax refund of $216 in 6 to 8 weeks. How much can he borrow for 2 months at 1.5% per month simple interest so as to be able to pay it and the interest back when he gets his check?
8. Find the time required for $1000 to grow to $1200 at 5% simple interest. State your answer to the nearest day, figuring 365 days to a year.
9. Doug would like to get a cash advance of $300 on his credit card, which he will pay back in 20 days. He's willing to pay up to $20 for the use of this money; i.e., he's willing to pay back up to $320 at the end of 20 days.

a. What is the maximum daily simple rate he is willing to pay?

b. If interest is not compounded, what does this translate to as a yearly rate? Figure 365 days in a year.

10. With what simple interest rate will $1 accumulate to $1.0642 in 1 year?

11. If the daily periodic rate on a credit card is .05425%, what is the annual percentage rate? Round your answer to the nearest hundredth of a percent.

12. If the monthly periodic rate on a credit card is 1.625%, what is the annual percentage rate?

13. If the annual percentage rate on a credit card is 16.4%, find each of the following.

a. The monthly periodic rate (State your answer as a percentage, rounded to three decimal places.)

b. The daily periodic rate (State your answer as a percentage, rounded to five decimal places.)

14. The average daily balance on a credit card charging a monthly periodic rate of 1.40% is $100.21. What is the finance charge?

15. The average daily balance on a credit card charging a daily periodic rate of .04082% is $312.49. What is the finance charge for a 30-day billing period?

16. A certain person typically has an average daily balance of $300. How much per year does the person save in interest charges on a card that charges an annual percentage rate of 13.9% over a card that charges an annual percentage rate of 19.8%?

17. In Figure 10.1 the daily periodic rate is .060246 and the annual percentage rate is 21.99. How many days in a year are used in this calculation?

WRITTEN ASSIGNMENTS 10.3

1. From the formula $I = Prt$, if you know any three of I, P, r, and t, the fourth can be determined. Suppose you know I, P, and t, where I and P are in dollars and t is in years. Describe a step-by-step procedure that a sixth-grader with a calculator could follow to calculate r.

2. Some banks on some loans take the interest charge out of the loan amount when the loan is made. In such a case the bank is said to have *discounted* the note, and the interest charge is called the *discount*. Write a more detailed description of discounted notes, including an example.

3. Argue that for a credit card finance charge, the charge is often higher than a month's simple interest at the monthly periodic rate on the new end-of-the-month balance. When would the charge be exactly this amount and when would it be higher?

10.4 COMPOUND INTEREST

GOALS

1. To follow the development of the formula for compound interest, calculate the compound amount, describe the exponential growth of money at compound interest, and calculate present value.
2. To calculate effective rates.
3. To calculate the effect of inflation on prices.
4. To calculate the time for money to double at a fixed rate with compound interest and the rate needed for money to double in a fixed amount of time.

Compound Amount

Most interest that you might earn or pay is *compound interest*. That is, you earn or pay interest on the accumulated amount in the account repeatedly rather than earning or paying interest just once on the original principal. Because the interest is calculated on the accumulated amount, the amount in the account or owed grows faster than with simple interest at the same annual rate. The following example illustrates this.

Invest In Your Future Today.

Buy U.S. Savings Bonds

EXAMPLE 10.11

$1000 is invested at $3\frac{1}{2}$%, compounded yearly. How much is there in the account at the end of 3 years?

SOLUTION At the end of 1 year:

$$
\begin{aligned}
\text{amount} &= \text{principal} + \text{interest} \\
&= \$1000 + Prt \\
&= \$1000 + \$1000(.035)1 \\
&= \$1000 + \$35 \\
&= \$1035 \qquad \text{(new principal)}
\end{aligned}
$$

At the end of the second year:

$$
\begin{aligned}
\text{amount} &= \text{principal} + \text{interest} \\
&= \$1035 + \$1035(.035)(1)
\end{aligned}
$$

The new principal is the amount at the start of the second year.

$$
\begin{aligned}
&= \$1035 + \$36.22 \\
&= \$1071.22
\end{aligned}
$$

At the end of the third year:

$$
\begin{aligned}
\text{amount} &= \text{principal} + \text{interest} \\
&= \$1071.22 + \$1071.22(.035)(1) \\
&= \$1071.22 + \$37.49 \\
&= \$1108.71
\end{aligned}
$$

Therefore, at the end of 3 years, $1108.71 is in the account.

As a check note that $3\frac{1}{2}$% simple interest for 3 years would produce interest $I = Prt = \$1000(.035)3 = \$105$ for a total amount of $1105. The extra $3.71 is the effect of compounding.

The procedure of Example 10.11 allows the calculation of the compound amount for any investment, at any interest rate, for any time. However, if the investment were for 30 years, this calculating process would become quite tedious. To cut down on the time required, we will develop a formula for the compound amount. Much of the notation will be the same as for simple interest (Section 10.3). Let

P = principal

r = rate per compounding period (stated as a decimal or proportion)

n = number of periods

A = accumulated amount

If money is invested or borrowed at 6% compounded yearly, then $r = .06$ and n is the number of years; if interest is at 6% compounded semiannually, then $r = .06/2 = .03$ and n is the number of half-years; if interest is at 6% compounded quarterly, then $r = .06/4 = .015$ and n is the number of quarters, etc.

After one period:

$$
\begin{aligned}
\text{amount} &= \text{principal} + \text{interest} \\
&= P + Pr\,(1) \\
&= P + Pr \\
&= P(1 + r)
\end{aligned}
$$

CHAPTER 10 CONSUMER MATHEMATICS

402

After two periods:

$$\text{amount} = \text{principal at start of period 2} + \text{interest}$$
$$= P(1 + r) + P(1 + r)r\,(1)$$
$$= P(1 + r) + P(1 + r)r$$
$$= P(1 + r)(1 + r)$$
$$= P(1 + r)^2$$

After three periods:

$$\text{amount} = \text{principal at start of period 3} + \text{interest}$$
$$= P(1 + r)^2 + P(1 + r)^2 r\,(1)$$
$$= P(1 + r)^2 + P(1 + r)^2 r$$
$$= P(1 + r)^2(1 + r)$$
$$= P(1 + r)^3$$

Continuing in this manner, it can be seen that A, the amount after n periods, is as follows.

$$A = P(1 + r)^n \qquad\qquad (10.1)$$

If this formula is applied to the situation of Example 10.11,

$$A = \$1000(1 + .035)^3 = \$1108.717875 \approx \$1108.72$$

Use the x^y key, the y^x key, or the $\wedge$ key on your calculator.

The difference between this result and that of Example 10.11 was caused by rounding in our computations.

EXAMPLE 10.12

$320 is invested at 3.07% (annual rate) compounded monthly. How much is in the account after 3 years?

SOLUTION

$$A = P(1 + r)^n$$
$$= \$320\left(1 + \frac{.0307}{12}\right)^{36}$$

3 years = 36 months

$$\approx \$350.83$$

As a check, at simple interest $I = Prt = \$320(.0307)(3) \approx \$29.47$, so the compound amount should be a little more than $320 + \$29.47 = \$349.47$, which it is.

Exponential Growth

If everything else remains the same, the more often money is compounded, the greater is the compound amount. To see an example of this, consider $1000 invested for 4 years at 3% annual interest, compounded over different time intervals, as shown in ● Table 10.1.

TABLE 10.1	
Effect of the Frequency of Compounding $1000 at 3% for 4 Years	
Compounded	Compound Amount
Not at all (simple interest)	$1120.00
Yearly	1125.51
Semiannually	1126.49
Quarterly	1126.99
Monthly	1127.33
Daily (take 365 days/yr)	1127.49

Notice that in the compound-amount formula (equation 10.1), if principal P and rate r are fixed, then the compound amount is an exponential function of n. For example, if $10,000 is invested at 5% interest, compounded annually, then

$$A = \$10,000(1 + .05)^n$$

where n is the number of years the money is invested. ● Table 10.2 and ► Figure 10.2 show the exponential growth of that $10,000 over 40 years.

TABLE 10.2	
Growth of Money at Compound Interest	
n (years)	A
5	$12,762.82
10	16,288.95
15	20,789.28
20	26,532.98
25	33,863.55
30	43,219.42
35	55,160.15
40	70,399.89

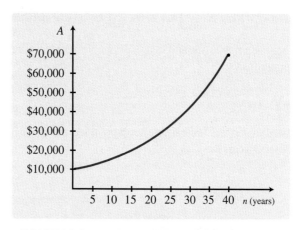

► FIGURE 10.2 Growth of money at compound interest

Notice how the amount grows by increasing amounts over fixed time intervals. For example, during the first 5 years the amount increased by $2762.82, during the second 5 years the amount increased by $3526.13, during the third 5 years the amount increased by $4500.33. Another way to say this is that the amount increases at an increasing rate.

Let's now compare the result of compounding with the result of simple interest. If $10,000 is invested at 5% simple interest, then the interest is $500 each year, or $2500 each 5 years. The amounts available after different periods of time are shown in ► Figure 10.3. The amounts with simple interest are shown on the straight line, while the amounts with annual compounding are shown on the curve.

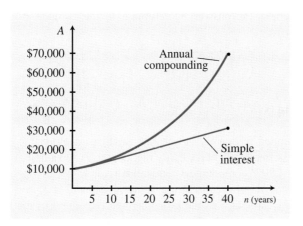

▶ FIGURE 10.3 Effect of annual compounding

Present Value

In the compound-amount formula, $A = P(1 + r)^n$, we can solve for P, as follows.

$$P = \frac{A}{(1 + r)^n} = A(1 + r)^{-n} \qquad (10.2)$$

The amount P is called the *present value* of the amount A. (When P is known, A is called the *future value* of P.) The present value is the amount that needs to be invested now in order to have the amount A after n periods of time when money is compounded each period at a rate r per period. The next example illustrates a present-value calculation.

EXAMPLE 10.13

How much money needs to be invested now so that $20,000 will be available in 4 years? Assume the investment is at 4%, compounded monthly.

SOLUTION From the compound-amount formula (equation 10.1), we have

$$A = P(1 + r)^n$$

$$\$20,000 = P\left(1 + \frac{.04}{12}\right)^{48}$$

and
$$P = \$20,000\left(1 + \frac{.04}{12}\right)^{-48} = \frac{\$20,000}{(1 + .04/12)^{48}}$$

If we use the present-value formula (equation 10.2), we go directly to the same result:

$$P = \$20,000\left(1 + \frac{.04}{12}\right)^{-48} \approx \$17,047.41$$

So $17,047.41 needs to be invested now so that $20,000 will be available in 4 years.

LIGHTEN YOUR LOAD

In Example 10.13, why go through the longer method (from the compound-amount formula) to calculate P when we can get that result directly through the present-value formula? The main reason is that it requires memorizing only one formula rather than two. If you have to keep formulas in your head, then remembering one formula and using the longer method is preferable. You need to know the basic compound-amount formula and how to solve for one variable in terms of another. If you have access to written formulas, then the present-value formula is the way to go.

We can check the answer by returning to the compound-amount formula (equation 10.1):

$$A = \$17,047.41\left(1 + \frac{.04}{12}\right)^{48} \approx \$20,000.00$$

Effective Rate

Money compounded frequently at one rate might earn more interest than money compounded less frequently at a slightly higher rate. For example, which is the better investment: 4.80% compounded semiannually or 4.76% compounded monthly? We can determine this by considering some investment, say, $1 for 1 year. The amount of the investment, i.e., the principal, and the length of the investment will determine the future value. But for comparison purposes all that matters is that they are the same. At 4.80% compounded semiannually, the $1 grows over 1 year to

$$\$1\left(1 + \frac{.048}{2}\right)^2 = \$1.048576 \approx \$1.04858$$

and the $1 at 4.76% compounded monthly grows to

$$\$1\left(1 + \frac{.0476}{12}\right)^{12} \approx \$1.04865$$

We conclude that the 4.76% interest rate compounded monthly is slightly better.

Rather than go through the preceding kind of analysis to compare interest rates and compounding periods, a standard rate called the *effective rate* is often used. The effective rate is the simple interest rate that would produce the same interest in a year as the nominal rate (the rate named or specifically stated). For example the nominal rate of 4.80% compounded semiannually produces interest on $1 of $.04858 for a year. Let r denote the simple interest rate that would also do this.

$$\text{interest} = Prt$$
$$\$.04858 = \$1(r)(1)$$
$$\$.04858 = \$r$$
$$.04858 = r$$

The effective rate is 4.858%, for a nominal rate of 4.80% compounded semiannually. Similarly, for a nominal rate of 4.76% compounded monthly, the effective rate is 4.865%. Notice that in both of these cases the effective rate is higher than the nominal rate. Because of the power of compounding, this will always be the case unless the nominal rate is for simple interest. The effective rate gives a standard way to compare interest rates with different compounding periods.

●EXAMPLE 10.14

A money market fund is earning 5.75% annually. It compounds monthly. Calculate the effective rate.

SOLUTION If $1 is invested for 1 year,

$$A = P(1 + r)^n$$
$$= \$1\left(1 + \frac{.0575}{12}\right)^{12} \approx \$1.05904$$

Now let r denote the simple interest rate that would earn the same interest of $.05904.

$$interest = Prt$$
$$\$.05904 = \$1(r)(1)$$
$$\$.05904 = \$r$$
$$.05904 = r$$

The effective rate is 5.904%.

YOUR FORMULATION

Notice in Example 10.14 the relation of the effective rate 5.904% to the future value $1.05904. There is a similar relationship in the two effective rates calculated in the paragraphs before Example 10.14. If you can identify and use this relationship, you can short-cut the calculation of an effective rate.

Inflation

Another way that compounding affects the consumer is through inflation. Inflation drives up the price of an item in the same way that interest drives up the value of an investment. For example, with a 3% annual rate, the price of an item goes up by 3% each year over the previous year, and the effect of the inflation compounds in the same way that the effect of interest compounds. In particular, to calculate what a book that cost $50 in 1996 would cost in the year 2000, assuming no other changes in costs, we could apply the basic compound-amount formula (equation 10.1).

$$A = P(1 + r)^n = \$50(1 + .03)^4 \approx \$56.28$$

The price in the year 2000 would be $56.28.

EXAMPLE 10.15

Inflation from 1960 to 2000 will average about 5% per year. If a new compact car cost $3000 in 1960, what would the comparable new car cost in 2000 because of inflation?

SOLUTION Use the compound-amount formula (equation 10.1).

$$A = P(1 + r)^n$$
$$= \$3000(1 + .05)^{40} \approx \$21,000$$

The comparable car in the year 2000 would cost about $21,000.

Rule of 72

A useful rule of thumb in working with compound interest is known as the *Rule of 72*. This rule deals with the length of time it takes money to double when interest is compounded.

INFLATION

Inflation, defined as a general increase in prices, is the world's greatest robber. A covert thief, inflation steals from widows, orphans, bondholders, retirees, annuitants, beneficiaries of life insurance, and those on fixed salaries, decreasing the value of their incomes. Inflation extorts more wealth from the public than do all other thieves, looters, embezzlers, and plunderers combined.

Inflation, a Jekyll and Hyde character, is not only a great robber but also a great benefactor. Inflation is the world's greatest giver, doling out benefits to debtors, hoarders of goods, owners of property, government (for which it reduces the burden of the public debt), and, over time, owners of common stocks. The largesse thus bestowed on the debtor class and owners of property exceeds the combined total of all charities, contributions, and donations.

Don Paarlberg, *An Analysis and History of Inflation*, Wesport, CT: Praeger, 1993, p. xi.

> ### RULE OF 72
>
> The interest rate r as an annual percentage times the time d (in years) it takes money to double at the rate r, compounded annually, is about 72.
>
> $$rd \approx 72$$

EXAMPLE 10.16

If money is invested at an interest rate of 5%, compounded yearly, how long will it take to double?

SOLUTION Let d be the doubling time, in years. By the Rule of 72, $5d \approx 72$, so $d \approx 72/5 = 14.4$. It will take about 14 years to double. Notice in Table 10.2 that after 15 years at 5%, money had more than doubled.

Keep in mind that the Rule of 72 is only an approximation. However, it does give a quick estimate of doubling time if the rate is known or a quick estimate of the rate needed to double money in a fixed number of years. It is also a useful check on calculations with compound interest. For example, in Example 10.15, with 5% inflation, prices should double in about 14 years (see Example 10.16). so to look at prices in 40 years we could say:

For Original Price of $3,000
Price after 14 years: about $6,000
Price after 28 years: about $12,000
Price after 42 years: about $24,000

Therefore, the price after 40 years should be close to $24,000. This is a reasonable approximation to the calculated amount of $21,000.

Summary

In summary, with compound interest, the basic result is the compound-amount formula. It can be used to calculate present value and effective rates. Inflation is a form of compound interest. The Rule of 72 is used with doubling times and to check results when compounding.

EXERCISES 10.4

In Exercises 1–4, find the compound amount for the investments listed below.

1. $1000 for 2 years at 4.2% compounded yearly
2. $800 for 4 years at 4.9% compounded semiannually
3. $600 for $4\frac{1}{2}$ years at 5.5% compounded quarterly
4. $400 for $5\frac{1}{2}$ years at 6.2% compounded monthly
5. A certain U.S. Savings Bond earns a guaranteed interest rate of 6.0% if held for 30 years. For a bond that will pay out $10,000 in 30 years, what should a person pay now?

6. A U.S. Savings Bond purchased on March 1, 1983, and cashed on March 1, 2003, would earn interest of 7.75% compounded semiannually.
   a. If a bond were purchased for $5000 on March 1, 1983, what would it be worth on March 1, 2003?
   b. If you have a bond that will pay you $10,000 on March 1, 2003, what is the maximum amount the purchaser should have paid on March 1, 1983?
7. Frank earnestly pledged $100 to his college alumni association, to be paid in 4 years. He wants to set aside that money

now in his bank account, which earns $2\frac{1}{2}\%$ annual interest, compounded quarterly. How much should he set aside now?

8. How much should be invested in a money market fund (interest is compounded monthly) so that $11,000 will be available in 3 years? Assume the annual interest rate is 5.75%.

9. A certain money market fund pays 4.59%. It compounds monthly. What is the effective rate?

10. Certain U.S. Savings Bonds pay 7.0%, compounded semiannually. What is the effective rate?

11. In 1960 many college students worked at summer jobs that paid $1.00 per hour. The average inflation rate from 1960 to 1997 was about 5% per year. What would the comparable hourly wage be in 1997?

12. If the inflation rate holds at 3% per year, what will the comparable wage be 35 years from now to a current wage of $4.35 per hour?

13. A new high school teacher in a school system in Missouri started in 1966 at a salary of $3600. What would be the comparable starting salary in 1997 if the average inflation rate from 1966 to 1997 was 5% per year?

14. A certain college graduate starts at age 23 at a salary of $25,000. If that person gets only cost-of-living increases averaging 3.5% per year until retirement at age 65, what will the person's salary be at retirement?

15. In 1803 the United States made the Louisiana Purchase from France for about $15,000,000.
   a. If inflation averaged 2% a year since that time, what would its cost be on its bicentennial in 2003?
   b. What if inflation averaged 3% a year?

16. If money earns 4% interest, approximately how long will it take for it to double?

17. Heidi has $4000 and would like it to grow to $8000 in 8 years. What interest rate does she need to seek?

18. If inflation holds at about 3%, how long will it take for prices to double?

19. Prices more than doubled from 1970 to 1980. What does this say about the average annual rate of inflation during that time? Be as specific as possible.

WRITTEN ASSIGNMENTS 10.4

1. Assume you have been hired by a financial institution to write to customers encouraging them to save their money (invest with this financial institution). What you prepare will go to current customers as part of a quarterly report on the status of their accounts. You are to emphasize the power of compounding using the ideas of Table 10.2 and Figure 10.3. Write a half a page to a page promotion.

2. The Rule of 72 is an approximation rather than an exact formula. Investigate how close to being accurate it is by looking at some different interest rates, predicting the doubling time from the Rule of 72, then seeing what happens to the future value after that period of time. For example, if the annual interest rate is 5%, the predicted doubling time is about 14 years. After 14 years, $1 has increased in value to $1.98, which is close to, but not quite, double the original amount. Comment on what rates of interest give the best results with the Rule of 72.

3. With compound interest, money "grows exponentially" over time. Write a paragraph or two to a high school junior describing what this means.

4. Talk to a retiree of age 70 or more about the effect of inflation on a fixed income. Summarize that person's comments.

10.5 ANNUITIES

GOALS
1. To describe what an annuity, an ordinary annuity, and an annuity due are.
2. To use a formula to calculate the future value of an ordinary annuity.
3. To use the formula for the future value of an ordinary annuity to calculate the future value of an annuity due.
4. To use a formula to calculate the present value of an ordinary annuity.

In the preceding two sections dealing with interest we have looked at lump-sum payments: How much money will be paid out (is due) over n years if an amount P is invested (borrowed) now? How much money do we need to invest now so that a

What is the value of an annuity?

given amount A is available after n years? Now we look at a sequence of payments needed to reach a savings objective or to pay off a loan. For example, if, to save for a college education, a parent pays $100 at the end of each month into an account earning 3.0% annual interest, compounded monthly, how much will be available after 5 years?

This type of investment is an example of an *annuity:* a sequence of equal payments made at equal periods of time. If the frequency of payments is the same as the frequency of compounding (every month, quarter, year, etc.) and the payments are made at the *end* of each time period, then the annuity is called an *ordinary annuity.* If the frequency of payments is the same as the frequency of compounding but the payments are made at the *start* of each time period, then the annuity is called an *annuity due.*

Ordinary Annuity

To illustrate the way money accumulates with an ordinary annuity we take a more tractable example: If a parent puts $1200 at the end of each year into an account earning 3% interest, compounded annually, how much will be available after 5 years? The answer is shown in ▶ Figure 10.4. The future value of the annuity is $6370.96. This is a reasonable answer, since $6000 was invested (five payments of $1200 each) and some interest was earned.

If this procedure were tried on the earlier example ($100 a month for 5 years), there would be 60 payments to trace. Rather than do this, let

R = payment made each period

n = number of periods

r = interest rate per period as a decimal or proportion (e.g., 3% = .03)

S = future value (value of the annuity after the n payments)

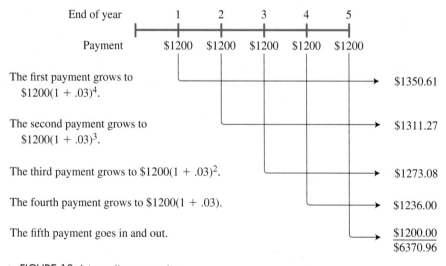

▶ FIGURE 10.4 An ordinary annuity

Then it can be shown that the following is a formula for the future value of an ordinary annity.

$$S = R\left[\frac{(1+r)^n - 1}{r}\right] \qquad (10.3)$$

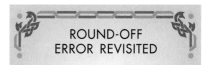

ROUND-OFF
ERROR REVISITED

As in the calculation of present value, no intermediate calculations need be written down when calculating future value. All the calculations can be left in the calculator. This both saves time and avoids round-off error.

To illustrate the use of this formula for the future value of an annuity consider the annuity of Figure 10.4. The formula for the future value of an annuity (equation 10.3) gives

$$S = \$1200\left[\frac{(1+.03)^5 - 1}{.03}\right] \approx \$6370.96$$

The next example returns to the example of saving for college.

EXAMPLE 10.17

A parent who has a child entering eighth grade considers setting aside some money each month to help provide for the child's college education. The parent pays $100 at the end of each month into an account earning 3.0% annual interest, compounded monthly. How much will be available after 5 years (when the child is ready to enter college)?

INCORRECT "SOLUTION" We're first going to do an incorrect "solution" to illustrate how a check will detect that an error was made.

$$S = R\left[\frac{(1+r)^n - 1}{r}\right]$$

$$= \$100\left[\frac{(1+.03)^{60} - 1}{.03}\right]$$

$$\approx \$16{,}305.34$$

Is this reasonable? The parent makes 60 payments of $100 for total payments of $6000. At 3% (compounded annually) it would take $6000 about 24 years to double to $12,000 (by the rule of 72). By our "incorrect solution," the money more than doubled in 5 years. Surely the effect of the monthly compounding is not enough to get as much as $16,305. There must be a mistake. Did you see where it was?

CORRECT SOLUTION The rate per period (month) is .03/12, so

$$S = R\left[\frac{(1+r)^n - 1}{r}\right]$$

$$= \$100\frac{(1 + .03/12)^{60} - 1}{.03/12}$$

$$\approx \$6464.67$$

Is this reasonable? Since it's a little more than the $6000 made in payments, it is reasonable. After 5 years, $6464.67 will be available.

cathy®

Annuity Due

The ordinary annuity is a little strange, in two ways. First, you set up the annuity but don't make a deposit until the end of the first time period. Second, the final payment is made, then taken right back out of the account on the same day. The annuity due, with payments made at the start of each period, overcomes both of these quirks.

To calculate the future value of an annuity due, we will use a procedure (rather than a formula) based on the formula for the future value of an ordinary annuity. We will describe the procedure in ▶ Figure 10.5. The same notation as for an ordinary annuity will be used:

$$R = \text{payment made each period}$$

$$n = \text{number of periods}$$

$$r = \text{interest rate per period}$$

$$S = \text{future value}$$

Think of adding one period onto the start and making one more payment at the end, as in ▶ Figure 10.6.

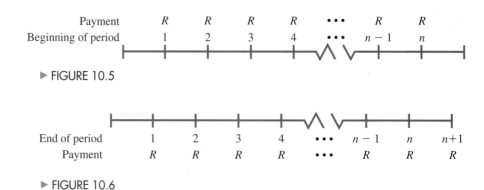

▶ FIGURE 10.5

▶ FIGURE 10.6

This "thought" annuity is an ordinary annuity with $n + 1$ periods, so

$$S = \left[R\frac{(1 + r)^{n+1} - 1}{r} \right] - R$$

We added a last payment of R in the first term of our "thought" annuity. Since it was not actually made, we subtract it from the first term.

In summary, to find the future value of an annuity due, find the future value of an ordinary annuity with one more period of time, then subtract the value of the one extra payment. The following example applies this procedure.

EXAMPLE 10.18

Suppose Jennie decides to create her own Christmas Club account by depositing $50 on the 15th of each month from January—November in a money market fund that pays 4.78%, compounded monthly. How much will she have on December 15?

SOLUTION Notice that this is an annuity due (since the payments are made at the start of each period) for 11 months. Consider it as an ordinary annuity for 12 months, then subtract one payment.

$$S = R\left[\frac{(1 + r)^{12} - 1}{r} \right] - R = \$50\left[\frac{(1 + .0478/12)^{12} - 1}{.0478/12} \right] - \$50 \approx \$563.32$$

Jennie will have $563.32 on December 15. This is reasonable, since she makes 11 payments of $50, for a total of $550, and earns some interest.

YOUR FORMULATION

If you had money available to make the same payments (dollar amount and total number of payments) into either an ordinary annuity or an annuity due, both of which earn the same interest and have the same compounding periods, which would have the larger future value, and why?

Present Value of an Ordinary Annuity

An annuity generates a future value after a certain amount of time (the S of our preceding calculations). The value of that annuity at the present time is called the *present value of an ordinary annuity*. In other words, the present value of an annuity is the amount of money that if invested today, using the same rate and compounding of the annuity, would provide the same amount of money after n periods of time as the annuity provides. If we let P be the present value of an ordinary annuity of n payments of amount R with interest compounded at a rate of r per period, the following can be shown.

PRESENT VALUE OF AN ORDINARY ANNUITY

$$P = R\left[\frac{1 - (1 + r)^{-n}}{r} \right] \tag{10.4}$$

We illustrate the use of this formula by applying it to Example 10.17.

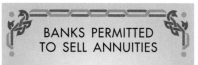
•EXAMPLE 10.19

Consider the annuity of Example 10.17: A parent saves for a college education by paying $100 at the end of each month into an account earning 3.0% annual interest, compounded monthly, for 5 years. What is the present value of the annuity?

SOLUTION By the formula for the present value of an ordinary annuity (equation 10.4), we have

$$P = \$100\left[\frac{1 - (1 + .03/12)^{-60}}{.03/12}\right] \approx \$5565.24$$

The present value is $5565.24. This is reasonable, since with 60 monthly payments of $100 the parent planned to invest $6000. The present value is less than that and the interest will make it grow.

Let's check more exactly: How much will $5565.24 grow to after 5 years at 3.0%, compounded monthly?

$$S = P(1 + r)^n = \$5565.24\,(1 + .03/12)^{60} \approx \$6464.68$$

This is just one penny different than the future value of the annuity.

Summary

In summary, an annuity is a way to invest through a sequence of regular payments. It might be in the form of an ordinary annuity or of an annuity due. Both of these can be calculated from the same formula. This formula can also be used to develop a formula for the present value of an ordinary annuity.

EXERCISES 10.5

1. A faculty member is 6 years from retirement. To add to her retirement benefits she plans to deposit $200 at the end of each month into an account that pays 5¾% interest, compounded monthly. How much will she have on deposit after 6 years?

2. A grandfather plans on depositing $1000 in his granddaughter's savings account on each of her first 18 birthdays. If the account earns 7% interest, compounded annually, how much will she have on her 18th birthday?

3. What if the parent of Example 10.17 had begun saving at the same rate of $100 at the end of each month in an account earning 3.0% annual interest, compounded monthly, when the child was born. How much would have been available after 18 years.

4. A student "stops out" of school for a year to earn money for the rest of his schooling. He plans to deposit $1500 at the end of each quarter into an account paying 6¼%, compounded quarterly. How much will he have at the end of a year?

5. Impressed by the power of compounding, a student vows upon graduation to set aside $100 twice a year into an Individual Retirement Account (IRA) paying 5½%, compounded semiannually. How much will she have in her IRA upon her retirement in 40 years?

6. Kathy wants to save for a car by putting $160 per month into a money market fund that pays 4.8%, compounded monthly. If she starts right away and makes 30 payments, how much will she have at the end of 30 months?

7. Mike is considering buying a whole life insurance policy. As an option he looks at what he would have in 40 years if he invested $150 at the start of each semiannual period at 7%, compounded semiannually. How much would he have?

8. If an investor had put $2000 a year in the Scudder Short-Term Bond Fund at the beginning of each year from 1988 through 1992, the investor would have accumulated $13,366 by the end of 1992. What would the investor have accumulated by the end of 1992 by putting $2000 a year into a savings account

paying 3.0% per year, compounded annually, at the *beginning* of each year from 1988 through 1992?

9. Calculate the present value of the annuity of Example 10.18.

10. Brandon sets up a supplemental retirement annuity (SRA) and pays $1000 into his SRA every 6 months. His SRA earns 8% annual interest, compounded semiannually. He will make the payments at the end of each 6-month period for 15 years. What is the present value of Brandon's annuity?

11. A grandfather said he would set up an ordinary annuity for a newly born granddaughter and pay $500 a quarter, with the last payment on her 18th birthday. Assume the money would earn annual interest of 7.5%, compounded quarterly. The grandmother said they should just give the granddaughter a trust with an amount of money now that would grow to the same amount as the annuity by her 18th birthday. If this is done, how much should be set aside?

12. Teachers often have a choice of being paid every 2 weeks on a 20-pay or a 26-pay basis. The 20-pay basis covers 40 weeks, or a little over nine months (the academic year), while the 26-pay distribution helps the teacher spread the salary throughout the year. For example, suppose that after deductions a teacher's salary was $16,900. The teacher could choose between receiving $845 every 2 weeks for 20 weeks (for $16,900) from roughly early September to early June or receiving $650 every 2 weeks for 26 weeks (for $16,900) from roughly early September for 1 year. With the 20-pay option, the teacher could take the difference of $195 between the two rates of pay, invest this amount each 2 weeks, then withdraw the total accumulated amount after 40 weeks (in early June) for use in the summer. Assume the account into which these payments are made earns 4.5% annual interest, compounded daily (for a 365-day year). How much extra money will the teacher who chooses 20 pays earn over the teacher who chooses 26 pays?

1. Describe the difference between an ordinary annuity and an annuity due.

2. Write a page to a recent college graduate pointing out the value of an annuity in providing for *either* of the following.

a. A child's college education

b. The graduate's retirement

10.6 LOAN PAYMENTS

GOALS
1. To find the size of payments necessary to pay off a loan.
2. To construct an amortization schedule.

In the preceding section, methods were developed to find the future value or the present value of an annuity. In this section, the focus is still on equal payments made at equal periods of time, but with concern on the amount of a payment required. We will consider this in the context of a *fixed-installment loan*—a loan that the borrower pays by making fixed payments (installments) at fixed points in time. If the fixed payments are all the same and if the fixed points in time are equally spaced, then the borrower makes equal payments at equal periods of time, which means the borrower is doing the same thing as buying an annuity. Fixed-installment loans are commonly used to fund major purchases, such as houses, cars, appliances, and furniture.

How's the car payment calculated?

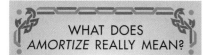
EXAMPLE 10.20

Perissa wants to buy a car that will cost $7250. She will pay 20% down, or $1450, and finance the remaining $5800 through a bank at an annual percentage rate (APR) of 6.9% with monthly payments for 3 years.

a. How large are the monthly payments?
b. What is the total amount paid on the loan (principal plus interest)?
c. How much interest is paid on the loan?

SOLUTION

a. Call the payment amount R. It will be made at the end of each month. The $5800 can be regarded as the present value of an ordinary annuity. Use the formula for the present value of an ordinary annuity (equation 10.4).

$$P = R\left[\frac{1 - (1 + r)^{-n}}{r}\right]$$

$$\$5800 = R\left[\frac{1 - (1 + .069/12)^{-36}}{.069/12}\right]$$

To solve for R, divide both sides of the equation by the number in brackets:

$$R = \frac{\$5800}{\dfrac{1 - (1 + .069/12)^{-36}}{.069/12}} \approx \$178.82$$

Perissa will need to make monthly payments of $178.82.

Is this reasonable? She will make 36 payments of $178.82 for a total payment of $6437.52. This will be for the $5800 owed plus interest, so it looks reasonable.

b. Perissa will make 36 payments of $178.82 for a total payment of $6,437.52.
c. Of the $6437.52 paid, $5800 is the principal, so the interest is $6,437.52 − $5,800 = $637.52.

Amortization Schedule

A loan is *amortized* provided both the principal and the interest are paid by a sequence of payments. In Example 10.20 the amount of the monthly payments needed to amortize a fixed-installment loan was calculated. In the next example we look at a table, called an *amortization schedule,* for paying off a loan.

SHOE by Jeff MacNelly

Reprinted by permission: Tribune Media Services.

EXAMPLE 10.21

Tom borrows $6000 for some home improvements. He will pay off the loan in 2 years, with four semiannual payments. The money is borrowed at $9\frac{1}{2}\%$ per year on the unpaid balance. Construct an amortization schedule.

SOLUTION The table to be filled in is the following. The values of a, b, c, etc. will be calculated.

Payment Number	Amount of Payment	Interest for Period	Portion to Principal	Principal Owed at End of Period
0	—	—	—	$6000
1	a	b	c	d
2	a	e	f	g
3	a	h	i	j
4	m	k	l	n

Step-by-step instructions for filling in the table follow. We suggest that as soon as possible you stop reading the instructions and complete the table yourself. Be careful on the line for payment number 4.

a. The amount of the payment is figured as in Example 10.20.

$$P = R\left[\frac{1 - (1 + r)^{-n}}{r}\right]$$

$$\$6000 = R\left[\frac{1 - (1 + .095/2)^{-4}}{.095/2}\right]$$

$$R = \frac{\$6000}{\frac{1 - (1 + .095/2)^{-4}}{.095/2}} \approx \$1682.26$$

Enter this in the table at a (three places).

b. The interest for the first semiannual period is

$$I = Prt = \$6000\left(\frac{.095}{2}\right)(1) = \$285.00$$

Enter this in the table at b.

c. Of the first payment of $1682.26, $285.00 goes toward interest and the rest ($1682.26 − $285.00 = $1397.26) goes toward the principal. Enter this at c.

d. The principal that was $6000 is now reduced by $1397.26 to $4602.74. Enter this at d.

e. The interest for the second semiannual period is

$$I = Prt = (4602.74)\left(\frac{.095}{2}\right)(1) \approx \$218.63$$

Enter this at e.

f. $1682.26 − $218.63 = $1463.63

g. $4602.74 − $1463.63 = $3139.11

h. $I = Prt = (\$3139.11)(.095/2)(1) \approx \$149.11$

i. $1682.26 − $149.11 = 1533.15

j. $3139.11 − $1533.15 = $1605.96 (Notice that for the line with payment 4 we employ a different order of calculation.)

k. $I = Prt = (\$1605.96)(.095/2)(1) \approx \$76.28$

l. To pay off the loan the remaining principal of $1605.96 must be paid.

m. The payment is the sum of interest and principal: $76.28 + $1605.96 = $1682.24.

n. $1,605.96 − $1,605.96 = $0 (This is what is needed to have the loan paid off.)
The amortization schedule is as follows:

Payment Number	Amount of Payment	Interest for Period	Portion to Principal	Principal Owed at End of Period
0	—	—	—	$6000
1	$1682.26	$285.00	$1397.26	4602.74
2	1682.26	218.63	1463.63	3139.11
3	1682.26	149.11	1533.15	1605.96
4	1682.24	76.28	1605.96	0

YOUR FORMULATION

1. In the amortization schedule for Example 10.21 the interest payments decreased from one period to the next. Explain why, without doing the exact calculation.
2. Will such a decrease in interest payments from one period to the next happen in every amortization schedule? Why or why not?

Long-Term Loans

Recall that with money borrowed at compound interest the amount due is an exponential function of time. Picture what the graph of an exponential function looks like. The difference between the amount due and the principal (the amount borrowed) is the interest. Consequently, for a long-term loan the interest is the major portion of the amount paid. To see this consider the next example.

EXAMPLE 10.22

A person wants to buy an $87,000 house by paying 20% down and financing the rest on a 30-year mortgage at 9.0% per year, with monthly payments. Display the amortization schedule for payments 0, 1, 2, and 3.

SOLUTION The person pays 20% down, which is $17,400, and finances the remaining $69,600.

Payment Number	Amount of Payment	Interest for Period	Portion to Principal	Principal Owed at End of Period
0	—	—	—	$69,600
1	$560.02[a]	$522.00[b]	$38.02	69,561.98
2	560.02	521.71[c]	38.31	69,523.67
3	560.02	521.43	38.59	69,485.08

[a]
$$P = R\left[\frac{1-(1+r)^{-n}}{r}\right] \quad \text{so} \quad \$69{,}600 = R\left[\frac{1-(1+.09/12)^{-360}}{.09/12}\right]$$

$$R = \frac{\$69{,}600}{\dfrac{1-(1+.09/12)^{-360}}{.09/12}} \approx \$560.02$$

[b]$I = Prt = \$69{,}600(.09/12)(1) = \$522.00$
[c]$I = Prt = \$69{,}561.98(.09./12)(1) \approx \$521.71$ (Notice the huge fraction of the first payments that goes toward interest. This is typical for loans taken out over a long period of time.)

Taking Out a Loan

Looking at the annual percentage rates (APRs) on the loans is one way to compare them, but there are often other costs to consider. On automobile loans, consider rebates (see Exercise 8) and costs such as credit life insurance. On some mortgages, "points" may be added to the mortgage cost along with closing costs. Carefully calculate these as you compare loans. Since financial institutions are regularly marketing their products in different ways, consider checking the advice of consumer groups before finalizing your purchase. These can be found by searching a database covering magazines under "loans," "installment buying," or some similar key words. Even after reading such a source, you'll probably want to do some calculating—some consumer mathematics—to find the best deal.

EXERCISES 10.6

1. a. Find the monthly payments necessary to pay off a $1000 loan in 2 years if the interest rate is 7.2%, compounded monthly.
   b. Find the total amount paid (principal plus interest).
   c. Find the total amount of interest paid.

2. Perissa of Example 10.20 decides that monthly payments of $178.82 are more than she can afford. She really likes the car, so she considers a 4-year repayment schedule. What will be her monthly payments?

3. A bank offers a bi-weekly mortgage program (payments every 2 weeks) and argues its advantages over monthly payments through the following calculations: Suppose $74,700 is financed.
   a. What is the amount of a monthly payment if payments are made at an APR of 9.0684% for 15 years?
   b. What is the total amount of all payments made?
   c. What is the amount of a biweekly payment if payments are made at an APR of 9.0527% (slightly lower than in part a) for 12 years and 10 months (335 biweekly payments)?
   d. What is the total amount of all payments made in part c?
   e. How much is saved in total payments by using the biweekly mortgage program?

4. Suppose Raul has a $760 credit card bill that he is determined to pay off before making other charges on his card. The interest rate on his credit card is 16.9%, compounded monthly. If he pays off his bill in five equal payments, how much will each monthly payment be?

5. Consider Example 10.20. The start of the amortization schedule for the loan is shown. Fill in the next (blank) line of the table.

Payment Number	Amount of Payment	Interest for Period	Portion to Principal	Principal Owed at End of Period
0	—	—	—	$5800.00
1	$178.82	$33.35	$145.47	$5654.53
2				

6. Chippwood Swim Club borrowed $8000 for some repairs, with the money to be repaid in four semiannual payments with interest at 9.9% a year, compounded semiannually. Construct the amortization schedule.

7. You buy a house for $75,000 and pay 20% down. You finance the rest with a 30-year mortgage at 9.25% annual interest with monthly payments. Construct the amortization schedule for payments 0, 1, 2, and 3.

8. You want to buy a particular car for which you need to pay $9000 beyond the down payment and trade-in you have. You have two options. One is to borrow through the car dealer at 6.7% annual interest, compounded monthly. The other is to get a $1200 rebate through the dealer and finance the remaining $7800 through a bank at 9.2%, compounded monthly. Assuming that either financing arrangement lasts for 3 years (36 months), which is the better deal? Answer by doing the following calculations.
   a. How much is a monthly payment with dealer financing?
   b. How much in total ($9000 + interest) do you pay through the dealer?
   c. How much is a monthly payment through the bank?
   d. How much in total ($7800 + interest) do you pay through the bank?
   e. Which is the better deal, and how much in total do you save by financing your loan that way?

9. You want to compare purchasing a new car with leasing the same car. The lease is for $269 per month for 36 months. The purchase price is $14,700. Assuming you can borrow the $14,700 purchase price at 9.1%, compounded monthly, what would your payments be for loans of each of the following terms?
   a. 36 months   b. 48 months   c. 60 months

10. An investment quiz contains the following true-false question. A 15-year fixed-rate mortgage saves you nearly 60% of the total interest costs over the life of the loan when compared to a 30-year fixed-rate mortgage.

Explore the answer as follows. Assume a $100,000 loan is borrowed on a fixed-rate mortgage at 8% annual interest with monthly payments.

a. If the mortgage is a 15-year mortgage, how much are the monthly payments?

b. What is the total amount paid over the 15 years?

c. How much of the amount in part b is interest?

d. If the mortgage is a 30-year mortgage, how much are the monthly payments?

e. What is the total amount paid over the 30 years?

f. How much of the amount in part e is interest.

g. How many dollars in interest does the 15-year mortgage save over the life of the loan when compared with the 30-year mortgage?

h. (*Fill in the blank*) For a $100,000 loan, the 15-year 8% mortgage saves you nearly _____% of the total interest costs over the life of the loan when compared to a 30-year 8% mortgage.

i. How would your answer in part h change if the amount borrowed were $200,000? Would it be larger, the same, or smaller?

j. To see what would happen to your answer in part h if the interest rate were to change, repeat parts a–i for an annual interest rate of 9%.

k. Answer the original true-false question.

WRITTEN ASSIGNMENTS 10.6

1. The size of a payment on a fixed-installment loan depends on three variables: the amount borrowed, the interest rate, and the number of payments. The dependence on the amount borrowed is direct. For example, if the amount borrowed doubles, then the payments double if the other two variables remain constant.

   a. Explore the dependence on the interest rate by considering a $1000 loan to be paid back in 36 monthly payments at various interest rates. Describe what you've found, and display it in both a table and a graph.

   b. Explore the dependence on the number of payments by considering a $1000 loan at 9% annual interest to be paid back with 24 monthly payments, 36 monthly payments, 48 monthly payments, or 60 monthly payments. Describe what you've found, and display it in both a table and a graph.

2. Consider a car that you might be interested in and find an approximate price on it. Find out what some possible interest rates are that you might pay. Explore several different options using combinations of different rates and different numbers of payments. Write this up in a summary to present to someone who might be interested, such as a parent, a spouse, a boyfriend or girlfriend, a brother or sister, or some other friend.

3. A survey, *Adult Literacy in America,* released by the Department of Education and reported by *Time,* September 20, 1993,

FIXED RATE • FIXED TERM

Home Equity Loans **14.25%**
Annual Percentage Rate
Ten Year Term

SAMPLE MONTHLY REPAYMENT SCHEDULE

Amount Financed	Monthly Payment
$10,000	$156.77
$25,000	$391.93
$40,000	$627.09

120 Months 14.25% APR

was based on interviews with 26,000 American adults. One question follows: You need to borrow $10,000. Look at the ad and explain to the interviewer how you would compute the total amount of interest charges you would pay under this loan plan. Write your explanation.

4. Go to a local financial institution (bank or savings and loan) and get information on its loan rates of a particular type (home mortgage, automobile loan, college loan, etc.). Select a type of loan for which this financial institution has more than one rate, such as a 30-year mortgage or a 20-year mortgage. Compare and contrast the loans. Under what conditions would you recommend one over the other?

CHAPTER 10 REVIEW EXERCISES

1. Describe three distinct areas of consumer mathematics.

2. State three facts about the history of interest.

3. Pepsi is available in a 12-pack of 12-oz cans for $3.00, a 2-liter bottle (67.6 oz) for 99¢, or an eight-pack of 20-oz bottles for $2.59. Rank these from best buy to poorest buy based on cost per ounce.

4. A gas utility has a monthly customer charge of $7.50 and a usage charge of $.380590 per 100 cubic feet (CCF). The state

has a 6% sales tax on the total. If a customer uses 45 CCF of gas in a month, what will the monthly bill be?

5. An electric utility has an electric use charge of $.075210 per kilowatt-hour (kWh) and a monthly customer charge. If the May bill before sales tax for 709 kWh is for $58.32, how much is the monthly customer charge?

6. An electric utility charges $.076511 per kilowatt-hour (kWh) for the first 680 kWh and a higher rate for the next chunk of

kilowatt-hours. It has no monthly customer charge. In September a customer used 1369 kWh and had a bill, before sales tax, of $119.01. What is the per-kWh charge on the second chunk of kilowatt-hours?

7. If the millage rate is 32 mills, what is the tax on a property assessed at $72,500?

8. If the millage rate is 22 mills and the tax on a property is $1348.60, what is the assessed value of the property?

9. If a property assessed at $84,700 pays a tax of $3472.70, what is the millage rate?

10. A catalog for a major retailer says "Save 30%."
   a. If the original price of a top was $22.00, what is the sale price?
   b. If the sale price of a sweater is $13.99, what was the original price?

11. Some shoes are on sale for $39.88 that were regularly $55. What percentage does the customer save?

12. A shoe store advertised 50% off on a pair of shoes when another pair of equal or greater value is purchased at the regular price. What is the percentage off on the two pairs?

13. A coupon for A-1 Steak Sauce is good for 50¢ off on a bottle of any size. The choices are a 10-oz bottle for $3.29 and a 15-oz bottle for $4.49.
   a. With a coupon, which is the better buy?
   b. With double coupons, which is the better buy?

14. a. Find the simple interest for 8 months on $250 at 5.15% per year.
   b. Find the amount to be repaid after 8 months.

15. a. How much must a person invest at 0.01411% per day simple interest for 14 days in order to earn the price of a 32¢ stamp?
   b. What daily rate of simple interest must a person earn for it to be worthwhile to pay 32¢ to mail a $150 investment to be invested for 14 days?
   c. How many days must a $150 investment be deposited at 0.01411% per day simple interest in order to earn 32¢ or more?

16. The monthly periodic rate on a credit card is 1.40%. What is the annual percentage rate?

17. The annual percentage rate on a credit card is 19.15%.
   a. What is the monthly periodic rate? State your answer as a percentage, rounded to three decimal places.
   b. What is the daily periodic rate? State your answer as a percentage, rounded to five decimal places.

18. a. The average daily balance on a credit card charging a monthly periodic rate of 1.40% is $235.10. What is the finance charge?
   b. The average daily balance on a credit card charging a daily periodic rate of 0.060081% is $476.63. What is the finance charge for a 29-day billing cycle?

19. Describe the difference between simple interest and compound interest. Use an example to illustrate.

20. In developing the formula for the compound amount, we can find that the amount after three periods is $P(1 + r)^2 + P(1 + r)^2 r$. This is then written in the following form: $P(1 + r)^2 ($ _____ $)$. Fill in the missing factor.

21. The formula for the compound amount is $A = P(1 + r)^n$. Find the compound amount for $2000 invested for 3 years at 5.8% compounded quarterly.

22. Compound interest makes money grow exponentially. Sketch a graph to show exponential growth.

23. A student organization wants to set aside money now in order to have $2000 available 1 year from now. It can put money in a money market fund earning 5.75%, compounded monthly. How much should the organization set aside now?

24. What is the effective rate for a money market fund that earns 5.75%, compounded monthly?

25. During the 1980s inflation averaged about 4.7% per year. If an item cost $100 in 1980, what would it be expected to cost in 1990?

26. If money were invested at 6%, about how long would it take to double?

27. If you wanted an investment to double in 5 years, what rate of interest would you need to seek?

28. Describe what an annuity is.

29. What is the difference, if any, between an ordinary annuity and an annuity due?

30. The formula for the future value of an ordinary annuity is

$$S = R \left[\frac{(1 + r)^n - 1}{r} \right]$$

where R is the payment made each period, n is the number of periods, and r is the interest rate per period as a decimal or proportion. If a person puts $100 at the end of each month in a money market fund that pays 5.8%, compounded monthly, how much will the person have at the end of 3 years?

31. Suppose the investor in Review Exercise 30 moves the deposits up a month by making each of 36 deposits at the beginning of each month. Under this scheme how much will the person have at the end of 3 years?

32. The formula for the present value P of an ordinary annuity of n payments of amount R with interest compounded at a rate of r per period is

$$P = R \left[\frac{1 - (1 + r)^{-n}}{r} \right]$$

If an investor puts $100 at the end of each month into an account paying 6.0%, compounded monthly, what is the present value of the annuity?

33. The formula for the present value of an ordinary annuity is given in Review Exercise 32. Assume you want to borrow $10,000 from a bank at an annual percentage rate (APR) of 7.75% with monthly payments for 4 years.
   a. How large are the monthly payments?
   b. What is the total amount paid on the loan (principal plus interest)?
   c. How much interest will be paid on the loan?

34. A friend borrows $1000 from you. Your friend will repay the loan in three monthly payments starting in 1 month at 7.0% annual interest, compounded monthly. Construct the amortization schedule.

Hall, Arthur P., and Wheat, J. Marc. "Do Credit Cards Need Interest Rate Caps?" *Consumers' Research,* February 1992, pp. 28–31.

Havighurst, Clark C., ed. *Consumer Credit Reform.* Dobbs Ferry, NY: Oceana Publications 1970.

Homer, Sidney. *A History of Interest Rates.* New Brunswick, NJ: Rutgers University Press, 1963.

Krafft, Susan. "Discounts Drive Women to Clothes," *American Demographics,* July 1991, v. 13, n. 7, p. 11.

Pearlberg, Dan. *An Analysis and History of Inflation.* Wesport, CT: Praeger, 1993.

Shepherdson, Nancy. "Credit Card America," *American Heritage,* November 1991, pp. 125–132.

U.S. News & World Report. "How Do They Charge Thee? Let Us Count the Ways," June, 1992, p. 66.

CHAPTER 11

DISCRETE MATHEMATICS

PROLOGUE

One of the great achievements of humanity was the development of calculus in the 1670s by Sir Isaac Newton and Gottfried Wilhelm Leibniz. A key idea in calculus is the idea of continuity. Intuitively, we can say that the graph of a function is continuous if we can draw it without lifting our pencil off the paper. Think of the set of all real numbers on the number line. They form a continuous set; There are no holes or gaps. By comparison, the set of integers, {. . . , −3, −2, −1, 0, 1, 2, 3, . . .}, is not a continuous set; it has (big) holes between any two numbers in the set. Mathematics advanced significantly with ideas related to continuity.

Discrete mathematics is in some sense the antithesis of continuous mathematics. The set of integers is a discrete set. Any finite set is a discrete set. Discrete sets have the property that we can count the elements in the set—we can go from one object to the next. Discrete mathematics has been

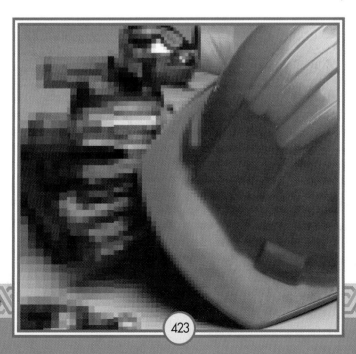

around since the beginning of mathematics. Counting 1, 2, 3, and so on is an activity in discrete mathematics. Although topics in discrete mathematics have been studied throughout time, the advent of the computer has prompted the growth of discrete mathematics. Computers can be made to count an enormous number of objects and can perform an enormous number of arithmetic and logical operations in a short amount of time.

The finest television pictures today are produced on digital TVs. The screen is broken into a large number of cells, each with a number associated to tell brightness, color, and other characteristics of the picture in that cell. The finest musical recording is done with digital equip-

ment. Even more, compact discs have error-correcting capability that enables them, even if scratched with a knife, to reproduce the sound originally recorded, and the scratched CD plays with the quality of the original. The error-correcting codes are also a part of discrete mathematics. They are advanced forms of the error-detecting codes discussed in Chapter 2.

Isn't it ironic that a picture, which is a continuous visual image, can today be reproduced best as a collection of discrete images. Similarly, sound, which we think of as a continuous phenomenon, can be reproduced best as a collection of discrete sounds. We live in a world of continuous production and discrete reproduction.

Chapter opening photo: A collection of discrete images blend to give a continuous-appearing image.

11.1 INTRODUCTION

GOALS
1. To describe the nature of discrete mathematics.
2. To describe what has prompted today's focus on discrete mathematics.

Discrete mathematics is a part of mathematics, just like geometry and logic are parts of mathematics. *Discrete mathematics* deals with sets of objects that can be counted or processes that consist of a sequence of individual steps. This contrasts with *continuous mathematics,* which deals with infinite sets containing an interval and infinite processes other than those composed of a sequence of individual steps. Continuous mathematics includes calculus.

Your previous experience with mathematics includes experience with both discrete and continuous mathematics. You learned to count 1, 2, 3, 4, and so on, which deals with discrete mathematics. You draw a line segment ●——●, which has an infinite number of points on it and contains an interval; it is in the province of continuous mathematics. In algebra, the equation $F = (9/5)C + 32$ for converting a Celsius temperature to Fahrenheit can be regarded as in the province of discrete mathematics if only whole-number temperatures ($0°$, $1°$, $2°$, etc.) are involved or in the province of continuous mathematics if any temperature on a nondigital thermometer is possible.

Section 9.4 contains material on the distinction between discrete and continuous random variables. The same distinction applies more broadly between discrete and continuous mathematics.

Several topics covered earlier in this book fall in the province of discrete mathematics:

Check digits (Section 2.5)	Sets (Chapter 3)
Logic (Chapter 4)	Counting (Section 8.3)

Discrete mathematics has been around for a long time. Since counting falls in the province of discrete mathematics, many of the first mathematical concepts were part of what we now call discrete mathematics. Discrete mathematics has become more prominent in recent years because of developments with computers, which are machines

Times can be regarded as part of continuous mathematics or as part of discrete mathematics.

with a finite number of states corresponding to switches that are open or closed. Computers make it possible to attack problems involving a large number of calculations. Consequently, discrete mathematics has evolved as the computer has developed.

In addition to the problems in the sections and chapters just mentioned, we now present one other problem that's in the province of discrete mathematics.

Consider a convenience store whose layout is as shown in ▶ Figure 11.1. Is it possible to enter the store at the door and travel in each aisle exactly once (once and only once) and leave by the door? Give some thought to this. Either find such a path or provide an argument why none is possible.

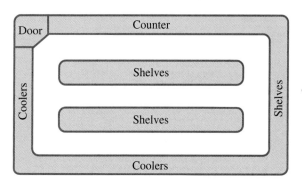

The blank areas denote aisles.

▶ FIGURE 11.1 Convenience store layout

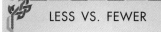

WRITTEN ASSIGNMENTS 11.1

1. Earlier we said that since counting falls in the province of discrete mathematics, many of the first mathematical concepts were part of what we now call discrete mathematics. We didn't say that *all* of the first mathematical concepts involved discrete mathematics, because some of the first concepts were geometric—such as lines and line segments—and involved continuity. Which do you believe came first—the discrete mathematics of counting or the notion of a line or curve that embodies the concept of continuity? Argue your point based on research regarding primitive peoples or research or observations on how infants learn.

2. If your college or university has one or more courses in discrete mathematics, look at a catalog description, a syllabus, or a textbook's table of contents, and describe what topics from these are included in this textbook.

3. Describe the difference between *discrete* and *discreet*. Assume your audience is students in English 101.

4. Answer the question posed earlier about the convenience store layout (Figure 11.1).

11.2 • EULER CIRCUITS

GOALS
1. To answer the convenience store question of the previous section and to explain the answer.
2. To use the concept of degree of a vertex of a graph and the theorem that every vertex on a graph with an Euler circuit has even degree.
3. To determine whether or not a graph is connected.
4. To work with Euler circuits in graphs and application problems.

The example problem in Section 11.1 about the convenience store layout can be formulated in several different ways. ▶ Figure 11.2 presents a diagram of the convenience store with the labels removed.

1. Suppose the diagram represents a small subdivision, with the blank areas showing streets. Can a police car patrol the subdivision by traveling exactly once on each street of the subdivision?
2. Suppose the diagram represents a floor in a classroom building, with the blank areas showing hallways. Is it possible to deliver a message to each classroom by going down each hallway exactly once?
3. Suppose the diagram is of a network of water pipes, with the blank areas being pipes. Is it possible to run your hand over each pipe exactly once without lifting your hand from a pipe and without going over some pipe a second time?

These three situations and the convenience store layout example of Section 11.1 can all be modeled using the diagram of Figure 11.2.

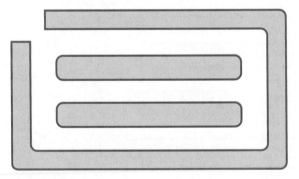

▶ FIGURE 11.2 Unlabled layout

Now let's picture the situation with the even less detailed diagram in ▶ Figure 11.3.

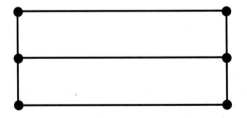

▶ FIGURE 11.3 Model of Figure 11.2

This diagram is an example of a *graph*. Each dot is called a *vertex* (plural *vertices*), and each line segment is called an *edge*. The graph of Figure 11.3 can be related to the diagram of Figure 11.2 as shown in ▶ Figure 11.4.

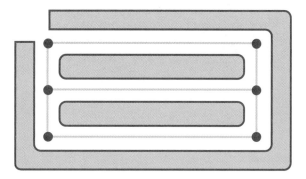

▶ FIGURE 11.4 Using a graph

In the problems illustrated with Figure 11.3 the goal is to start at a vertex, travel on each edge exactly once, and return to the starting vertex. Such a path is called an *Euler circuit*, in honor of the famous Swiss mathematician Leonhard Euler (1701–1783). (His last name is pronounced "oiler." A historical sketch of Euler is found in Section 3.4.) The question we've been asking can now be rephrased: Does the graph of Figure 11.3 have an Euler circuit?

We hope you have formulated an answer to this question after reading Section 11.1. Here's one argument: Label the vertices of the graph *A, B, C, D, E,* and *F,* as shown in ▶ Figure 11.5. Suppose we attempt to start an Euler circuit at vertex *A*. We must first go to either vertex *B* or vertex *F*. Let's say we go to *F* first. From *F* we must go to *C* or *E*. With the first option (*AF* then *FC*), we must eventually traverse edge *EF* in the direction from *E* to *F*. When (eventually) this is done, there's no way to get back to *A* without using an edge a second time. With the second option (*AF* then *FE*) a similar problem arises. We must eventually traverse edge *CF* in the direction from *C* to *F*. When (eventually) this is done, there's no way to get back to *A* without using an edge a second time. Therefore, if we go to *F* first, we cannot complete a circuit back to *A* without covering some edge twice.

▶ FIGURE 11.5 Labeled graph

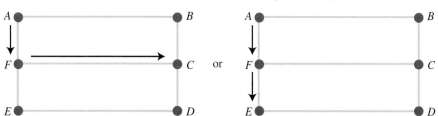

If we don't go to *F* first—i.e., if we start out traversing *AB first*—then we'll have to go through *F* on the way back to *A* and we'll have a similar problem to that just described: we take an edge into *F,* another edge out of *F,* then eventually another edge back into *F,* and we're stuck. (Notice that a similar problem occurs at *C:* we come in on one edge, go out on a second, eventually come back to *C* on a third, then we're stuck.)

We have shown that this graph has no Euler circuit. The answer to the convenience store problem in Section 11.1 (Is it possible to travel in each aisle of the convenience store exactly once and leave by the door?) is no, and the answers to our three reformulations of it are also no.

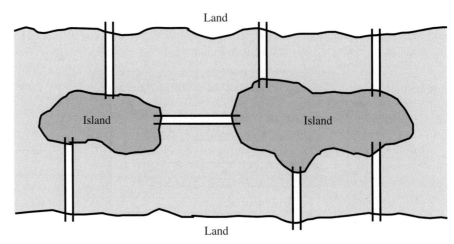

Königsberg Bridge Problem

The problem Euler solved that led to his name being attached to Euler circuits dealt with bridges in the Prussian city of Königsberg and is called the *Königsberg bridge problem*. (Königsberg, now called Kaliningrad, is located in a part of Russia between Lithuania and Poland.) The story is that the residents of Königsberg liked to take a Sunday afternoon stroll around a part of the city that included two islands in the Pregel River. The islands and mainland were connected by seven bridges, as shown in ▶ Figure 11.6. The efficient stroller wonders if there is a way to take a stroll through the city that crosses each bridge once and only once.

Euler answered the question by modeling the situation with a graph. Since the main focus is on the bridges and since the size of the islands and land masses is immaterial, the graph will have the land masses as vertices and the bridges as edges. A graph of Königsberg is shown in ▶ Figure 11.7.

Use the argument stated earlier or an argument like you used in Section 11.1 to answer the Königsberg bridge problem. Before we present an answer, we will examine in general when a graph has an Euler circuit.

When Does a Graph Have an Euler Circuit?

Suppose a graph has an Euler circuit. Let *V* be a vertex of the graph that is on an edge. Consider edges that come into or go out from *V*. We consider two cases according to whether or not *V* is the starting vertex of an Euler circuit.

Case 1 Suppose *V* is a starting vertex of an Euler circuit. Then we start on the circuit by some edge going out from *V*. To complete the circuit there must be another edge coming into *V*. Thus, there are at least two edges on *V*.

If there is a third edge on *V*, it will be used to go out from *V*. To complete the circuit there must be a fourth edge coming back to *V*.

If there is a fifth edge on *V*, it will be used to go out from *V*. To complete the circuit there must be a sixth edge coming back to *V*.

In general, for each edge going out from *V* there must be an edge coming into *V*. Therefore, the edges on *V* came in pairs; i.e., there is an even number of edges on *V*.

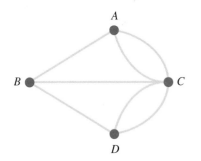

▶ FIGURE 11.7 Graph of Königsberg

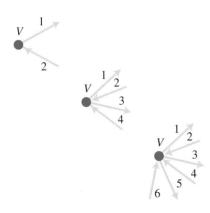

Case 2 Suppose *V* is not a starting vertex of an Euler circuit but is on an edge. Since *V* is on an edge, the Euler circuit must use an edge, call it 1, coming into *V* and a second edge, call it 2, to leave *V*. Therefore, there are two edges on *V*.

If there is a third edge on *V*, it will be used to come into *V*. Since the circuit does not end at *V* (an Euler circuit starts and ends at the same vertex and *V* is not a starting vertex), there must be a fourth edge going out from *V*.

In general, for each edge coming into *V* there must be an edge going out from *V*. Therefore, the edges on *V* come in pairs; i.e., there is an even number of edges on *V*.

In either case 1 or case 2, if V is a vertex on an edge of a graph with an Euler circuit, then V has an even number of edges on it. In a graph, the number of edges on a vertex is called the *degree* of the vertex. For example, in Figure 11.7, *A* has degree 3, *B* has degree 3, *C* has degree *5,* and *D* has degree 3. We have proved the following:

> ## THEOREM 11.1
>
> Every vertex on a graph with an Euler circuit has an even degree.

YOUR FORMULATION

1. Is the converse of Theorem 11.1 true? In other words, if every vertex of a graph has even degree, must the graph have an Euler circuit? Experiment with a couple of graphs and see what you think.
2. If all your graphs give a yes answer to the question in part 1, the answer to the question still might be either yes or no. However, if some graph in which every vertex has even degree does not have an Euler circuit, then the answer is no. Did you find such a graph? If not, try again to construct one.

In Figure 11.7 there are vertices without even degree. In fact, every vertex has odd degree. This is what Euler observed: It is not possible to take a stroll through Königsberg that crosses each bridge once and only once.

One example of a graph in which every vertex has an even degree but there is no Euler circuit is shown in ► Figure 11.8. Notice that the graph comes in two parts. It is said to be *not connected*. The graph in Figure 11.7 is *connected*. This means that there is a path in the graph from any one vertex to any other vertex. In Figure 11.8 there is no path from *A* to *D,* so the graph is not connected. Intuitively, a graph is not connected if it has two or more distinct pieces; otherwise, it is connected.

Here is one form of the converse of Theorem 11.1 that is true.

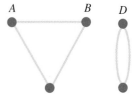

► FIGURE 11.8 A graph where every vertex has even degree and there is no Euler circuit

> ## THEOREM 11.2
>
> If in a connected graph every vertex has an even degree, then the graph has an Euler circuit.

We will omit proof of this theorem.

The combination of Theorems 11.1 and 11.2 tells us exactly when a graph has an Euler circuit. Therefore, a graph has an Euler circuit exactly when both of the following are true.

1. The graph is connected.
2. Every vertex has an even degree.

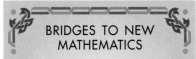

EXAMPLE 11.1

Which of the following graphs has an Euler circuit?

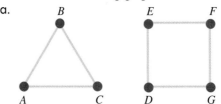

a.

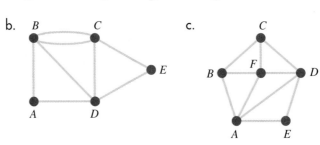

b.   c.

SOLUTION

a. This graph does not have an Euler circuit, since it is not connected.
b. This graph has an Euler circuit, since it is connected and every vertex has even degree.
c. This graph does not have an Euler circuit, since vertices *B* and *C* both have odd degree (3).

When a graph does have an Euler circuit, that circuit can usually be found fairly easily by trial and error. For example, for the graph of Example 11.1b, see if you can write down an Euler circuit starting at *A*. One possibility is *AB, BC, CB, BD, DC, CE, ED, DA*. (We list edges since an Euler circuit is a path that traverses each *edge* exactly once). Another possibility is *AB, BC, CD, DB, BC, CE, ED, DA*. There are other possibilities.

The two conditions telling whether or not a graph has an Euler circuit were applied to solve the Königsberg bridge problem. They can be applied to solve the convenience store problem of Section 11.1. That store layout can be modeled by Figure 11.5. Since vertices *C* and *F* have odd degree (3), there is no Euler circuit. The three reformulations of the convenience store problem show other kinds of problems that can be solved with Theorem 11.1.

EXERCISES 11.2

1. For the graph pictured below, find the degree of each of the four vertices.

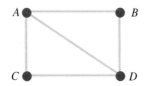

2. For the labeled graph of Figure 11.5, state the degree of each vertex.
3. Construct a graph with four vertices, each with degree 3.
4. Construct a graph with three vertices, each with degree 4.
5. a. Is the graph of Exercise 1 connected?
   b. If you answered yes, state a path from *B* to *C*. If you answered no, state two vertices that have no path connecting them.

In Exercises 6–8, tell whether or not the graph has an Euler circuit. If it does, list one. If not, tell why not.

6.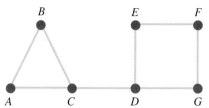

7.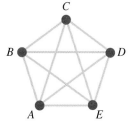

8.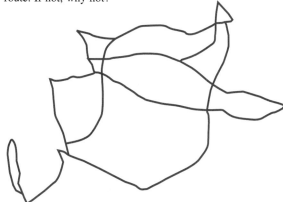

9. The Laurel Highlands is in southwestern Pennsylvania. Assume you were hired by its tourist bureau. As part of your job you would like to drive on each road shown on the accompanying map. You'd like to do this as efficiently as possible. Is it possible to do this by driving on each stretch of road exactly once and returning to where you started? If so, show such a route. If not, why not?

10. An area park has a series of trails for hiking or jogging, as shown here. Is it possible to start at one dot, go on each section of trail exactly once, and return to the starting point? If so, show such a route. If not, why not?

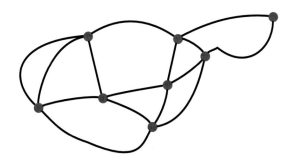

11. Can you trace the star with a pencil without lifting the pencil from the paper or tracing a line more than once? Assume you start and stop at the same point. If you can, show the tracing. If not, why not?

12. Can you trace the "house" shown with a pencil without lifting the pencil from the paper or tracing a line more than once? Assume you start and stop at the same point. If you can, show the tracing. If not, why not?

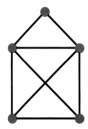

13. The floor design of the first floor of a house is shown. Is there a way to start outside the house or in one of the rooms, make a trip through the house that passes through each doorway exactly once, and return to the starting spot? If so, describe it. If not, why not?

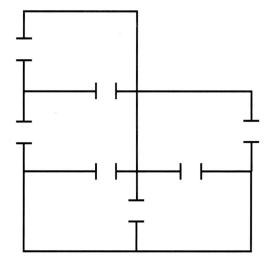

14. An Euler circuit starts and ends at the same vertex. What if we weaken this restriction by permitting a path to start at one vertex and end at another one? Such a path is sometimes called an *Euler trail*—it starts at some vertex, traverses each edge exactly once, and stops at a different vertex.
    a. Make a conjecture, like Theorem 11.1, about when a graph has an Euler trail.

b. Try your conjecture on several graphs, including the one in Exercise 12.

c. Prove that if a graph has an Euler trail, your conjecture is true.

In each of Exercises 15–17 refer to the stated exercise. (a) Tell whether or not (exactly) one edge can be removed and leave a graph with an Euler circuit. If so, find one. If not, why not? (b) Tell whether or not (exactly) two edges can be removed and leave a graph with an Euler circuit. If so, find one. If not, why not?

15. Exercise 1
16. Exercise 6
17. Exercise 8

WRITTEN ASSIGNMENTS 11.2

1. Four different problems have been presented that go with Figure 11.2. Three are listed near the figure, and the other (the convenience store problem) is presented in Section 11.1. Describe another situation that can be modeled with the diagram of Figure 11.2.

2. Modeling a problem with a graph involves mathematical ideas that are not just using numbers and are not Euclidean geometry. List some other mathematical ideas that don't rely heavily on numbers or Euclidean geometry.

11.3 ▸ HAMILTONIAN CYCLES

GOALS
1. To determine whether or not a graph has a Hamiltonian cycle, to find one if it does, and to do applications involving Hamiltonian cycles.
2. To calculate the number of routes you need to examine to attack the traveling salesperson problem (TSP) by the brute force method, to find TSP routes using the nearest-neighbor and sorted-edges algorithms, and to summarize what is known about the traveling salesperson problem.
3. To describe the difference between an exact method and a heuristic method.

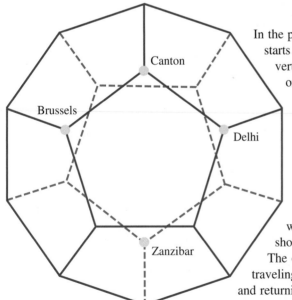

In the previous section we looked at an Euler circuit in a graph—a path that starts at a vertex, traverses each edge exactly once, and returns to the starting vertex. Let's reverse the role of vertices and edges and explore whether or not a graph has a path that visits each vertex exactly once and ends at the original vertex. Such a path is called a *Hamiltonian cycle*, in honor of the Irish mathematician Sir William Rowan Hamilton (1805–1865).

The attachment of Hamilton's name to such a path came from his invention of a puzzle called "Traveler's Dodecahedron" or "A Voyage Round the World," which consisted of a regular dodecahedron (a 3-dimensional solid with 12 faces, each a regular pentagon) in which each of the 20 corners represented an important place in the world. A regular dodecahedron with some of those places labeled is shown in ▸ Figure 11.9.

The objective was to find a route that visited each place exactly once by traveling along the edges of the dodecahedron (not necessarily all of them) and returning to the starting city. Hamilton's puzzle had nails in each corner of the dodecahedron and a string to wrap around the nails. A flattened version of the dodecahedron is shown in ▸ Figure 11.10.

▸ FIGURE 11.9 A regular dodecahedron

Sir William Rowan Hamilton

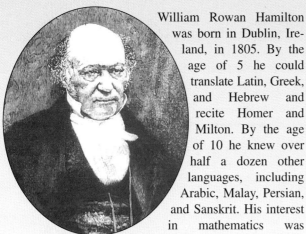

William Rowan Hamilton

William Rowan Hamilton was born in Dublin, Ireland, in 1805. By the age of 5 he could translate Latin, Greek, and Hebrew and recite Homer and Milton. By the age of 10 he knew over half a dozen other languages, including Arabic, Malay, Persian, and Sanskrit. His interest in mathematics was spurred by a meeting in 1820 with an American "calculating boy" who did mental calculations amazingly fast. At Trinity College in Dublin he took highest honors in both mathematics and English verse and earned respect for a paper he wrote on optics. At the age of 22, while still an undergraduate, he was appointed Professor of Astronomy and Royal Astronomer of Ireland.

Hamilton's broad scholarly interests expanded to poetry. He was friends with the English writers Coleridge and Wordsworth. However, Hamilton was not as adept at poetry as at mathematics, and Wordsworth advised him to stick to writing mathematics.

Hamilton's most significant mathematical achievement was developing the system of quaternions. The quaternions can be regarded as an extension of the system of complex numbers. He had been working on a system of triples of real numbers. The story goes that while walking in a park in Dublin he realized he needed quadruples instead of triples. Their basic units he called 1, i, j, k, and he supposedly carved in a bridge the following fundamental relations:

$$i^2 = j^2 = k^2 = ijk = -1$$

A plaque commemorating this can be found in Dublin today. Although quaternions are sometimes used by physicists, their main significance is as an example of an algebraic system in which the multiplication is not commutative—e.g., $ij \neq ji$.

Even though the problem of finding a Hamiltonian cycle is much like the problem of finding an Euler circuit, it seems to be much harder. In fact, no solution as simple and complete as Theorem 11.1 is known. In this book, a trial-and-error approach will be used to determine whether or not a graph has a Hamiltonian cycle.

Can you find a Hamiltonian cycle in the dodecahedron graph of Figure 11.10? One such cycle follows the edges joining the following sequence of vertices: 1, 6, 7, 8, 9, 10, 11, 12, 13, 19, 18, 17, 16, 20, 15, 14, 5, 4, 3, 2, 1.

●EXAMPLE 11.2

Does the graph shown have a Hamiltonian cycle? If so, state one. If not, explain why not.

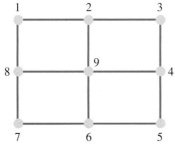

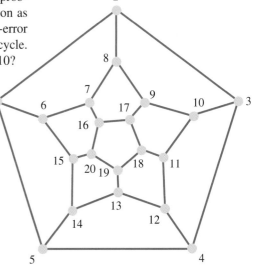

▶ FIGURE 11.10 Flattened dodecahedron

SOLUTION Because of the symmetry of the graph, there are three possible starting points: a corner (1, 3, 5, 7), the midpoint of an outside edge (2, 4, 6, 8), or the center (9). Let's consider each of these in order.

Start at a corner—say, 1. Because of the symmetry it doesn't matter whether we go to 2 or 8 next, so let's go to 2 (1 → 2). From 2 there are two choices: Go to 3 or go to 9. This can be illustrated as in ▶ Figure 11.11. Via this type of illustration, all the possible paths formed by starting at 1 can be listed, as in ▶ Figure 11.12, which also presents reasons why each path is not Hamiltonian. We have shown that there is no Hamiltonian cycle when we start at a corner.

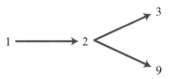

▶ FIGURE 11.11 Starting a Hamiltonian cycle from vertex 1

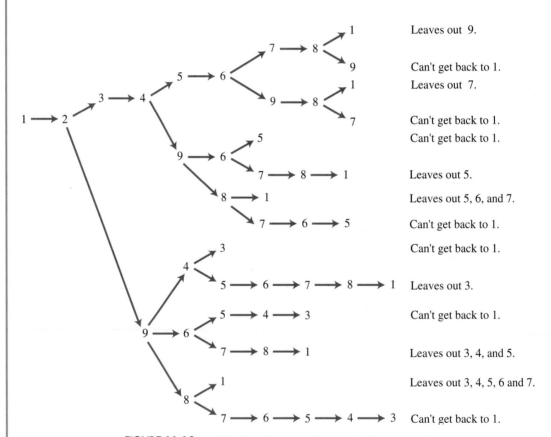

▶ FIGURE 11.12 Possible Hamiltonian cycles starting from vertex 1

Next, suppose we start at the midpoint of an edge. Because of the symmetry, let's start at 2. There are three possible next vertices—1, 3, and 9. Because of the symmetry, 1 and 3 are alike. We'll go to 3. From this start, possible circuits are as displayed in ▶ Figure 11.13. As you can see, none of these routes produces a Hamiltonian cycle.

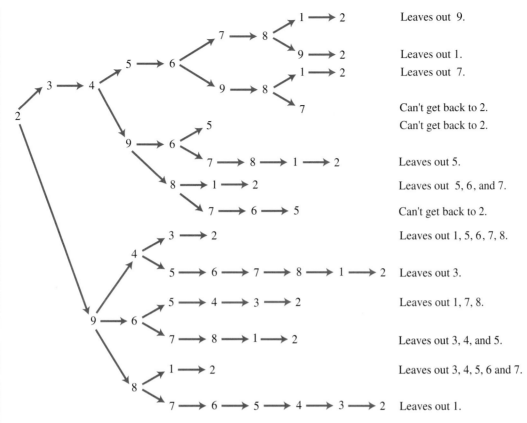

Leaves out 9.

Leaves out 1.
Leaves out 7.

Can't get back to 2.
Can't get back to 2.

Leaves out 5.

Leaves out 5, 6, and 7.

Can't get back to 2.

Leaves out 1, 5, 6, 7, 8.

Leaves out 3.

Leaves out 1, 7, 8.

Leaves out 3, 4, and 5.

Leaves out 3, 4, 5, 6 and 7.

Leaves out 1.

▶ FIGURE 11.13 Possible Hamiltonian cycles starting from vertex 2

Finally, consider starting in the middle at 9. By the symmetry of the graph, going to 2, 4, 6, or 8 produces the same result, so let's go to 2. Again, 1 and 3 are symmetrically placed, so we'll go to 3. This yields the following route:

$$9 \rightarrow 2 \rightarrow 3 \rightarrow 4 \rightarrow 5 \rightarrow 6 \rightarrow 7 \rightarrow 8 \rightarrow 1$$

By this route we can't get back to 9. Therefore, there is no Hamiltonian cycle starting at vertex 9.

This rather tedious approach has exhausted all possible circuits without producing a Hamiltonian cycle. Thus, the graph does not have a Hamiltonian cycle.

Many applications of Hamiltonian circuits are similar to applications of Euler circuits. One application is shown in the next example.

EXAMPLE 11.3

The traffic lights in a small city are shown by dots in the diagram on the next page. The lines represent main streets.

a. After an electrical storm a city official wants to drive through each intersection where there is a street light to see if it is working. Is there a way to do this starting at City Hall, using only main streets, going through each intersection exactly once, and returning to City hall?

b. If the light at 1 is omitted, is there such a route?

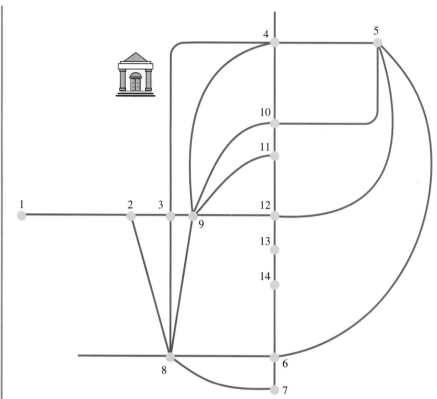

SOLUTION

a. No. Once the official gets to 1 she is stuck.
b.

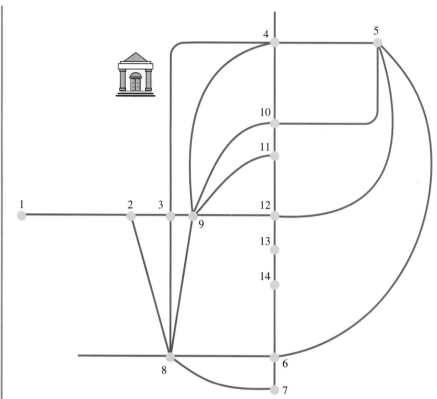

$$\text{🏛} \to 3 \to 2 \to 8 \to 7 \to 6 \nearrow 5 \to (\text{can't get 13 and 14})$$
$$\searrow 14 \to 13 \to 12 \to 9 \to 11 \to 10 \to 5 \to 4 \to \text{🏛}$$

We have found a Hamiltonian circuit.

Traveling Salesperson Problem (TSP)

There are times when a particular Hamiltonian cycle is of interest. One such problem is known as the *traveling salesperson problem* (TSP): A salesperson starts at a city, wants to visit several cities without repeating any, and then return to the starting city. Also, this is to be done in the most efficient way (least distance, least time, least cost, etc.). In other words, we want to find a Hamiltonian cycle that is most efficient.

┌─ **EXAMPLE 11.4**

A college student wants to visit the four western cities shown. Driving distances between them are shown. What is the shortest trip starting in Denver that visits each city exactly once and returns to Denver?

SOLUTION Let's look at all the possible routes that visit each city exactly once. Since a route, such as D → SF → LA → P → D, and its reverse both give the same mileage, we will consider them as the same route.

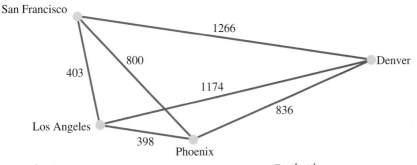

Route	Total Mileage
D → P → LA → SF → D	836 + 398 + 403 + 1266 = 2903
D → P → SF → LA → D	836 + 800 + 403 + 1174 = 3213
D → LA → P → SF → D	1174 + 398 + 800 + 1266 = 3638

The shortest route, at 2903 miles, is Denver to Phoenix to Los Angeles to San Francisco and back to Denver, or its reversal.

YOUR FORMULATION

1. In Example 11.4 there were four cities and we started in Denver. From there, the number of possible routes is:

$$\underline{3} \cdot \underline{2} \cdot \underline{1} = 6$$

— Number of cities that could be visited first after Denver
— Number of of cities that could be visited second after Denver and the first city
— Number of of cities that could be visited third after Denver and the first two cities

Since half the routes are reversals of others, only 6/2 = 3 possible routes need checking.

Apply this same argument to the case where there are five cities. How many distinct routes need to be checked?

2. In general, given n cities to be visited, pick one city as the starting point (call it A). The number of possible routes is then (fill in the blanks):

$$\underline{n-1} \cdot \underline{n-2} \cdot \underline{n-3} \cdots \underline{} = \underline{}$$

— Number of cities that could be visited first after A
— Number of cities that could be visited second after A and the first city
— Number of cities that could be visited third after A and the first two cities
⋮
— Number of cities that could be visited $(n-1)^{\text{th}}$ after A and the first $(n-2)$ cities

However, _____ (What proportion?) of the routes are reversals of the others, so only _____ possible routes need to be checked.

In the traveling salesperson problem, it is usually possible to travel from any one city to any other city. Consequently, there are Hamiltonian cycles that can be created by choosing, at each city, any city not previously visited as the next choice of city to visit. The goal of the TSP is to find an efficient (shortest-length) cycle from among all of the possible Hamiltonian cycles.

Brute Force Method The method of Example 11.4 is a systematic way to attack the traveling salesperson problem: List all the distinct routes and find the least-cost route. We will call this method of attacking the traveling salesperson problem the *brute force method*.

With the brute force method, for how many routes do we need to calculate a cost (in time, distance, or dollars)?

In the Your Formulation, you have just shown the following.

BRUTE FORCE METHOD

To attack the traveling salesperson problem for n cities by the brute force method requires examining $(n-1)!/2$ routes.

In Example 11.4, n was 4, so the number of routes was

$$\frac{3!}{2} = \frac{6}{2} = 3$$

If $n = 5$, the number of routes is

$$\frac{4!}{2} = \frac{24}{2} = 12$$

Notice that to use the brute force method for only five cities requires the calculation of costs for 12 routes. If six cities are involved, the number of routes to check on is

$$\frac{5!}{2} = \frac{120}{2} = 60$$

Somewhere around this point it's easy to lose interest in doing the calculations by hand and to want to involve the computer. How quickly can the computer do the calculations?

Suppose you wanted to visit all the U.S. metropolitan areas with populations over 2 million. As of the 1990 census, there were 21 of these. The number of routes to consider is $20!/2 \approx 1.216 \times 10^{18}$. If a computer generated 1 million of these a second, it would take about 1.216×10^{12} seconds, or over 38,000 years. If the computer's speed increased by a factor of 1000 to calculating the cost of 1 billion routes a second, then the time to calculate the costs of the routes for 21 cities would be reduced to between 38 and 39 years.

The brute force method for attacking the traveling salesperson problem becomes unwieldy as the number of cities increases. As the preceding paragraph shows, with 21 cities even a fast computer cannot solve the problem in a lifetime. Consequently, we need a faster method to solve the traveling salesperson problem.

Unfortunately, no method is currently known that will guarantee a minimum-cost route in a reasonable amount of time for problems of moderate size. This is an unsolved problem in mathematics. What is typically done for a traveling salesperson problem that is too large to do efficiently by the brute force method is to use a method that gives a reasonably good solution but not necessarily a minimum-cost solution—in other words, a method that gives a low-cost solution but not necessarily a lowest-cost solution. Such a method, called a *heuristic method,* makes intuitive sense.

Heuristic Methods: Nearest-Neighbor Algorithm   One heuristic method for the traveling salesperson problem is called the *nearest-neighbor algorithm.* An *algorithm* is a step-by-step method for solving a problem. For example, we are using a division algorithm when we do long division to divide 187 by 11:

$$\begin{array}{r} 17 \\ 11\overline{)187} \\ \underline{11} \\ 77 \\ \underline{77} \\ 0 \end{array}$$

In the nearest-neighbor algorithm, at each step we go to the nearest city that has not yet been visited. For example, consider in Example 11.4, in which a student wants to start in Denver and visit each city (shown here again).

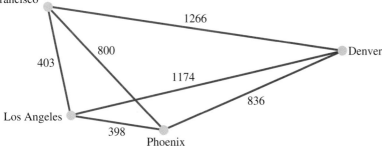

D ➡ P   If the student starts at Denver, the nearest city is Phoenix.

P ➡ LA   The nearest city to Phoenix is Los Angeles.

LA ➡ SF   The nearest city to Los Angeles that has not yet been visited is San Francisco.

SF ➡ D   This completes the trip. Therefore, the nearest-neighbor algorithm produces the route Denver to Phoenix to Los Angeles to San Francisco and back to Denver. This is also the minimum-distance route.

The nearest-neighbor algorithm seems like a good procedure since at each step we pick the lowest-cost alternative. However, it does not always produce a minimum-cost route. The following example illustrates this.

EXAMPLE 11.5

For Example 11.4, use the nearest-neighbor algorithm to find a route starting in Los Angeles.

$$\begin{array}{ccccc} & 398 & 800 & 1266 & 1174 \\ \text{SOLUTION}\quad \text{LA} \longrightarrow & \text{P} \longrightarrow & \text{SF} \longrightarrow & \text{D} \longrightarrow & \text{LA} \end{array}$$

This route has a total mileage of 3638. Notice (see Example 11.4) that this is in fact the longest of the three possible routes.

In the four-city situation of Example 11.4, the nearest-neighbor algorithm sometimes gives the lowest-mileage route and sometimes doesn't. If we start in Denver (see Example 11.4), the nearest-neighbor algorithm gives the lowest-mileage route; but if we start in Los Angeles (see Example 11.5), it does *not*. And if we start in either Phoenix or San Francisco, the nearest-neighbor algorithm again gives the lowest-mileage route. The nearest-neighbor algorithm is an efficient heuristic method to produce traveling salesperson routes, but it does not always give the minimum-cost solution.

Heuristic Methods: Sorted-Edges Algorithm A second heuristic method for designing a route in the traveling salesperson problem is the *sorted-edges algorithm*. Since we want a minimum-mileage route, we first sort the distances (edges of a graph), from the smallest to the largest. In the situation of Example 11.4 they are 398, 403, 800, 836, 1174, and 1266.

Let's start by choosing edges 398 and 403. If we next choose 800, we form a loop without including Denver. Consequently, we bypass 800.

Next we try edge 836.

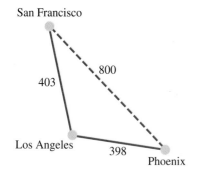

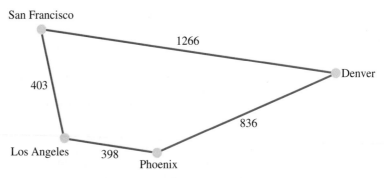

To complete the circuit, we fill in the edge between Denver and San Francisco.

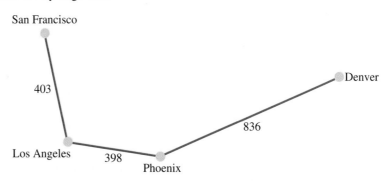

So the cycle starting at Denver is Denver to Phoenix to Los Angeles to San Francisco then back to Denver. Notice (see Example 11.4) that this is the minimum-mileage tour.

The sorted-edges algorithm does not begin by looking at a particular vertex. Instead it considers edges from smallest to largest and uses them as necessary. Two conditions cause an edge to be eliminated from consideration before a cycle is completed:

1. If including the edge creates a loop (cycle) that does not include all the vertices
2. If including the edge puts three edges on some vertex (this would lead to a path that would visit that vertex more than once)

Once a cycle containing all the vertices is found, it is the route suggested. Intuitively (heuristically), it should be a good route, since we've chosen the shortest edges we could to build it. However, it need not be the best route. This is illustrated in the next example.

EXAMPLE 11.6

Pick an efficient route for a traveling salesperson to move among Cincinnati, Cleveland, Indianapolis, and Nashville. Mileages are as shown.

	Ci	Cl	I	N
Cincinnati	—	244	110	275
Cleveland	244	—	318	513
Indianapolis	110	318	—	283
Nashville	275	513	283	—

a. Use the sorted-edges algorithm.
b. Use the nearest-neighbor algorithm starting in Nashville.
c. Use the brute force method.

SOLUTION Let's first construct a graph that models the situation.

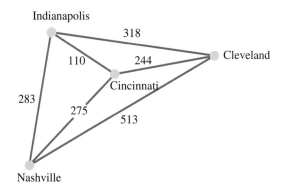

a. The edges, in order, are 110, 244, 275, 283, 318, 513.
   First use 110 and 244. If we use the next edge, 275, this gives three edges at Ci, so we do not use this edge.

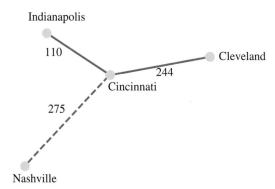

Next we try 283. If we try the next edge, 318, there is a premature loop: I → Ci → Cl → I. Therefore, we do not include 318.

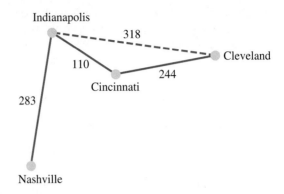

Finally, we include 513. This sorted-edge route is N → I → Ci → Cl → N, with length 1150 miles.

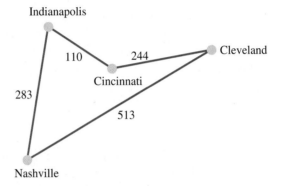

b.
$$N \xrightarrow{275} Ci \xrightarrow{110} I \xrightarrow{318} Cl \xrightarrow{513} N$$

The total mileage is 1216 miles.

c. With four cities there are $3!/2 = 6/2 = 3$ routes. Two of them are examined in parts a and b. The third is

$$N \xrightarrow{283} I \xrightarrow{318} Cl \xrightarrow{244} Ci \xrightarrow{275} N$$

The total mileage is 1120. This is the best route.

Notice in the preceding example that the sorted-edges algorithm and the nearest-neighbor algorithm gave different results. Neither result was the lowest-mileage route. However, the sorted-edges and nearest-neighbor algorithms usually give an efficient route. We just can't be sure the suggested route is the best. In fact, it might be the worst.

The merit of the two heuristic methods we've examined is their speed. In fact, they give a way to attack the TSP in cases where the brute force method takes too long. To see how this works, consider the following example, with 10 cities taken from the upper left corner of a mileage chart.

EXAMPLE 11.7

Find an efficient route for visiting the following 10 cities.

	Alba	Albu	At	Ba	Bil	Bir	Boi	Bos	Buf	C
Albany, NY	—	2041	1010	339	2098	1071	2518	166	301	2434
Albuquerque, NM	2041	—	1404	1890	991	1254	940	2220	1773	1554
Atlanta, GA	1010	1404	—	654	1799	153	2223	1108	907	2369
Baltimore, MD	339	1890	654	—	1916	771	2406	427	401	2309
Billings, MT	2098	991	1799	1916	—	1775	606	2197	1755	563
Birmingham, AL	1071	1254	153	771	1775	—	2065	1226	902	2256
Boise, ID	2518	940	2223	2406	606	2065	—	2685	2214	794
Boston, MA	166	2220	1108	427	2197	1226	2685	—	465	2636
Buffalo, NY	301	1773	907	401	1755	902	2214	465	—	2148
Calgary, AB	2434	1554	2369	2309	563	2256	794	2636	2148	—

SOLUTION First we use the nearest-neighbor algorithm starting from Albany.

$$\text{Alba} \xrightarrow{166} \text{Bos} \xrightarrow{427} \text{Ba} \xrightarrow{401} \text{Buf} \xrightarrow{902} \text{Bir} \xrightarrow{}$$

$$\text{At} \xrightarrow{1404} \text{Albu} \xrightarrow{940} \text{Boi} \xrightarrow{606} \text{Bil} \xrightarrow{563} \text{C} \xrightarrow{2434} \text{Alba}$$

The total mileage is 7996 miles. We can see fairly quickly that the nearest-neighbor algorithm starting at any other city would produce a different route.

Now we apply the sorted-edges algorithm. The edges, from smallest to largest, are 153, 166, 301, 339, 401, 427, 465, 563, 606, 654, 771, 794, 902, 907, 940, 991, 1010, 1071, 1108, 1226, 1254, 1404, 1554, 1755, 1773, 1775, 1799, 1890, 1916, 2041, 2065, 2098, 2148, 2197, 2214, 2220, 2223, 2256, 2309, 2369, 2406, 2434, 2518, 2636, 2685. We insert edges one at a time, starting with the smallest, to produce the following diagram. When an edge is omitted, the description of why appears in the accompanying list.

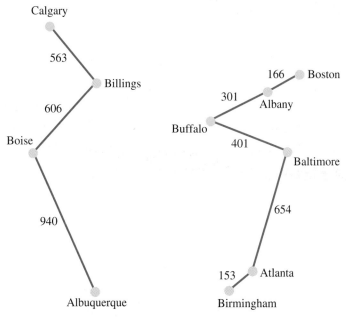

339 is omitted—it would be a third edge at Albany.

427 is omitted—it would prematurely complete a loop

465 is omitted—it would be a third edge at Buffalo and would prematurely complete a loop.

771 is omitted—it would be a third edge at Baltimore and would prematurely close a loop.

794 is omitted—it would prematurely close a loop.

902 is omitted—it would prematurely close a loop.

907 is omitted—it would be a third edge at Buffalo and would prematurely close a loop.

Notice that at this point all 10 cities are on the diagram. To complete it we need to hook up Calgary with Birmingham or Boston and we need to hook up Albuquerque with Birmingham or Boston. The distances are, respectively, 2256, 2636, 1254, 2220. The sorted-edges algorithm chooses the smallest of these distances, 1254, which hooks up Albuquerque with Birmingham. (You might want to pencil this in on the diagram.) To complete the cycle, we connect Calgary and Boston. The tour described from Albany is as follows.

$$\text{Alba} \xrightarrow{301} \text{Buf} \xrightarrow{401} \text{Ba} \xrightarrow{654} \text{At} \xrightarrow{153} \text{Bir} \xrightarrow{1254}$$

$$\text{Albu} \xrightarrow{940} \text{Boi} \xrightarrow{606} \text{Bil} \xrightarrow{563} \text{C} \xrightarrow{2636} \text{Bos} \xrightarrow{166} \text{Alba}$$

The total mileage is 7674. Notice that this is shorter than the cycle given by the nearest-neighbor algorithm starting from Albany. The brute force method would require looking at $9!/2 = 181{,}440$ cycles. We only have 181,438 more to examine.

Other Heuristic Algorithms The traveling salesperson problem has been studied seriously since the 1930s (Lawler et al., 1985, p. 5). However, some related problems have origins dating back over 200 years. An 1832 German book, "Von einem alten Commis-Voyageur" (by a veteran traveling salesman) stated in the last chapter:

> By a proper choice and scheduling of the tour, one can often gain so much time that we have to make some suggestions. . . . The most important aspect is to cover as many locations as possible without visiting a location twice.

Other places where the TSP arises (besides visiting cities) are in scheduling for a variety of chemical manufacturing facilities, X-ray crystallography (up to 14,000 "cities"), circuit board assembly, and the study of protein conformations.

One of the most widely known heuristic algorithms is by Shen Lin and Brian Kernahan of Bell Labs. This algorithm gives results that are often within a few percent of optimal in a relatively short amount of time. For example, on a 532-city problem the Lin–Kernahan solution was within 1.7% of the optimal solution, and it was calculated in 7% of the time taken for the optimal solution (which was reached by a method other than the brute force technique). An algorithm that gives an exact solution to the TSP for certain classes of problems with up to 5000 cities has been developed by Donald Miller of Du Pont and Joseph Pedny of Purdue University. An algorithm developed by David Johnson and Jon Bentley of AT&T Bell Laboratories in New Jersey and Lyle McGeach of Amherst College can compute, in under 3 hours, a solution that is within 2% of the minimum-cost solution to a TSP with a million cities. Work continues on TSP algorithms that will give either exact solutions to larger classes of problems or, in a reasonable time, approximate solutions with very good accuracy that improve either in speed or accuracy on known results.

A significant part of discrete mathematics involves finding algorithms to solve problems. Sometimes the algorithm gives an exact result and sometimes the solution is only approximated by means of a heuristic algorithm.

In Exercises 1–6, tell whether or not the graph has a Hamiltonian cycle. If it has one, describe it.

1.

2.
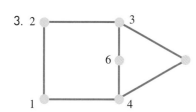

3.
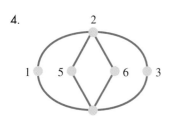

4.
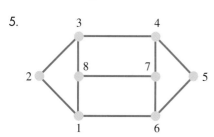

5.

6.
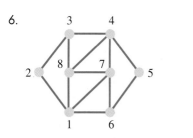

7. Some cities joined by interstate highways in Illinois are shown. Is there a way to start in some city, visit each city exactly once, and return to the starting city by traveling only on the interstate highways shown? If so, state a route. If not, why?

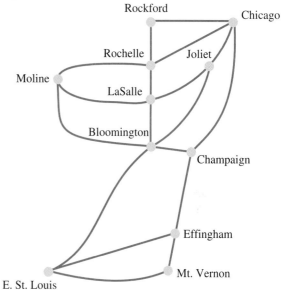

8. In this graph presenting South America, dots represent countries and an edge is drawn between two dots if the countries share a border. Is it possible to start in one of these countries, visit each country exactly once, and return to the starting country by always going from one country to an adjacent one? If so, state the route. If not, tell why not.

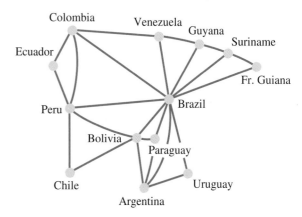

9. Major league baseball's American League and National League each have 14 teams. If you wanted to start in one major league city, visit the other 13 cities in that league exactly once, and return to the starting point, how many different routes would you need to check?

11.3 HAMILTONIAN CYCLES

445

10. If the brute force method is used to calculate the number of traveling salesperson problem tours for 13 cities on a computer that can calculate the cost of 1 million tours per second, how long will the calculation take?

11. Does the total mileage in the best route for a traveling salesperson problem (as determined by brute force) depend on the city where the salesperson starts?

12. Does the total mileage in a route for the traveling salesperson problem determined by the nearest-neighbor algorithm depend on the city where the salesperson starts?

In Exercises 13 and 14, find a low-mileage route through the cities mentioned by means of (a) the nearest-neighbor algorithm starting at the first city mentioned alphabetically, (b) the sorted-edges algorithm, and (c) the brute force method.

13.

	B	D	H	St. L
Birmingham, AL	—	653	692	508
Dallas, TX	653	—	242	638
Houston, TX	692	242	—	799
St. Louis, MO	508	638	799	—

14.

	A	Bos	Buf	P
Albany, NY	—	166	301	237
Boston, MA	166	—	465	115
Buffalo, NY	301	465	—	538
Portland, ME	237	115	538	—

15. For the 10 cities of Example 11.7 find the nearest-neighbor circuit starting at Boise.

16. Find an efficient route for visiting the 10 cities of Example 11.7 plus Minneapolis, MN. Mileage from Minneapolis to the other cities is as follows.
   a. Use the nearest-neighbor algorithm starting at Albany.
   b. Use the sorted-edges algorithm.

	Alba	Albu	At	Ba	Bil
M	1250	1219	1121	1113	826

	Bir	Boi	Bos	Buf	C
M	1068	1476	1390	949	1244

   c. With the brute force method, how many cycles need to be considered?

17. Pick any four or more cities for which you know mileages between each pair of them.
   a. Find an efficient route to visit all of these cities. Use each of the three methods of this section: nearest-neighbor algorithm, sorted-edges algorithm, and brute force method.
   b. Did the heuristic methods produce the shortest route? What was the overall result in the class on this question?

For Exercises 18 and 19, apply the technique described in the margin note "The TSP and Hamiltonian Cycles" to the graph shown. Do each of the following steps.
   a. Label the given edges with a weight of 0. Add edges so that each pair of vertices is joined by an edge. What edges have you added?
   b. Label the added edges with a weight of 1. The new graph has three TSP cycles. List them and find the total weight of each.
   c. Does any cycle have minimum total weight 0?
   d. What does your answer to part c say about the existence of a Hamiltonian cycle in the original graph?

18. 19.

1. The traveling salesperson problem as we've described it involves a salesperson traveling to a number of cities at minimum cost. An alternate formulation might ask for the quickest route in the manufacture of a circuit board for a laser to drill holes. Write two other formulations of the traveling salesperson problem.

2. Look up the word *heuristic* in the dictionary. Describe a situation, other than the traveling salesperson problem, in which a heuristic method might be used to "solve" it. Describe what the heuristic method is in this situation.

3. Summarize what is known about solving the traveling salesperson problem. Assume the reader has read this chapter and knows what the TSP problem is and what the different solution methods are.

4. (*For chess fans*) Look up and report on a knight's tour on a chessboard.

5. The margin note "TSP Competition" says that the record is a 3038-city problem. Has that record been beaten? (Check your periodical database under "traveling salesman problem.") If so, give a reference to your article and write a summary of it telling who, how many cities, and how long it took.

11.4 THE MARRIAGE PROBLEM

GOALS
1. To determine whether or not a set of marriages is stable and to tell why.
2. To determine how many sets of marriages are possible with a set of *n* men and *n* women.
3. To follow the proof of the stable marriage theorem.
4. To use the deferred acceptance algorithm and to describe other situations to which the deferred acceptance algorithm applies.

"Matchmaker, Matchmaker, make me a match. Find me a find. Catch me a catch."

—From *Fiddler on the Roof*

There are a variety of situations that involve matching people or organizations with each other. One such instance is usually called *the marriage problem*.

Stable Marriages

Suppose there are *n* men and *n* women. Each person ranks the *n* people of the opposite sex according to their attractiveness as a potential spouse. Is there a way to pair the men and women so that the set of marriages will be stable? A set of marriages will be defined as unstable if there are a man and a woman who are not married to each other but who both prefer to be married to each other rather than their current spouses. Otherwise, the set of marriages is defined as stable.

An unstable set of pairings.

EXAMPLE 11.8

There are three women, A, B, and C, and three men X, Y, and Z, who are to be paired in marriages. Their preferences for marriage partners are as follows: A(XZY) (meaning A prefers X to Z and Z to Y), B(XZY), C(ZYX), X(ACB), Y(CAB), Z(BCA).

a. How many possible sets of marriages are there?
b. How many of these, if any, are stable?

Before proceeding to the solution, we will introduce a matrix method for representing the preferences. A *matrix* is a rectangular array of objects, such as

	X	Y	Z
A	(1,1)	(3,2)	(2,3)
B	(1,3)	(3,3)	(2,1)
C	(3,2)	(2,1)	(1,2)

The first entry in each pair of a row (rows are horizontal) is the position the woman at the left of the row rates the corresponding man. For example, A rates X first, Y third, and Z second. The second entry in each pair of a column (columns are vertical) is the position the man at the top of the column rates the corresponding woman. For example, Y rates A second, B third, and C first.

SOLUTION

a. Consider the number of pairings for the three women, A, B, and C.

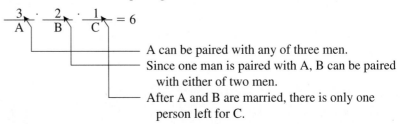

$$\frac{3}{A} \cdot \frac{2}{B} \cdot \frac{1}{C} = 6$$

— A can be paired with any of three men.
— Since one man is paired with A, B can be paired with either of two men.
— After A and B are married, there is only one person left for C.

Thus, six possible sets of marriages exist.

b. We'll look at each of the six possible sets of marriages and pick the stable ones.

i. First, we consider AX, BY, CZ. This notation means A and X are married, B and Y are married, and C and Z are married. We circle the corresponding pairs in the matrix:

	X	Y	Z
A	⟨(1,1)⟩	(3,2)	(2,3)
B	(1,3)	⟨(3,3)⟩	(2,1)
C	(3,2)	(2,1)	⟨(1,2)⟩

The pair BY—(3,3)—is problematic. Both B and Y are married to their last choice of partner. Notice that in the B row the pair (2,1) has a smaller entry in the first component than the pair (3,3), so B prefers Z to Y. How does Z feel? He rates B as his first choice, so he prefers B to his spouse, C. Therefore, B and Z are not married to each other but would both prefer to be married to each other rather than to their current spouse, and so this set of marriages is unstable.

ii. Now we consider AX, BZ, CY. Again, circle the corresponding pairs.

	X	Y	Z
A	(1,1)	(3,2)	(2,3)
B	(1,3)	(3,3)	(2,1)
C	(3,2)	(2,1)	(1,2)

(In this table: A–X pair (1,1) circled, B–Z pair (2,1) circled, C–Y pair (2,1) circled.)

Is this set of marriages stable? Notice that each man is married to his first choice, so no man would prefer to be married to anyone other than his current spouse. Even though some women would prefer to be married to other men, there is no man–woman pair who would prefer to be married to each other rather than to their current spouse. Therefore, this set of marriages is stable.

iii. Next we consider AY, BX, CZ:

	X	Y	Z
A	(1,1)	(3,2)	(2,3)
B	(1,3)	(3,3)	(2,1)
C	(3,2)	(2,1)	(1,2)

(In this table: A–Y pair (3,2) circled, B–X pair (1,3) circled, C–Z pair (1,2) circled.)

In the A row, since A is married to her last choice, the AY marriage is a likely candidate for instability. Consider A and X. They are not married to each other but would both prefer to be married to each other than to their current spouse. Consequently, this set of marriages is unstable. This leaves the following three possible sets of marriages.

iv. AY, BZ, CX
v. AZ, BX, CY
vi. AZ, BY, CX

Exercise 1 asks you to show that each of these sets of marriages is unstable.

In Example 11.8 there is exactly one set of stable marriages. This need not always happen, as the following example shows.

EXAMPLE 11.9

The marriage preferences for three women, A, B, and C, and three men, X, Y, and Z, are shown here. How many of the six possible sets of marriages are stable?

A: YZX   X: ABC
B: ZXY   Y: BCA
C: XYZ   Z: CAB

	X	Y	Z
A	(2, 1)	(3, 1)	(1, 2)
B	(1, 2)	(2, 2)	(3, 1)
C	(3, 3)	(1, 3)	(2, 3)

SOLUTION The same six sets of marriages as in Example 11.8 are possible. We will consider them in a different order.

AY, BZ, CX. This is stable, since each woman has her first choice of partner. So no woman would prefer to move to another man.

AX, BY, CZ. This is stable, since each man has his first choice of partner.
AZ, BX, CY.

	X	Y	Z
A	(3,1)	(1,3)	(2,2)
B	(2,2)	(3,1)	(1,3)
C	(1,3)	(2,2)	(3,1)

The only man A would prefer to Z is Y, but Y is happier with C than with A. The only man B would prefer to X is Z, but Z is happier with A then with B. The only man C would prefer to Y is X, but X is happier with B than with C. Therefore, this set of marriages is stable. Notice that in this set of marriages each partner gets her or his second choice of mate.

AX, BZ, CY; AY, BX, CZ; AZ, BY, CX. Each of these is unstable. (Exercise 2 asks for the details.) Therefore, three of the six possible sets of marriages are stable.

In the two marriage problem examples we have examined, one has exactly one stable set of marriages while the other has several stable sets of marriages. This raises two questions: (1) Does every marriage problem with n men and n women have a stable set of marriages? (2) If a marriage problem has more than one stable set of marriages, which one is best?

Deferred Acceptance Algorithm

The answer to the first question is yes. A technique for finding a stable set of marriages known as the *deferred acceptance algorithm* was first described by David Gale and Lloyd Shapley in 1962 in the proof of the following theorem.

> ### THEOREM 11.3
> There is always a stable set of marriages.

Proof of Theorem 11.3 Given a set of n men and n women, we describe a procedure (an algorithm) for arranging a set of marriages in which each member of one sex proposes to that person's most preferred member of the opposite sex. We will first follow the traditional pattern, where men propose. Then we will see what happens if women do the proposing. Is there a difference between these two stable sets of marriages?

Each man proposes to his more preferred woman. If every woman receives a proposal (this is equivalent to supposing that no two men propose to the same woman), then the process concludes. This set of marriages is stable, since each man has his first choice of partner and no man would prefer to be married to someone else. (This would happen In Example 11.8.b.ii: X proposes to A, Y proposes to C, and Z proposes to B.)

If the process does not conclude in one step, then there are some women who have not received a proposal and some women who have more than one proposal.

The women who have more than one proposal choose their favorite proposer, put that man on hold (maybe someone better will come along), and reject the other proposers. The women who have only one proposer put that man on hold. The women who have no proposals wait patiently. To help you follow the proof, we will illustrate the steps of the proof with the following example.

EXAMPLE 11.10

The marriage preferences for three men, X, Y, and Z, and three women, A, B, and C, are shown here.

X: ABC   A: ZXY
Y: ABC   B: XYZ
Z: BAC   C: YZX

	X	Y	Z
A	(2, 1)	(3, 1)	(1, 2)
B	(1, 2)	(2, 2)	(3, 1)
C	(3, 3)	(1, 3)	(2, 3)

SOLUTION We will set up the following chart to keep track of the process.

	Proposals by: X Y Z	A B C holds	A B C rejects
1.	A A B	X Z —	Y — —

Proof of Theorem 11.3 (cont.) In the second step any men who were rejected in the first step propose to their second choices. Proposals on hold continue as proposals. The women who have more than one proposal again choose their favorite proposer, put that man on hold, and reject the other proposers. The women who have only one proposal put or keep the proposer on hold. The women who have no proposals continue to wait.

EXAMPLE 11.10 (cont.)

We repeat the chart, with a second step added.

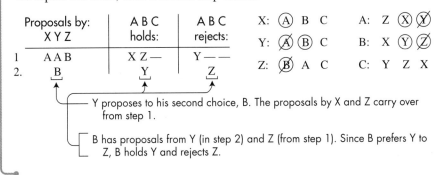

Proof of Theorem 11.3 (cont.) If at the end of the second step, every woman has received a proposal, then the process concludes. Notice that when this happens each woman is holding a proposal and no man is rejected at this step. However, if at the end of the second step some woman has not received a proposal, then some man's proposal has been rejected and the process continues to a third step.

EXAMPLE 11.10 (cont.)

We complete the process by repeating the chart, but with additional steps.

Proposals by: X Y Z	A B C holds:	A B C rejects:
1. A A B	X Z —	Y — —
2. B	Y	Z
3. A	Z	X
4. B	X	Y
5. C	Y	

X: Ⓧ Ⓑ C   A: Ⓩ Ⓧ Ⓩ̸
Y: Ⓐ Ⓑ̸ Ⓒ   B: Ⓧ̸ Ⓨ̸ Ⓩ̸
Z: Ⓑ̸ Ⓐ C   C: Ⓨ Z X

The process is now complete, since each woman is holding a proposal. The set of marriages is AZ, BX, CY. It is stable since each woman has her first choice of partner. (Notice that even though the men did the proposing, the women in this case ended up with their first choice of partner while the men did not.)

Proof of Theorem 11.3 (conclusion)   At the third step we repeat the process of the second step: Men who were rejected in the previous step propose to their next choices and proposals on hold continue as proposals. The steps continue to be repeated until every woman has received a proposal. This must happen at some step in the process. (The argument for this is asked for in Exercise 5.) At this point each woman is holding a proposal from a man, which she now accepts. This produces the set of marriages.

Is this set of marriages stable? If not, then there is a man, call him Ted, and a woman, call her Carol, who are not married to each other but who would both prefer to be married to each other rather than their current spouses. Carol, then, did not receive a proposal from Ted; otherwise, she would have held him over her current spouse. Since Ted did not propose to Carol, he must have preferred his current spouse to Carol; otherwise, he would have proposed to Carol before his current spouse. Therefore, a pair such as Ted and Carol cannot exist. Consequently, the set of marriages is stable. This concludes the proof.

The algorithm for finding a stable set of marriages in the proof of the Theorem 11.3 is known as the *deferred acceptance algorithm* because women receiving proposals defer accepting them until every woman is holding exactly one proposal. The next example illustrates the deferred acceptance algorithm with four men and four women.

EXAMPLE 11.11

The marriage preferences for four men, W, X, Y, and Z, and four women, A, B, C, and D, are shown here. We apply the deferred acceptance algorithm to arrive at a stable set of marriages.

W: BDCA   A: XYWZ
X: BACD   B: XZYW
Y: BACD   C: WYXZ
Z: CBAD   D: XZWY

	W	X	Y	Z
A	(3, 4)	(1, 2)	(2, 2)	(4, 3)
B	(4, 1)	(1, 1)	(3, 1)	(2, 2)
C	(1, 3)	(3, 3)	(2, 3)	(4, 1)
D	(3, 2)	(1, 4)	(4, 4)	(2, 4)

Proposals by: W X Y Z	A B C D holds:	A B C D rejects:
1. B B B C	— X Z —	W,Y
2. D A	Y W	

W: Ⓑ Ⓓ C A   A: X Ⓨ W Z
X: Ⓑ A C D   B: Ⓧ Z Y W
Y: Ⓑ Ⓐ C D   C: W Y X Ⓩ
Z: Ⓒ B A D   D: X Z Ⓦ Y

The process is complete, since every woman is holding a proposal (and no man is currently rejected). The set of marriages is AY, BX, CZ, DW.

	W	X	Y	Z
A	(3, 4)	(1, 2)	(2, 2)	(4, 3)
B	(4, 1)	(1, 1)	(3, 1)	(2, 2)
C	(1, 3)	(3, 3)	(2, 3)	(4, 1)
D	(3, 2)	(1, 4)	(4, 4)	(2, 4)

YOUR FORMULATION

Choose four men and four women in the class, assign them the preferences shown in Example 11.11, and have them go through the steps described in the proof of Theorem 11.3. Also choose someone to be the "marriage broker," whose job it is to let everyone know when one step has been concluded and the next step is to begin. At the first step each man goes to the woman to whom he wishes to propose (the woman who is first on his list). Each woman receiving a proposal (B and C) holds her favorite proposer and rejects the others. The marriage broker announces that this concludes step 1. Then step 2 takes place, followed by as many steps as necessary for every woman to be holding a proposal. Details are as given in Example 11.11, but we suggest that you try acting out the procedure from its description.

Our description of the deferred acceptance algorithm involves the idea that the men propose. The algorithm works equally well if the women propose. Let's consider the deferred acceptance algorithm applied to Example 11.8, but with women proposing.

EXAMPLE 11.12

Apply the deferred acceptance algorithm to Example 11.8, but with women proposing.

SOLUTION

A: XZY   X: ACB
B: XZY   Y: CAB
C: ZYX   Z: BCA

Proposals by: A B C	X Y Z holds:	X Y Z rejects:
1. X X Z	A — C	B
2. Z	B	C
3. Y	C	

A: Ⓧ Z Y   X: Ⓐ C B
B: Ⓧ Ⓩ Y   Y: Ⓒ A B
C: Ⓩ Ⓨ X   Z: Ⓑ Ⓒ A

The stable set of marriages is AX, BZ, CY. Notice that this is the same set of marriages that was attained with men proposing. That had to happen, since there is only one stable set of marriages for that example. Notice, however, that the steps are not identical.

Let's try the deferred acceptance algorithm again with the women proposing, but in a situation where there is more than one stable set of marriages.

EXAMPLE 11.13

Apply the deferred acceptance algorithm to Example 11.9, but with women proposing.

SOLUTION

A: YZX   X: ABC
B: ZXY   Y: BCA
C: XYZ   Z: CAB

	X	Y	Z
A	(3, 1)	(1, 3)	(2, 2)
B	(2, 2)	(3, 1)	(1, 3)
C	(1, 3)	(2, 2)	(3, 1)

Proposals by: A B C	X Y Z holds:	X Y Z rejects:
Y Z X	C A B	— — —

A: Ⓨ Z X   X: A B Ⓒ
B: Ⓩ X Y   Y: B C Ⓐ
C: Ⓧ Y Z   Z: C A Ⓑ

The algorithm is done in one step with the set of marriages AY, BZ, CX. Notice that this time each woman got her first choice of mate. Exercise 7 compares this result with what happens when men propose.

Unequal Numbers of Men and Women

So far we have stated the marriage problem with the same number of men as women. However, the concept of a stable set of marriages still applies if the numbers of men and women are different. Also, the deferred acceptance algorithm still applies. If there are fewer proposers than potential receivers, then the process ends when no proposer is rejected. If there are more proposers than potential receivers, then the process ends when every potential receiver has received a proposal. We illustrate the first case with the following example. Exercises 12 and 13 illustrate the second case.

EXAMPLE 11.14

Three men, X, Y, and Z, and five women, A, B, C, D, and E, have the following marriage preferences:

X: EACDB   A: XZY
Y: CEADB   B: XZY
Z: EADCB   C: YXZ
            D: XZY
            E: XYZ

Apply the deferred acceptance algorithm, with men proposing.

Proposals by: X Y Z	A B C D E holds:	A B C D E rejects:
1. E C E	Y X	Z
2. A	Z	

X: Ⓔ A C D B
Y: Ⓒ E A D B
Z: Ⓔ Ⓐ D C B

A: X Ⓩ Y
B: X Z Y
C: Ⓨ X Z
D: X Z Y
E: Ⓧ Y Ⓩ

The stable set of marriages determined by the algorithm is AZ, CY, and EX. B and D remain unmarried.

Best Set of Stable Marriages

Earlier we asked two questions: (1) Does every marriage problem with n men and n women have a stable set of marriages? (2) If a marriage problem has more than one stable set of marriages, which one is best? Theorem 11.3 gives an affirmative answer to the first question. The second question is more difficult to answer and ultimately depends on what we mean by "best." Notice that the deferred acceptance algorithm favors the proposers—they generally get higher choices than do those receiving proposals. Written Assignments 1 and 4 pursue this second question.

The Marriage Problem in Other Contexts

The marriage problem has been stated in terms of marriages, but the concept applies in many other contexts. Three related situations would be dating possibilities, roommate selection, and the assignment of partners in mixed doubles for tennis, bowling, bridge, or other games. House sellers and house buyers are in a similar situation. See Written Assignment 2 for further exploration of this.

A slightly different situation would be in a fraternity or sorority rush in which potential pledges prioritize the Greek organizations while the groups in turn prioritize individuals. Usually in this situation the groups are choosing several members instead of just one, so some modification of the deferred acceptance algorithm would be necessary.

Another situation like the fraternity like the fraternity or sorority rush would be when students are choosing graduate or professional schools and those schools are choosing students. A matching procedure is used by the National Intern and Resident Matching Program to place medical interns with participating hospitals. Also, in a job placement situation, employers and prospective employees have a similar relationship. A version of this job placement is the basis for Exercise 13.

Strategy

If a person is a participant in a matching procedure like the marriage problem, is it ever to the person's advantage to state preferences other than the true ones in the hope of getting a better outcome? Yes, it has been shown that no stable matching procedure exists that gives all the participants an incentive for revealing their true preferences (Roth, 1982, p. 622). This produces another side to a matching problem.

STRATEGY

The question of strategy and the application of the marriage problem to economics are addressed in an article by Jinpeng Ma in the *Journal of Economic Theory,* August 1995, pp. 352–369:

Once any mechanism [such as the deferred acceptance algorithm] is instituted, a classical incentive problem arises. Could it be that players may sometimes have incentive to strategically misrepresent their true preferences? This motivates the search for "strategy-proof" mechanisms.

1. Show that each of the given sets of marriages from Example 11.8 is unstable. In each case, name a man and woman who are not married to each other but who would both prefer to be married to each other than to their current spouse.

   a. Set iv: (AY, BZ, CX)

   b. Set v: (AZ, BX, CY)

   c. Set vi: (AZ, BY, CX)

2. Show that each of the given sets of marriages from Example 11.9 is unstable. In each case, name a man and a woman who are not married to each other but who would prefer to be married to each other than to their current spouse.

   a. AX, BZ, CY   b. AY, BX, CZ   c. AZ, BY, CX

3. There are three women, A, B, and C, and three men, X, Y, and Z, who are to be paired in marriages. Their preferences in marriage partners are as follows: A(XYZ), B(XZY), C(XZY), X(ABC), Y(CAB), Z(CAB). For each of the six possible sets of marriages, tell whether it is stable or unstable. In each unstable set, name a man and a woman who are not married to each other but who would both prefer to be married to each other than to their current spouse.

4. Three women, A, B, and C, and three men, X, Y, and Z, are to be paired in marriage. Their preferences in marriage partners are as follows: A(XYZ), B(XZY), C(XZY), X(ABC), Y(CAB), Z(CAB). For each of the six possible sets of marriages, tell whether it is stable or unstable. In each unstable set, name a man and a woman who are not married to each other but would both prefer to be married to each other rather than to their current spouse.

5. In the proof of the theorem that there is always a stable set of marriages (Theorem 11.3), argue that the process must reach a point where every woman has a proposal.

6. If there are four men and four women, how many sets of marriages are possible?

7. Apply the deferred acceptance algorithm to Example 11.9, with men proposing. Compare your result with the result of Example 11.13, in which women proposed.

8. Apply the deferred acceptance algorithm to Example 11.10, with women proposing.

9. Apply the deferred acceptance algorithm to Example 11.11, with women proposing.

10. The marriage preferences for four men W, X, Y, and Z, and four women, A, B, C, and D, are shown. Apply the deferred acceptance algorithm under each of the following conditions.

W: BCDA	A: WXYZ
X: CBAD	B: ZXWY
Y: BCAD	C: ZWYX
Z: ABCD	D: YZXW

    a. Men proposing   b. Women proposing

11. Apply the deferred acceptance algorithm to Example 11.14, with women proposing.

12. The marriage preferences for five men, V, W, X, Y, and Z, and four women, A, B, C, and D, are shown.

V: ABCD	A: VYZWX
W: DCAB	B: WYVXZ
X: ABDC	C: ZYWXV
Y: DABC	D: VWZYX
Z: DBAC	

    a. How many sets of marriages are possible?

    b. Apply the deferred acceptance algorithm with men proposing.

    c. Apply the deferred acceptance algorithm with women proposing.

13. Four faculty members, A, B, C, and D, are each going to have a student work for her. Six students have applied for these jobs. The pairing will be done by means of the deferred acceptance algorithm. The preferences are as follows.

A: UVWXYZ	U: ABCD
B: UYVWZX	V: ACBD
C: VXUZYW	W: BACD
D: YWZUVX	X: CABD
	Y: ABDC
	Z: BADC

    a. How many pairings are possible? (One pairing is AU, BV, CW, DX.)

    b. Apply the deferred acceptance algorithm with faculty "proposing."

    c. Apply the deferred acceptance algorithm with students "proposing."

1. When the marriage problem has more than one set of stable marriages, suggest a definition for which set is the "best." Illustrate your definition with at least two examples, and argue why it is a reasonable definition of "best."

2. In the subsection headed "The Marriage Problem in Other Contexts" several other situations are mentioned in which the matching concept of the marriage problem can be used. Describe another situation that is similar to the marriage problem but in a different context. Give an example of this situation and solve it by means of the deferred acceptance algorithm.

3. In working with the marriage problem, you are doing mathematics with very little use of numbers and no use of shapes.

Numbers are involved only for ordering preferences.

a. Describe two other places in mathematics that use neither numbers nor shapes.

b. In the marriage problem and the two places you named, what is it about the situations or problems that makes them mathematical even though they don't involve numbers or shapes?

4. In Exercise 11, different pairings occur depending on whether faculty "propose" or students "propose." Both results give stable pairings ("marriages").

a. Present reasons why having faculty "propose" gives the better pairing.

b. Present reasons why having students "propose" gives the better pairing.

11.5 ▶ VOTING METHODS

GOALS

1. To calculate the number of possible rankings of a set of alternatives.

2. To use the voting methods of majority rule, the plurality method, binary voting, Condorcet winner, and the Borda method, and to describe how the voting method used by a group can affect the outcome of an election.

3. To describe the following results in the theory of voting: the paradox of voting (Condorcet's paradox); certain voting methods don't satisfy the independence of irrelevant alternatives condition; and Arrow's theorem and its impact.

4. To describe some of the people involved in the development of voting theory.

There are many situations in which a group of people make choices among alternatives. Political parties choose their nominees for offices. Voters choose a president and a variety of other officials. A group of faculty decide on which courses to require on a major. Five people decide on which two videos to rent for the weekend. A family decides where to go on vacation. Sportswriters decide who the top 20 college football teams are. And the list continues. In this section we will look at different methods that a group might use to aggregate individual choices into a group choice. This area of study is called *social choice theory*.

Political parties choose their nominees.

Probably the simplest method is to have a dictator. Who will be president? The dictator decides. What courses are required on a major? The dictator (maybe the department head) decides. Where will the family go on vacation? The dictator (maybe dad, maybe mom) decides. The method, though simple, is probably not desirable to most people. We generally believe in participatory decisions making.

 YOUR FORMULATION

Describe different ways, besides having a dictator, that a group can aggregate individual choices into a group choice. In other words, describe several different voting methods.

When a group votes to pick among choices, the idea of *majority rule* comes quickly to mind. We need to be careful about using that term correctly. A majority is a fraction larger than $\frac{1}{2}$. If there are 11 voters, 6 or more constitute a majority. If there are 12 voters, 7 or more are required for a majority. If there are three alternatives—call them A, B, and C—and 11 voters—who vote 5 for A, 3 for B, and 3 for C—then no alternative has a majority. If the procedure is to choose A in this case, then the *plurality method* is involved. With plurality voting, the alternative with the most votes (a plurality) is the winner.

If there are only two alternatives for voters to choose between and all the voters actually vote, then the plurality method and majority rule become the same. The main questions are procedural. What happens if there is a tie vote? What if not everybody votes? For an alternative to be chosen, does it require a majority of those voting or a majority of those eligible to vote? Once such procedural questions are settled, if there are only two alternatives, then majority rule is usually preferred.

What if there are three or more alternatives? Then do the voters want to select one alternative, select several alternatives, or rank the alternatives? For example, the voters in a presidential election elect one person; a search committee for president of a college might submit to the board of trustees three names selected from a large number of alternatives; and sportswriters rank the top 20 teams in some sports. Let's focus on the case in which the voters select one alternative (a president) rather than when the voters select several alternatives (three candidates, the top 20 teams).

Even though the voters select one alternative, each voter may have in mind some ranking of all the alternatives. The following notation will be used to indicate rankings of alternatives: Given three alternatives A, B, and C, we indicate that a voter prefers A to B and B to C by writing

$$\text{(ABC),} \qquad \text{(A, B, C),} \qquad \text{or} \qquad \begin{matrix} A \\ B \\ C \end{matrix}$$

Notice that if there are three alternatives—say, A, B, and C—then the number of rankings of them, with ties excluded, is the number of permutations of A, B, and C, which is 3!, or 6. This counting can be viewed as filling in a priority chart:

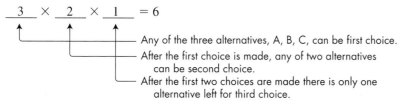

(_____ _____ _____)
    First   Second   Third
   choice   choice   choice

How many ways are there to fill in this priority chart?

___3___ × ___2___ × ___1___ = 6

— Any of the three alternatives, A, B, C, can be first choice.
— After the first choice is made, any of two alternatives can be second choice.
— After the first two choices are made there is only one alternative left for third choice.

In particular, the six choices are (ABC), (ACB), (BAC), (BCA), (CAB), and (CBA).

One way for the group to select its alternative is to use the plurality method. Whichever alternative gets the most votes is selected. However, this may yield a group selection that, in some sense, does not reflect the group's overall feeling. This is illustrated in the following example.

EXAMPLE 11.15

Let's consider the earlier-mentioned situation with 11 voters and three alternatives, A, B, and C, in which 5 favor A, 3 favor B, and 3 favor C. Furthermore, let's assume the individual rankings are as follows.

	A	A	B	B	C	C	or	A	B	C
Ranking	B	C	A	C	A	B		B	C	B
	C	B	C	A	B	A		C	A	A
Number of votes	5	0	0	3	0	3		5	3	3

a. Which alternative is the winner under the plurality method?
b. In what sense does the plurality winner not reflect the group's overall feeling?

SOLUTION

a. A is the plurality winner because more voters favor A (5) than favor either B or C.
b. Two ways that A is not the group's overall choice are that 3 + 3 = 6 voters (a majority) like B better than A and that A is the last choice of a majority (6) of the voters.

In Example 11.15, the group's feeling seems to favor B over A, yet A is the plurality winner. Is there a better way than plurality?

Binary Voting

In a tournament with three competitors, one method of selecting a winner would be to give one competitor a bye, have the other two compete against each other, then have that winner compete against the holder of the bye. We will call this *binary voting*, since the alternatives are considered in pairs. If such binary voting were used in Example 11.15, then there would be three ways to award the bye (to A, to B, or to C). Each of these is shown in ▶ Figure 11.14.

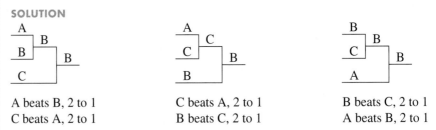

B beats A, 6 to 5
B beats C, 8 to 3

C beats A, 6 to 5
B beats C, 8 to 3

B beats C, 8 to 3
B beats A, 6 to 5

▶ FIGURE 11.14 Binary voting in Example 11.15

In each of the ways of assigning the bye, B was the alternative selected. This supports our earlier analysis that the group seems to favor B over A. Let's consider binary voting in another example.

EXAMPLE 11.16

Use binary voting to select the group choice if the three members of the group have one with a rating (ABC), one with a rating (BCA), and one with a rating (CAB).

SOLUTION

A beats B, 2 to 1
C beats A, 2 to 1

C beats A, 2 to 1
B beats C, 2 to 1

B beats C, 2 to 1
A beats B, 2 to 1

The group choice depends entirely on the pairings. Whichever alternative gets the bye wins.

Example 11.16 illustrates a major problem with binary voting: The outcome can be significantly influenced by the order in which the pairs are presented. The tournament director or a committee chair who sets the agenda wields significant influence if a group choice is to be made by pairwise voting. This problem is sometimes called the *agenda effect:* Different agendas may produce different winners.

Although binary voting was problematic in Example 11.16, it worked fine in the situation of Example 11.15, where B beats both A and C in pairwise voting. Such an alternative is called a *Condorcet winner*. More specifically, a Condorcet winner is one that beats all other alternatives in pairwise competition. Thus, in the situation of Example 11.15, B is a Condorcet winner, while in Example 11.16 there is no Condorcet winner.

The group of Example 11.16—three voters with ratings (ABC), (BCA), and (CAB)—illustrates another situation in which Condorcet's name is applied. We consider the individual ratings to be *transitive:* in the rating (ABC), since A is preferred to B and B is preferred to C, A is preferred to C. However, if the group votes between two alternatives, A beats B 2 to 1, B beats C 2 to 1, and C beats A 2 to 1. The group's choice is not transitive; if it were transitive, once it is known that A beats B and B beats C, then it would follow that A beats C. This phenomenon is known as the *paradox of voting* or *Condorcet's paradox.* Whereas the individual (transitive) preferences line up, the group (nontransitive) preference cycles around, as illustrated in ▶ Figure 11.15.

ABC　　　　　　BCA　　　　　　CAB

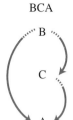
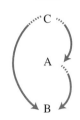
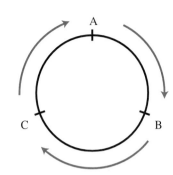

► FIGURE 11.15 Voting cycles

The next example uses binary voting with more than three alternatives.

EXAMPLE 11.17

A group has five voters, whose rankings of four alternatives, A, B, C, D, are as follows.

Voter 1	Voter 2	Voter 3	Voter 4	Voter 5
A	A	C	C	D
B	B	D	D	A
C	D	A	B	C
D	C	B	A	B

a. Is there a Condorcet winner?
b. Which alternative would win with binary voting?

SOLUTION

a. How many pairs of alternatives are there? There are

$$C(4, 2) = \frac{4 \cdot 3}{2 \cdot 1} = 6$$

A 4		A 3		A 2		B 2		B 2		C 3	
	A		A		D		C		D		C
B 1		C 2		D 3		C 3		D 3		D 2	

Notice that there is no Condorcet winner. There is what we might call a Condorcet loser—every alternative beats B in a pairwise contest. The remaining three alternatives, A, C, and D, have the group preference cycle A > C (read this "A is preferred to C"), C > D, and D > A. Therefore, there is no alternative that beats all the others in pairwise contests, and so there is no Condorcet winner.

b. There are two ways to implement binary voting. One way is via a tournament format, and the other way is via a game show format starting with a pair of alternatives and having the winner compete against successive challengers.

Tournament Format

Notice that B never wins (B can't win any individual contest) but any of A, C, or D can win, depending on the pairings. This is another illustration of the agenda effect. Notice also that whichever alternative started out against B ended up winning.

Game Show Format

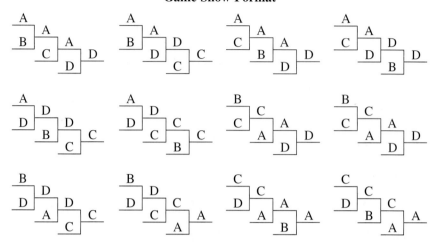

Notice, again, that B never wins but that any of A, C, or D can win. Once more the agenda effect is illustrated. Also notice that one alternative in the original pairing never wins (for example, C or D in the lower right format). And whichever alternative enters last ends up winning (except for B, which can never win).

Runoff Elections

When there is no majority winner, a form of majority rule can be attained by eliminating one or more alternatives and voting again on the remaining choices. The alternative eliminated might be either the one with the fewest first-place votes or the one with the most last-place votes. In the final vote, a majority of those voting select one alternative over the other.

Runoff elections often occur when there is a large number of voters. They happen in some statewide elections for political office, although usually in a two-step form where the first step produces the top two vote-getters and the second step chooses between those two.

Let's examine a runoff election in the three example situations presented so far in this section. Suppose the voting is done by eliminating the alternative with the fewest first-place votes and then voting on the remaining alternatives. In the situation of Example 11.15,

	A	B	C
	B	C	B
	C	A	A
Number of votes	5	3	3

B and C are tied with the fewest first-place votes at 3 each. To break the tie we look at which of these two has the fewer second-place votes, which is C, by a 3-to-8 count. Then C is eliminated and a run-off election is held between A and B. We assume the voters' preference scales are compressed as follows:

	A	B	C		A	B	B
	B	C	B	→	B	A	A
	C	A	A				
Number of votes	5	3	3		5	3	3

Then B wins 6 to 5. In other words, we assume that the group choice is made by treating the individual choices as if the removal of alternative C does not affect the voters relative to choices A and B. If a voter prefers A to B when C was a possible choice, we assume that the voter still prefers A to B when C is not a possible choice. This assumption is an example of the principle of *independence of irrelevant alternatives* (the irrelevant alternative, C, did not affect the outcome).

In the situation of Example 11.16, each alternative gets one first-place vote and one second-place vote, so a runoff election cannot be implemented by eliminating the alternative with the fewest first-place votes. Likewise, a runoff election cannot be implemented by eliminating the alternative with the most last-place votes. In this example, there are only three voters.

In Example 11.17, B is eliminated since it has no first-place votes. The ratings are compressed, by means of the principle of independence of irrelevant alternatives, to yield the following:

Voter 1	Voter 2	Voter 3	Voter 4	Voter 5
A	A	C	C	D
C	D	D	D	A
D	C	A	A	C

There is no change in first-place votes: A gets 2, C gets 2, and D gets 1. No alternative has a majority, so D, the alternative with the fewest first-place votes, is eliminated. The compressed ratings are:

Voter 1	Voter 2	Voter 3	Voter 4	Voter 5
A	A	C	C	A
C	C	A	A	C

Now A wins, 3 votes to 2.

Borda Count

Another voting method that is particularly useful in ranking a set of alternatives is called the *Borda count:* Each voter rates the n alternatives from 1 (first choice) to n (last choice). Then n points are assigned to the first choice, $n - 1$ points to the second choice, and so on down to 1 point for the last choice. The points are then totaled for all the voters. The alternative with the most points is the group's first choice, the alternative with the second-most points is the group's second choice, etc. We'll apply this method to each of our three examples.

In Example 11.15 [5 voters ranked the alternatives (ABC), 3 ranked them (BCA), and 3 ranked them (CBA)], A gets 5 first-place votes at 3 points each and 6 third-place votes at 1 point each for a total of $(5 \cdot 3) + (6 \cdot 1) = 21$ points. B gets 3 first-place votes and 8 second-place votes for a total of $(3 \cdot 3) + (8 \cdot 2) = 25$ points. C gets 3 first-place, 3 second-place, and 5 third-place votes for a total of $(3 \cdot 3) + (3 \cdot 2) + 5 = 20$ points. The group ranking by the Borda count is (BAC).

In Example 11.16 [1 voter has the rating (ABC), 1 has the rating (BCA), and 1 has the rating (CAB)], A gets 6 points, B gets 6 points, and C gets 6 points. There is a three-way tie among A, B, and C.

JEAN-CHARLES DE BORDA

Jean-Charles de Borda lived in France about the same time as Condorcet 1733–1799). Like Condorcet he was a member of the Academy of Sciences. During the last half of the eighteenth century, the time of the American and French Revolutions, the time was ripe in France for discussions of a mathematical theory of elections. Borda presented to the Academy of Sciences in 1784 a paper on elections in which he pointed out that the plurality method might lead to a choice that was not the real preference of the group (Black, 1968, p. 179). His example was similar to Example 11.15.

His method of "marks," what we've called the Borda method, was adopted by the Academy of Sciences for electing its members and used until 1800, when a new member opposed it.

Although Borda and Condorcet achieved distinction as mathematicians in their time, they are not considered great mathematicians today. Nonetheless, Borda was perhaps the first person to start discussion of a mathematical analysis of voting.

In Example 11.17, in which the following holds,

Voter 1	Voter 2	Voter 3	Voter 4	Voter 5
A	A	C	C	D
B	B	D	D	A
C	D	A	B	C
D	C	B	A	B

A gets 2 first-place, 1 second-place, 1 third-place, and 1 fourth-place vote for a total of $(2 \cdot 4) + (1 \cdot 3) + (1 \cdot 2) + (1 \cdot 1) = 14$ points. B gets 0 first-place, 2 second-place, 1 third-place, and 2 fourth-place votes for a total of $(0 \cdot 4) + (2 + 3) + (1 \cdot 2) + (2 \cdot 1) = 10$ points. C gets 2 first-place, 0 second-place, 2 third-place, and 1 fourth-place vote for a total of $(2 \cdot 4) + (0 \cdot 3) + (2 \cdot 2) + (1 \cdot 1) = 13$ points. D gets 1 first-place, 2 second-place, 1 third-place, and 1 fourth-place vote for a total of $(1 \cdot 4) + (2 \cdot 3) + (1 \cdot 2) + (1 \cdot 1) = 13$ points. The group ranking is (A(CD)B) where "(CD)" indicates that C and D tie.

Notice that the Borda count will produce a group's first-choice alternative as well as the group's ranking of all the alternatives. This is the method sportswriters and coaches use when voting on the top 10 or top 20 teams in various sports. It is also the one coaches or tournament officials use when seeding teams or individuals in a tournament. It is also the one committees frequently use when ranking alternatives. A variation of it is used when scoring swimming and track meets where first place may earn a team 5 points, second place 3 points, and third place 1 point.

A problem with the Borda method is that a majority winner, which is therefore a Condorcet winner, might not be the Borda winner. In other words, the Borda method can violate majority rule. The following example illustrates this.

EXAMPLE 11.18

There are five voters with three alternatives, A, B, and C. Three voters have the rating (ABC) and two have the rating (BCA). Find the winner by each of the voting methods discussed so far:

a. Majority rule
b. Plurality method
c. Binary voting
d. Condorcet winner
e. Runoff election
f. Borda count

SOLUTION

a. Under majority rule, A wins 3 votes to 2.
b. Since A has a majority, A has a plurality.
c. Under binary voting in any format, since A beats any other alternative, A will be the winner.
d. A is a Condorcet winner.
e. In a runoff election, C would be dropped for having the fewest (0) first-place votes or the most (3) last-place votes. Then A beats B 3 to 2, so A is the winner.
f. Under the Borda count, A has $(3 \cdot 3) + (2 \cdot 1) = 11$ points, B has $(2 \cdot 3) + (3 \cdot 3) = 12$ points, and C has $(2 \cdot 2) + (3 \cdot 1) = 7$ points. The Borda winner is B and the Borda ranking is (BAC).

Notice that voting systems (a) through (e) all make A the winner while the Borda count makes B the winner.

An attractive feature of the Borda method is that a voter has an opportunity to express not only a first choice but also second, third, and lower choices. Some alternatives might rate very highly with a few voters but be toward the bottom of the priority lists of a number of voters.

Arrow's Theorem

Each of the voting methods we've described has some flaws. Majority rule may not produce a winner, and even if it does, the winner may not represent the overall group attitude (see Example 11.18). The plurality method can produce a winner that does not represent the overall group attitude (see Example 11.15). Binary voting has the agenda effect. Selecting a Condorcet winner is often inconclusive—there may not be a Condorcet winner. A runoff election requires two or more elections, which in a national, state, or city election is time consuming and costly, or else it requires that each voter order the alternatives completely. The Borda method might produce a winner that is not a majority winner.

The *independence of irrelevant alternatives (IIA) condition* applied to the group choice states that if one alternative is removed from consideration (an irrelevant alternative) and the voters compress their rankings, then the group ranking is also the compressed ranking. Aside from the majority method, none of the methods satisfies this condition. Majority rule satisfies IIA since if some alternative—say, X—gets a majority of the votes and some other alternative is withdrawn, then X still gets a majority of the votes, so X is still the majority winner. To see that the plurality method does not satisfy IIA, let's consider Example 11.15, in which A wins. Let's assume alternative B withdraws. The compressed rankings become:

A	C	C
C	A	A
Number of votes 5	3	3

C now wins, 6 to 5. To show that runoff voting does not satisfy IIA, consider the following example.

EXAMPLE 11.19

There are five voters, with the following rankings of four alternatives:

Number of Voters:	2	2	1
	A	B	C
	B	C	A
	C	A	B
	D	D	D

a. Which alternative wins under runoff voting in which the alternative with the fewest first-place votes is dropped at each stage?
b. Show that an irrelevant alternative can be dropped that will lead to a different winner's being selected.

SOLUTION

a. D has the fewest first-place votes, so drop D. The resulting ratings are:

Number of Voters:	2	2	1
	A	B	C
	B	C	A
	C	A	B

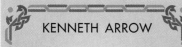

Now C has the fewest first-place votes, so drop C to get the following resulting ratings:

Number of Voters:	2	2	1
	A	B	A
	B	A	B

Consequently, A wins 3 to 2.

b. Assume alternative B is not considered. The resulting ratings are:

Number of Voters:	2	2	1
	A	C	C
	C	A	A
	D	D	D

Again, D has the fewest first-place votes and so is eliminated. The resulting ratings are:

Number of Voters:	2	2	1
	A	C	C
	C	A	A

Now, C wins 3 to 2. Since there is a new winner, this shows that IIA is not satisfied.

Given that all the methods we've discussed have flaws, you should suspect that people interested in social choice theory have been seeking to develop flawless methods for group decision making. What flaws do we want to avoid? Here are four characteristics that many people consider desirable for a voting method.

Unrestricted domain. Any set of individual rankings of the alternatives is possible. Thus, if there are n alternatives, any of the $n!$ rankings is possible for each voter. All the methods we have considered have unrestricted domain.

Decisiveness. Given any set of individual ratings, the method produces a winner. Majority rule and selecting a Condorcet winner are not decisive. The other methods need some tie-breaking rules, but with these they are decisive.

Independence of irrelevant alternatives (IIA). As noted earlier, only majority rule among the methods we've examined satisfies this condition.

Pareto principle. If everyone in the group prefers alternative X to alternative Y, then under its voting method the group should prefer X to Y. This principle generally holds for all the methods we've discussed, although its application to the case where X is not the first choice requires more consideration for some voting methods.

A summary of whether the methods we've discussed have these characteristics is given in ● Table 11.1.

Is there a voting method that has all four characteristics? There's good news and bad news. The good news is that there is a method—in fact, exactly one method, which was proved by Kenneth Arrow in 1951. The bad news is that the method is to have a dictator. To put Arrow's theorem differently, if a fifth desirable characteristic of a voting method is not to have a dictator, then no voting method has all five characteristics.

Essentially, Arrow's theorem says there is no perfect voting method. Where do we go from there? There are two approaches. One is to change the description of a

TABLE 11.1
Comparison of Voting Methods

	Characteristic			
Method	Unrestricted Domain	Decisiveness	IIA	Pareto Principle
Majority rule	Yes	No	Yes	Yes
Plurality	Yes	Yes	No	Yes
Binary voting	Yes	Yes	No	Yes
Condorcet winner	Yes	No	Yes*	Yes
Runoff election	Yes	Yes	No	Yes
Borda count	Yes	Yes	No	Yes

*If there is a Condorcet winner, then IIA is satisfied. If there is not a Condorcet winner, then the withdrawal of an alternative might produce a Condorcet winner.

perfect voting method from the five characteristics Arrow suggested to some other set of reasonable characteristics. The other is to live with imperfect methods and to describe their strengths and flaws. Efforts of people doing research in social choice theory have been directed toward both approaches.

Attempts to describe conditions for a desirable voting method have led to a variety of theorems like Arrow's theorem that say that there is no such voting method. Some of these are summarized in a book entitled *Arrow Impossibility Theorems* by Jerry S. Kelly (1978). Some relatively new voting schemes have been developed. Two of these are *approval voting* (Brams and Fishburn, 1983) and *acceptability voting* (Paine, 1990). However, like other voting systems, they have weaknesses.

A number of articles dealing with voting methods, including their strengths and weaknesses, have been published in the popular press and in journals related to economics, political science, psychology, and mathematics. We hope it is clear that the voting method a group uses in deciding among alternatives can affect the outcome.
● Table 11.2 summarizes the voting methods and examples presented in this section.

We conclude with one further caveat. All of our discussions of voting methods have assumed that the voters state their true preferences. That is, we have assumed that they vote sincerely rather than insincerely or strategically. That frequently is a

TABLE 11.2
Summary of voting methods and examples

	Winner in:				
Voting Method	Example 11.15	Example 11.16	Example 11.17	Example 11.18	Example 11.19
Majority rule	None	None	None	A	See Exercise 4
Plurality	A	A, B, and C tie	A and C tie A beats C	A	See Exercise 4
Binary voting	B	A or B or C	A or C or D	A	See Exercise 4
Condorcet winner	B	None	None	A	See Exercise 4
Runoff election (first place)	B	A, B, and C tie	A	A	A
Borda count	B	A, B, and C tie	A	B	See Exercise 4

faulty assumption. For example, in the 1996 presidential election some people who favored H. Ross Perot chose not to vote for him because they feared they were, in some sense, throwing their vote away. Consequently, they voted strategically for Bill Clinton or Bob Dole. Such insincere voting further complicates the analysis of voting methods.

EXERCISES 11.5

1. If there are two alternatives and ties are permitted, how many rankings of the alternatives are there?

2. If there are three alternatives and ties are permitted, how many rankings of the alternatives are there?

3. If there are four alternatives and ties are *not* permitted, how many rankings of the alternatives there?

4. In Example 11.19, which alternative, if any, is the winner under each of the following?
   a. Majority rule
   b. Plurality method
   c. Binary voting using the tournament method
   d. Selecting the Condorcet winner
   e. Borda method (give the group ranking of all four alternatives)

5. Seven people on a committee are considering three alternatives, A, B, and C. One person has the ranking (ABC), three have (ACB), one has (BAC), and two have (CAB). Which alternative, if any, is the winner under each of the following?
   a. Majority rule
   b. Plurality method
   c. Binary voting using each of the three possibilities (A gets a bye, B gets a bye, C gets a bye)
   d. Selecting the Condorcet winner
   e. Runoff election (drop the alternative with the fewest first-place votes)
   f. Borda method (give the group ranking of all three alternatives)

6. There are five voters with five alternatives, A, B, C, D, and E. Their individual rankings are (DBECA), (CABED), (BCAED), (BEDCA), and (AECDB). Which alternative, if any, is the winner under each of the following?
   a. Majority rule
   b. Plurality method
   c. Selecting the Condorcet winner
   d. Borda method (give the group ranking of all five alternatives)

7. Suppose a group of 55 voters has the following preferences on five alternatives, A, B, C, D, and E: 18 have the ranking (ADECB), 12 have (BEDCA), 10 have (CBEDA), 9 have (DCEBA), 4 have (EBDCA), and 2 have (ECDBA). Which alternative, if any, is the winner under each of the following? (The results of the voting in this example are dramatized in the *For All Practical Purposes* series of television program

videotapes, Part III, Social Choice, Show One, "The Impossible Dream: Election Theory.")
   a. Majority rule   b. Plurality method
   c. Selecting the Condorcet winner
   d. Runoff method with the alternative, with the fewest first-place votes eliminated at each stage
   e. Runoff method, with the runoff between the top two vote-getters initially
   f. Borda method

8. In the situation of Example 11.15, binary voting leads to the group choice (BCA). In the situation of Example 11.16, binary voting leads to the nontransitive group preference, where A > B, B > C, and C > A, i.e., the preferences cycle. If there are three alternatives, a group of any size, and binary voting is used, is there any other possibility for the group choice besides a transitive choice or a cycle? If so, give an example. If not, explain why not.

9. In the Borda voting in Example 11.15, the total points for the three alternatives were $21 + 25 + 20 = 66$. In Example 11.16, the total points were $6 + 6 + 6 = 18$. In Example 11.17, the total points were $14 + 10 + 13 + 13 = 50$.
   a. If there were 5 voters and 5 alternatives, how many total points would there be?
   b. If there were 7 voters and 5 alternatives, how many total points would there be?
   c. If 100 sportswriters rank the top 10 college football teams, how many total points would there be?

10. Show that binary voting does not satisfy IIA by using Example 11.17 under the following conditions.
   a. The tournament format with A and B paired. The withdrawal of which irrelevant alternative will change the outcome? Which alternative is the new winner?
   b. The tournament format with A and C paired. The withdrawal of which irrelevant alternative will change the outcome? Which alternative is the new winner?
   c. The tournament format with A and D paired. The withdrawal of which irrelevant alternative will change the outcome? Which alternative is the new winner?
   d. The game show format with the pairing

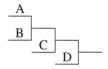

The withdrawal of which irrelevant alternative will change the outcome? Which alternative is the new winner?

11. Explain why the Borda method does not satisfy IIA by using the rankings in Example 11.19. The withdrawal of which irrelevant alternative will change the outcome? Which alternative is the new winner?

12. a. Show that in voting by choosing a Condorcet winner, if there is one, the method does satisfy IIA.
    b. Use Example 11.17, in which there is no Condorcet winner, to show that the withdrawal of one alternative might produce a Condorcet winner. Which alternative withdrawing produces which winner?

Exercises 13–29 are True or False.

13. If there is a majority winner, it will be a plurality winner.
14. A plurality winner will be a majority winner.
15. If there is a majority winner, it will be a winner under binary voting of every format.
16. If there is a majority winner, it will be a Condorcet winner.
17. If there is a Condorcet winner, it will be a majority winner.
18. If there is a majority winner, it will be a Borda winner.
19. A Borda winner will be a majority winner.
20. A plurality winner will be a winner under binary voting of every format.
21. A winner under binary voting will be a plurality winner.
22. A plurality winner will be a Condorcet winner.
23. A Condorcet winner will be a plurality winner.
24. A plurality winner will win every runoff election.
25. A plurality winner will be a Borda winner.
26. A winner under binary voting will be a Condorcet winner.
27. A Condorcet winner will win under binary voting of every format.
28. A Condorcet winner will be a Borda winner.
29. A Borda winner will be a Condorcet winner.

WRITTEN ASSIGNMENTS 11.5

1. Write a paragraph or two describing majority rule and the plurality method and the difference between them. Assume your audience is a group of fourth-, fifth-, or sixth-graders.

2. Nontransitive preferences often occur in athletic competition: A beats B, B beats C, and C beats A. Might an individual preference be nontransitive? Write a paragraph to a page on this question. If you answer yes, give an example. If you answer no, describe why you believe an individual preference must be transitive.

3. We stated that a problem with the Borda method is that a majority winner might not be the Borda winner. Is that a problem with the Borda method or a problem with majority rule? Take a position and support it in a page or two. Give at least one example.

4. One principle we used is independence of irrelevant alternatives. Some people object to it. Show a reason to object by describing an individual preference rating (ABC) for you or someone else that has the property that if one of the alternatives is deleted, the ratings of the other two alternatives reverse. That is, A is deleted and then you prefer C to B; B is deleted and then you prefer C to A; or C is deleted and then you prefer B to A.

5. (*Research project*) Borda and Condorcet lived at the same time and were both members of the French Academy of Sciences. What was their relationship? Keith Michael Baker (1975, p. 44) implies that Borda, "for whom Condorcet savored a special resentment" and Condorcet were not close. However, Duncan Black (1968, p. 179) says, "Throughout their lives the two were close friends, and during Borda's prolonged absences from Paris they corresponded much on scientific matters." Find some other sources that comment on the relationship between Borda and Condorcet and report on your findings. Give detailed bibliographic references. (The books by both Baker and Black have substantial bibliographies.)

6. Discuss strategic voting with majority rule.

7. (*Research project*) Write a two- to five-page paper on approval voting. The book by Brams and Fishburn (1983) is a good starting point. Also check your periodical database. Name some organization or groups of voters that use approval voting. Would you support its use in a presidential primary election in your state?

8. Describe a voting method that has not been discussed in this section. It might be one mentioned in the Your Formulation at the start of this section. Add this method to Table 11.1; i.e., tell whether or not this method has unrestricted domain, is decisive, satisfies IIA, and satisfies the Pareto principle.

9. Suppose the people in the class for which you are using this book were to vote to select one of four possible times to have an exam. Of the voting methods discussed in this section, which one would you support? Write a one-page position paper with the intent of convincing your classmates that the method you support should be adopted.

1. Write a paragraph describing the nature of discrete mathematics. How does it differ from other areas of mathematics?

2. What has produced the focus on discrete mathematics today? What is there today that prompts chapters in books, entire books, and entire courses on discrete mathematics that was not present a century ago?

3. A convenience store has the layout shown here.

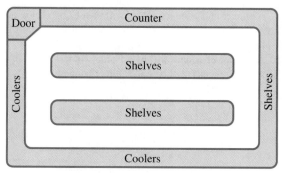

The blank areas denote aisles.

   a. Is it possible to enter the store at the door and travel in each aisle exactly once (once and only once) and leave by the door?

   b. Why or why not? Answer in terms that a typical convenience store customer could understand.

4. Construct a graph with three vertices, one of degree 1, a second of degree 2, and a third of degree 3.

5. Prove the theorem that every vertex on a graph with an Euler circuit has even degree.

6. Consider the following graph.

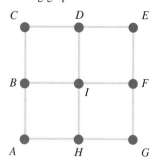

   a. What is the degree of vertex *A? B? I?*

   b. Is the graph connected? If so, state a path from *A* to *I*. If not, state two vertices that have no path joining them.

   c. Is there an Euler circuit? If so, state one. If not, explain why not.

   d. If you answered yes to part c, add one or more edges but no additional vertices so that the resulting graph does *not* have an Euler circuit. If you answered no to part c, add one or more edges but no additional vertices so that the resulting graph has an Euler circuit.

   e. Does the graph have a Hamiltonian cycle? If it has one, state it. If not, add one or more edges between existing vertices so that the graph has a Hamiltonian cycle, and state the Hamiltonian cycle.

7. The people who live in a subdivision with the layout shown would like a police car occasionally to enter their subdivision and drive on each street.

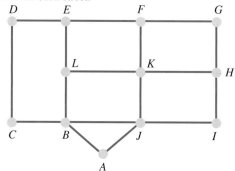

   a. Is it possible to do this by entering at point *A,* driving on each street exactly once, and exiting at *A?*

   b. If so, describe the route. If not, explain why not.

8. a. Represent the following map of the southeastern United States as a graph, with each state being a vertex and with an edge joining two states exactly when the two states have a border in common.

   b. Is it possible to start in some state, go from each state to an adjoining state, and return to the starting state without passing through any state except the starting state more than once? Why or why not?

9. Santa Claus wants to visit each of the seven continents by starting in North America, traveling to each of the six other continents exactly once, and returning to North America. How many tours are possible? (Count NA-SA-E-As-Af-Au-G-NA as the same tour as NA-G-Au-Af-As-E-SA-NA.)

10. By means of the following find a low-mileage route through the cities shown.
   a. The nearest-neighbor algorithm
   b. The sorted-edges algorithm
   c. The brute force method

	C	Den	Det	E
Chicago, IL	—	1050	279	1439
Denver, CO	1050	—	1310	652
Detroit, MI	279	1310	—	1696
El Paso, TX	1439	652	1696	—

11. Describe the difference between an exact method and a heuristic method for solving a problem.

12. Write a paragraph on what is known about solutions to the traveling salesperson problem. Can the most efficient TSP route always be found? Is it easy to find? What is the best method(s)?

13. There are three women, A, B, and C, and three men X, Y, and Z, who are to be paired in marriages. Their preferences in marriage partners are as follows: A(XYZ), B(XZY), C(YXZ), X(BAC), Y(BCA), Z(CBA).
   a. For the set of marriages AX, BY, CZ, tell whether it is stable or unstable. If it is unstable, name a man and a woman who are not married to each other but who would both prefer to be married to each other rather than to their current spouse.
   b. Repeat part a for the set of marriages AY, BZ, CX.
   c. Use the deferred acceptance algorithm to find a stable set of marriages. Leave evidence of your use of the algorithm.

14. The musical *Seven Brides for Seven Brothers* centers on seven men and seven women. How many sets of marriages are possible?

15. Answer the following questions regarding the proof that the deferred acceptance algorithm produces a stable set of marriages in a set of *n* men and *n* women.
   a. The algorithm begins by having each man, say, propose to his most preferred woman. If every woman receives a proposal, then the process concludes. Why is this set of marriages stable?
   b. If the process does not conclude in one step, which of the following *must* be true? (There might be more than one correct answer.)
      i. There are some men who have not made a proposal.
      ii. There are some women who have not received a proposal.
      iii. Some men have made more than one proposal.
      iv. Some women have received more than one proposal.

16. Describe a situation, other than one with marriages, to which the deferred acceptance algorithm applies.

17. Are majority rule and plurality voting the same? If so, why are different terms used? If not, explain the difference and illustrate with an example.

18. Six noted artists are Audubon, Botticelli, Chagall, Dali, Eakins, and Feuerbach. In how many ways could these six be ranked from most favorite to least favorite?

19. In the 1992–93 National Basketball Association season, the five players named to the All-League team were Charles Barkley, Michael Jordan, Karl Malone, Akeem Olajuwon, and Mark Price. Their rankings in seven statistical categories were:

Points	Points/ Game	Rebounds/ Game	Field Goal %	Minutes Played	Free Throw %	Assists
J	J	O	M	O	P	P
M	M	B	O	M	J	J
O	O	M	B	J	O	B
B	B	J	J	B	B	M
P	P	P	P	P	M	O

Treat each statistical category as a voter, and pick the winner among these five players by means of the following.
   a. Majority rule (if there is no winner, drop the player with the most last-place votes and vote again)
   b. Plurality method (if there is no winner, drop the player with the most last-place votes and vote again)
   c. Binary voting (you decide the format)
   d. A Condorcet winner
   e. Borda method (give a complete ranking of all five players

20. Review Exercise 19 mentions five voting methods.
   a. Describe the strengths of any two of them. Make sure the grader knows which two methods you are describing.
   b. Describe the weaknesses of any two of them.

21. Describe the paradox of voting (Condorcet's paradox).

22. Use Review Exercise 19 to show that the Borda method does not satisfy the independence of irrelevant alternatives condition by comparing the relation between Malone and Olajuwon if either Barkley or Price is not considered. State which one is not considered and what the effect is.

23. Name two people involved in the development of voting theory. For each person tell the century in which he or she lived and at least one fact about the person's contribution to voting theory.

24. Describe Arrow's theorem.

25. Describe how the voting method used by a group can affect the outcome of an election.

Baker, Keith Michael. *Condorcet: From Natural Philosophy to Social Mathematics.* Chicago: University of Chicago Press, 1975.

Biggs, N. L., Lloyd, E. K., and Wilson, R. J. *Graph Theory 1736–1936.* Oxford: Clarendon Press, 1976.

Black, Duncan. *The Theory of Committees and Elections.* Cambridge: Cambridge University Press, 1968.

Boyer, Carl B., and Merzbach, Uta C. rev. ed. *A History of Mathematics,* 2nd ed. New York: Wiley, 1991.

Brams, S. J., and Fishburn, P. C. *Approval Voting.* Cambridge, MA: Birkhauser Boston, 1983.

Calinger, Ronald, ed. *Classics of Mathematics.* Englewood Cliffs, NJ: Prentice Hall, 1995.

Eves, Howard. *An Introduction to the History of Mathematics,* 6th ed. Chicago: Saunders College, 1990.

Gale, D., and Shapley, L. S. "College Admissions and the Stability of Marriage," *The American Mathematical Monthly 69* (1962):9–15.

Gittleman, Arthur. *History of Mathematics.* Columbus, Merrill, OH: 1975.

Kelly, Jerry S. *Arrow Impossibility Theorems.* New York: Academic Press, 1978.

Lawler, E. L., Lenstra, J. K., Rinnooy Kan, A. H. G., Shmoys, D. B., ed. *The Travelling Salesman Problem.* New York: Wiley, 1985.

Ma, Jinpeng. "Stable Matchings and Rematching—Proof Equilibria in a Two-Sided Matching Market," *Journal of Economic Theory* (August 1995):352–369.

Miller, Donald L., and Pekny, Joseph F. "Exact Solution of Large Asymmetric Traveling Salesman Problems," *Science 251* (1991):754–761.

Paine, Neil R. "Acceptability Voting: A Modification," *British Journal of Political Science 20* (1990):142.

Roth, Alvin E. "The Economics of Matching: Stability and Incentives," *Mathematics of Operations Research 7* (1982):617–628.

Sangalli, Arturo. "Short-circuiting the Travelling Salesman Problem, *New Scientists 134* (June 27, 1992):16.

CHAPTER 12

MATHEMATICS AND COMPUTERS

PROLOGUE

One characterization of the time period in which we live is that it is a period of increasing mathematization of knowledge. The physical sciences have traditionally relied heavily on mathematics, and they continue to do so. The biological and social sciences also rely on mathematics. Technical areas are by their nature based on mathematical ideas. But the arts and humanities also interact with mathematics. The increasing mathematization of knowledge is prompted by both the effectiveness of mathematical modeling and the ease with which complex calculations can now be performed.

Through time, developments in mathematics and technology have made calculation easier and increased the use of calculations and other

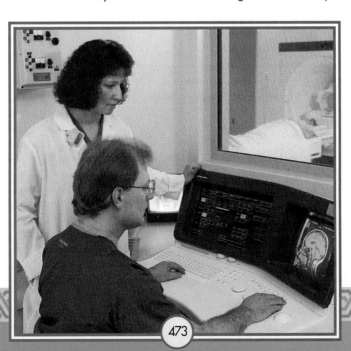

mathematical ideas in applications of the mathematical sciences. As a consequence, computer science has opened up as a new branch of the mathematical sciences, and experimentation in the mathematical sciences has grown.

In this chapter we will explore some of the significant developments in mathematics and technology that have made calculation easier, and we'll look at some of the uses of technology in mathematics and some recent developments in mathematics.

Chapter opening photo: A mathematical algorithm, implemented by computer technology, generates a picture of an internal organ.

12.1 MECHANICAL AND MATHEMATICAL AIDS TO COMPUTATION

GOALS
1. To describe the dependence of algorithmic calculations on a writing medium.
2. To become familiar with the abacus, to use it for addition, and to describe its history.
3. To become familiar with logarithms: to find logs, to multiply numbers, and to describe today's uses of logarithms.
4. To describe a slide rule, its use, and its precursor.

If you were faced with a moderately complex arithmetical calculation—such as multiplying two five-digit numbers—you would probably reach for a calculator. What has taken place in mathematics and technology that led to the development of the calculators and computers of today?

FRANK & ERNEST by Bob Thaves

With permission of Bob Thaves

Before calculating can be done, a system of numeration must be in place. (Ancient numeration systems are discussed in Section 2.1.) The first ideas of counting probably involved matching objects to be counted—say, sheep—with counting objects such as pebbles. Such a system would make addition easy. For example,

can be represented by

The pebbles representing the sheep are simply lumped together and counted.

Our method of writing down a simple calculation such as $3 + 2 = 5$ or a more complex calculation such as

$$\begin{array}{r} 1 \\ 28 \\ + 35 \\ \hline 63 \end{array}$$

is dependent on the availability of a medium on which to write (for us, paper) and, in the more complex example, on the development of the place-value system of numeration. The ancient Egyptians prior to 600 B.C.E. developed a way of transforming fibers from the papyrus plant into a form of paper. The ancient Chinese made a form of paper from rice. Around 100 B.C.E. papyrus was replaced in Rome and Greece by parchment, a material made from the skins of sheep, goats, or other animals. These early writing materials were both relatively scarce and expensive. They were not available for routine arithmetical calculations.

Paper as we know it was invented in China around the year 100. Its use had spread to Arab countries by about 800 and later spread into Europe. Making paper from wood pulp, the way it is made in modern paper mills, is only a little over a hundred years old.

Because of the relative scarcity and expense of writing materials, other mediums were adopted for arithmetic calculations. One of these was the abacus.

The Abacus

The abacus can be regarded as the earliest mechanical computing device. It appeared in many forms in parts of the ancient world. Its exact origin is not known. Early implementations of the ideas of the abacus involved lines drawn in the sand or in the dust on a dustboard or on a counting board. Here, *abacus* will refer to an instrument for calculating like the one shown in ▶ Figure 12.1, and *counting boards* will mean either an abacus or lines drawn in the sand or dust.

To appreciate how the counting board was an aid to computation, we will do an addition involving Roman numerals. (Roman numerals are discussed in Section 2.1.) ▶ Figure 12.2 shows how to add MDCCLXXIV and MMCXLVIII.

▶ FIGURE 12.1 Four calculators: author Doug Nance, graphing calculator, abacus, and author Tom Miles. One is very old and one is very new.

1.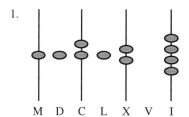

M D C L X V I

Draw four vertical lines and label them, from left to right, M, C, X, and I; label the spaces between the lines D, L, and V.
Place counters on and between the lines to represent the first number, MDCCLXXIV.

2.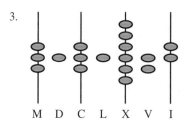

M D C L X V I

Now insert the counters for the second number, MMCXLVIII.

3.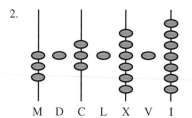

M D C L X V I

Replace five of the markers in the I column with one marker in the V space.

▶ FIGURE 12.2 Adding MDCCLXXIV and MMCXLVIII on an abacus

Evidence of the use of abacus-like devices goes back at least as far as the ancient Persians and Egyptians of 600 B.C.E. They were also found in ancient Greece. Joseph Dauben says (Dauben, 1992, p. 7):

> Already by the seventh century B.C., the use of such counting boards must have been generally familiar, for the Greek lawmaker Solon (according to Diogenes Laertius) made a political reference to those who had influence with tyrants as being like pebbles on a counting board, their values standing sometimes for more and sometimes for less.

The abacus is often associated with Chinese or Japanese cultures. However, it is not clear whether it was invented by the Chinese or borrowed by them from Roman or Arabic sources. In its present form, it was employed in China by about 1200 and went from there to Korea and to Japan. However, the Chinese of about 300 B.C.E. used a system of "rod numerals." Calculation was done with them in much the same way that it is done with the abacus.

The Incas of Central America prior to 1600 apparently had developed a counting board. Other civilizations have also been advanced enough that they might have used counting boards.

Two factors led to the demise of the abacus. Hindu-Arabic numerals gained in popularity because calculation with them was easier than with numbers written in an additive system of numeration. Their increased usage led to disputes between the abacists and the algorists, who used algorithms like our procedures for addition, subtraction, multiplication and division.

4.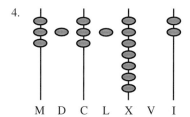

Replace the two markers in the V space by a single marker in the X column.

5.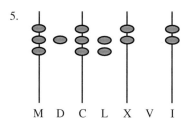

Replace five of the markers in the X column by a single marker in the L space.

6.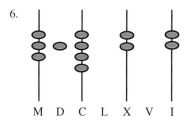

Replace the two markers in the L space by a single marker in the C column.

▶ FIGURE 12.2 continued

The second factor was the appearance of rag paper about 1300. This permitted the kind of algorithmic calculation we do, such as in adding 287 and 469.

By the 1500s the use of the abacus in Italy had died out. It continued to be employed in northern Europe until the 1700s and in Russia into the 1800s. It was used, with great dexterity, in the Orient until only recently.

Although its use has died out, the abacus played a prominent role in the development of calculation by being the first mechanical calculational device. As a tool, it has been replaced by more powerful devices.

Logarithms

A development in mathematics—the creation of logarithms—enabled multiplication of large numbers to be carried out more quickly than the usual hand calculation. Logarithms can help to multiply because exponents are added when numbers with the same base are multiplied. For example, if we wanted the product

$$2^3 \cdot 2^5$$

we could write $$2^3 + 2^5 = 2^{3+5} = 2^8$$

This product could have been found from ● Table 12.1 by noting that the sum of the exponents, 3 and 5, is 8 and then selecting 256 as the answer because it is below the 8 in the top row.

TABLE 12.1
Powers of 2

Exponent	1	2	3	4	5	6	7	8	9	10	11	12
Number	2	4	8	16	32	64	128	256	512	1024	2048	4096

Now we change the base from 2 to 10. ● Table 12.2 matches up exponents and numbers. Notice that

TABLE 12.2
Powers of 10

Exponent	1	2	3	4	5	6	7	8
Number	10	100	1,000	10,000	100,000	1,000,000	10,000,000	100,000,000

(the mth number) $\times$ (the nth number) $= 10^m \times 10^n = 10^{m+n} =$ the $(m + n)$th number

The exponents are called *logarithms,* from a Greek phrase meaning "ratio number." We write log 10 = 1, log 100 = 2, log 1000 = 3, etc., and read these, respectively, as "the log of 10 equals 1," "the log of 100 equals 2," etc.

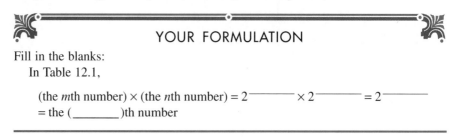

YOUR FORMULATION

Fill in the blanks:
In Table 12.1,

(the mth number) $\times$ (the nth number) $= 2^{\underline{\hspace{1cm}}} \times 2^{\underline{\hspace{1cm}}} = 2^{\underline{\hspace{1cm}}}$
= the (_____)th number

TABLE 12.3					
Nonpositive Powers of 10					
Exponent	0	−1	−2	−3	−4
Number	10^0	10^{-1}	10^{-2}	10^{-3}	10^{-4}
	1	.1	.01	.001	.0001

Table 12.2 can be extended to include all integers as exponents. This is indicated in • Table 12.3. We write log 1 = 0, log .1 = −1, log .01 = −2, etc.

An extension of this process allows us to write all real numbers as decimal powers of 10. For example, $2 \approx 10^{0.30103}$. This means that log 2 ≈ .30103. This extension is what allowed logarithms to be developed as a computational aid. In general, they simplified computation because "product" was replaced by "sum" and "exponentiation" was replaced by "product."

The person credited with discovering the concept of logarithms is John Napier, a Scot born in 1550 who lived most of his life at the family castle near Edinburgh, where he died in 1617. He was a fervent Protestant and a landowner in a time of religious and political controversies. He wrote on many topics, including religion and mathematics, and though not a professional mathematician, is credited with several developments in mathematics.

Napier's development of logarithms was based only in part on the reasoning underlying our description of logarithms in this section. One invention that assisted the development of logarithms was Simon Stevin's introduction of decimal fractions in the late 1500s (see Section 2.3). Napier thought about the concept of logarithms for many years before publishing in 1614 a small book entitled *Mirifici Logarithmorum Canonis Descriptio* (A Description of the Wonderful Law of Logarithms). This book contained a table of logarithms.

Napier's publication did not use base-10 logarithms. About a year after its publication, the English mathematician Henry Briggs visited Napier. Napier and Briggs agreed that base-10 logarithms would be an improvement over those in Napier's book. In 1624, 7 years after Napier's death, Briggs published his *Arithmetica Logarithmica* with a table of logarithms of numbers to 14 decimal places. The base-10 logarithms are sometimes called *common logarithms* or *Briggsian logarithms*.

A prominent driving force in the development of logarithms was the need to multiply large numbers. The demands of astronomy, in particular, prompted the desire for a relatively fast way to multiply large numbers. The key to multiplying large numbers via logarithms is to generalize the property stated earlier:

(the *m*th number) × (the *n*th number) = the (*m* + *n*)th number

We will state the generalization as a theorem.

CAUGHT BLACK-HANDED

John Napier achieved a reputation as an eccentric and as a dealer in magic. One story, perhaps unfounded, maintains that he announced that his coal black rooster would identify which of his servants was stealing from him. Each servant was sent into a darkened room with the instruction to pet the rooster. Unknown to all of them, Napier had smeared lampblack on the rooster's back. The guilty servant, afraid to touch the rooster, established his guilt by emerging with clean hands.

John Napier

THEOREM 12.1
────●────

If A and B are positive real numbers, then log (AB) = log A + log B.

Proof: Let A and B be positive real numbers and let log $A = a$ and log $B = b$. Then

$$10^a = A \quad \text{and} \quad 10^b = B$$

So

$$AB = 10^a 10^b = 10^{a+b}$$

This says that

$$\log (AB) = a + b,$$

So

$$\log (AB) = \log A + \log B.$$

TABLE 12.4
Five-place Logarithms

N.		0	1	2	3	4	5	6	7	8	9
1000	000	0000	0434	0869	1303	1737	2171	2605	3039	3473	3907
1001		4341	4775	5208	5642	6076	6510	6943	7377	7810	8244
1002		8677	9111	9544	9977	*0411	*0844	*1277	*1710	*2143	*2576
1003	001	3009	3442	3875	4308	4741	5174	5607	6039	6472	6905
1004		7337	7770	8202	8635	9067	9499	9932	*0364	*0796	*1228
1005	002	1661	2093	2525	2957	3389	3821	4253	4685	5116	5548
1006		5980	6411	6843	7275	7706	8138	8569	9001	9432	9863
1007	003	0295	0726	1157	1588	2019	2451	2882	3313	3744	4174
1008		4605	5036	5467	5898	6328	6759	7190	7620	8051	8481
1009		8912	9342	9772	*0203	*0633	*1063	*1493	*1924	*2354	*2784
1010	004	3214	3644	4074	4504	4933	5363	5793	6223	6652	7082
1011		7512	7941	8371	8800	9229	9659	*0088	*0517	*0947	*1376
1012	005	1805	2234	2663	3092	3521	3950	4379	4808	5237	5666
1013		6094	6523	6952	7380	7809	8238	8666	9094	9523	9951
1014	006	0380	0808	1236	1664	2092	2521	2949	3377	3805	4233
1015		4660	5088	5516	5944	6372	6799	7227	7655	8082	8510
1016		8937	9365	9792	*0219	*0647	*1074	*1501	*1928	*2355	*2782
1017	007	3210	3637	4064	4490	4917	5344	5771	6198	6624	7051
1018		7478	7904	8331	8757	9184	9610	*0037	*0463	*0889	*1316
1019	008	1742	2168	2594	3020	3446	3872	4298	4724	5150	5576
1020		6002	6427	6853	7279	7704	8130	8556	8981	9407	9832
1021	009	0257	0683	1108	1533	1959	2384	2809	3234	3659	4084
1022		4509	4934	5359	5784	6208	6633	7058	7483	7907	8332
1023		8756	9181	9605	*0030	*0454	*0878	*1303	*1727	*2151	*2575
1024	010	3000	3424	3848	4272	4696	5120	5544	5967	6391	6815
1025		7239	7662	8086	8510	8933	9357	9780	*0204	*0627	*1050
1026	011	1474	1897	2320	2743	3166	3590	4013	4436	4859	5282
1027		5704	6127	6550	6973	7396	7818	8241	8664	9086	9509
1028		9931	*0354	*0776	*1198	*1621	*2043	*2465	*2887	*3310	*3732
1029	012	4154	4576	4998	5420	5842	6264	6685	7107	7529	7951
1030		8372	8794	9215	9637	*0059	*0480	*0901	*1323	*1744	*2165
1031	013	2587	3008	3429	3850	4271	4692	5113	5534	5955	6376
1032		6797	7218	7639	8059	8480	8901	9321	9742	*0162	*0583
1033	014	1003	1424	1844	2264	2685	3105	3525	3945	4365	4785
1034		5205	5625	6045	6465	6885	7305	7725	8144	8564	8984
1035		9403	9823	*0243	*0662	*1082	*1501	*1920	*2340	*2759	*3178
1036	015	3598	4017	4436	4855	5274	5693	6112	6531	6950	7369
1037		7788	8206	8625	9044	9462	9881	*0300	*0718	*1137	*1555
1038	016	1974	2392	2810	3229	3647	4065	4483	4901	5319	5737
1039		6155	6573	6991	7409	7827	8245	8663	9080	9498	9916
1040	017	0333	0751	1168	1586	2003	2421	2838	3256	3673	4090
1041		4507	4924	5342	5759	6176	6593	7010	7427	7844	8260
1042		8677	9094	9511	9927	*0344	*0761	*1177	*1594	*2010	*2427
1043	018	2843	3259	3676	4092	4508	4925	5341	5757	6173	6589
1044		7005	7421	7837	8253	8669	9084	9500	9916	*0332	*0747
1045	019	1163	1578	1994	2410	2825	3240	3656	4071	4486	4902
1046		5317	5732	6147	6562	6977	7392	7807	8222	8637	9052
1047		9467	9882	*0296	*0711	*1126	1540	*1955	*2369	*2784	*3198
1048	020	3613	4027	4442	4856	5270	5684	6099	6513	6927	7341
1049		7755	8169	8583	8997	9411	9824	*0238	*0652	*1066	*1479
1050	021	1893	2307	2720	3134	3547	3961	4374	4787	5201	5614
N.		0	1	2	3	4	5	6	7	8	9

To illustrate how this theorem enables large numbers to be multiplied, we will consider the product of 10,192 and 10,197, using the portion of a logarithm table shown in ● Table 12.4. First we find the log of each number. Recall that log 10,000 = 4 and log 100,000 = 5. Since both 10,192 and 10,297 are between 10,000 and 100,000, both have logs between 4 and 5. The part of the log after the decimal point is described in the table. To find log 10,192, look at the row headed 1019 and then go to the column headed 2 to get 0082594. Knowing that log 10,192 is between 4 and 5, we have log 10,192 = 4.0082594. Similarly, log 10,297 = 4.0127107. By Theorem 12.1,

$$\log [(10,192)(10,297)] = \log 10,192 + \log 10,297$$
$$= 4.0082594 + 4.0127107 = 8.0209701$$

We now seek the number whose log is 8.0209701 by looking in the table for the part to the right of the decimal. It is not there exactly, but it is between .0209411 and .0209824 and is closest to .0209824, which is the log of 10,495. The 8 to the left of the decimal point says that the number is between

$$10^8 = 100,000,000 \qquad \text{and} \qquad 10^9 = 1,000,000,000$$

The product is therefore approximately 104,950,000.

We know that this is only an approximation since the true value has a units digit of 4 (from $2 \times 7 = 14$). How close is it to the actual value? The actual value is 104,947,024, which if rounded to five significant digits is 104,950,000.

This process seems complicated at first, but it is easily learned. It speeds up the multiplication by converting multiplication to addition, which is easier to do. If you need more convincing, try multiplying 10,192 by 10,297 without the aid of a calculator.

Until a decade ago most textbooks on college algebra and trigonometry included tables of logarithms. Today many don't because even inexpensive calculators have a logarithm key that produces logarithms with more accuracy than the five-place values in Table 12.4. To see how this works, use your calculator to find the following logs that have been discussed in this section.

n	$\log n$
10	1
100	2
1,000	3
1	0
0.1	−1
0.01	−2
2	0.301019995664
10,192	4.00825941499
10,297	4.01271071274

If your calculator has a [LOG] key, it also will have an [LN] key. *LN* stands for "natural logarithm." It has a base of e ($e \approx 2.71828$) rather than 10.

Napier's development of logarithms was influenced by a number of activities that involved calculating with large numbers. These included astronomy, navigation, trade, engineering, and warfare. The pressures of these activities gave impetus to computational developments.

As with most inventions and mathematical developments, the climate was ripe for the invention of logarithms. A Swiss watchmaker, Jobst Bürgi (1552–1632), published a table of logarithms in 1620, 6 years after Napier published his book. Bürgi's work was done independent of Napier's work.

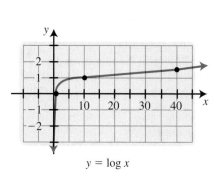

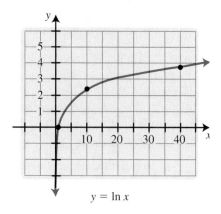

$y = \log x$ $\qquad\qquad$ $y = \ln x$

▶ FIGURE 12.3 Graphs of logarithmic functions

Since calculators can now multiply large numbers with more speed and accuracy than can be attained with logarithms, do logarithms still have any value? Yes, they are used in at least two ways. The logarithmic functions (base 10, base e, etc.) are inverses of exponential functions [$f(x) = 10^x$, $f(x) = e^x$, etc.] in the sense that the logarithmic function undoes what the exponential function has done (in the same way that taking the square root undoes what squaring does). This idea allows us to solve equations with a variable in the exponent.

A second use of logarithmic functions is to model certain phenomena, such as the loudness of a sound. The graphs of two logarithmic functions are shown in ▶ Figure 12.3. As the graphs show, the logarithmic functions can help to describe some phenomena that are always growing, but growing at a slower rate as x increases.

CALCULATOR NOTE

Graph the logarithm function on a graphing calculator by graphing $y(x) = \log x$. Set the range on x to go from 0 to 40 and on y to go from −2 to 2. You should see a graph like the first one in Figure 12.3.

Graph $y = \ln x$ and compare your result with the second graph in Figure 12.3. Then graph both the log function and the ln function on the same set of axes.

To use the numbers we referred to in connection with Table 12.4, change the range on x from 10,100 to 10,300 and on y from 4 to 4.02. Graph $y = \log x$ and use your trace key to estimate log 10,192. To how many significant digits is it correct?

The applications of exponential functions in Section 5.3 are places where logarithms can solve equations with a variable in the exponent. Also, the Richter scale can be described most directly by means of logarithms. (A description of the Richter scale without the use of logarithms is given in Section 5.3.)

Logarithms originally and for over 350 years served a very important purpose by giving a relatively fast and accurate way of multiplying large numbers. Howard Eves, (1990, p. 312) quotes the French mathematician Pierre-Simon Laplace

(1749–1827) as stating that the invention of logarithms, "by shortening the labors doubled the life of the astronomer." Michael R. Williams (1985, p. 111) writes:

> It is hard to imagine an invention that has helped the process of computation more dramatically than has logarithms, the one exception being the modern digital computer. During a conference held in 1914 to celebrate the three hundredth anniversary of the publication of the *Descriptio,* it was estimated that, of all the calculation done in the previous three hundred years, the vast majority had been done with the aid of logarithms.

Today logarithms help to solve equations with variables in an exponent and assist in modeling certain phenomena.

The Slide Rule

The calculators students carry to classes in mathematics, engineering, physics, certain business areas, and other disciplines have replaced a mechanical instrument called a slide rule that was the badge of distinction for students in those classes in the first three-quarters of the twentieth century. The slide rule is an adaptation of the idea of logarithms.

Edmund Gunter (1581–1626), a colleague of Henry Briggs' at Gresham College in London, worked with logarithms because of the many calculations required in his work with astronomy, navigation, and sundials. He noted adding together two logarithms (which is done to multiply two numbers) can be accomplished by making a logarithmic scale on a piece of wood (see ▶ Figure 12.4), then adding the logarithms by means of a compass (what you use to draw a circle—also called a pair of compasses or a pair of dividers). To multiply 2 × 3 on this scale, take the distance for 3 (the distance from 1 to 3) and add this to the distance for 2. If you don't

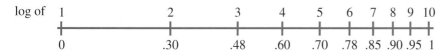

▶ FIGURE 12.4 A logarithmic scale

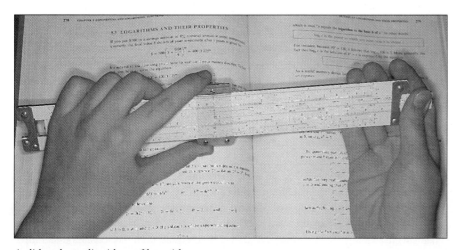

A slide rule applies ideas of logarithms.

CALVIN AND HOBBES ©1992 Watterson. Dist. by UNIVERSAL PRESS SYNDICATE. Reprinted with permission. All rights reserved.

have a compass handy, try this with a piece of paper. Make marks on your paper at 1 and 3 (or put one end of your compass at 1 and the other end of 3). Put the mark that was at 1 at 2. The mark that was at 3 tells the product. It should be at 6.

Gunter's line of numbers was produced on a piece of wood about 2 feet long. Its use spread quickly throughout Europe.

The next step was taken on William Oughtred (1574–1660), an English cleric and mathematician. He noted that to apply Gunter's Line of Numbers, a compass was needed (or a scrap of paper). He observed that if two logarithmic scales were available side by side, one could be slid along to add two distances and the compass would not be needed. This is the idea of the slide rule.

Although Oughtred had the idea for a slide rule around 1622, he did not describe it in print until 1632. It evolved to its current form as a movable sliding piece between two fixed pieces of wood by the 1650s, but very little use was made of the device for about 200 years. A number of people contributed to its development, including James Watt, who worked on the steam engine. The impetus for its widespread acceptance in the 1850s seemed to come with the addition of a movable cursor, a piece that fit on top of the two fixed pieces of wood and moving sliding piece of wood and served to line up the scales. The cursor was invented by Amedee Mannheim (1831–1906), who at the time was a 19-year-old French artillery officer.

The slide rule was a mark of an advance in technology. It was fast and relatively accurate. Although its calculations were only accurate to three or four places, few calculations required more accuracy at that time. It was to the student or professional of the first three-quarters of the twentieth century what a calculator is to the student or professional of today.

Summary

In this section we have looked at some mechanical and mathematical devices that have been invented to aid calculation: an efficient system of numeration, the abacus, logarithms, and the slide rule. All of these require some action by the user other than entering the numbers. In the next section we will look at the development of mechanical calculating machines and computers that require no intermediate human input.

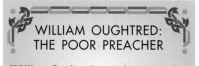

WILLIAM OUGHTRED: THE POOR PREACHER

William Oughtred served as rector in two parishes in England. Several of his parishioners complained that he was a poor preacher because he studied nothing but mathematics.

In Exercises 1 and 2, show the steps in adding the two numbers given in Roman numerals. Use diagrams such as in Figure 12.2.

1. MMDCLXXIII + MDCXXXIV
2. MCCCLVII + MMDCCI
3. Consult Table 12.4 to find the following.
   a. log 10,001   b. log 10,032   c. log 10,093

In Exercises 4 and 5, consult Table 12.4 to estimate the indicated product.

4. 10,001 × 10,032
5. 10,040 × 10,400

In Exercises 6–9, use a calculator to find the log of the stated number.

6. 0.13
7. 1.3
8. 13
9. 130

In Exercises 10 and 11, use the logarithmic scale of Figure 12.4 to find the indicated products. Describe, in a manner like the discussion accompanying Figure 12.4, how you've calculated these.

10. 2 × 4
11. 3 × 3
12. Consult the graph of the logarithm function (see Figure 12.3 or use a graphing calculator) and other facts about logarithms to answer this exercise. For what values of x is log x each of the following?
    a. positive   b. zero   c. negative
13. The number e is sometimes defined as the number such that ln $e = 1$. Graph $y = \ln x$, and use the trace key on a graphing calculator to approximate the x value so that ln $x = 1$. This will be an approximation to e. Approximate e to the nearest hundredth.

1. One federal agency that has *bureau* in its name is the Bureau of Management and Budget. Name two other agencies with *bureau* in the name.
2. In the Middle Ages new numerals called Hindu-Arabic numerals were introduced into Europe.
   a. Write a page promoting the use of Hindu-Arabic numerals.
   b. Write a page opposing the use of Hindu-Arabic numerals. Assume your audience consists of citizens of a French town during the 1400s. (You need not write in French.)
3. (*Library research*) A technique to do multiplication that was common around the time of Napier was called *Gelosia multiplication*. Describe it (one reference is Aspray, 1990, pp. 17–18), and illustrate its use in multiplying two three-digit numbers and a three-digit number times a four-digit number.
4. Howard Eves (1990) says that demands that mathematical calculations be performed more quickly and accurately were met by four remarkable inventions: the Hindu-Arabic notation, decimal fractions, logarithms, and the modern computing machine. Write a page or two describing how decimal fractions enable mathematical calculations to be performed more quickly or accurately.
5. On your calculator, try finding log 0. Describe the output on your calculator. Do you believe this is right? Describe why this happens.
6. (*Library research for those who know some trigonometry*) About 25 years before Napier published his book on logarithms, a technique was practiced that was a precursor of log-

arithms, in that it found the product of the sines of two angles by doing an addition. This *method of prosthaphaeresis* used the following formula:

$$(\sin a) \times (\sin b) = \frac{1}{2}[\cos (a - b) - \cos (a + b)]$$

Write a page describing the method of prosthaphaeresis. Include some historical background and at least one example of its use. Aspray (1990) and Boyer (1991) contain relevant information.

7. (*Verbal report*) Find a slide rule and demonstrate to your class its use in multiplying two numbers. If possible, describe another mathematical operation that can be performed on the slide rule.
8. The invention of logarithms was made independently by Napier in Scotland and Bürgi in Switzerland.
   a. Name another mathematical invention or discovery that was made independently by two or more people.
   b. Why do you believe that creations like logarithms or like what you named in part a often are made independently by two or more people?
9. In teaching mathematics to younger elementary-age children, say, grades K–4, how could an abacus be helpful? What properties of mathematics could it illustrate?
10. Write a paragraph to a page describing the connection between logarithms and the slide rule.

12.1 MECHANICAL AND MATHEMATICAL AIDS TO COMPUTATION

12.2 AUTOMATIC CALCULATING MACHINES AND COMPUTERS

GOALS
1. To describe developments with mechanical calculating machines: the motivation for them, the mathematical ideas behind them, and some of their developers.
2. To describe the early development of computers: the advent of programming, the connections between logic and circuits, and people involved with these developments.
3. To describe the development of tabulating machines and electromechanical calculators.
4. To describe software and hardware developments since the 1940s.

In the preceding section we looked at some mechanical and mathematical devices that aid calculation. All of them require some action by the user in addition to entering the numbers. In this section we look at the development of mechanical calculating machines and computers that require no intermediate human input between entering the numbers and reading the output.

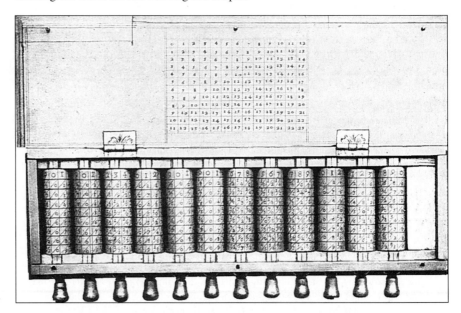

A cylindrical form of Napier's bones

One of Pascal's calculating machines

Mechanical Calculating Machines

Although the slide rule is a mechanical device for carrying out calculations, it depends on the operator to perform a number of tasks. The first calculating machine to do addition with a mechanism that would carry numbers was invented in about 1620 by Wilhelm Schickard (1592–1635). Essentially, Schickard created the first calculating machine.

Schickard was a professor of Hebrew, Oriental languages, mathematics, astronomy, and geography and also a Protestant minister in Tübingen, Germany. He was influenced in the development of his machine by logarithms and another invention of Napier's called *Napier's bones,* rectangular pieces of bone, metal, wood, or cardboard used to multiply large numbers. Napier described this device in a publication in 1617, the year of his death and 3 years after his publication on logarithms.

The next attempt to construct a calculating machine was by Blaise Pascal (1623–1662), a French mathematician and theologian. He created his first machine in 1642, at the age of 19, and during his lifetime produced a total of about 50 machines, all based on the same idea. Some of these are still in existence.

Pascal's machine consisted of a sequence of dials, each numbered from 0 through 9. The machine was built so that when one dial turned from 9 to 0, the dial to its left would increase by 1. This built a carry mechanism into the machine.

Since the dials could turn in only one direction, subtraction was accomplished by adding the first number to what is called the *nines complement* of the second number, then making two adjustments. The nines complement of a number is the number that when added to it will produce a sum that is all nines. For example, the nines complement of 17 is 82 because $17 + 82 = 99$. Similarly, the nines complement of 723 is 276 because $723 + 276 = 999$. Here is how nines complements can replace subtraction with addition: Suppose the problem is 83–28. View this as

$$83 + (71 - 100 + 1) = 154 - 100 + 1$$
$$= 155 - 100$$
$$= 55$$

$\uparrow$ nines complement of 28

We can describe this process as: add 83 and 71 (the nines complement of 28) to get 154, then delete the left-most 1 (subtract 100) to get 54, and add 1 to get 55.

EXAMPLE 12.1

Find $641 - 478$ by adding the nines complement.

SOLUTION The nines complement of 478 is 521. $641 + 521 = 1162$. Drop the left-most 1 to get 162, then add 1 to get 163. So $641 - 478 = 163$.

 YOUR FORMULATION

You have seen that the process of subtracting by adding the nines complement works for $83 - 28$ and $641 - 478$. Describe in general why it works. Let the description of $83 - 28$ guide your thinking.

Gottfried Wilhelm Leibniz (1646–1716), a German mathematician best known as one of the two developers of calculus, was aware of Pascal's mechanical adding machine. He suggested an attachment to Pascal's machine to enable it to do multiplication. He constructed a mechanical multiplier in 1674.

In the mid-1660s Samuel Morland (1625–1695), an English diplomat, probably influenced by Pascal's machine, invented three different calculating machines. One was an adding machine small enough to be carried in a pocket. Another was a machine to do multiplication via Napier's bones.

We have discussed four of the inventors who produced calculating machines in the 1600s: Schickard, Pascal, Leibniz, and Morland. There were at least three reasons for the creation of calculating machines:

- Need to do calculations with large numbers
- Intellectual curiosity. Could the idea of Napier's bones be mechanized? Could a previous calculating machine be improved?
- Toys for the rich. A calculating machine was a conversation piece for the wealthy. It was in such a setting that Samuel Pepys first saw a calculating device. (See the margin note "My, What a Pretty Calculator!")

Although these four inventors and others produced a number of calculating machines from 1620 to the mid-1800s, the machines did not gain widespread acceptance. They were slow and prone to errors.

The first machine to overcome these problems and prove to be relatively fast, reliable, and useful was produced by Thomas de Colmar of France in the early 1820s. It was a revision of a Leibniz model using advances in engineering and design. His "arithmometer" was manufactured until the early 1900s.

The next major advance occurred in the 1870s with the use of a variable-toothed gear patented by Frank S. Baldwin in the United States in 1875 and developed independently by Willgodt T. Odhner, a Swede working in Russia. Machines containing this gear were the first ones that could perform the four fundamental operations of arithmetic—addition, subtraction, multiplication, and division—without resetting the machine. The variable-toothed gears increased reliability and ease of use and allowed for a more compact construction. The need for these machines was prompted by developments of the Industrial Revolution. In turn, techniques for the mass production of these calculating machines were made possible by the Industrial Revolution.

Early Development of Computers

We now turn from calculating machines to computers. The ancestral roots of the computer are traced to Charles Babbage (1791–1871). Babbage has been called the father of the computer, the grandfather of the computer, and the godfather of the computer. He was trained as a mathematician at Cambridge University in England.

Although Babbage worked some as a mathematician, he had an interest in a wide range of intellectual activities and a special interest in mechanical things. This mechanical interest, combined with his attention to detail and concomitant frustration with inaccuracies in the printed mathematical tables of his day, spurred his ambition to mechanize calculation. He desired a machine that would calculate without error and would bypass copying and typesetting errors by printing the results directly.

In 1822 Babbage built a prototype of a calculating machine called a Difference Engine. It was so named because it was based on the *principle of finite differences:* If a polynomial function [a function of the form $f(x) = a_0 + a_1x + a_2x^2 + \cdots + a_nx^n$] is evaluated at successive integral values and the differences between consecutive

TABLE 12.5
Differences for a polynomial function

x	$f(x) = x^3 + 2x + 4$	First difference	Second difference	Third difference
0	4			
		3		
1	7		6	
		9		6
2	16		12	
		21		6
3	37		18	
		39		6
4	76		24	
		63		
5	139			

function values are then examined, and the differences of the differences are then examined, etc., eventually all the differences will be constant. For example, the differences for the function $f(x) = x^3 + 2x + 4$ are shown in ● Table 12.5. Notice that the third differences are all the constant 6.

Table 12.5 permits us to find values of the polynomial by doing only additions. This is illustrated by the arrows in ● Table 12.6. We can consult either table to calculate $f(6)$, but first we will explain how to use such a table to calculate $f(3)$ by referring to the part of Table 12.6 above the top boxed diagonal. The boxed diagonal can be filled in by working from right to left. The entry under "Third Difference" is 6. In fact, all the entries in this column will be 6. We get the next entry to the left in the diagonal by adding 6 + 6 to get 12. The next entry to the left on the diagonal is 12 + 9, or 21, and the next entry to the left, which is $f(3)$, is 21 + 16, or 37. We can calculate $f(4)$ by filling in the diagonal between the boxed diagonals, doing repeated additions to get $f(4) = 76$. The bottom boxed diagonal shows the additions that give us $f(5) = 139$.

TABLE 12.6
Using finite differences to calculate function values

x	$f(x) = x^3 + 2x + 4$	First difference	Second difference	Third difference
0	4			
		3		
1	7		6	
		9		6
2	16		12	
		21		6
3	37		18	
		39		6
4	76		24	
		63		
5	139			

1. In Table 12.6, try the diagonal method, using only additions, to calculate $f(6)$.
2. Compare your answer from part 1 with a direct calculation of $f(6)$.

Babbage's initial efforts brought him the first Gold Medal of the Astronomical Society of London and the interest of the British government. At that time, astronomical tables to aid navigation at sea had to be recalculated yearly. The consequences of errors in calculation could be serious. Therefore, for a nation such as Great Britain that relied heavily on sea trade, there was substantial interest in a fast, accurate calculating machine.

After 11 years of work by Babbage and a number of assistants on developing the actual Difference Engine, over 12,000 parts had been produced and it was partly assembled. In 1833 a dispute arose between Babbage and his chief engineer. By 1834 the British government had stopped supporting the project and work on it ceased. A full-sized Difference Engine was not constructed until the Science Museum in London made one in 1991, 200 years after Babbage's birth, from Babbage's original design. The completed Difference Engine is 7 feet long, 3 feet wide, and 8 feet high and weighs about 3 tons. And it works!

Babbage's credentials as the progenitor of the computer rest on a second calculating machine he conceived, in 1834, called the Analytical Machine. He intended it as a programmable computing machine that would take information from a stack of punched cards and perform any sequence of arithmetic operations. Babbage's Analytical Engine was also not physically completed, but it was described fully in diagrams.

Doron Swade, Senior Curator in Computing and Control at the Science Museum, London, operates the Charles Babbage-designed Difference Engine.

During the later 1800s and early 1900s difference engines and analytical engines were designed by several other people, but Babbage was regarded as the person who made the first major intellectual contribution toward the development of a computer. However, the effective realization of a computer did not take place until over 100 years later. Babbage was ahead of his peers in two other ways. His book *Economy of Manufactures and Machinery* is regarded by some people as containing the original seeds of the discipline in the mathematical sciences known as operations research, although it did not bloom as a discipline until around World War II. Curiously, that's about the time that Babbage's ideas on computers also came to fruition. Babbage was also one of the first people to have government support for his research, although both he and the British government were unhappy about what happened with the research funding. The British government was unhappy that his Difference Engine was not completed after a dozen years, and Babbage was unhappy that the government withdrew the funding of his project.

ADA LOVELACE: COMPUTER PIONEER

Among the people who worked with Charles Babbage was Ada Lovelace (1816–1852), the daughter of the poet Lord Byron. She translated a French paper on some of Babbage's work into English and added extensive notes that included ideas on programming the computer. Some call her the world's first programmer. In 1979 the U.S. Department of Defense assigned the name Ada to a programming language it commissioned.

Logic Machines

Up to this point this chapter has talked about advances in calculation. Today, only a small amount of what computers do is numerical calculation. A majority of students use a computer mainly for word processing. Another major application of computers is data management via databases, such as library databases on books and periodicals. Computers are effective in carrying out both computational and other tasks because they can make logical decisions. For example, to calculate $(2 + 3) \times 4$, the computer's logic must tell it first to add 2 and 3, then to multiply the sum by 4. In alphabetizing a set of words, the computer's logic must enable the machine to sort the words in alphabetical order.

The computer was not the first mechanical device designed to make logical decisions. The first logic machine was designed in 1800 or before by Charles, third Earl of Stanhope (1753–1816). He called it the Demonstrator. It was meant to solve a certain type of logical argument (called a syllogism) mechanically and to solve elementary probability problems. William Aspray (1990, pp. 106–107) describes it:

> It consists of a $4'' \times 4'' \times 0.75''$ mahogany block with a brass top, having carved out of it a window $1'' \times 1'' \times 0.5''$. Slots were grooved in three sides of the block to allow transparent red and gray slides to enter and cover a portion of the window. On the brass face, along three sides of the window, integer calibrations from 0 to 10 were marked.

Consider its application to the following numerical syllogism:

> Seven of ten *A*s are *B*s.
> Five of ten *A*s are *C*s.
> Therefore, at least two *B*s are *C*s.

First, Venn diagrams tell us this is valid reasoning, as shown in ▶ Figure 12.5, in which the nonoverlapping parts are labeled I, II, . . . , VIII. Let n (I) denote the number of objects in region I, n (II) the number of objects in region II, etc. In the syllogism, the first and second statements both say that n (A) = 10, so

$$n\,(\mathrm{I}) + n\,(\mathrm{II}) + n\,(\mathrm{IV}) + n\,(\mathrm{V}) = 10 \qquad (a)$$

The first statement says that

$$n\,(\mathrm{II}) + n\,(\mathrm{V}) = 7 \qquad (b)$$

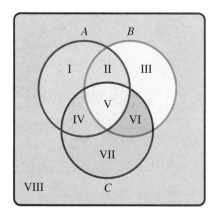

▶ FIGURE 12.5 Venn diagram for a numerical syllogism

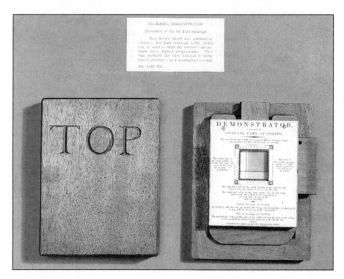

Lord Stanhope's Demonstrator

The second statement says that

$$n\,(\text{IV}) + n\,(\text{V}) = 5 \qquad\qquad\qquad (c)$$

To reach the conclusion, we are interested in $n\,(\text{B} \cap \text{C}) = n\,(\text{V}) + n\,(\text{VI})$. Since we have no information about $n\,(\text{VI})$, we will focus on $n\,(\text{V})$. Add equations (b) and (c) to get

$$n\,(\text{II}) + n\,(\text{V}) + n\,(\text{IV}) + n\,(\text{V}) = 12 \qquad\qquad (d)$$

From equation (a), $\qquad n\,(\text{II}) + n\,(\text{IV}) + n\,(\text{V}) = 10 - n\,(\text{I}).$

Putting this in equation (d) gives

$$10 - n\,(\text{I}) + n\,(\text{V}) = 12$$

or
$$n\,(\text{V}) = 2 + n\,(\text{I})$$

Therefore, $n\,(\text{V}) \geq 2$. So there are at least two objects in $A \cap B \cap C$, and at least two objects in $B \cap C$, or at least two Bs are Cs.

With Lord Stanhope's Demonstrator, the syllogism can be solved by pushing the red (top in the picture) slide (representing B) seven units across the window (representing A) and pushing the gray slide (representing C) five units across the window from the opposite direction (bottom in the picture) . The two slides overlap in two units. These two units represent the minimum number of Bs that are Cs.

The Demonstrator could also calculate certain probabilities. We will not go into detail except to comment that in Chapter 8's discussion of probability, we relied on arguments related to Venn diagrams, which are in turn related to Lord Stanhope's Demonstrator. Lord Stanhope's device was limited to logical arguments with two premises and probability problems with two events. It remained essentially unknown until much later than 1800. However, in the 1860s it inspired William Stanley Jevons (1835–1882), a professor of logic and political economy in England, to construct a "logic piano." Jevons was a student of Augustus De Morgan. (Section 3.5 contains De Morgan's laws for sets, and Section 4.7 contains De Morgan's laws for logic.) According to William Aspray (1990, p. 110):

> The logic piano was a box approximately three feet high. A faceplate above the keyboard displayed the entries of the truth table. Like a piano, the keyboard had black-

and-white keys, but here they were used for entering premises. As the keys were struck, rods would mechanically remove from the face of the piano the truth-table-entries inconsistent with the premises entered on the keys.

In effect, it was a mechanical way to construct a truth table. (Truth tables are discussed in Section 4.3.)

Another English logician and contemporary of Jevons was John Venn (1834–1923). His technique for diagramming logical arguments bears his name—Venn diagrams (see Section 3.4). He also described a machine for diagramming logical arguments and wrote about logic machines in general.

Through the years several other people have invented logic machines or improved on those of Lord Stanhope, Jevons, and Venn. None of these were able to handle the level of logical argument that a modern computer can. However, they did show that the use of logic could be mechanized.

In 1938 Claude Shannon, in his master's thesis at MIT, showed how logic could be modeled by switching circuits and how switching circuits could be described by means of logical connectives.

A year earlier, "On Computable Numbers," a paper by Alan Turing (1912–1954), an English mathematician, was published. In it, Turing described which functions were effectively computable. They were described by means of a theoretical machine now known as a Turing machine. The Turing machine is a theoretical model of a digital stored-program computer.

Turing was also involved in developing an actual computer called the Colossus, which became operational in 1943. It was designed for and used by the British during World War II in decoding intercepted German radio messages. This decoding made a significant contribution to the Allied victory in World War II. Winston Churchill described the work of Turing's team as "my secret weapon."

Both logic machines and mechanical calculating machines influenced the development of computers, with the calculating machines having the bigger impact. Nonetheless, logic machines provided insights into what could be done mechanically with ideas from logic. Today's computers can do much more than what was envisioned for the logic machines. In addition, Shannon's paper served as the impetus for the theory of switching circuits and Turing's work began the area of computer science known as automata theory.

Tabulating Machines

A major contribution toward computer development came in response to a counting problem. The U.S. Constitution requires a census every 10 years. By 1880 the size of the U.S. population and the number of questions asked on the census form caused the compiling of the results of the census to take 7 years. Herman Hollerith (1860–1929), a Census Bureau employee, won a contest to find a faster way of processing the data for the 1890 census. His system was to record information by means of holes punched in a card. It provided the thrust for automatic data processing. His system reduced the time required to compile the results of the 1890 census to a little over 2 years.

Electromechanical Calculators

In the first electromechanical calculators, electricity both drove the motor and activated relays. A relay is the same type of switch involved in sending Morse code over telegraph lines and in telephone transmissions. The relay was developed in the

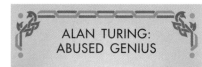

Stibitz model K (for "kitchen")

1930s by at least three people: Konrad Zuse (1910–1995) in Germany and George Stibitz (1904–1995) and Howard Aiken (1900–1973) in the United States.

It appears that the world's first operational programmable digital computer was developed by Zuse, a German engineer, in 1941, during World War II. Zuse, like Babbage before him in England, had trouble convincing the German government of the merit of his machine. The Nazi military personnel pursued technological advances in rockets and atomic weapons, but they seemed less interested in Zuse's computing technology. They were apparently unaware of the computing machine's potential military significance in areas ranging from calculating missile trajectories to coding and decoding messages. (It was only a few years later that the British, under the guidance of Turing, constructed the Colossus computer to decode German messages.)

In 1937 George Stibitz, a mathematician at Bell Labs, wired some relays together to form a device that could add two single-digit binary (base 2) numbers. His work helped show the value of base-2 arithmetic in the design of computers. He went on to help build several calculating machines at Bell Labs.

About the same time that Stibitz was working at his kitchen table, Howard Aiken, at Harvard, was looking for ways that calculating machines could help him with research problems in physics. His ideas, which were supported by IBM, involved relays that were a standard part of IBM tabulating machines. A team of engineers from IBM built an electromechanical calculating machine called the Harvard-IBM Automatic Sequence Controlled Calculator, more commonly known as the Mark I. It was revealed publicly in 1944. It was 51 feet long, 8 feet tall, and 2 feet deep and weighed 5 tons.

Exactly how does the development of the Mark I fit into the history of computing? Most things it did were predated by the machine of Zuse and the British Colossus. However, Zuse's machine was probably unknown outside of Germany. Even inside Germany it was not fully appreciated. Knowledge of Colossus was controlled by

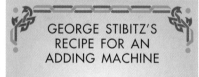

GEORGE STIBITZ'S RECIPE FOR AN ADDING MACHINE

In 1937, working in his kitchen, Stibitz cobbled together a primitive adding device out of dry-cell batteries, metal strips from a tobacco can, flashlight bulbs and telephone wires. Many consider it the earliest antecedent to the digital computer.

Obituary for George Stibitz in *Time*, Feb. 13, 1995.

The Mark I

government security, since it was involved in the war effort. In fact, as of 1996 some details about Colossus are still veiled in secrecy. This made the Mark I the first well-known electronically operated computing machine. Technically, it was not what most people now call a computer, since it was electromechanical rather than purely electronic and it did not employ a stored program. In any case, the Mark I was a notable achievement in computing. Aiken went on to develop the Mark II, III, and IV, which were sophisticated computing machines.

The Mark I contained arithmetic units designed to do arithmetic operations (addition, subtraction, multiplication, and division) and to calculate 10^x and log x when x is given. Aiken estimated that calculation in the Mark I was about 100 times faster than calculation on desk calculators of the time (Moreau, 1984, p. 31). However, to calculate 10^x required over 61 seconds and to calculate log x required over 68 seconds. Still, the computations had 23-place accuracy (Williams, 1985, p. 245).

Even before the Mark I was publicly unveiled it was doing classified work for the U.S. Navy. One of the problems it was enlisted for was to calculate the blast effects of the first atomic bomb. One of the naval officers assigned to work with the Mark I was Grace Murray Hopper (1906–1992), who had been a mathematics professor at Vassar College. She is credited with developing several programming languages, including COBOL. Her efforts have established her as one of the pioneers in software development.

Electronic Computers

The electromechanical computers of Zuse and Aiken and the Colossus, although many times faster than other available calculating machines, were relatively slow. Their relay switches also were prone to breakdowns. Much higher speeds were attained with electronic circuits instead of electromechanical relays. The story of the development of electronic computers is like that of several other mathematical

Grace Murray Hopper: Grandmother of the Computer Age

Grace Murray Hopper has been called the Ada Lovelace of the Mark I. She also popularized the term *bug* as applied to a computer program when she found that a moth trapped in a computer was causing it to malfunction. She taped the moth to a page of the logbook and wrote, "First actual case of bug being found."

The poster honoring Hopper in the National Women's Hall of Fame says:

Mathematics genius, computer pioneer, inventor, teacher—Grace Hopper's accomplishments encompass a wide range of achievements that have helped transform society.

The woman who became known as "Amazing Grace" and as "The Grandmother of the Computer Age" was born before Ford launched his first Model T and before women had the right to vote. Educated at Vassar and Yale, Hopper joined the Navy during World War II

because of a desire to help win the war. Her formidable skills in mathematics helped propel her into the brand new world of "computing machines" and she loved the opportunities to innovate that this brought. She worked on the early UNIVACs, and soon began to create computer "languages"—mathematical equations computers could understand. Recognizing the need for a more "user-friendly" language in English to enable more people to work with computers, she pioneered COBOL, a computer language that promoted easier access. A leader and pioneer in the technology that has transformed information flow forever, Hopper was also the first woman to attain the rank of Rear Admiral in the U.S. Navy. In 1991, she won the National Medal of Technology.

Grace Hopper lived out the charge she gave to others:

A ship in port is safe, but that is not what ships are for. Be good ships. Sail out to sea and do new things.

Grace Murray Hopper

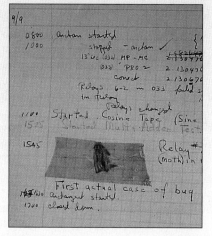

She found the first actual case of a bug in a computer program.

developments: Substantial work set the stage for their development, several people developed the idea at about the same time, and controversy arose over who developed it first.

From 1939–1942 John Atanasoff (1903–1995), a professor of physics at Iowa State College (now Iowa State University), and a graduate student, Clifford Berry, built an electronic computer to help solve systems of linear equations. The machine, limited in its ability, was left behind when Atanasoff and Berry left Iowa State College. In Germany, Helmut Schreyer saw that electronic circuits could replace the mechanical and electromechanical circuits in Zuse's machines, but World War II brought his efforts to a halt. The British Colossus machines (there were probably 10 in use by the end of World War II) incorporated a combination of vacuum tubes and electromechanical relays. However, the Colossus did not perform ordinary arithmetic. It focused on logical operations for decoding German messages.

What a number of people have described as the world's first general-purpose electronic digital computer was developed in 1946 by two professors of electrical engineering at the University of Pennsylvania, John Mauchly (1908–1980) and J. Presper Eckert, Jr. (1919–1995). Their machine, the ENIAC (Electronic Numerical Integrater and Calculator), was housed in a 30- by 50-foot room, contained over 18,000 vacuum tubes, and weighed 30 tons. It could do arithmetic calculations about 1000 times as fast as the Mark I and 100,000 to 1 million times as fast as a desk calculator of that time. Its development was funded by the U.S. Army to speed up the calculation of ballistic tables.

Years later, the question of who should be credited with the invention of the computer was an essential part of a federal court case between computer giants Honeywell and Sperry Rand. The judge ruled in 1973: "Eckert and Mauchly did not themselves first invent the automatic electronic digital computer, but instead derived that subject matter from . . . Atanasoff." Historians of computing still argue over who should be credited with inventing the computer, but all or almost all agree that Atanasoff, Berry, Eckert, and Mauchly all made significant contributions to computing.

The ENIAC, with some modifications, was in service from 1946 until 1955. According to Michael Williams (1985, p. 287): "It has been conjectured that, during the ten years of ENIAC's useful life at the Aberdeen Proving Grounds, this machine did more arithmetic than had been done by the whole human race prior to 1945."

Stored-program Electronic Computers

Electronic computers such as the ENIAC offered huge increases in the capability of mechanized arithmetic over relay machines or mechanical machines. However, a design challenge was the blending of the speeds of arithmetic, input, output, and programming. This challenge was met with the development of the stored program.

The concept of a stored program is usually attributed to the brilliant Hungarian American mathematician John von Neumann (1903–1957). This concept was incorporated into the design of the EDVAC (Electronic Discrete Variable Automatic Computer) at the University of Pennsylvania. This machine was completed in 1951. The stored program allowed the machine to have instructions in its memory rather than feeding instructions to it on paper tape or punched cards or by reconfiguring cable connections.

Before the EDVAC was completed, Mauchly and Eckert left the University of Pennsylvania to form their own computer company. This company was taken over by Remington-Rand, which produced the first commercial computer, the UNIVAC I (Universal Automatic Computer) in 1951. This machine was purchased by the U.S. Census Bureau.

Technological Changes

Calculations have been speeded up through the years by mathematical developments (systems of numeration, logarithms), design changes, and technological developments. Technological changes have included going from strictly mechanical

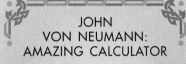

JOHN VON NEUMANN: AMAZING CALCULATOR

John von Neumann was born and educated in Hungary before moving to the United States in 1930. He taught first at Princeton University, and in 1933 became a member of the Institute for Advanced Study at Princeton. He had amazing calculational abilities and made significant contributions to a large number of diverse areas in the mathematical sciences. In addition to writing on different aspects of pure mathematics, including set theory and logic, he wrote on quantum theory, developed and applied game theory to economics in particular and social sciences more broadly (see Section 13.2), studied the functioning of the human brain and related it to the computer, and was involved in construction of the hydrogen bomb. His enormous reputation helped popularize computers.

machines such as Babbage's to electromechanical relays (Zuse, Stibitz, Aiken) to vacuum tubes (the Colossus, Atanasoff and Berry, Mauchly and Eckert) to transistors to integrated circuits on silicon chips. Developments in both computer hardware and software have aided the user. Immense changes have been made in the last 60 years and significant developments are likely in the next 60 years.

1. Find the nines complement of 381.
2. Find the nines complement of 652.
3. Find 83 − 76 by adding the nines complement.
4. Find 695 − 128 by adding the nines complement.
5. Find 4032 − 2164 by adding the nines complement.
6. Find 92,345 − 36,789 by adding the nines complement.

In Exercises 7–12, construct a table of differences (like Table 12.5).

7. $f(x) = x^2$ 8. $f(x) = x^3$
9. $f(x) = x^4$ 10. $f(x) = x^5$
11. $f(x) = 2x^3 - 4x^2 + 5x - 6$ 12. $f(x) = -2x^3 - 3x^2 - 4x + 5$
13. Predict what the constant difference is for the function $f(x) = x^n$, where n is a positive integer.
14. Let $f(x)$ be a polynomial function that ends up with a constant difference $d \neq 0$. Given the function $g(x) = f(x) + r$, where r is a real number, what will be the eventual nonzero constant difference for $g(x)$? Explain why.
15. Consider the following numerical syllogism:

    Six of ten As are Bs.
    Three of ten As are Cs.
    Therefore, at least two Bs are Cs.

    a. Is this valid reasoning?
    b. Support your answer by describing the solution to the syllogism using Lord Stanhope's Demonstrator.
    c. Support your answer with a Venn diagram, as in Figure 12.5.
16. Consider the following syllogism:

    No B is A.
    All C is A.
    Therefore, no B is C.

    a. Is this valid reasoning?
    b. Support your answer by describing the solution to the syllogism using Lord Stanhope's Demonstrator.
    c. Support your answer with a Venn diagram.

17. Stibitz' model K circuit could add two single-digit binary numbers. Write all possible sums of two single-digit binary numbers.
18. Match the development or description in the first column with the person or persons in the second column and the dates in the third column. A development may have more than one associated person.

Development or Description	People	Date
Ancestor of the computer	Aiken	before 1600
Connection between logic and switching circuits	Atanasoff	1600s
	Babbage	1700s
Electromechanical calculator	Berry	1800s
First electronic computer	de Colmar	1900–1935
Helped develop Colossus	Eckert	after 1935
Logic machine	Hollerith	
Mechanical calculating machine in the 1600s	Hopper	
	Jevons	
One of the first programmers	Leibniz	
Pioneer in software development	Lovelace	
Stored-program computer	Mauchly	
Successful commercial calculating machine	Morland	
	Odhner	
Tabulating machine	Pascal	
	Schickard	
	Shannon	
	Stanhope	
	Stibitz	
	Turing	
	Venn	
	von Neumann	
	Zuse	

1. (*Library research*) Write a one- to three-page paper describing Napier's bones. Include an example of how to use them to multiply a three-digit number and a four-digit number and an example of how to use them to multiply two four-digit numbers.

Aspray (1990) and Eves (1990) contain information on Napier's bones.
2. (*Library research*) Write a two- to five-page paper on whether or not Ada Lovelace was the world's first programmer. Some

possible sources are Aspray, 1990; Moreau, 1984; Williams, 1985; Dorothy Stein, *Ada: Life and Legacy* (Cambridge, MA: MIT Press, 1987); Betty Alexandra Toole, *Ada, Enchantress of Numbers,* (Mill Valley, CA; Strawberry Press, 1992); and Raymond Kurweil, *The Age of Intelligent Machines* Cambridge, MA: MIT Press, 1990).

3. Sometimes a government agency funds a research project that, like Babbage's, goes on for years without achieving the desired result. Write a one- or two-page paper either supporting or opposing the position that if after 5 years a government-supported research project has not reached its goal, then that project should no longer be supported by the government.

4. (*Library research*) Write a three- to five-page paper on one of the following accounts of Alan Turing's life.
   a. *Breaking the Code* by Hugh Whitemore
   b. *Alan Turing: The Enigma* by Andrew Hodges.
   Include:
   • A summary of the mathematics or computer science done by Turing
   • A description of Turing's personal life, with at least one anecdote about him
   • Comments from a review of the play or the book

5. Hollerith's development of the forerunner of the punched-card machine was based on a counting problem. Name at least two

(other) mathematical developments that were based on a counting problem. Describe the counting problem, the mathematical development, and the historical situation surrounding the development.

6. In 1944 the Mark I required over 61 seconds to calculate 10^x and over 68 seconds to calculate log x. Today your calculator does those same calculations instantly. Write a paragraph to a page commenting on the impact of that change in capability.

7. (*Library research*) Write a two- to five-page paper on John von Neumann. Include at least one anecdote about his computational ability, at least one additional comment (beyond what is in this section) on his development of the stored program for a computer, and additional information about his work in some other area.

8. Outline the developments in calculating that have been described in sections 12.1 and 12.2.

9. List reasons for the development of calculators and computers and improvements in them.

10. In this section, Babbage was described as the father of the computer and Hopper was described as the grandmother of the computer. How can this be reconciled with the fact that Babbage died before Hopper was born?

12.3 ⬥ PROGRAMMING CONCEPTS

GOALS
1. To follow the execution of program segments involving assignment statements, if-then segments, if-then-else segments, nested conditional statements, while loops, and for-next loops.
2. To describe, give examples of, and recognize iterated procedures.
3. To construct a trace table for an iterated procedure.
4. To modify program segments to produce related results.

Languages

When two people talk to each other, they use a language that both understand. It might be English, Spanish, sign language, pig Latin, or something else. When we want a computer to perform a procedure, we must communicate with it in a language that it understands. Some common languages that computers understand are: BASIC, FORTRAN, Pascal (named after Blaise Pascal, see Sections 8.1 and 12.2), PL/1, C, and Ada (named after Ada Lovelace, see Section 12.2). In this chapter we will write instructions for the computer in a kind of pseudocode combining elements of all those languages. This pseudocode can easily be translated into other languages. However, it is not our intent that you learn to write a computer program, but rather that you learn how a computer program functions.

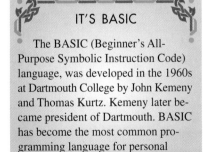

IT'S BASIC

The BASIC (Beginner's All-Purpose Symbolic Instruction Code) language, was developed in the 1960s at Dartmouth College by John Kemeny and Thomas Kurtz. Kemeny later became president of Dartmouth. BASIC has become the most common programming language for personal computers.

Algorithms

An *algorithm* is a procedure for solving a problem, especially a computational procedure, such as dividing 3302 by 13 by means of long division:

$$
\begin{array}{r}
254 \\
13\overline{)3302} \\
\underline{26} \\
70 \\
\underline{65} \\
52 \\
\underline{52}
\end{array}
$$

This procedure is called a division algorithm. Computers are ideally suited to perform algorithms since they can do repetitive tasks with great speed and accuracy. Some algorithms are built into computers, while others are entered as a program to be performed.

Variables and Expressions

In the pseudocode that we will use, the term *variable* will mean a specific storage location in a computer's memory. If we say that the variable x has the value 2, this means that the memory location corresponding to x contains the number 2. A given memory storage location can only hold one value at a time, so if a variable is given a new value during the execution of a program, then the new value is stored in the storage location and the old value is gone.

Variables can take on values from many different sets, including integers, real numbers, and character strings (words or other sequences of symbols from the keyboard).

An algebraic expression involving variables is evaluated by the computer by substituting the current value of all the variables (stored in memory) for the variables in the expression and performing the indicated operations. For example, if the current value of x is 3, then the algebraic expression

$$2 \cdot x + 4 \qquad \text{is evaluated as} \qquad 2 \cdot 3 + 4, \text{ or } 10$$

We will use the following symbols of operation:

Symbol	Operation
+	Addition
−	Subtraction
*	Multiplication
/	Division
^	Exponentiation

Here's an example using the exponentiation symbol:

$$5^3 = 5*5*5 = 125$$

Assignment Statements

We assign a value to a variable through an *assignment statement*. To assign a value to variable x, we will write

```
x := <value>
```

where <value> stands for a particular value. For example, we could assign 3 to x by writing

$$x := 3$$

It is also possible to assign values from one (or more) variables to another variable. For example,

$$x := y + z$$

first adds the values in y and z and then assigns that value to x.

EXAMPLE 12.2

Find the value of x when the following program segment is executed.

```
i := 2
j :=-3
x := i^4 - 5*j + i/j
```

SOLUTION The value of x is

$$2^4 - 5*(-3) + 2/(-3) = 16 + 15 - \frac{2}{3}$$
$$= 31 - \frac{2}{3}$$
$$= \frac{93}{3} - \frac{2}{3}$$
$$= \frac{91}{3} = 30\frac{1}{3}$$

Notice that the evaluation depends on the usual conventions about the order in which operations are performed: exponentiation, multiplication, and division are done before addition or subtraction.

Conditional Statements

The statements just described are executed one after the other. This sequential order can be altered by means of conditional statements. A *conditional statement* in logic is one of the form "if p, then q." It says that if a condition (p) is satisfied, then another condition (q) must also be satisfied. In our pseudocode one form of a conditional statement is:

```
if        (condition)
  then s₁
  else s₂
```

Here, *condition* is a statement that is either true or false. The symbols s_1 and s_2 represent statements or groups of statements. If is often the case that *condition* is a relational test such as $x < y$ or $x = y$. In general, indentation indicates that statements are grouped together.

EXECUTION OF AN *IF-THEN-ELSE* STATEMENT

1. *Condition* is evaluated by substituting the current values of the variables and determining the truth or falsity of the resulting statement.

2. If *condition* is true, then s_1 is executed and execution moves to the next program statement after the **if-then-else** statement.

   If *condition* is false, then s_2 is executed and execution moves to the next program statement after the **if-then-else** statement.

EXAMPLE 12.3

Consider the following program segment.

```
if (j < 5 or j > 9)
  then y := 1
  else y := 0
```

What is the value of *y* after execution of this program segment for the following values of *j?*

a. $j = 0$ b. $j = 5$ c. $j = 10$

SOLUTION

a. When $j = 0$, the statement $j < 5$ or $j > 9$ is true, so we go to the **then** part and get $y = 1$.

b. When $j = 5$, the statement $j < 5$ or $j > 9$ is false, so we go to the **else** part and get $y = 0$.

c. When $j = 10$, the statement $j < 5$ or $j > 9$ is true, so we go to the **then** part and get $y = 1$.

The program segment in Example 12.3 can be represented by the flowchart in ▶ Figure 12.6.

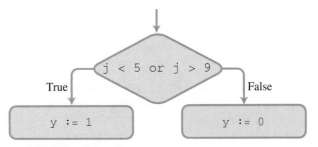

▶ FIGURE 12.6 Flowchart for Example 12.3

EXAMPLE 12.4

Consider the following program segment.

```
if Hrs < 30
  then Cl := Freshman
  else Cl := Not freshman
```

What is the value of Cl after execution of this program segment for the following values of Hrs?

a. Hrs = 15   b. Hrs = 30   c. Hrs = 45

a. Since Hrs < 30, Cl is Freshman.
b. Since the statement "Hrs < 30" is false, Cl is Not freshman.
c. Since the statement "Hrs < 30" is false, Cl is Not freshman.

The program segment in Example 12.4 can be represented by the flowchart in ▶ Figure 12.7.

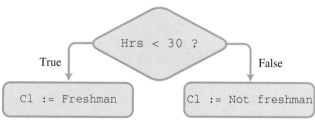

▶ **FIGURE 12.7** Flowchart for Example 12.4

A program can have one conditional statement nested within another one. The next example illustrates this.

EXAMPLE 12.5

Given the following program segment, find the value of q when the program segment is executed for the stated values of x and y.

```
if x > 0
        then if y > 0
                then q := 1
                else q := 4
        else if y > 0
                then q := 2
                else q := 3
```

a. $x = 1$ and $y = 2$
b. $x = -1$ and $y = -2$
c. $x = 1$ and $y = -2$
d. $x = -1$ and $y = 2$

SOLUTION

a. Since "$x > 0$" is true, we are in the top **(then)** branch of the program segment. Since "$y > 0$" is true, $q = 1$.
b. Since "$x > 0$" is false, we are in the bottom **(else)** branch of the program segment. Since "$y > 0$" is false, $q = 3$.
c. Since "$x > 0$" is true, we are in the top branch of the program segment. Since "$y > 0$" is false, $q = 4$.
d. Since "$x > 0$" is false, we are in the bottom branch of the program segment. Since "$y > 0$" is true, $q = 2$.

Notice that for a point (x, y) not on one of the coordinate axes, the program segment tells in which quadrant q the point is located. For a point on one of the coordinate axes, the program segment still assigns a q-value to it (see Exercise 7).

While Loops

A **while** loop is used when a sequence of statements in a program is to be executed over and over again. It has the following form.

while *(condition)*
[statements that make up the body of the loop]
end while

Condition is a statement involving variables in the program. A particular example follows.

EXAMPLE 12.6

Describe the execution of the following *while* loop.

```
n := 1, f := 1
while (n < 4)
      n := n + 1
      f := n * f
end while
```

SOLUTION First, *n* and *f* are both assigned the value 1. Since $n < 4$, the **while** loop is entered. Now *n* is set to 2 and *f* is set to 2*1. Since *n* is still less than 4, the **while** loop is repeated. This time *n* is set to 3 and *f* is set to 3*2*1. Since *n* is still less than 4, the **while** loop is repeated again. This time *n* is set to 4 and *f* is set to 4*3*2*1. Now the statement "$n < 4$" is false, so execution progresses to **end while.** After execution of the **while** loop, the value of *n* is 4 and the value of *f* is 4! = 4*3*2*1 = 24.

YOUR FORMULATION

1. Modify the **while** loop of Example 12.6 to produce the value 13!.
2. Modify the **while** loop of Example 12.6 to produce the value *i*! for any positive integer *i*. In other words, given a positive integer *i*, the desired output is *i*!.

A **while** loop is an example of an *iterated* process. Each pass through the loop is called an *iteration* of the loop. Another iterated process is a **for-next** loop, described later in this section.

The description of the execution of the loop in Example 12.6 can be summarized in the following table, called a *trace table*. The values of the variables are their values after the given iteration. "After iteration 0" means the same thing as "before iteration 1."

Trace Table

		Iteration Number			
		0	1	2	3
Variable Name	*n*	1	2	3	4
	f	1	2	6	24

EXECUTION OF A *WHILE* LOOP

1. *Condition* is evaluated by substituting the current values of all the variables in *condition* for the variables. The resulting statement is determined to be true or false.
2. If the statement is true, all the statements that make up the body of the loop are executed in the order in which they appear. Execution then returns to step 1.
3. If the statement is false, execution moves to the next program statement after the **while** loop (the next statement after **end while**).

A **while** loop can be represented by the flowchart in ▶ Figure 12.8.

A second type of iterated process is the **for-next** loop, which has the following form:

for variable := initial expression **to** final expression
                    [statements that make up the body of the loop]
next (same) variable

A particular example follows.

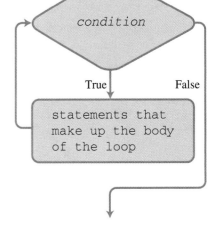

▶ FIGURE 12.8 Flowchart for **while** loop

EXAMPLE 12.7

Construct a trace table to describe the execution of the following program segment.

```
f := 1
for i := 1 to 4
        f := i*f
next i
```

SOLUTION

		Iteration Number				
		0	1	2	3	4
Variables	*i*		1	2	3	4
	f	1	1	2	6	24

Notice that this program segment produces 4!. This was also done in Example 12.6 using a **while** loop.

EXECUTION OF A *FOR-NEXT* LOOP

1. The "variable" is set equal to the value of "initial expression."
2. If the value of "variable" is less than or equal to the value of "final expression," then the statements in the body of the program are executed.
3. If the value of "variable" is greater than the value of "final expression," then execution moves to the next program statement after the **for-next** loop [the next statement after **next** (same) variable].

For-next loops are used when it is known in advance exactly how many times a process is to be repeated. A **while** loop is for when it is not known how many times a loop is to be repeated.

In summary, we communicate with a computer via a programming language, many of which are available. Numerical values or keyboard characters like "Freshman" can be assigned to variables. Programs are algorithms that can execute logical commands, like **if-then-else** statements, and can harness the ability of a computer to perform many arithmetic operations rapidly in order to do complicated calculations.

In Exercises 1 and 2, find the value of x when the following program segment is executed.

1.
```
a := 1
b := 2
x := 3*a - b/a + b^5
```

2.
```
m := 4
n := m/2
x := m^n
```

In Exercises 3–6, find the value of z for each given value of i after execution of the given program segment.

3.
```
if i < 10
        then z := 1
        else Z := 2
```
a. $i = 2$ b. $i = -7$ c. $i = 11$

4.
```
if i > 0
    then z := i
    else z := -i
```
a. $i = 5$ b. $i = 7$ c. $i = 9$
d. Describe z as a function of i.

5.
```
if (i > 12 and i < 20)
  then z := teenager
  else z := nonteenager
```
a. $i = 9$ b. $i = 15$ c. $i = 20$ d. $i = 30$

6.
```
if i = 12
  then z := chapter 12
  else z := not chapter 12
```
a. $i = 2$ b. $i = 12$ c. $i = 21$

7. In the program segment of Example 12.5 find the value of q for a point on each of the following.
   a. The positive x-axis
   b. The negative x-axis
   c. The positive y-axis
   d. The negative y-axis
   e. $(0, 0)$

8. Find the value of s in the following program segment for the given values of a and b.
```
if a > 0
    then if b > 0
            then s := a + b
            else if a > -b
                    then s := a - (-b)
                    else s := -(-b - a)
    else if b < 0
            then s := - (-a + -b)
            else if -a > b
                    then s := -(-a - b)
                    else s := b - (-a)
```
   a. $a = 2$ and $b = 5$ b. $a = 2$ and $b = -5$

c. $a = 2$ and $b = -1$ d. $a = -3$ and $b = 2$
e. $a = -3$ and $b = 6$ f. $a = -3$ and $b = -4$
g. $a = -3$ and $b = 3$
h. Describe what this program segment does.

9. Find the value of p in the following program segment for the given values of a and b.
```
if a > 0
        then if b > 0
                then p := a*b
                else p := -(a*(-b))
        else if b > 0
                then p := -((-a)*b)
                else p := (-a)*(-b)
```
   a. $a = 2$ and $b = 3$ b. $a = 2$ and $b = -3$
   c. $a = 2$ and $b = 0$ d. $a = 0$ and $b = 1$
   e. $a = -2$ and $b = 3$ f. $a = -2$ and $b = 0$
   g. $a = -2$ and $b = -4$
   h. Describe what this program segment does.

10. a. The **for-next** loop in Example 12.7 calculates 4!. Modify it to produce a **for-next** loop that calculates 100!.
    b. Modify the **for-next** loop in Example 12.7 to produce a **for-next** loop that will calculate $n!$ for any positive integer n.

In each of Exercises 11 and 12 construct a trace table to describe the execution of the program segment.

11.
```
for i := 1 to 5
        s := i^2
next i
```

12.
```
for k := 1 to 6
        n := 10^k
        x := (1 + (1/n))^n
next k
```

13. a. Construct a trace table to describe the execution of the following program segment.
```
a := 1
b := 1
for i := 3 to 7
c := a + b
a := b
b := c
next i
```
    b. State a general description, rather than a particular number, for what this program segment produces.

14. Write a program segment that will produce the twentieth Fibonacci number (the twentieth number in the Fibonacci sequence). Recall that the Fibonacci sequence is the sequence $1, 1, 2, 3, 5, \ldots$.

1. If we want to use a computer to calculate 4!, we can simply write 4*3*2*1. Why then might someone want to use a **while** loop like that in Example 12.6 to calculate a factorial? Be specific.

2. Some people believe that by writing a computer program to perform a process, the author of the process learns how to do the process. Do you believe that a junior high school student could learn rules of adding signed numbers better by writing a computer program to do it than by traditional methods? Write a paragraph to a page supporting your answer.

3. (*For those familiar with a computer spreadsheet*) Describe how some of the program segments listed in this section (including the exercises) could be implemented by means of a spreadsheet.

4. Assume a faculty member in the philosophy department at your school has stated in the campus newspaper that she doesn't know what an algorithm is. Write a letter to the editor explaining what an algorithm is.

12.4 SPEED AND ALGORITHMS

GOALS
1. To describe the increase in computing speed since the 1950s.
2. To use the two algorithms given for determining whether or not a number is prime and to use the factoring algorithm.
3. To know and explain why a computer's ability is based on both its hardware and its software.
4. To interpret simple BASIC programs.

Sections 12.1 and 12.2 chronicle some advances in computing, especially those that involve mathematics rather than just technology. The advances were prompted by necessity, technological developments, and curiosity. Development of a writing medium and its improvement to paper encouraged the creation and evolution of a written numeration system and algorithms for doing the arithmetic operations (addition, subtraction, multiplication, and division). Logarithms were invented, at least in part, in response to the need for calculation with large numbers in astronomy, navigation, trade, engineering, and warfare. The demands of business transactions, the census, and warfare encouraged the development of computing machines. Technological development with vacuum tubes, transistors, and integrated circuits have made more computing power available. Section 12.3 examines ways a computer uses processes of logic and iterative processes. Pervasive throughout these developments is inventive (mathematical and technical) curiosity and genius.

In this section we examine the increase in the speed of calculation that computing devices have made possible and the nature of algorithms used in computing.

Speed

The mathematical and technological developments that have been described in the preceding sections have served to make calculations faster. ● Table 12.7 gives an indication of the immense reductions in calculating times that have been achieved. The table is consistent with what R. Moreau (1984, p. 188) describes as "the 100-fold increase in speed every 10 years that is generally reckoned to be typical."

Another indication of the increase in speed of computers is the time it takes a computer to show that a large number is or is not prime. The time it took various machines to show that the number $2^{8191} - 1$ (a 2416-digit) is not prime is shown in

DILBERT by Scott Adams

Reprinted by permission of United Feature Syndicate, Inc.

TABLE 12.7
Increase in Computer Speed

Machine	Date	Number of Multiplications per Second
Mark I	1944	1/3
Zuse Z4	1945	1/3
ENIAC	1946	360
UNIVAC I	1951	400
IBM 701	1953	2,000
NORC	1954	30,000
PDP 1	1959–63	50,000
IBM 7094	1959–63	100,000 – 250,000
LARC	1959–63	125,000
CDC	1959–63	170,000

● Table 12.8. This increase in computing speed represents more than a 100-fold increase every 10 years from the late 1950s to the late 1970s (see Exercise 1) and about a 100-fold increase from 1961 to 1971.

In *Business Week 1994 Annual,* Peter Coy (p. 54) comments on increases in speed:

Very, ultra, super, extremely: The rapid devaluation of superlatives is a hallmark of the Information Revolution. Progress is so rapid that what is powerful today will be passé tomorrow. Million-dollar vacuum-tube computers with fearsome names like ENIAC that awed people in the 1950s couldn't keep up with a 1990s pocket calculator.

TABLE 12.8
Decrease in Time to Show that $2^{8191} - 1$ is Not Prime

Computer	Approximate Date	Time
ILLIAC-I	Late 1950s to early 1960s	100 hours
IBM System 7090	1961	5.2 hours
ILLIAC-II	1963	49 minutes
IBM System 360/91	1971	3.1 minutes
CRAY-1	1979	10 seconds

Some Algorithms

Section 12.3 has some examples of computer algorithms. To see another example of how a computer algorithm works we'll consider an algorithm that determines whether or not a positive integer is prime.

Recall that a positive integer greater than one is *prime* if and only if it has no divisors other than 1 and itself. For example, 7 is prime and 6 is not prime. The idea for determining whether or not a number is prime is to start with 2 and see if it divides the number. If it does, the number is not prime (unless the number is 2). If 2 is not a divisor of the number, try 3 as a divisor. This process is continued until a divisor is found or it is determined that the number is prime. Based on this idea of successive trial divisors, here is a very crude algorithm.

CRUDE ALGORITHM FOR DETERMINING WHETHER A NUMBER IS PRIME

1. Get a number N.
2. If $N = 1$, write a special message. (1 is not a prime.)
3. If $N = 2$, then N is prime (and you are done). Write "N is prime."
4. Initialize the trial divisor to 2.
5. Repeat steps 6 to 8 until you determine whether or not N is prime.
   6. If the trial divisor is a divisor of N, then N is not prime (and you are done). Write "N is not prime."
   7. If the trial divisor is not a divisor of N, then increase the trial divisor by 1.
   8. If the trial divisor equals N, then N is prime (and you are done). Write "N is prime."

This algorithm guarantees you will reach a termination stage because eventually the trial divisor will either divide N or be equal to N. The following example looks at the steps of this algorithm of $N = 5$.

EXAMPLE 12.8

Write out the steps of the preceding algorithm for $N = 5$.

SOLUTION
1. $N = 5$
2. $N \neq 1$, so go to step 3.
3. $N \neq 2$, so go to step 4.
4. The trial divisor is 2.
5. First time
   6. 2 is not a divisor of 5. Go to step 7.
   7. Increase the trial divisor to 3.
   8. Since 3 does not equal 5, go back to step 6.
   Second time.
   6. 3 is not a divisor of 5. Go to step 7.
   7. Increase the trial divisor to 4.
   8. Since 4 does not equal 5, go back to step 6.
   Third time.
   6. 4 is not a divisor of 5.
   7. Increase the trial divisor to 5.
   8. Therefore, 5 is prime.

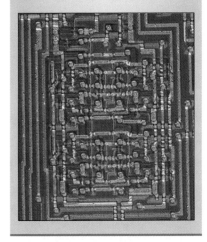

Advances in Computer Technology

Advances in computer technology can be measured not only by speed but also by memory, cost, and the power of a machine. The following comparisons over time were supplied by Dale Jarman, Manager, Computer Center, Central Michigan University.

Computing technology has changed at an astounding rate during the last 20 years. One illustration of such change is the amount of memory available in a typical personal computer (PC). In 1975, less than 1K of memory was available; by 1985, available memory had increased to 512K. Eight years later, in 1993, PCs typically had 16 Meg (16,000K) of memory, and industry

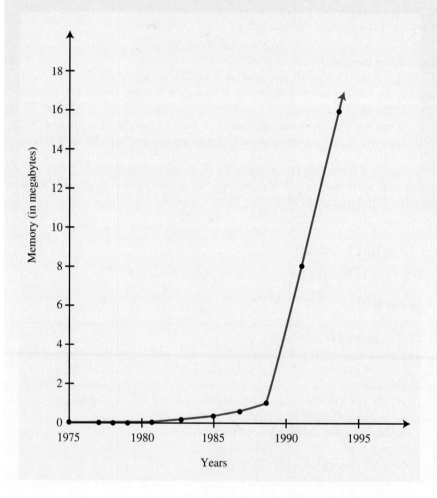

projections indicate 64 Meg of memory will be available on PCs by 1997. This growth is shown in the following graph.

Another method of considering advances in computing technology is to compare cost to power. The graph following shows that the cost (in actual dollars) of a personal computer has decreased steadily since 1975. However, the power of the PC has grown tremendously. In this comparison, the Norton power rating, which assigns a power rating of 1 to the 64K machine of 1981 with an 8088 chip, has been used.

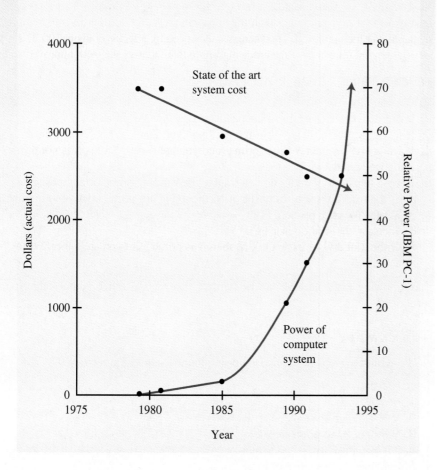

Although the crude algorithm works, it is not very efficient. Two improvements can be made to reduce drastically the number of trial divisors that need to be examined before you can determine whether the number is prime or composite. First, consider the trial divisor 2. If 2 is not a divisor of N, then no other even integer will divide N because N is odd. Thus, there is no need to check the divisors 4, 6, 8, ... This means that about half of the trial divisors have been eliminated.

Let's next consider how large the trial divisors have to be before we can quit. In our crude algorithm, step 8 said we could quit when the trial divisor equaled N. Actually you can quit as soon as the trial divisor exceeds the square root of N.

 YOUR FORMULATION

Explain why we can quit in the crude prime-number algorithm as soon as the trial divisor exceeds the square root of N.

This means that if you are checking a prime number N near 10,000, you only need to look for divisors less than or equal to $\sqrt{N}$ (about 100). Since you are checking for odd divisors, you can determine that the number is prime by using approximately 50 trial divisors rather than the approximately 10,000 trial divisors required by the first crude algorithm. This increases the speed by a factor of about 200. Here is a fairly easy modification of the crude algorithm that incorporates both improvements.

REVISED PRIME-NUMBER ALGORITHM

1. Get a number N.
2. If $N = 1$, write a special message and stop. (1 is not a prime.)
3. If $N = 2$ or $N = 3$, then N is prime (and you are done). Write "N is prime."
4. If 2 is a divisor, then N is not prime (and you are done). Write "N is not prime."
5. Initialize the trial divisor to 3.
6. Repeat steps 7 to 9 until you determine whether N is prime or composite.
    7. If the trial divisor is a divisor of N, then N is not prime (and you are done). Write "N is not prime."
    8. Increase the trial divisor by 2.
    9. If the trial divisor exceeds $\sqrt{N}$, then N is prime (and you are done). Write "N is prime."

The following example looks at the steps of this revised prime-number algorithm for $N = 97$.

EXAMPLE 12.9

Write out the steps of the revised prime-number algorithm for $N = 97$.

SOLUTION
1. $N = 97$
2. $N \neq 1$, so go to step 3.
3. $N \neq 2$ or 3, so go to step 4.
4. 2 is not a divisor, so go to step 5.
5. Initialize the trial divisor to 3.
6. First time.
    7. 3 is not a divisor of 97, so go to step 8.
    8. Change the trial divisor to 5.
    9. 5 does not exceed $\sqrt{97}$ ($\sqrt{97} \approx 9.8$), so go back to step 7.

Second time.

7. 5 is not a divisor of 97, so go to step 8.
8. Change the trial divisor to 7.
9. 7 does not exceed $\sqrt{97}$, so go back to step 7.

Third time.

7. 7 is not a divisor of 97, so go to step 8.
8. Change the trial divisor to 9.
9. 9 does not exceed $\sqrt{97}$, so go back to step 7.

Fourth time.

7. 9 is not a divisor of 97, so go to step 8.
8. Change the divisor to 11.
9. 11 exceeds $\sqrt{97}$. Therefore, 97 is prime.

Exercise 7 asks you to compare the number of trial divisors used in Example 12.9 with the number that would have been used in determining that 97 is prime by the crude prime-number algorithm.

Actual computer code for this algorithm in the programming language Pascal is given in ▶ Figure 12.9.

```
FUNCTION PrimeCheck (Num : integer) : boolean;
VAR
    TrialDivisor : integer;
    Prime : boolean
    Bound : integer;
BEGIN
    Prime := true;
    IF Num MOD 2 = 0 THEN
        Prime := false
    ELSE
        BEGIN
            TrialDivisor := 3;
            Bound := trunc (sqrt(Num));
            WHILE (TrialDivisor <= Bound) AND Prime DO
                IF Num MOD TrialDivisor = 0 THEN
                    Prime := false
                ELSE
                    TrialDivisor := TrialDivisor + 2
        END; { of ELSE option }
    PrimeCheck := Prime
END; { of FUNCTION PrimeCheck }
```

▶ FIGURE 12.9 Pascal code for prime-number algorithm

We have looked at two algorithms for determining whether or not a given positive integer is prime. The revised algorithm is considerably more efficient than the crude algorithm. This illustrates a more general rule of thumb that better algorithms increase computing power and speed. A computer's ability to perform a task and the speed with which it completes the task are dependent on both its hardware and its software (programs).

Some people believe that since mathematics teachers in elementary and secondary schools consider it important to know whether or not (small) numbers are

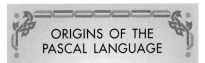

ORIGINS OF THE PASCAL LANGUAGE

The programming language Pascal is named after French mathematician Blaise Pascal (1623–1662), in honor of his work with calculating machines (see Section 12.2). It was created by Niklaus Wirth in Zurich, Switzerland, and was first released in 1971.

prime, research mathematicians must spend their time determining whether or not really big numbers are prime. We hope that you do not buy wholeheartedly into that belief. However, it is true in the following sense. In order to code data for national security purposes or for privacy purposes one scheme that has become popular, called *public key encryption,* encrypts (codes) a message using a large composite number that is made public. To decode the message you need to know the factors of the number. The success of the method relies on the difficulty of factoring a large number. Consequently, one area of mathematical research focuses on finding faster algorithms to factor large numbers.

Our revised prime-number algorithm can be modified to produce a factoring algorithm—an algorithm that will factor a positive integer into its prime factors. The idea is that to factor a number *N,* first find a trial divisor *d,* write $N = d(N/d)$, then repeat the procedure for *N/d* in place of *N.*

FACTORING ALGORITHM

1. Get a number *N.*
2. If $N = 1$, then write a message that *N* has no prime factors, and stop.
3. If $N = 2$ or 3, then *N* is prime. Write "*N* (fill in its value) is a factor" and stop.
4. If 2 is a divisor, then let $d = 2$, write "2 is a factor" and go on to step 10.
5. Initialize the trial divisor to 3.
6. Repeat steps 7 through 9 until you find a divisor of *N.*
7. If the trial divisor is a divisor of *N,* then let $d =$ trial divisor. Write "*d* (fill in its value) is a factor" and go to step 10.
8. Increase the trial divisor by 2.
9. If the new trial divisor exceeds $\sqrt{N}$, then *N* is prime (and you are done).
10. Let $N = N/d$ and repeat steps 1–9.

Notice that the procedure will stop when *N/d* is prime. The number *N* will be the product of the factors written out. The following example looks at the steps of this factoring algorithm for $N = 12$.

EXAMPLE 12.10

Write out the steps of the factoring algorithm for $N = 12$.

SOLUTION

	Printed information
First pass through the steps:	
1. $N = 12$.	
2. $N \neq 1$, so go to step 3.	
3. $N \neq 2$ or 3, so go to step 4.	
4. 2 is a divisor, so let $d = 2$ and go to step 10.	`2 is a factor`
10. Let $N = 12/2 = 6$ and go back to step 1.	
Second pass through the steps:	
1. $N = 6$	
2. $N \neq 1$, so go to step 3.	
3. $N \neq 2$ or 3, so go to step 4.	
4. 2 is a divisor, so let $d = 2$ and go to step 10.	`2 is a factor`
10. Let $N = 6/2 = 3$ and go back to step 1.	

Third pass through the steps:

1. $N = 3$
2. $N \neq 1$, so go to step 3.
3. $N = 3$ `3 is a factor`

The prime factorization of 12 is $2 \cdot 2 \cdot 3$.

Next we look at interpreting some algorithms. In ▶ Figure 12.10 is an algorithm written in the BASIC language.

```
10 A = 1
20 C = 1
30 B = A + 1
40 C = C + B
50 A = B
60 if B < 4 go to 30
70 print C
80 end
```

▶ FIGURE 12.10 A BASIC program

The numbers at the left in the figure are line numbers, which give a way to refer to a particular command. The following are the steps executed by this program.

1. Line 10 sets the value of A at 1.
2. Line 20 sets the value of C at 1.
3. Line 30 sets the value of B at $1 + 1$, or 2.
4. Line 40 sets the value of C at $C + B = 1 + 2 = 3$.
5. Line 50 resets the value of A at 2.
6. Since $B < 4$, go to line 30.
7. Line 30 resets the value of B at $2 + 1 = 3$.
8. Line 40 resets the value of C at $C + B = 3 + 3 = 6$.
9. Line 50 resets the value of A at 3.
10. Since $B < 4$, go to line 30.
11. Line 30 resets the value of B at $3 + 1 = 4$.
12. Line 40 resets the value of C at $C + B = 6 + 4 = 10$.
13. Line 50 resets the value of A at 4.
14. Since $B \not< 4$, go to line 70.
15. Line 70 prints 10.
16. Line 80 ends the program.

What has the program done? It has added the positive integers through 4 to give $1 + 2 + 3 + 4 = 10$ and printed the sum (10).

Summary

In this section we have looked at the increase in the speed of calculation made through different computing devices and the nature of algorithms used in computing. Increases in computer ability and speed depend on both the hardware and the software used.

1. Table 12.8 shows an increase the speed of computing a particular calculation from 100 hours in the late 1950s to 10 seconds in the late 1970s.
   a. (*Fill in the blank.*) This is a _____-fold increase in computer speed over a 20-year period.
   b. (*Fill in the blank.*) This is a _____-fold increase in computer speed every 10 years.

In Exercises 2–5, use the crude prime-number algorithm and write out the steps for the given value of N.

2. 3
3. 4
4. 6
5. 7

6. In the revised prime-number algorithm, step 9 deals with the trial divisor greater than $\sqrt{N}$. What happens when the trial divisor equals $\sqrt{N}$? Describe what happens for N = 9 and N = 25, then generalize.

7. In Example 12.9, with the revised prime-number algorithm we tried six trial divisors (2, 3, 5, 7, 9, 11) before concluding that 97 is prime. If the crude prime-number algorithm had been used to determine that 97 is prime, how many trial divisors (starting with 2) would have been tried?

In Exercises 8–11, use the revised prime-number algorithm and write out the steps for the given value of N.

8. 6
9. 7
10. 102
11. 103

12. Tell the number of trial divisors that must be tried in the revised prime-number algorithm to show that 3613 is prime.

13. Table 12.8 presents the times it has taken various computers to show that $2^{8191} - 1$ is not prime. Suppose the revised prime-number algorithm were used to show that $2^{8191} - 1$ is not prime.
    a. What is the approximate maximum number of trial divisors that would be needed?
    b. (*Multiple choice*) If a computer tried a billion trial divisions per second, how long would it take for it to try all the trial divisors?
       i. less than 1 year
       ii. between 1 and 10 years
       iii. between 10 and 100 years
       iv. between 100 and 1000 years
       v. over 1000 years

In Exercises 14–21, write out the steps of the factoring algorithm for the given value of N.

14. 8
15. 9
16. 10
17. 15
18. 20
19. 63
20. 65
21. 121

22. What number would the BASIC program in Figure 12.10 print if the 4 in line 60 were replaced by 6?

23. What happens with the program in Figure 12.10 if lines 40 and 50 are transposed i.e., if the present line 40 becomes line 50 and the present line 50 becomes line 40?

24. How could the program in Figure 12.10 be modified to print the sum of the first 200 positive integers?s

1. Following Table 12.7, R. Moreau is quoted as estimating a 100-fold increase in speed every 10 years. What was the increase in speed (as an *n*-fold increase) from 1944 to 1954? from 1946 to 1956? from 1951 to 1961? Do some other calculations of this type. Do they support or contradict Moreau's estimate? Write a paragraph or two stating your position.

2. Four examples of algorithms have been mentioned: a division algorithm, the deferred acceptance algorithm for the marriage problem, and the nearest-neighbor and sorted-edges algorithms for the traveling salesperson problem. State two other algorithms. Give enough detail that other people in your class know what problem the algorithm is designed to solve and how it works.

3. Write an algorithm for getting from your classroom to your house.

4. Write an algorithm in the style of the prime-number algorithms or the factoring algorithm that does each of the following.
   a. Takes an 11-digit number used by the U.S. Postal Service (10 digits plus a check digit) and determines whether or not a single-digit error has been made (see Section 2.5)
   b. Take an amount *A* of money invested at an interest rate *i*, compounded *n* times a year, for *m* years and prints out at the end of each compounding period the interest earned during that period and the compound amount at the end of that period (see Section 10.4).

12.5 SOME USES OF COMPUTING IN MATHEMATICS

GOALS

1. To name and describe four ways that mathematicians use the computer.
2. To work with some concepts from number theory: to state an unsolved problem in mathematics, to give examples of twin primes and identify pairs of numbers as twin primes, and to describe and give examples of Mersenne primes.
3. To identify computer proof and computer-assisted proof and to distinguish between them.
4. To describe the four-color theorem.

In the previous sections of this chapter we have looked at the influence of mathematics in the development of calculating machines and computers. Developments in both hardware and software have enabled computations and logical processes to be performed at increasing speeds. The resulting speed and computing power have, in turn, affected mathematics. We will look at four ways that mathematicians use the speed and the power of the computer:

- Exploration
- Visualization
- Simulation
- Proof

Exploration

The speed and power of the computer enable mathematicians to collect data, make conjectures, and test those conjectures with examples produced on a computer. For example, it has been known for over 2000 years that there is an infinite number of prime numbers. Primes that differ by 2 are called *twin primes*. For example, 3 and 5 are twin primes, 5 and 7 are twin primes, 11 and 13 are twin primes, and 617 and 619 are twin primes. Is there an infinite number of pairs of twin primes? Answer that question with proof to support your position and you will become famous in mathematical circles. The answer is not known, but it is expected to be yes.

The computer is ideally suited to searching for twin primes. However, no matter how fast a computer is or how long it runs, it can never prove that there is an infinite number of pairs of twin primes by searching for twin primes. Also, it cannot prove that there is only finitely many twin primes.

SOME COMPUTER-DISCOVERED RESULTS ABOUT TWIN PRIMES

- There are 224,376,048 pairs of twin primes less than 10^{11} (published by Brent in 1976 (Ribenboim, 1988, p. 202). In other words, there are 224,376,048 pairs of twin primes in which each prime has 11 or fewer digits.
- The 1040-digit numbers $256,200,945 \times 2^{3426} \pm 1$ are twin primes (discovered by Atkin and Rickert in 1980 (Ribenboim, 1988, p. 203).
- The 2259-digit numbers $107,570,463 \times 10^{225} \pm 1$ are twin primes (discovered by Dubner in 1985 and verified by Atkin (Ribenboim, 1988, p. 203).
- The largest-known twin primes as of mid-1996 were $4,650,828 \times 1001 \times 10^{3429} \pm 1$ (discovered by Dubner).

> ### "NICELY DONE" CALCULATIONS TURN UP ERROR IN A COMPUTER CHIP
>
> In October 1994 the newly introduced Intel Pentium processor, a computer chip found in some personal computers, was announced to be flawed. Intel then replaced the flawed chip in many of the computers that had incorporated it. The announcement of the problem to the public was made as the result of research by Thomas Nicely, a mathematics professor at Lynchburg College in Virginia, research involving the adding of the reciprocals of twin primes. Discrepancies in his results pointed to the error in the chip. Because computations in number theory—such as calculations with prime numbers or with twin primes—are known to be a good way to reveal computer errors, such calculations are sometimes used to test new computer hardware or software.

MARIN MERSENNE

Marin Mersenne (1588–1648) was a French mathematician and monk whose life overlapped the lives of Galileo, Descartes, and Fermat. In the seventeenth century, ideas were communicated by mail between people working in a particular area, rather than through journals or e-mail as today. Mersenne served as a focal point for much communication in mathematics. In disseminating information, he at times went perhaps too far by sharing ideas that the originator had communicated with Mersenne in confidence. (See Written Assignment 2.) He is best known in mathematics for having his name attached to Mersenne primes and for a bad conjecture about them.

In working with twin primes the computer is clearly employed in discovering new ones and in pointing the way toward conjectures about them.

Another area of number theory that uses the computer for exploration is the area of Mersenne primes. A *Mersenne prime* is a prime of the form $2^p - 1$. In order for $2^p - 1$ to be prime, p must be a prime. (Exercise 6 outlines a proof.) • Table 12.9 examine this for some small values of p.

Some primes of the form $2^p - 1$ were known since the time of Euclid (around 300 B.C.E.). Two others, for $p = 17$ and $p = 19$, were discovered by Cataldi in 1588, the year of Mersenne's birth. Mersenne conjectured that $2^p - 1$, where p is any one of 55 primes between 2 and 257 inclusive, is prime only for $p = 2, 3, 5, 7, 13, 17, 19, 31, 67, 127,$ and 257. Table 12.9 shows that for $p = 2, 3, 5,$ and 7, $2^p - 1$ is prime. Cataldi had shown that for $p = 17$ and $p = 19$, $2^p - 1$ is prime. It has subsequently been shown (but prior to computers) that Mersenne made five mistakes in his list: $p = 67$ and $p = 257$ in his list do not make $2^p - 1$ prime, and three primes not in his list—61, 81, and 107—do make $2^p - 1$ prime.

The advent of the computer made feasible the search for larger Mersenne primes. • Table 12.10 contains information on the 34 Mersenne primes known as of mid-1996. This table exemplifies the effects of increasing computer speed and improved algorithms.

The Mersenne primes with $p \leq 127$ were discovered prior to computers. The first attempt to employ a computer to find Mersenne primes was by Alan Turing in 1951, but he was unsuccessful. In 1952, Raphael Robinson, with the help of D. H. Lehmer and E. Lehmer and a SWAC computer at the National Bureau of Standards, discovered two Mersenne primes on January 30 and three more later in the year.

Notice the increase in size from the twelfth to the thirteenth Mersenne prime: the twelfth has 39 digits while the next one has 157 digits. That might be a big enough gap to lead you to believe that the twelfth Mersenne prime is the largest one. Imagine your thoughts if you are looking for the next Mersenne prime after the twelfth one. You know the twelfth one has 39 digits. You find that there are no 40-digit Mersenne primes, then that there are no 41-digit Mersenne primes, then that there are no 42-digit Mersenne primes, etc. It's not until you get to the 157-digit numbers that the next Mersenne prime appears. In terms of the values of the prime p that you try, there are 31 primes between 2 and 127, inclusive, and 12 of these produce Mersenne primes. The prime 521, which is used to produce the thirteenth Mersenne

TABLE 12.9 $2^p - 1$			
p	Is p prime?	$2^p - 1$	Is $2^p - 1$ prime?
1	No	1	No
2	Yes	3	Yes
3	Yes	7	Yes
4	No	15	No, $15 = 3 \cdot 5$
5	Yes	31	Yes
6	No	63	No, $63 = 3^2 \cdot 7$
7	Yes	127	Yes
8	No	255	No, $255 = 3 \cdot 5 \cdot 17$

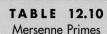

TABLE 12.10
Mersenne Primes

Order	Value of p for Which $2^p - 1$ is Prime	$2^p - 1$	When Proved Prime	By Whom	Machine Used
1	2	3	by 300 BCE	—	
2	3	7	by 300 BCE	—	
3	5	31	by 300 BCE	—	
4	7	127	by 300 BCE	—	
5	13	8,191	1461	—	
6	17	131,071	1588	Cataldi	
7	19	524,287	1588	Cataldi	
8	31	2,147,483,647	1750	Euler	
9	61	19 digits	1883	Pervouchine	
10	89	27 digits	1911	Powers	
11	107	33 digits	1914	Powers	
12	127	39 digits	1876	Lucas	
13	521	157 digits	1952	Robinson	SWAC
14	607	183 digits	1952	Robinson	SWAC
15	1,279	386 digits	1952	Robinson	SWAC
16	2,203	664 digits	1952	Robinson	SWAC
17	2,281	687 digits	1952	Robinson	SWAC
18	3,217	969 digits	1957	Riesel	BESK
19	4,253	1,281 digits	1961	Hurwitz	IBM-7090
20	4,423	1,332 digits	1961	Hurwitz	IBM-7090
21	9,689	2,917 digits	1963	Gillies	ILLIAC-II
22	9,941	2,993 digits	1963	Gillies	ILLIAC-II
23	11,213	3,376 digits	1963	Gillies	ILLIAC-II
24	19,937	6,002 digits	1971	Tuckerman	IBM 360
25	21,701	6,533 digits	1978	Nickel, Noll	CDC-CYBER
26	23,209	6,987 digits	1979	Noll	CDC-CYBER
27	44,497	13,395 digits	1979	Nelson, Slowinski	Cray-1
28	86,243	25,962 digits	1982	Slowinski	
29	110,503	33,265 digits	1988	Colquitt, Welsh	
30	132,049	39,751 digits	1983	Slowinski	
31	216,091	65,050 digits	1985	Slowinski	Cray XMP
32?	756,839	227,832 digits	1992	Slowinski, Gage	Cray-2
33?	859,433	258,716 digits	1994	Slowinski, Gage	Cray C90
34?	1,257,787	378,632 digits	1996	Slowinski, Gage	Cray T94

prime, is the sixty-seventh prime after 127. Therefore 12 of the first 31 primes produce Mersenne primes, while none of the next 66 primes produces a Mersenne prime. This kind of exploration is tremendously tedious by hand but is routine, albeit very time-consuming, for the computer.

The twenty-fifth Mersenne prime was discovered by high school students, Laura Nickel and Curt Noll, on a CDC computer. The twenty-eighth Mersenne prime was discovered after the thirtieth and thirty-first Mersenne primes. In fact, it was believed for a while that there was no Mersenne prime between those for $p = 86,243$ and $p = 132,049$ (Ribenboim, 1988, p. 79).

The question marks after 32, 33, and 34 for the thirty-second, thirty-third, and thirty-fourth Mersenne primes are because, as of late 1996, it was unknown whether there is a Mersenne prime for certain values of p between 216,091 and 1,257,787. A cooperative Internet project is seeking to determine this. The 34th? Mersenne prime is also the largest known prime of any type. The proof of the primality of the largest-known Mersenne prime was done by a fast method for primality testing called the Lucas-Lehmer test (rather than by the methods of the previous section). It took about 6 hours on a Cray T94 supercomputer. Cray Research and other supercomputer companies run programs such as this as a way to test hardware and software.

An illustration of how the computer can aid exploration outside the area of number theory is with the traveling salesperson problem (TSP). The TSP is to find a minimum-cost route to visit n cities and to return to the starting city. (See Section 11.3.) Suppose we conjecture that the sorted edges algorithm (one method to attack the TSP) is at least as good as the nearest-neighbor algorithm (another method to attack the TSP); i.e., the route given by the sorted edges algorithm has either a lower cost or the same cost as the route given by the nearest-neighbor algorithm. To explore this conjecture we can feed a mileage chart for, say, 30 cities into the computer, then have it pick, say, 20 cities at random and calculate a route by both the sorted edges algorithm and the nearest-neighbor algorithm. The times of the two routes are compared. This is repeated for a number of sets of cities.

If for some set of cities, the cost (mileage) of the route produced by the sorted edges algorithm is greater than the cost of the route produced by the nearest-neighbor algorithm, then the conjecture is disproved. If for all the sets of cities examined by the computer, the cost of the route produced by the sorted edges algorithm is less than or equal to the cost of the route produced by the nearest-neighbor algorithm, then there is evidence that the conjecture might be true and we might seek to prove it.

The TSP illustration is typical of computer exploration, in that the computer might be help to disprove a conjecture or give evidence for a conjecture. However, a computer analysis of examples can prove a conjecture only if all examples can be considered. Since there is often an infinite number of examples, the computer, by looking at examples, usually cannot prove a conjecture. In spite of this, the computer is increasingly relied on for exploration in mathematics to search for patterns.

Visualization

The graphical capability of a computer enables us to test the old adage that a picture is worth a thousand words, for example, the graphical presentation of data in descriptive statistics. Consider the histogram in ▶ Figure 12.11, showing the percentage change in the U.S. population, by state, from 1980 to 1990. (For example, eight states increased in population by between 4.95% and 9.95% from 1980 to 1990.) In this pictorial form the data are much easier to interpret than if they had been presented as 50 percentages arranged alphabetically by state. Many statistical computer packages are available that will draw histograms, pie charts, stem-and-leaf displays, and other pictorial representations of data.

One common problem in algebra is to solve an equation; for example, find x such that $2x^2 - 11x + 12 = 0$. This can be solved by factoring or by the quadratic

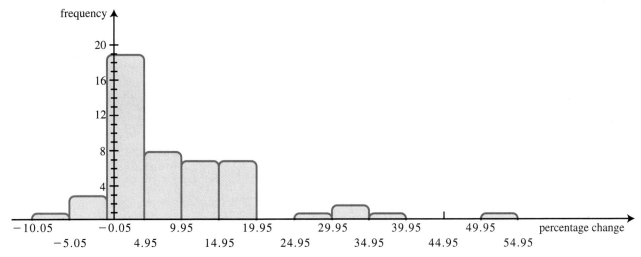

▶ FIGURE 12.11 Percentage change in U.S. population, by state, 1980–1990

formula, but the graph of $f(x) = 2x^2 - 11x + 12$ in ▶ Figure 12.12 pictures the solutions. This graph can easily be drawn on a graphing calculator or on a computer. The trace key on a graphing calculator will show, with some manipulation, that when $x = 1.5$, $y = 0$ and when $x = 4$, $y = 0$, so the solutions to the equation $2x^2 - 11x + 12 = 0$ are 1.5 and 4.

CALCULATOR NOTE

On your graphing calculator graph the function of Figure 12.12. Use the trace key to get the solutions to the equation $2x^2 - 11x + 12 = 0$.

In a similar way, three-dimensional graphing programs will enable the user to depict surfaces in three dimensions. For example, the graph of $f(x, y) = y^2 - x^2$ is shown from three different angles in ▶ Figure 12.13.

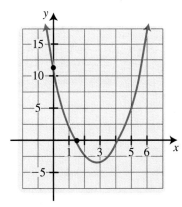

▶ FIGURE 12.12 Graph of $f(x) = 2x^2 - 11x + 12$

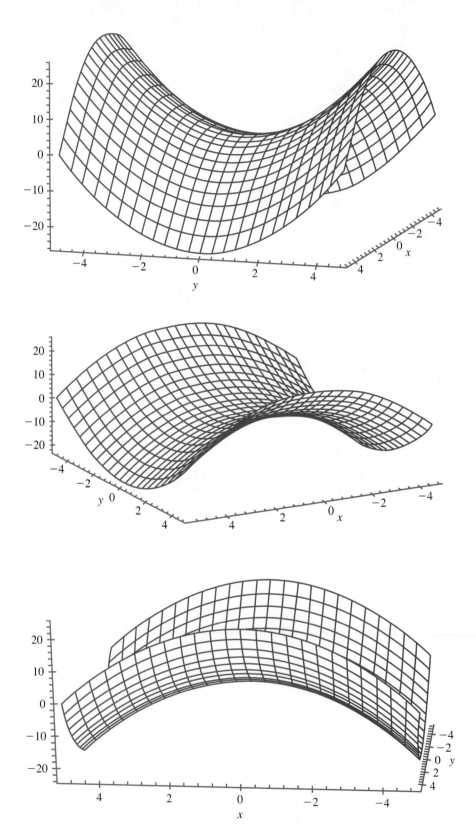

▶ FIGURE 12.13 Graph of $f(x, y) = y^2 - x^2$ from three different angles

▶ FIGURE 12.14 Picture of soap bubble minimal surface

Many computer programs have the capability of rotating the perspective on a three-dimensional graph. Instead of seeing just three views as in Figure 12.13, we can see a motion picture of the surface as we change perspective.

Another area where the visualization of mathematical ideas via computer has proved very helpful is in the search for minimal surfaces. If you took the wand that comes with a bottle of bubble solution (for blowing bubbles) and dipped it into the bubble solution, the surface that forms across the wand—a disk—is an example of a minimal surface. It has the smallest surface area of all the surfaces that go across the wand. This is illustrated in ▶ Figure 12.14.

If we take two wands, dip them both in the bubble solution, and ease them apart and nearly parallel, the minimum surface that forms is an hourglass-shaped surface called a *catenoid*. It is an example of what is called a *natural surface*—it does not intersect itself. Until fairly recently there were only three known natural minimal surfaces: the catenoid, the plane, and the helicoid. These are pictured in ▶ Figure 12.15.

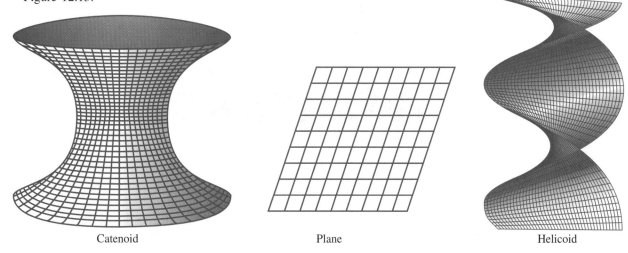

Catenoid   Plane   Helicoid

▶ FIGURE 12.15 Three natural minimal surfaces
Source: David Hoffman and William H. Meek, III, "Minimal Surfaces Based on the Catenoid," *Am. Math. Monthly,* Oct. 1990, p. 708.

A mathematician found a set of equations that represented a new natural minimal surface and two other mathematicians represented these equations graphically on a computer. By observing the surface pictured on the computer screen from different angles, they created a proof that it was in fact a natural minimal surface. (This is described and pictured in the videotape "Computer Science Overview," which is part of the television series *For All Practical Purposes*.) In the process of doing this they also used the computer to explore various hypotheses. Their proof led to ideas that, with computer visualization, gave clues to finding other natural minimal surfaces. The computer visualization helped develop new mathematics.

Simulation

Simulation is a depiction on the computer of a real-world situation. You have probably seen computer games that simulate landing an airplane or playing basketball. The flow of traffic on a street network might be simulated on a computer in order to predict how to set traffic lights for the most efficient traffic flow.

Many probability questions can be answered approximately by simulation. For example, suppose we wish to calculate the probability that a bridge hand (13 cards out of a 52-card deck) contains exactly three aces. We could ask the computer to "deal," say, 1000 bridge hands. Suppose that in 39 of the hands there were exactly three aces. It would be reasonable to estimate that about 39/1000 = .039 of all bridge hands have exactly three aces.

The design of automobiles relies heavily on computers. For example, the effect of windshield designs, with different amounts of curvature and at different angles to the hood, on the fuel efficiency of a car can be simulated on the computer before the more expensive physical models of the car are constructed and tested in wind tunnels. The simulation requires substantial amounts of mathematical knowledge to set it up. Oftentimes, the simulation creates a need for new mathematical ideas.

In medical technology, the CAT scan is in effect a simulation. A machine produces pictures of part of the human body. From these pictures, mathematics and the computer construct a model of what the actual human body part looks like. The situation is like knowing a few points on a curve, such as in ▶ Figure 12.16, and trying to predict what the curve looks like. Three possibilities for the curve from which these points came are shown in ▶ Figure 12.17. The information in Figure 12.16 is not sufficient to choose which of the curves in Figure 12.17 is the original one. More data would help. The developers of the CAT scan found a way to collect enough data (from the pictures) to make a reasonably accurate model of part of the particular human body.

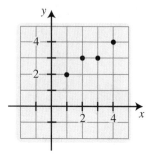

▶ FIGURE 12.16 Points from a curve

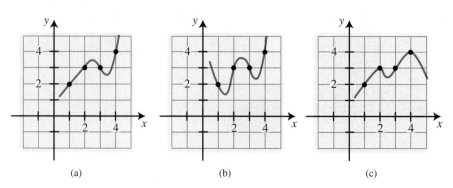

(a)　　　　　　(b)　　　　　　(c)

▶ FIGURE 12.17 Curves from which points in Fig. 12.16 might have come

YOUR FORMULATION

Refer to the three curves in Figure 12.17.

1. If it were known that the point (2.5, 2.5) is on the curve, which of the three curves is the original one?
2. If it were known that the curve is decreasing (from left to right) as it goes through the point (1, 2), which of the three curves is the original one?

In economics, the Leontief input-output model can simulate the national economy. With it, a large matrix (rectangular array of numbers) describes the interaction of various sectors of the economy. For example, one entry in the matrix might be .015, to show that 1.5% of the output of the forestry industry goes into the soft drink beverage industry. This model enables us to simulate the economy by predicting what will happen if a tree disease cuts the output of the forestry industry by 10% or what will happen if a hot summer increases the demand for soft drinks by 20%.

The availability of computers makes possible the analysis of the national economy in much smaller sectors. Over 100 sectors requires a matrix with over $(100)^2 =$ 10,000 entries. Clearly, hand calculation with such a matrix is infeasible. Hand calculation would require analyzing the economy with the Leontief model based on only three or four or five sectors.

In addition to traffic flow, probability, automobile design, medical technology, and economics, computer simulation is widely applied in all sciences and engineering. Sophisticated mathematical tools are often employed to describe and evaluate the computer models.

Proof

We will distinguish between *proof by computer* and *computer-assisted proof.* Computer programs exist that enable the user to enter the definitions and axioms of an axiomatic system, such as Euclidean geometry (this includes high school geometry), and have the computer prove theorems in the system. In particular, all the theorems typically proved in a high school geometry textbook can be proved by means of such a computer program. The computer employs properties of logic (see Chapter 4) to arrive at a conclusion in a step-by-step argument. Such proofs can verify conclusions in some axiomatic systems. (In geometry, the conclusions are quite well known and have been verified many times by many people.) At their current stage of development, these computer programs to do proofs are of more interest to people working in the area of artificial intelligence, an area of computer science that borders on psychology, than they are to creating or checking proofs in mathematics.

Computers can assist with proofs to check cases that are too numerous or too time-consuming to check by hand. Perhaps the most noteworthy instance of this is in the 1976 proof of the four-color theorem by Kenneth Appel and Wolfgang Haken. The question deals with coloring maps so that two countries (states) that share a border of more than a point have different colors. To do so, at least four colors are required.

Are four colors sufficient in all cases, or is there a map that might require five or more colors? Appel and Haken proved that four colors are sufficient in all cases.

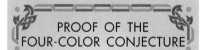

PROOF OF THE FOUR-COLOR CONJECTURE

The four-color conjecture was over 100 years old when it was finally proved by Kenneth Appel and Wolfgang Haken in 1976. The conjecture apparently originated in 1852 with an observation by Francis Guthrie, a university student in London, that a map of the counties of England could be colored with only four colors. Could every map be colored with only four colors? He asked his brother Frederick, who in turn mentioned it to his teacher Augustus DeMorgan (see Section 3.5). DeMorgan mentioned it to Sir William Rowan Hamilton (see Section 11.3), but Hamilton never attempted to solve the problem.

In 1880, A. B. Kempe published a "proof" of the four-color theorem. His "proof" contained an error that was not detected until 1890. A number of other incorrect "proofs" have been proposed. The first correct proof was in 1976 by Appel and Haken, two mathematicians at the University of Illinois.

▶ FIGURE 12.18 A map requiring four colors

Argue that at least four colors are required for the map in ▶ Figure 12.18.

Appel and Haken's technique involved classifying any map as belonging to one of several classes, then showing that a map in any of these classes requires no more than four colors. They used over 6 months of computer time to exhaust all the cases. Without the computer, the proof cannot be done by their technique.

Some people claim that what Appel and Haken did is not a proof. They argue that a proof is something that can be checked in logic and detail by a person. Since, because of time constraints, no person can check the details in Appel and Haken's argument, these people claim that Appel and Haken's argument does not constitute a proof that four colors suffice.

Whether or not computer-assisted proofs are really proofs, they do serve a useful purpose in mathematics. A similar style of proof by Clement Lam and a team of computer scientists was completed in 1988. It showed, after about 3000 hours of computer time, that there is no projective plane (a type of geometric diagram) of order 10 (that is, with exactly 10 points on it).

Summary

Computers are intimately connected to mathematics in their historical development, in their physical design, and in the algorithms that run them. Many applications with computers require significant amounts of mathematics. In turn, mathematicians use the computer as a tool for exploration, visualization, simulation, and proof.

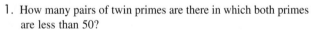

EXERCISES 12.5

1. How many pairs of twin primes are there in which both primes are less than 50?

2. 101 is prime. Is it in a pair of twin primes?

3. 113 is prime. Is it in a pair of twin primes?

4. Complete Table 12.9 for $p = 9$.

5. Show that even though 11 is prime, $2^{11} - 1$ is not prime by producing a factorization of $2^{11} - 1$. (*Hint:* One of the factors is 23.)

6. Prove that "for $2^p - 1$ to be prime, p must be prime" by showing that if p is not prime—say, $p = mn$, then

$$2^p - 1 = 2^{mn} - 1$$
$$= (2^m - 1)(2^{(n-1)m} + 2^{(n-2)m} + \cdots + 2^{2m} + 2^m + 1)$$

7. Use Exercise 6 to find a factor of $2^{12} - 1$.

8. An illustration of using the computer for exploration called for selecting 20 cities at random from 30 cities. How many different sets of 20 cities can be chosen from a fixed set of 30 cities?

9. The curve from which the points in Figure 12.16 come contains the points (1, 2), (2, 3), (3, 3), and (4, 4).

   a. Which one(s) of the following curves go(es) through all four of these points?

   i. $y = 3$

   ii. $y = x + 1$

   iii. $y = -\frac{1}{2}x^2 + \frac{5}{2}x$

   iv. $y = \frac{1}{3}x^3 - \frac{5}{2}x^2 + \frac{37}{6}x - 2$

   v. $y = \frac{7}{24}x^4 - \frac{31}{12}x^3 + \frac{185}{24}x^2 - \frac{101}{12}x + 5$

   b. The graphs of options iv and v in part a are shown. Describe a characteristic of one of the curves that would distinguish the two curves.

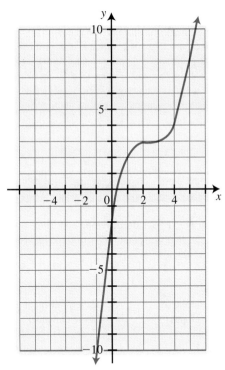

Option iv

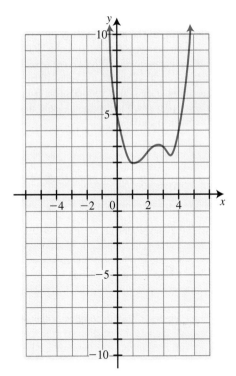

Option v

10. Construct, if possible, a map with six or more countries requiring exactly the given number of colors. If it is not possible, so state.

a. 2   b. 3
c. 4   d. 5

WRITTEN ASSIGNMENTS 12.5

1. (*Library research*) As of January 1, 1995, the largest-known twin primes were $4,650,828 \times 1001 \times 10^{3429} \pm 1$. Is a larger pair of twin primes now known? Write a one-paragraph summary of what you find. Document your answer by photocopying your source or giving a reference to your source.

2. Mersenne at times passed on to the mathematical community new ideas that others had shared with him in confidence. How do you judge Mersenne in this regard? On one hand, he betrayed the confidence of those who shared information with him, and perhaps blunted their announcement and development of new ideas. On the other hand, he stimulated mathematical thinking by quickly releasing the new ideas. Write a page stating and supporting your position. More on Mersenne can be found in Bell (1990).

3. (*Library research*) As of January 1, 1995, there were 32 known Mersenne primes. Have any been discovered since then? Write a one-paragraph summary of what you find. Document your answer by photocopying your source or giving a reference to your source.

4. Write a two- or three-paragraph summary of how the computer is used for exploration in mathematics. Include as one paragraph an example different from any of the three mentioned in this section of the kind of exploration that might be done.

5. Write a one- or two-paragraph summary of how the computer is used for visualization in mathematics.

6. Pick a concept in mathematics for which visualization is very helpful to you or for which you wish there were a way to visualize it. Describe the concept and either how visualization is helpful to you or how, if you were able to visualize it, that would be helpful to you.

7. Write a one- or two-page summary of how mathematics and the computer are used for simulation.

8. A computer program can help to prove all the theorems typically proved in a high school geometry textbook. Big deal!

Euclid proved these over 2000 years ago—by hand. So what's the use of having a computer prove these same theorems? Write one or two paragraphs answering this question.

9. Write a page on the issue of whether or not Appel and Haken's argument that four colors suffice is a proof. Write so as to persuade others in your class that your position is correct.

10. Write a paragraph or two summarizing how mathematicians use the computer for proofs.

11. The formula $p = n^2 - n + 41$, where n is a positive integer, generates some prime numbers. When $n = 1$, $p = 41$, which is prime. When $n = 2$, $p = 43$, which is also prime. Does this formula always produce a prime? Suppose a computer were enlisted to help find the answer to this question by entering values of n and calculating values of p. Outline the possible outcomes of this computer exploration.

CHAPTER 12 REVIEW EXERCISES

1. Even before the development of a place-value system of numeration or the abacus, another invention helped facilitate algorithmic calculation. What was that invention, and how did it help facilitate algorithmic calculation?

2. With diagrams, show how to add CCCLXXVII and CCLXIV on an abacus.

3. Write a paragraph describing the origins of the abacus or abacus-like devices. Include information on where and when.

4. What was the original purpose of logarithms? Be specific.

5. Let y be a positive real number such that $4 < 10^{2/3} < y < 10^{3/4} < 6$. If $x = \log y$, which one of the following is true?
   a. $x < 2/3$ b. $x = 2/3$ c. $2/3 < x < 3/4$ d. $x = 3/4$
   e. $3/4 < x < 4$ f. $x = 4$ g. $4 < x < 6$ h. $x = 6$
   i. $x > 6$

6. Prove that the log of a product of two numbers is the sum of their logs. In particular, prove that if x and y are positive real numbers, then $\log xy = \log x + \log y$.

7. Use Table 12.4 of logarithms to estimate $10{,}077 \times 10{,}382$. Leave evidence of your reasoning.

8. Logarithms are seldom, if ever, utilized today to do the kind of calculation in Exercise 7. Describe a use of logarithms today.

9. With the logarithmic scale in Figure 12.4, multiply 4×2. Describe your procedure.

10. Describe Gunter's Line of Numbers. For what was it used?

11. Describe a slide rule.

12. For what was a slide rule used?

13. Listed below as items a through h are several developments with calculating machines or computers. Place them in chronological order, from the earliest to the most recent.
   a. Electromechanical calculator
   b. First electronic computer
   c. Colossus
   d. Programming work of Ada Lovelace
   e. Software development by Grace Murray Hopper
   f. Stored-program computer
   g. Successful commercial calculating machine
   h. Tabulating machine

14. Listed here are people who have made significant contributions in mathematics or computing, along with a list of those developments. After each development listed in the right-hand column, name one person from the two left-hand columns

who contributed to that development. (For some developments more than one person contributed, but you need name only one of them.)

Development

Aiken	Hollerith	a. Called father (or grandfather) of the computer
Archimedes	Hopper	
Aristotle	Jevons	b. Connection between logic and switching circuits in the 1930s
Arrow	Leibniz	
Atanasoff	Lovelace	c. Developed one of the first two or three electronic computers
Babbage	Mauchly	
Baldwin	Morland	d. Electromechanical calculators
Banzhaf	Napier	e. First successful commercial calculating machine
Bernoulli	Newton	
Berry	Odhner	f. Helped develop the Colossus; aided the British effort in WWII
Borda	Pascal	
Cantor	Plato	
Condorcet	Pythagoras	g. Invented a device made out of rectangular pieces of bone, metal, wood, or cardboard for multiplying large numbers
de Colmar	Schickard	
Descartes	Shannon	
Eckert	Stanhope	h. Invented a mechanical calculating machine in the 1600s
Euclid	Stibitz	
Euler	Turing	i. Logic machines
Fermat	Venn	
Fibonacci	von Neumann	j. Pioneer in programming
Galileo	Zuse	k. Pioneer in software development
Gauss		l. Stored-program computer
		m. Tabulating machines; helped speed the census

15. a. Why was the technique of subtracting a number by adding its nines complement developed?
   b. Find $423 - 278$ by adding the nines complement. Leave evidence of your reasoning.

16. List two reasons for the development of calculating machines in the 1600s.

17. Construct a table of differences for the function
$$f(x) = 2x^3 - 4x.$$

18. Write one or two paragraphs describing the development of logic machines and their relation to computers.

19. State a fact that quantifies the increase in speed of computers through the years.
20. Define *algorithm*.
21. What is the value of *x* when the following program segment is run?

```
i := 6
j := 2
x := 3*j - i/j + 4^j
```

22. Find the value of *y* for each given value of *i* after execution of the following program segment.

```
if i > 10
  then y := 25
  else y := 22
```

    a. *i* = 10   b. *i* = 22   c. *i* = 25

23. Find the value of *z* for each given value of *i* after execution of the following program segment.

```
if (i < -3 or i > 3)
    then z := 0
    else z := 1
```

    a. *i* = −4 b. *i* = −3 c. *i* = 0 d. *i* = 1 e. *i* = 3 f. *i* = 4

24. The following program segment tells the amount of tax *t* (in dollars) a single person owed (in 1993) when the adjusted income is *a* (in dollars). What would be the tax for each of the given adjusted incomes (*a* values)?

```
if (a < 22,100 or a = 22,100)
    then t := 0.15·a
    else if (a < 53,500 or a = 53,500)
      then t := 3315.00 + 0.28*(a - 22,100)
      else if (a < 115,000 or a = 115,000)
        then t := 12,107.00 + 0.31*
              (a - 53,500)
        else if  (a < 250,000 or
                 a = 250,000)
          then t := 31,172.00 + 0.36*
                (a - 115,000)
          else t := 79,772.00 + 0.396*
                (a - 250,000)
```

    a. $3000   b. $30,000   c. $115,000
    d. $1 million (dream on)

25. Consider the following program segment.

```
i := 0, p := 1
while  (i < 5)
  i := i + 1
  p := p*2
end while
```

    a. After this program segment has been executed, what is the value of *i*?
    b. What is the value of *p*?
    c. Modify the program segment to iteratively calculate 3^{20}.

26. Describe what is meant by an iterative process, and give an example of one.

27. a. Construct a trace table to describe the execution of the following program segment.

```
p := 100
for n := 1 to 4
  p := 0.9*p
next n
```

    b. In a Dutch auction sale, in each period of time (day, two-day period, week, etc.) the price of an item is a fixed percentage (maybe 90%) of its price the previous period. Modify the program segment to give the price of a $150 leather jacket in the tenth day of a Dutch auction sale in which the price each day is 90% of the price the previous day.

28. (*Multiple choice*) From the earliest computers like the Mark I and the ENIAC in 1944 to the present time, computing speed has increased so that the present speed is about how many times the earlier speed?
    a. 10 b. 100 c. 10,000 d. 10,000,000
    e. 10,000,000,000

29. Use the "very crude algorithm" for determining whether or not a number is prime (given prior to Example 12.8 in Section 12.4) and write out the steps for *N* = 8.

30. If the "very crude algorithm" for determining whether or not a number is prime (given prior to Example 12.8 in Section 12.4) were used to determine that 103 is prime, how many trial divisors (starting with 2) would be tried?

31. Use the revised algorithm for determining whether or not a number is prime (given prior to Example 12.9 in Section 12.4) and write out the steps for *N* = 35.

32. If the revised algorithm for determining whether or not a number is prime (given prior to Example 12.9 in Section 12.4) is used to show that 103 is prime, what trial divisors (starting with 3) will be used?

33. A computer's ability is based on both its hardware and its software. Write one or two paragraphs expanding on this idea.

34. Write out the steps of the factoring algorithm given in Section 12.4 prior to Example 12.10 for *N* = 18.

35. Consider the following BASIC program.

```
10 A = 1
20 C = 1
30 B = A + 2
40 C = C + B
50 A = B
60 if B < 8 go to 30
70 print C
80 end
```

    a. After this program is executed, what will be printed?
    b. How could the program be modified to print the sum of the first 100 odd positive integers?

36. Name three ways that mathematicians use the computer. Write a paragraph describing each of these three ways.

37. State an unsolved problem in mathematics.

38. Give an example of a pair of twin primes in which both primes are greater than 30.

39. a. Describe what a Mersenne prime is.
    b. Give an example of one.
40. Describe the distinction between *proof by computer* and *computer-assisted proof.*

41. Give an example of a result that has been proved via a computer-assisted proof.
42. What does the four-color theorem say? Be specific.

 BIBLIOGRAPHY

Aspray, William, (ed.). *Computing Before Computers.* Ames, IA: Iowa State University Press, 1990.

Baylor, Jill S. "Happenings," *IEEE Annals of the History of Computing 17,* (1) (1995):59–60. (An account of Grace Murray Hopper's induction into the National Women's Hall of Fame.)

Temple Bell, Eric, revised and updated by Underwood Dudley. *The Last Problem.* Providence, RI: The Mathematical Association of America, 1990.

Boyer, Carl, revised by Uta C. Merzbach. *A History of Mathematics,* 2nd ed. New York: Wiley, 1991.

Coy, Peter. "Faster, Smaller, Cheaper," *Business Week 1994 Annual,* pp. 54+.

W. Dauben, Joseph. "The King Was in His Counting House . . ." In *Historical Notes: Mathematics Through the Ages,* pp. 6–9. Lexington, MA: COMAP, 1992.

Eves, Howard. *An Introduction to the History of Mathematics,* 6th ed. Chicago: Saunders College, 1990.

Hodges, Alan. *The Enigma.* New York: Touchstone, 1983.

Kidder, Tracy. *Soul of a New Machine.* Boston: Little, Brown 1981.

Lee, J. A. N. "Unforgettable Grace Hopper," *Reader's Digest,* October 1994, pp. 181–185.

Moreau, R. translated by J. Howlett. *The Computer Comes of Age.* Cambridge, MA: MIT Press, 1984.

Ribenboim, Paulo. *The Book of Prime Number Records.* New York: Springer-Verlag, 1988.

Shapiro, Fred R. "The First Bug," *Byte,* April 1994, p. 308.

Time, January 13, 1992, p. 50.

Williams, Michael R. *A History of Computing Technology.* Englewood Cliffs, NJ: Prentice-Hall, 1985.

CHAPTER 13

MATHEMATICS AND THE OTHER LIBERAL ARTS

PROLOGUE

Mathematics is both a tool in other disciplines, including other liberal arts, and one of the liberal arts. In fact, historically, mathematics was at the core of the liberal arts.

Pythagoras found that harmony in music depended on a ratio of whole numbers. This helped fuel the Pythagoreans' belief that "all is number." They extended this omnipresence of number and its relation to harmony to the skies, where they spoke of "the harmony of the spheres" or "the music of the spheres." They assumed that the spheres must make a sound as they travel through the skies. Aristotle, in his work *On the Heavens*, says that the Pythagoreans believed that

the motion of bodies of that size must produce a noise, since on our earth the motion of bodies far inferior in size and speed of movement has that effect. Also, when the

531

sun and the moon, they say, and all the stars, so great in number and in size are moving with so rapid a motion, how should they not produce a sound immensely great? Starting from this argument, and the observation that their speeds, as measured by their distances, are in the same ratios as musical concordances, they assert that the sound given forth by the circular movement of the stars is a harmony.

From the time of Pythagoras onward, people recognized that music had mathematical properties and properties that affected the lives of people. Music joined arithmetic, geometry, and astronomy to form the fundamental liberal arts, the basis of study for educated people. This course of study was endorsed by Plato in the *Republic* and by the fifth century Christian philosopher Augustine, and in the Middle Ages it became known as the *quadrivium*. To these four were added the trivium of grammar, rhetoric, and logic (dialectic). These seven liberal arts were regarded as essential for educated persons. They have formed the basis for the liberal arts or general-education part of a college education up to the present.

Chapter opening photo: The title page of Gregor Reisch's Margarita Philosophica *(1503) depicts the seven liberal arts inside the circle with arithmetic seated in the middle holding a counting board.*

13.1 MATHEMATICS AND THE NATURAL SCIENCES

GOALS
1. To describe similarities and differences between mathematics and the natural sciences.
2. To describe some ways mathematics is used in astronomy, biology, chemistry, geology, medical sciences, meteorology, and physics.
3. To do applications involving distances.
4. To do applications involving scaling with areas, volumes, weights, and respiration rates.

Mathematics is often grouped with the natural sciences. People talk about the natural sciences and mathematics. In fact, Carl Friedrich Gauss described mathematics as the "queen of the sciences." The philosopher Immanuel Kant said, "I contend that each natural science is a real science insofar as it is mathematics."

Many of the early and great mathematicians were also scientists, including Archimedes, Copernicus, Galileo, Descartes, Newton, Leibniz, and Gauss. Today, the explosion of knowledge forces all but a few brilliant people to focus on a few areas of specialization, so very few people are classed as both scientists and mathematicians. However, also because of the explosion of knowledge, most quantitatively trained scientists know more mathematics than did any of those early greats.

Why is there such a close connection between mathematics and science? John Barrow, an astronomer and popular author, in his book *The World Within the World* (1988), asks "Why are the laws of Nature mathematical?" In response he answers (pp. 238–239):

We have grown so used to the extraordinary fact that we can reliably predict the course of natural phenomena by means of little squiggles on a sheet of paper that we may have ceased to be impressed by it. But we can see that something like mathematics— a language that possesses a form of built-in logical structure which constrains its form and direction—is necessary, although not sufficient, for the representation of the laws of Nature. Mathematics allows statements to be made unambiguously, in a manner that is value-free and culturally independent. Only when hypotheses can be

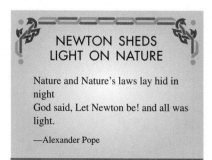

NEWTON SHEDS LIGHT ON NATURE

Nature and Nature's laws lay hid in night
God said, Let Newton be! and all was light.

—Alexander Pope

Liberal arts The arts appropriate for a free person, from the Latin *artes liberales*. Historically, the liberal arts included two groups of studies. One was the *trivium*: grammar (the correct use of language), logic (thinking clearly), and rhetoric (expressing ideas persuasively). The other was the *quadrivium*, which included the various branches of mathematics. Originally, these were arithmetic, geometry, astronomy, and harmony. Out of these traditional studies have come the groupings into natural sciences, social sciences, and humanities.

stated precisely in this way can they be compared and tested experimentally or observationally. Science has to use such a language if it is to progress in a meaningful way.

Astronomy is filled with basic geometric relationships, including the observation that the sun and moon are round. (We now know they are approximately spherical.) Planets travel in elliptical orbits. An object casts a shadow, and geometric relationships allow us to measure the height of the object from the length of the shadow and another measurement. Astronomy also involves a number of algebraic calculations.

Biology includes genetics, whose study involves probability. In fact, significant advances about probability grew out of the study of genetics. Mathematics helps in analyzing the effects of scaling up or scaling down a plant or animal species. For instance: Can a giant reptile have the same body shape as a small reptile? Why is the respiration rate of a hummingbird higher than the respiration rate of an eagle? Models of population growth are mathematical. Studies of the greenhouse effect, global warming, and other environmental phenomena are often done by means of mathematical modeling.

A major branch of chemistry is quantitative analysis. The structure of the atom is described through geometrical and probabilistic models. The accuracy of such models can be checked using high powered microscopes. Chemical reactions are described in a form that requires mathematics to balance them. The speed of chemical reactions is often influenced by temperature. The relationship between temperature and speed can be described by a functional equation, which is often presented graphically.

The mineralogy branch of geology has as its base the study of crystal structure. This involves examining the symmetries of three-dimensional objects. The exponential function plays a key role in dating old rocks and other geologic objects as well as in describing the Richter scale to measure the power of earthquakes.

The medical sciences are heavy users of statistics. The popular press frequently reports on articles from scientific journals on studies involving statistical analysis of data. CAT scans take pictures of parts of three-dimensional objects, then try to describe the objects. This involves mathematics that is beyond the scope of this course. Sophisticated medical diagnoses are often made by feeding symptoms and physical information into a mathematical model to predict what condition is present.

Meteorology studies the air and the weather, with a major goal of predicting the weather. These predictions are couched in terms of probabilities that are derived by means of statistical models.

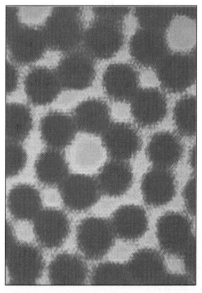

Silicon atoms viewed through a scanning tunneling microscope.

Much of physics can be regarded as applied mathematics. The study of mechanics (objects, forces, and motion), heat, light, electricity, sound, and the structure of matter all involve significant amounts of mathematics. Marie Curie helped discover radioactivity. Radioactive decay is described by an exponential function. Albert Einstein's theory of relativity uses concepts from non-Euclidean geometry.

We now examine some connections between mathematics and the sciences.

Mathematical Modeling and the Scientific Method

A real-world problem or situation can sometimes be solved or explained by a technique called *mathematical modeling*. (See Section 1.3.) ▶ Figure 13.1 depicts the mathematical modeling process. From a real-world problem or situation, we idealize or make assumptions to create a mathematical model. Then we apply logic or mathematical theorems to reach a mathematical conclusion. This conclusion is then translated into a real-world conclusion, which is compared with the real-world problem or situation to see if it is appropriate. If not, the model is refined.

Using mathematical modeling is very much like applying the scientific method. One description of the scientific method says that we begin with a preconception about a solution to a real-world problem or about an explanation for a real-world situation. We then proceed through the following four steps, which can be also described by a diagram such as the one in ▶ Figure 13.2.

THE SCIENTIFIC METHOD

1. *Observation.* If we want to describe planetary motion, we observe the planets. If we want to describe respiration, we observe some animals breathing.
2. *Synthesis.* We synthesize the results of observations into some common findings.
3. *Hypothesis.* From the synthesis of observations, we formulate a hypothesis.
4. *Prediction and testing.* The hypothesis gives us or leads us to a solution to the real-world problem or to an explanation of the real-world situation. This prediction we test by additional observation. If the predictions are consistently on target, then the hypothesis gains credibility; if not, then we formulate and test a new hypothesis.

Although the parts of the mathematical modeling process and the scientific method don't match up exactly, there are a number of similarities between them.

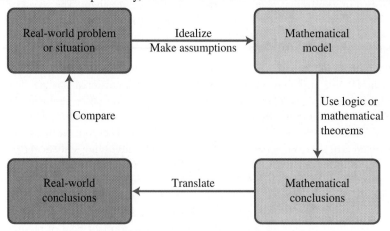

If necessary, refine the model

▶ FIGURE 13.1 Mathematical modeling

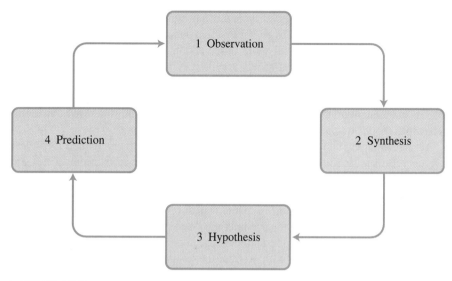

► FIGURE 13.2 The scientific method

The scientific method of going from observations through synthesis to a hypothesis is similar to the procedure of idealizing to a mathematical model and then using logic and theorems to reach a mathematical conclusion. This mathematical conclusion is like a hypothesis in the scientific method. Both procedures call for comparing the predicted results (prediction or real-world conclusion) with the observed behavior of what is being observed or modeled. In both cases, if the predictions don't match the observed behavior very well, then a refinement in the model or the hypothesis is made.

For example, biologists have asked if red dye #2 increases the incidence of cancer. They tried to answer this question by testing whether or not red dye #2 increased the incidence of cancer in laboratory rats. The rats were divided into two groups, a test group given red dye #2 and a control group fed a normal diet. All factors were identical in the two groups except the inclusion of red dye #2 in the test group's diet. Rats in the test group developed an unusually high incidence of cancer. Because there is a strong correlation between some causes of cancer in humans and in rats, the Food and Drug Administration (FDA) banned the use of red dye #2 as a food additive.

All disciplines try to establish truth or display beauty in the subject matter of their disciplines. The arts, including painting, sculpting, music, and literature, seek beauty or truths that come from adhering to a particular art form. The social sciences seek truths about social institutions and about people. The natural sciences seek truths about natural phenomena. Mathematics seeks truths about the objects of mathematics, including numbers, geometric forms, and patterns.

Although many disciplines seek truth, the means of establishing truth is different in mathematics than in other disciplines. In particular, it is different than the means of seeking truth in the natural sciences. In the natural sciences, a hypothesis is formulated through the scientific method. Can a hypothesis be proved via the scientific method? Think about your own answer to this question before reading the next paragraph.

Can a hypothesis be proved by means of the scientific method? In a word, no. The scientific method can provide a very strong reason to believe that a hypothesis

The scientific method and mathematical modeling are similar.

is true, but even one more observation could cause the hypothesis to be questioned. The tests with rats led the FDA to believe that red dye #2 contributes to causing cancer, but we can't be absolutely sure of the result. In the early twentieth century, the Bohr model of the atom was widely accepted. A number of observations seemed to confirm it, and it was successful in predicting a number of chemical results. However, by mid-century some problems with the Bohr model had been recognized, and a quantum theory model now serves to describe the structure of an atom. The Bohr model was not proved (and has now been disproved).

In mathematics, including mathematical modeling, statements (called *theorems*) are proved from definitions, postulates, and previously proved theorems. They are true relative to the truth of the postulates that are assumed. Thus, truth in mathematics is quite different than truth in science.

A word of caution about truth in mathematical modeling is in order. Mathematical modeling involves idealizing or making assumptions in going from the real-world problem or situation to the mathematical model. The "truth" of a mathematical model is thus dependent on the truth of the assumptions made. Because of this, the mathematical modeling process doesn't provide any surer form of truth than does the scientific method. However, in establishing mathematical theorems, there is a 100% guaranteed form of truth, relative to the postulates.

In summary, mathematical modeling is very similar to the scientific method. However, whereas we cannot in general prove a hypothesis by means of the scientific method, we can prove statements in mathematics. The truth established in mathematics is true relative to the axioms that make up the mathematical system in which we are working, and the truth established in mathematical modeling depends on the assumptions of the model.

Distance, Velocity, and Time

We now examine an application of mathematics to science that involves the following fact:

$$\text{distance} = \text{velocity} \times \text{time}$$

EXAMPLE 13.1

a. The planet Mercury averages about 29.75 miles a second in completing its solar orbit in 88.0 of our days. How long is Mercury's orbit?
b. Mercury's mean distance from the sun is 36,000,000 miles. If its orbit around the sun were circular, how long would that orbit be?

SOLUTION

a. $\dfrac{29.75 \text{ mi}}{\text{s}} \times \dfrac{60 \text{ s}}{\text{min}} \times \dfrac{60 \text{ min}}{\text{h}} \times \dfrac{24 \text{ h}}{\text{day}} \times 88 \text{ days} = 226{,}195{,}200 \text{ mi}$

Since the speed is given to four significant digits and the orbital time to three significant digits, we can't expect our answer to be accurate beyond three significant digits, so we say that Mercury's orbit is about 226,000,000 miles long.

b. If the orbit is circular, then its length is the circle whose circumference is

$$c = 2\pi r = 2\pi(36{,}000{,}000 \text{ mi}) \approx 226{,}194{,}671 \text{ mi}$$

Since the given mean distance from the sun has only two significant digits, we will report the answer to just two significant digits: 230,000,000 miles.

Notice that to two significant digits, the answer in part a is the same as this. The fact that the two results are the same to two significant digits says that Mercury's orbit is very nearly circular. In fact, it is elliptical but very nearly circular.

In many places in the sciences, we need to convert from one unit to another. For instance, in Example 13.1 a conversion is made from miles per second to miles per hour. The procedure for doing this is to deal with units as we deal with fractions.

The following example deals with conversion of units in another situation from astronomy.

● **EXAMPLE 13.2**

The star Cassiopeia is 45 light-years from the Earth. How many miles is that?

SOLUTION A light-year is the distance light travels in a year. Light travels at about 186,000 miles per second. We again convert units via fractions.

$$45 \text{ light-years} = 45 \text{ yr} \times \frac{365 \text{ days}}{\text{yr}} \times \frac{24 \text{ h}}{\text{day}} \times \frac{60 \text{ min}}{\text{h}} \times \frac{60 \text{ s}}{\text{min}} \times \frac{186,000 \text{ mi}}{\text{s}}$$

$$= 263,956,320,000,000 \text{ mi}$$

Since the number of light years to Cassiopeia is given to only two significant digits, we will round the answer to two significant digits and say it is about 2.6×10^{14} miles to Cassiopeia.

CALCULATOR NOTE

Some calculators have a conversion key to convert units. For example, on the TI-85 calculator, the conversion in Example 13.2 can be made as follows:

(45) (2nd) (CONV) (LNGTH) (lt-yr) (mile) (ENTER)

The calculator indicates 2.64532491636E14, meaning $2.64532491636 \times 10^{14}$ miles. We again round the answer to two significant digits: 2.6×10^{14} miles.

YOUR FORMULATION

The distances calculated in Example 13.2 and the Calculator Note differ from the third digit on. Why?

Problems of Scale

Science sometimes deals with the very small—such as cells, atoms, and electrons— and sometimes deals with the very large—such as planets, galaxies, and objects thousands of light-years away. Often, larger objects are enlarged versions of smaller objects, as shown in ▶ Figure 13.3. Such objects are called *similar.*

▶ FIGURE 13.3 Similar objects in nature

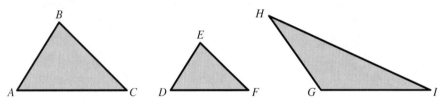

▶ FIGURE 13.4 Similar and nonsimilar triangles

Similar polygons were described in Section 6.4. Similar objects have the same shape but different sizes. For example, any two squares are similar and any two cubes are similar. However, not every two triangles are similar. In particular, triangles *ABC* and *DEF* in ▶ Figure 13.4 are similar while triangles *ABC* and *GHI* are not.

We often think of an adult human as a scaled-up version of a child. This is roughly correct, although several details are not exact. In fact, one rough estimate of the height a child will reach as an adult is to take the child's height at age 2 and double it. For example, if at age 2 a child is 34 inches tall, we can predict roughly that the child will grow to an adult height of about 68 inches (5 feet, 8 inches).

In our example of an adult as a scaled-up 2-year-old, the *scaling factor* is 2. In general, for similar objects the scaling factor is a number that tells how many times a length in the second object is of the corresponding length in the first object. For example, a scaling factor of 2/3 means that a length in the second object is 2/3 of the corresponding length in the first object.

EXAMPLE 13.3

One mouse is 3 inches long and a similar mouse is 4 inches long. What is the scaling factor?

SOLUTION In going from the smaller mouse to the larger mouse the scaling factor is 4/3. In going from the larger mouse to the smaller mouse, the scaling factor is 3/4.

Notice that the scaling factor is always a positive number. For scaling an object up (making it bigger), the scaling factor is greater than 1; for scaling an object down (making it smaller), the scaling factor is less than 1.

For a planar object such as a cell, a leaf, the wing of a bird, a cross section of a rock or tree, the surface of a pond, or the orbit of a planet, a change in a linear dimension (length or width) produces a change in the area of a scaled object. (For us the term *scaled* will include the possibilities that the scaled object could be larger, smaller, or the same size as the original object.) What is the effect of the scaling factor on the area of the scaled-up object?

Let's take the mathematical modeling approach by making a simplifying assumption and consider an object whose area is as simple as possible—the square, as shown in ▶ Figure 13.15. We'll call the length of a side of the square S and the scaling factor k. Then the length of a side of the scaled square is kS. The area of the original square is S^2, and the area of the scaled square is

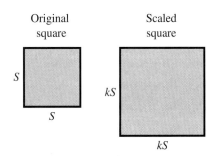

▶ FIGURE 13.5 Scaling a square

$$(kS)(kS) = k^2 S^2 = k^2 \times \text{area of original square}$$

Thus, scaling a square using a scaling factor of k changes the area of the square by a factor of k^2. Exercise 12 asks you to explore the effect of the scaling factor on the area of a rectangle.

In general, we measure the area of an object in square units, such as square millimeters, square meters, square inches, or square miles. An object can be pictured as containing a number of small squares as illustrated in ▶ Figure 13.6. Since the area of a square scaled by a factor of k changes by a factor of k^2, then if any planar object is scaled by a factor of k, the area changes by a factor of k^2. The same argument applies to the surface area of three-dimensional objects. The next example illustrates the effect of the scaling factor on area.

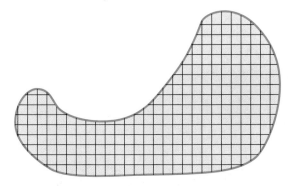

▶ FIGURE 13.6 The area of an object is measured in square units

●EXAMPLE 13.4

If a leaf is scaled up by a factor of 3, how does the area of the new leaf compare with the area of the original leaf?

SOLUTION The area of the new leaf is $3^2 = 9$ times the area of the original leaf.

Astronomers are interested in the surface area of a planet or star. Biologists study respiration and photosynthesis, in which oxygen and carbon dioxide are exchanged through surfaces. Heat loss in a bird or animal takes place through the surface of the skin. The strength of a bone is proportional to the cross-sectional

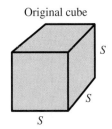

Original cube

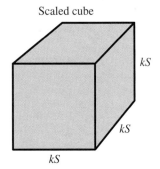

Scaled cube

▶ FIGURE 13.7 Scaling a cube

area. Chemists want to know the area of interface between two chemical substances because that affects the strength of their reaction. Geologists are interested in whether or not the rate of precipitation of a substance is dependent on the surface area where the substance is located. From physics, we know that the ability of one object to slide over another object is dependent on the surface areas they have in contact. In all of these instances, there may be an interest in what happens if an object is scaled.

A change in a linear dimension also produces a change in the volume of a three-dimensional object that is scaled. Again, let's start with an object whose volume is most basic—a cube.

Suppose a cube whose side length is S is scaled by a factor of k, as shown in ▶ Figure 13.7. The volume of the original cube is S^3. The volume of the scaled cube is

$$(kS)^3 = k^3 S^3 = k^3 \times \text{volume of original cube}$$

In general, the volume of an object is stated in cubic units, such as cubic centimeters, cubic meters, cubic inches, or cubic feet. The volume can be pictured as the number of small cubes that can be placed inside the object. Consequently, if an object is scaled by a factor of k, then the scaled object has a volume $k^3 \times$ the volume of the original object.

Since the weight of an object is proportional to its volume, then if an object is scaled by a factor of k, the scaled object has a weight $k^3 \times$ the weight of the original object. This assumes that the density of an object—the weight per unit volume (lead is denser than feathers)—stays constant.

⌐•EXAMPLE 13.5

A gorilla is the largest of the manlike apes. A large male gorilla might weigh 450 pounds. Standing on its hind legs, it might be 6 feet tall. If King Kong were a scaled-up gorilla 90 feet tall, how much would King Kong weigh?

King Kong

SOLUTION Going from 6 feet to 90 feet, the scaling factor is 90/6 = 15. King Kong's weight = $15^3 \times 450$ lb = 1,518,750 lb. Since there is only one significant digit in 6 feet and 90 feet, we'll state the answer to one significant digit as about 2 million pounds.

The foregoing use of the scaling factor in determining area, volume, and weight was approximate. Area and volume calculations depend on the objects' being similar geometrically; weight calculations depend on the objects' being similar geometrically and having the same densities. If objects are approximately similar and can be expected to have about the same densities, then the weight calculations based on the scaling factor are reasonably good approximations. They are almost always within a factor of 2 of the true value. In other words, the true weight is somewhere between half the predicted weight and twice the predicted weight. The next example illustrates what the magnitude of error might be.

EXAMPLE 13.6

The average weight of a 5-foot, 9-inch, 20–24-year-old American man is 163 pounds. Use this to predict the average weight of a 5-foot, 2-inch, 20–24-year-old American man and the average weight of a 6-foot, 4-inch, 20–24-year-old American man.

SOLUTION Going from 5 feet, 9 inches (69 inches) to 5 feet, 2 inches (62 inches), the scaling factor is 62/69. The predicted weight of the 5-foot, 2-inch man is

$$(62/69)^3 (163 \text{ lb}) \approx 118.25 \text{ lb}$$

Since the heights are given to two significant digits (62 and 69), we round the answer to two significant digits: 120 pounds.

The predicted weight of the 6-foot, 4-inch man is

$$163(76/69)^3 = 217.81$$

This rounds to 220 pounds.

cathy®   by Cathy Guisewite

In Example 13.6, the average weights actually are 130 and 198 pounds, respectively. Notice that the model underestimates the weight of the short man and overestimates the weight of the tall man. The actual weights have regressed toward the average weight of a man instead of being strictly height dependent.

All but a few living things take in oxygen and give off carbon dioxide. This process is called respiration. *Respiration rate* is measured in terms of a volume of oxygen per unit of body weight per unit of time. For example, a human uses 0.20 milliliters of oxygen per gram of tissue weight per hour. What is the effect of scaling on the respiration rate? Warm-blooded animals, including mammals and birds, give off heat in respiration. The heat energy produced maintains the body's temperature at a relatively constant level, which is usually higher than the surrounding environment. The heat loss, and the consequent need for oxygen consumption for respiration, is through the skin, whose area is proportional to the square of the scaling factor. The weight of the animal is proportional to the cube of the scaling factor. Therefore,

$$\text{respiration rate} \quad \text{is proportional to} \quad \frac{\text{surface area}}{\text{weight}}$$

That is,

$$\text{respiration rate} \quad \text{is proportional to} \quad \frac{k^2}{k^3} = \frac{1}{k}$$

where k is the scaling factor. Thus, the respiration rate is proportional to $1/k$, where k is the scaling factor.

As with our earlier analysis of the effect of scaling on weight, this analysis is dependent on the scaled animal's being similar to and having the same density as the original one. Generally, the results are good within a factor of 2.

EXAMPLE 13.7

Predict the respiration rate for a mouse.

SOLUTION The respiration rate for a human was stated earlier to be 0.20 milliliters of oxygen per gram of tissue weight per hour. An average human is about 5 feet, 6 inches tall (66 inches), while an average mouse is about 3 inches long. Therefore, the scaling factor in going from a human to a mouse is $3/66 = 1/22$. The respiration rate for a mouse should be about

$$\frac{1}{1/22} \times .20 = 4.4$$

The mouse's respiration rate is predicted to be about 4.4 milliliters of oxygen per gram of tissue weight per hour.

Actually, a mouse's respiration rate is 2.50 milliliters of oxygen per gram of tissue weight per hour. Notice that this is within a factor of 2 of the rate predicted in Example 13.7. Written Assignment 5 asks you to explain the discrepancy between the predicted and actual respiration rates for the mouse.

Summary

We've examined a few of the many places that mathematics and science are related. There are many others. At the start of this section we quoted Gauss's and Kant's strong statements about the role of mathematics in science. The philosopher and mathematician Alfred North Whitehead also made a strong statement when he said: "[T]he belief that the ultimate explanation of all things was to be found in Newtonian mechanics was an adumbration of the truth that all science, as it grows towards perfection, becomes mathematical in its ideas" (quoted in Rose, 1988, p. 137).

Mathematics is a tool in the sciences. The sciences in turn give rise to a number of problems that are the lifeblood of mathematics. In looking for solutions to these problems, more mathematics develops. Some of this new mathematics serves as a tool in the sciences. The cycle continues. Project 2061 of the American Association for the Advancement of Science puts it this way in *Benchmarks for Science Literacy* (Project 2061, 1993, p. 30):

> Much of mathematics is done because of its intrinsic interest, without regard to its usefulness. Still, most mathematics does have applications, and much work in mathematics is stimulated by applied problems. Science and technology provide a large share of such applications and stimulants. In doing their work, scientists and engineers may attempt to do some useful mathematics themselves, or may call on mathematicians for help. The help may be to suggest some already-completed mathematics that will suffice or to develop some new mathematics to do the job. On the one hand, there have been some remarkable cases of finding new uses for centuries-old mathematics. On the other hand, the needs of natural science or technology have often led to the formulation of new mathematics.

EXERCISES 13.1

1. The planet Venus moves about the sun at a mean distance of 67,000,000 miles in 225 of our days.
   a. If the orbit of Venus were circular (it's not, it's elliptical), what would be the length of its orbit?
   b. If the orbit were circular, what would be Venus' average speed, in miles per second?
2. The length of the Earth's equator is 24,901.55 miles. What is the Earth's diameter?
3. Mars' diameter is about 4200 miles. How long is its equator?
4. The star Canis Majoris (Sirius) is 8.7 light-years from Earth. How many miles is that?
5. In the Calculator Note, the distance to Cassiopeia was found on the calculator to 12 significant digits, while in Example 13.2 it was found to only eight significant digits. Ultimately, we rounded the distance off to two significant digits. Why does the calculator initially give more significant digits than the calculation in Example 13.2?
6. A star is 1.00 quadrillion miles from the Earth. How many light-years is that?
7. If a peregrine falcon is regarded as a scaled-down bald eagle and a peregrine falcon is 12 inches long and a bald eagle is 25 inches long, what is the scaling factor?
8. The smallest fish is the pygmy goby, which is about $\frac{1}{2}$ inch long. The largest fish is the whale shark, which can grow up to 60 feet. If we regard a whale shark as a scaled-up pygmy goby, what is the scaling factor?

9. A picture of a hermit thrush says "$\frac{2}{5}$ life size." The hermit thrush in the picture is $2\frac{7}{8}$ inches long. How long is a real hermit thrush?
10. If a cat is about 9 inches tall and we can make it the size of a bobcat by scaling up by $\frac{5}{3}$, how tall is a bobcat?
11. The tallest trees are the redwoods of California, which grow to a height of over 360 feet. If you make a scale model of a 360-foot redwood using a scaling factor of 1/50, how tall will the model tree be?
12. A rectangle that is *l* units long and *w* units wide is scaled by a factor of *k*.
    a. What is the area of the original rectangle?
    b. What is the area of the scaled rectangle?
    c. (*Fill in the blank*) The area of the scaled rectangle is _____ times the area of the original rectangle.
13. A triangle with a base *b* units long and a height *h* is scaled by a factor of *k*.
    a. What is the area of the original triangle?
    b. What is the area of the scaled triangle?
    c. (*Fill in the blank*) The area of the scaled triangle is _____ times the area of the original triangle.
14. A peregrine falcon is about 12 inches tall and a bald eagle is about 25 inches tall. How does the wing area of a peregrine falcon compare with that of a bald eagle if they are regarded as similar?

15. A tree that is 20 feet tall produces about 450 square feet of shade at a certain time of the day.
    a. When it grows to a height of 30 feet, how much shade will it produce at the same time of the day?
    b. What assumption do you make in stating your answer?
16. The area of a cell magnified about 1200 times is around 14 cm². What is the area of the cell (unmagnified)?
17. Female gorillas usually weigh about 200 pounds, whereas a large male gorilla might weigh 450 pounds. A large male gorilla stands about 6 feet tall.
    a. About how tall does a female gorilla stand?
    b. What assumption do you make in stating your answer?
18. An adult is about twice as tall as a 2-year-old. An average 2-year-old girl weighs 25 pounds, and an average 2-year-old boy weighs 26 pounds. Use this information to answer the following question: Is an adult human a scaled-up 2-year-old?
19. The average weight of a 5-foot, 5-inch, 20–24-year-old American woman is 130 pounds. Use this to predict the average weight of a 20–24-year-old woman whose height is each of the following.
    a. 4 feet, 10 inches (the actual average weight is 105 pounds)
    b. 6 feet, 0 inches (the actual average weight is 157 pounds)
20. A large cocker spaniel weighs about 28 pounds and stands 15 inches tall. A large St. Bernard stands about 30 inches tall. Which is the best estimate of its weight, in pounds? 28–40, 40–80, 80–160, 160–320, over 320
21. Cold-blooded animals have a body temperature approximately equal to that of their environment. Consequently, they don't need to produce much heat energy through respiration and have lower respiration rates than warm-blooded animals of the same size. A crayfish is about 5.5 inches long and has a respiration rate of 0.047 milliliters of oxygen per gram of tissue weight per hour. Use the fact that a human's respiration rate is 0.20 milliliters of oxygen per gram of tissue weight per hour to argue whether a crayfish is warm-blooded or cold-blooded.

WRITTEN ASSIGNMENTS 13.1

1. Describe an application of mathematics, beyond the ones mentioned in this section, to each of the following sciences.
    a. Astronomy   b. Biology
    c. Chemistry   d. Geology
    e. Medical sciences   f. Meteorology
    g. Physics
2. Write a paragraph or two describing similarities and differences between mathematical modeling and the scientific method.
3. A word of caution was given about truth in mathematical modeling. What development in mathematics established the principle that there is no absolute truth in mathematics; truth is relative to the axioms that describe a mathematical system. Describe how that development established the principle just stated.
4. Exercise 18 asks if an adult is a scaled-up child. Write one or more paragraphs describing differences in the shape of a child and the shape of an adult. A sketch of the differences is given in McMahon and Bonner (1983), p. 32.
5. In Example 13.7 the predicted respiration rate for a mouse was 4.4 milliliters of oxygen per gram of tissue weight per hour while the actual respiration rate is 2.50. Explain why this difference might occur.

13.2 MATHEMATICS AND THE SOCIAL SCIENCES

GOALS
1. To apply mathematical concepts discussed in earlier chapters to the social sciences.
2. To identify and give examples of two-player games and zero-sum games.
3. To determine, for a two-player, zero-sum game, (a) the best strategy for each player, (b) whether or not the game is a pure strategy game, and, if it is a pure strategy game, (c) the value of the game.
4. To apply dominant strategy analysis.

The social sciences deal with people in either their individual behavior (psychology) or their group behavior in institutions. This behavior is often described and analyzed by examining statistics. The social sciences are heavy users of statistics.

In fact, the study of statistics grew out of studies by John Graunt in the seventeenth century of patterns of human deaths due to the plague. His work is regarded as one of the first efforts in sociology. His methods and extensions of them were applied to economic data and called "political arithmetic." The original meaning of *statistics* is "a collection of data of interest to a statesman." More information on the origin of statistics is included in the introduction to Chapter 9.

Throughout this book we have presented a connection between mathematics and the social sciences by looking at historical developments of mathematical ideas. The direction of relationship also goes the other way. When we look at a society such as the Babylonian, the Mayan, the Igbo in Nigeria, the Yuki in California, or the ancient Chinese, relevant questions include the following ones about the society's mathematical development: What was the base of the system of numeration? Was the system of numeration additive, multiplicative, or place-value? Did the system have names for big numbers? Did it have a symbol for zero? Did it have the concept of fractions? Chapter 2 refers to the following numeration systems: Egyptian hieroglyphics, Roman numerals, Igbo in Nigeria, traditional African, Hindu, Arabic, Mayan, Yuki in California, western tribes of Torres Straits in Australia, Eskimos in Greenland, Babylonian, ancient Greek, and Mende in western Africa.

As civilizations and mathematics develop, each can influence the other. The ancient Greeks were known for their rational ways of analyzing the world. That style of thinking was influenced in its development by the mathematical thinking of people such as Pythagoras, and it in turn influenced the thinking of people such as Aristotle, Plato, and Euclid.

Aristotle

When mathematical developments have been slow (or fast) in coming, was it because of the nature of the mathematical systems in that society, or was it because of social considerations? For example, the development of fractions in ancient Greece was slowed by the Greek reliance on unit fractions (fractions whose numerators are 1). That is, the development of a mathematical idea (fractions) was slowed because of the way of dealing mathematically with that idea. In Europe in the Middle Ages mathematical development was slowed by the idea that knowledge comes directly from God through the hierarchy of the church and, therefore, humans should not dabble in developing new knowledge, mathematical or otherwise. In this case, mathematical development was slowed by religious considerations.

When social progress has been fast in coming, was it influenced by mathematical developments? For example, the political theory that flourished at the time of the American and French revolutions was influenced by an axiomatic approach to political theory similar to an axiomatic approach to geometry. Some places where topics in this book overlap with topics in the social sciences are listed in ⬤ Table 13.1.

The importance of the connection between mathematics and the social sciences has been recognized by professional organizations in both disciplines. A publication of the National Council for the Social Studies states: "As more and more information is formulated numerically, teachers and students ought to know about quantitative concepts that are useful in social analysis" (Laughlin et al., 1989, p. 7). The National Council of Teachers of Mathematics, in its curriculum and evaluation standards, has one standard titled "Mathematical Connections." It states: "[T]he curriculum should include deliberate attempts, through specific instructional activities, to connect ideas and procedures . . . with other content areas" (NCTM, 1989, p. 11).

The applications mentioned in Table 13.1 show quantitative concepts that are useful in social analysis. They connect mathematics with the social sciences.

Plato

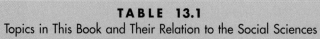

TABLE 13.1

Topics in This Book and Their Relation to the Social Sciences

Chapter or Section	Topic	Relation to social sciences
2	Systems of numeration	In describing a culture or civilization, what is the base of the system of numeration? Is it additive, multiplicative, or place-value? Does it have a symbol for zero? Does it have names for large numbers? Does it have the fraction concept? What happened to promote or to slow down the development of more sophisticated numeration?
4	Logic	Reliance of law on logic
5.3	Exponential functions	Inflation, effects of wage increases, population growth
7.1	Geometry as an axiomatic system	Model for the development of political systems as axiomatic systems
7.2	Non-Euclidean geometry	The idea of absolute truth is challenged.
8.1	Probability	The development of the science of probability in Europe was slowed by religious considerations.
8.2–8.6	Probability	Applications to conflict in a family, discrimination, jury selection, geography of tropical storms, and presidential elections
9 Prologue	Statistics	The word *statistics* refers to affairs of the state. The first study in statistics is also regarded as one of the first efforts in sociology.
9.1	Graphical presentation of data	Applications to population changes, sales tax, presidential performance, ages of U.S. Presidents, land areas of states, and public opinion
9.2	Measures of central tendency	Applications to income, sales tax, and property values
9.3	Measures of spread	Applications to taxes, ages of U.S. Presidents and their wives, property values, and ages of students and faculty
9.4	Normal distributions	Napoleon's government, car mileages, and grading on a curve
9.5	Statistical inference	Application to consumers' interests, public opinion, preelection polls, and margin of error
10	Consumer mathematics	Truth in Lending Act, Code of Hammurabi, millage, inflation, and financial matters
11.2, 11.3	Euler and Hamiltonian cycles	Public safety and application to map of South America
11.4	Marriage problem	Stable marriages and matchings
11.5	Voting methods	Political science and economics (Arrow of "Arrow's Theorem" is a Nobel-Prize–winning economist.)
12.1	Mechanical and mathematical aids to computation	Development of writing mediums
12.2	Automatic calculating machines and computers	The Industrial Revolution prompted the need for faster calculation, Alan Turing and World War II, German lack of interest in computer during W.W. II, and effect of U.S. Census on development of data-processing machines
12.5	Some uses of computers in mathematics	Leontief input-output model in economics

Game Theory

One additional application of mathematics to the social sciences is in the area of mathematics called game theory. Game theory provides a way to help people or institutions make decisions in certain situations.

The following problem is an example of a situation in which game theory could be useful. During World War II, Japan and the Allies battled over New Guinea. At one time, Allied intelligence reports indicated that the Japanese planned to send a convoy of ships from Rabaul on the eastern end of the island of New Britain to Lae on New Guinea (see ► Figure 13.8). This convoy had to travel either north of New Britain or south of it. Either route took 3 days, but the northern route had good visibility while the southern route had poor visibility. The American commander, General Kenney, had to decide on which route, north or south, to concentrate the reconnaissance aircraft. Once sighted, the ships could be bombed until they arrived at Lae. If the convoy took the northern route and General Kenney sent the aircraft north, then the convoy could be bombed for 2 days. If the convoy took the northern route and General Kenney sent the aircraft south, then the convoy could only be bombed for 1 day because of the loss of time seeking it out. If the convoy took the southern route and General Kenney sent the aircraft north, then the convoy could be bombed for 2 days. If the convoy took the southern route and General Kenney sent the aircraft south, then the convoy could be bombed for 3 days. What should the Japanese do? What should General Kenney do? This information is summarized in ➡ Table 13.2.

We look at things first from the point of view of the Japanese commander. If the convoy goes north (first column), it will be bombed for either 1 or 2 days. If the convoy goes south, it will be bombed for either 2 or 3 days. The worst thing on the northern route is 2 days of bombing, while the worst thing on the southern route is 3 days of bombing. So to minimize the losses, the Japanese commander would choose the northern route.

Now we analyze the situation from General Kenney's point of view, without knowing what the Japanese plan to do. If he sends the aircraft north (first row of the

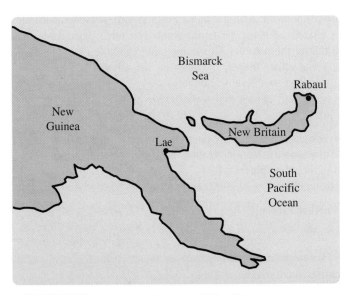

► FIGURE 13.8 Map of New Guinea and New Britain

Which route will optimize bombing?

TABLE 13.2
Estimated Number of Days of Bombing for Different Strategies
in the Battle of the Bismarck Sea

		Japanese Strategies	
		Northern Route	Southern Route
Kenney's strategies	Northern route	2	2
	Southern route	1	3

table), they will be able to bomb for 2 days. If he sends the aircraft south (row 2), they will be able to bomb for either 1 or 3 days, so the worst thing would be only 1 day of bombing. To maximize the amount of bombing when the Japanese commander is as clever as possible, General Kenney should send the planes on the northern route.

Historically this is what happened in the battle of the Bismarck Sea (see Haywood, 1954): The Japanese convoy traveled on the northern route, and General Kenney sent his aircraft on the northern route also. They bombed the convoy for 2 days.

This is an example of a *two-player, zero-sum game.* The two players are the Japanese commander and General Kenney. It is a zero-sum game since what one player wins, the other loses, and so the sum of amount won and amount lost is zero. Here, the Japanese lose 2: They are bombed for 2 days. General Kenney gains 2: His aircraft bomb for 2 days. The number 2 is called the *value* of the game. The rectangular array of four numbers in Table 13.2 is called the *payoff matrix.*

A basic assumption (axiom and postulate) about two-player games is that both players reveal their strategies at the same time. Here, the Japanese fleet started out and General Kenney's aircraft set out at essentially the same time. If you were General Kenney, would you change your strategy if you knew that the Japanese were traveling the northern route? No, since if you send your aircraft south, you only get to bomb for 1 day. If you were the Japanese commander, would you change your strategy if you knew that General Kenney was sending his aircraft on the northern route? If so, you will still be bombed for 2 days. Furthermore, if General Kenney found out about your change of plans, he could bomb you for 3 days. Therefore, you would stick with the northern route. This game is an example of a *pure-strategy game.* This means that either player, by knowing the other player's strategy, cannot change strategy to lead to a better outcome.

In the two-person, zero-sum games that we examine here, we assume that the following conditions are present.

1. There are two players, each intelligently pursuing his or her own self-interest.
2. Each player has two or more strategies from which to choose.
3. One player's gain is the other player's loss.
4. We use a worst-case analysis

We will examine another example.

EXAMPLE 13.8

In the following game, illustrated in ● Table 13.3:

a. Find the best strategy for each player.

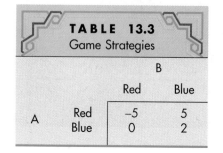

TABLE 13.3
Game Strategies

		B	
		Red	Blue
A	Red	−5	5
	Blue	0	2

b. Is the game a pure strategy game?

c. What is the value of the game?

The two players are called A and B. A plays the rows and B plays the columns. Each has two strategies, called red and blue. The entries in the table are the pay-offs to the row player, A. The 2 means that A gets 2 units from B. The −5 means that A gives up 5 units to B or, equivalently, B gets 5 units from A.

SOLUTION

a. Consider A's choices. If A plays red, the worst thing that can happen to A is −5. If A plays blue, the worst thing that can happen to A is 0. Among these two "worst things," the better for A is 0. Therefore, A should play blue.

Now consider B's choices. If B plays red, B can gain either 5 or 0. The worst thing for B when playing red is 0. If B plays blue, B might lose 5 or lose 2. For B, when playing blue, the worst thing is to lose 5. Among the two "worst things" for B, the better is to lose 0. Consequently, B should play red.

This analysis can be added to Table 13.3 to produce ● Table 13.4.

TABLE 13.4
Payoff Matrix with Worst Case Analysis

		B		
		Red	Blue	Worst Outcome
A	Red	−5	5	−5
	Blue	0	2	0
	Worst	0	5	

b. Is the game a pure-strategy game? Suppose A knows that B is going to play red. Can A change his strategy (rather than playing blue) to improve his situation? If A plays red, then A will lose 5, which is worse for A than losing 0. Therefore, A can't improve his payoff by changing his strategy.

Suppose B knows that A is going to play blue. Can B change her strategy (rather than playing red) to improve her situation? If B plays blue, then the payoff will be 2. This means that B loses 2, which is worse than losing 0. Therefore, B can't improve her payoff by changing her strategy.

Since neither player can improve his or her payoff by knowing the other player's strategy, the game is a pure-strategy game.

c. The value of the game is 0.

To help clarify the concept of a zero-sum game, consider a vote in Congress (either the United States Congress or a state's congress) on a tax increase. Assume that the two players are the Democrats and the Republicans and that the payoffs are the percentage changes in each party's approval rating in opinion polls. If both parties vote yes, one party might lose 2 percentage points and the other party might lose 3 percentage points. The "game" is likely not a zero-sum game; i.e., one party's gain is not necessarily the other party's loss.

To clarify the concept of a pure-strategy game, let's look at a game that is *not* a pure-strategy game.

A game has the following payoff matrix:

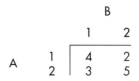

Is this a pure-strategy game? If so, describe the best strategy for each player. If not, describe why not.

SOLUTION To the payoff matrix, add the worst outcome for each strategy.

		B		
		1	2	Worst
A	1	4	2	2
	2	3	5	3
	Worst	4	5	

The better of the worst outcomes for A comes by playing row 2. The better of the worst outcomes for B comes by playing column 1. To see if the game is a pure-strategy game, we ask whether or not either player, by knowing the other player's strategy, can improve his or her outcome. Suppose B knows A is going to play the second row. If B plays the first column, B loses 3 (A gains 3). If B plays the second column, B loses 5. The better choice for B is to lose only 3, so B sticks with column 1. Suppose A knows that B is going to play the first column. If A plays the first row, then A gains 4. If A plays the second row, then A gains 3. Consequently, A is better to switch to the first row. Therefore, this is not a pure-strategy game, since if one of the players (here, A) knows the other player's strategy, that player (here, A) can switch strategies and obtain a better outcome.

The game in Example 13.9 is not a pure-strategy game. Think about how it's likely to be played. The analysis in the solution to Example 13.9 suggests that A minimizes the worst case by playing row 2 and B minimizes the worst case by playing column 1. We continue the analysis by writing A's thinking in plain text and B's thinking in italics:

A: Since B is likely to play column 1, I'll improve my outcome by switching to row 1.
B: *Since A is likely to switch to row 1, I can minimize my losses by switching to column 2.*
A: Since B is likely to switch to column 2, I can increase my outcome by switching to row 2.
B: *Since A is likely to switch to row 2, I can minimize my losses by switching to column 1.*

Notice that we're now back to the original starting strategies: A plays row 2 and B plays column 1. In this kind of game it can be shown that if the game is to be played a number of times, the best strategy for each player is to mix the two strategies available to the player. Such a game is called a *mixed-strategy game.* The analysis of how to mix strategies involves probabilities and expected values (topics in Chapter 8). We will not pursue this topic.

In some games, the number of strategies under consideration can be reduced using *dominant-strategy analysis*. One strategy *dominates* another if the first is always at least as good as the second. For example, consider the game whose payoff matrix is as follows:

$$\begin{bmatrix} 1 & -3 & 9 \\ 2 & 5 & 4 \end{bmatrix}$$

The third column is always worse for the column player than the first column (9 is worse than 1, 4 is worse than 2). Consequently, the column player will never consider the third column, so we remove it, producing the following payoff matrix:

$$\begin{bmatrix} 1 & -3 \\ 2 & 5 \end{bmatrix}$$

Can you see any dominance in this matrix? Neither column dominates the other, but the second row dominates the first (for the row player, 2 is better than 1 and 5 is better than −3). Thus, we remove row 1 to obtain the payoff matrix:

$$[2 \quad 5]$$

Which column dominates here? For the column player, 2 is better than 5, so we obtain the payoff matrix:

$$[2]$$

This is the value of the game. The row player plays the second row and the column player plays the first column.

Notice that this is a pure-strategy game and that the strategies found by dominant-strategy analysis are the same as if previous methods had been used. Dominant-strategy analysis doesn't always lead to a single number as in our example. But if one row dominates another or one column dominates another, that row or column can be eliminated to reduce the size of the payoff matrix.

 YOUR FORMULATION

1. Give an example of a payoff matrix with two rows and two columns in which dominant-strategy analysis does not lead to a single number, as it did in our example.
2. Is your example a pure-strategy game?

Historical Background Game theory was born as a branch of mathematics in 1944 with the publication of *The Theory of Games and Economic Behavior* by John von Neumann and Oskar Morgenstern. It is a key idea in economics, political science, psychology, and sociology. Because game theory ultimately deals with decision making, it is useful in the study of management.

We have made only an initial foray into the topic of game theory. We have touched on the "zero-sum game" that is in popular usage these days. Another term in popular usage that comes from game theory is the "win-win situation." This stems from games that are not zero-sum games but rather in which both players can win (rather than having one win what the other loses). Although game theory has been applied in biology and other places outside the social sciences, its usefulness in describing decision making makes it naturally applicable to the social sciences.

JOHN VON NEUMANN: A DEMIGOD?

John von Neumann had an amazing combination of almost-superhuman intellectual capabilities. He could do huge mental calculations at incredible speed. He had what appeared to be total recall. His charming personality endeared him to his friends. It was rumored that he was in fact a demigod, but he had made a detailed study of humans and could imitate them perfectly. More on von Neumann can be found in Section 12.2.

In Exercises 1–3, do the following.
a. State whether or not the game is a zero-sum game.
b. If the game is a zero-sum game, write the payoff matrix.
c. If the game is a zero-sum game, determine if the game is a pure-strategy game. If so, describe the best strategy for each player and the value of the game. If not, describe why not.

1. You park your car in a metered spot and decide whether to put 25¢ in the meter or to put no money in and risk a $5 parking ticket. The parking bureau makes a decision to send one of its employed officers to check the meters in this area. If an officer checks your meter and you've put in your 25¢, then the parking bureau gains 25¢. If an officer checks your meter and you've not put in your 25¢, then the parking bureau gains $5. If no officer checks your meter and you've put in your 25¢, then the parking bureau gains this amount. If no officer checks your meter and you've not put in your 25¢, then the parking bureau gains nothing.

2. Two contractors are planning to bid on a remodeling project. Contractor A plans to bid either $100,000 or $130,000; contractor B plans to bid either $110,000 or $120,000. If A bids $100,000 and B bids $110,000, A will get the bid and have a profit of $19,000 while B will be out the $1,000 cost of preparing the bid. If A bids $100,000 and B bids $120,000, A will get the bid and again make a profit of $19,000 while B will again be out $1,000. If A bids $130,000, then no matter whether B bids $110,000 or $120,000, B will get the bid and A will be out the $1,000 cost of preparing the bid. B will make a profit of $10,000 on a $110,000 bid and a profit of $11,000 on a $120,000 bid.

3. In a football game, on a particular play, the offensive coach may call a running play or a passing play. The defensive coach has a choice to defend against a run or a pass. Assume that on a running play the average gain is 2 yards if the defense expects it but is 6 yards if the defense expects a pass. Also assume that on a passing play the average gain is 12 yards if the defense expects a run and −7 yards (a loss of 7 yards) if the defense expects a pass.

4. In the game "Rock, Scissors, Paper," the following rules hold:

Rock breaks scissors.
Scissors cuts paper.
Paper covers rock.

The game is played by two players, call them A and B, who on the count of three both reveal their choice of rock, scissors, or paper. This is often done by showing a fist to indicate rock, the index and middle finger extended to indicate scissors, and a flat hand to indicate paper. The winner receives 1 point. For example, if A chooses scissors and B chooses paper, then A wins, so A gets 1 point and B loses 1 point. If both players show the same implement, then there is no winner and each player receives no points.
a. Why is this a zero-sum game?
b. Construct the payoff matrix.
c. Is this a pure-strategy game? Why or why not?

In Exercises 5–8, do the following.
a. Tell whether any row dominates another row.
b. Tell whether any column dominates another column.
c. If there is either row dominance or column dominance, use dominant-strategy analysis to reduce the size of the payoff matrix. If the game is a zero-sum game, determine if the game is a pure-strategy game. If so, describe the best strategy for each player and the value of the game. If not, describe why not.

5. The matrix of Table 13.2 on the battle of the Bismarck Sea
6. The matrix of Example 13.8
7. The matrix of Example 13.9
8. $\begin{bmatrix} 5 & 1 & 2 \\ 6 & 4 & 8 \\ 3 & 7 & 9 \end{bmatrix}$

1. For each of two different social sciences give an example of a relationship between mathematics and that social science.
2. Describe a situation (not included in this section) that can be analyzed by means of game theory. Include the payoffs, and analyze the game.
3. Shortly after the appearance of game theory in 1944, some people believed that game theory would become as important for the social sciences as calculus is for the natural sciences.

Although game theory is an important tool for the social sciences, few people, if any, would support the earlier claim of its importance. Why do you think game theory has not reached the status in the social sciences that some people first thought it might?
4. Describe a relationship between mathematics and the social sciences that has not been described in this section.

GOALS
1. To describe relationships between mathematics and art.
2. To describe relationships between mathematics and music.
3. To describe relationship between mathematics and literature.

In this section we look at places where mathematics is a tool in the humanities, but we emphasize other relationships between mathematics and the humanities. These relationships are manifested in the content of the two areas, the approach to them, and their effect. Should mathematics be considered as one of the humanities?

At first glance the humanities and mathematics may seem far apart, in that the humanities deal with human life and expression while mathematics deals with numbers, shapes, and patterns. But shapes and patterns are the essence of the visual arts, and patterns and ratios are the basis for music. Literature also deals with patterns. Furthermore, there is a connection between mathematics and the humanities in creativity. Mathematicians must be creative in discerning patterns and composing proofs of conjectures in the same way that artists, musicians, and writers depict patterns and compose new works.

What makes some art (painting, sculpture, music, poetry, literature) valued and other art not as valued? What makes some mathematics valued and other mathematics not as valued? All mathematical ideas, in order to be accepted, must be established logically. Some mathematical ideas are applied quickly, while others go for years and years, or perhaps forever, without being applied. Some mathematical ideas are elegant; others are less so. Take, for example, Euclid's fifth postulate and a postulate logically equivalent to it known as Playfair's axiom.

Euclid's fifth postulate: If a straight line falling on two straight lines makes the interior angles on the same side less than two right angles, the two straight lines, if extended indefinitely, meet on that side on which the angles are less than the two right angles.

Playfair's axiom: Given a line and a point not on the line, there is exactly one line through the given point parallel to the given line.

Which of these postulates do you regard as more elegant? Although elegance, like beauty, is in the eye of the beholder, most people choose Playfair's axiom.

The importance of elegance as a criterion in measuring the value of mathematical ideas is stated by Lynn Steen, a past president of the Mathematical Association of America (Steen, 1978, p. 10):

Despite an objectivity that has no parallel in the world of art, the motivation and standards of creative mathematics are more like those of art than of science. Aesthetic judgments transcend both logic and applicability in the ranking of mathematical theorems; beauty and elegance have more to do with the value of a mathematical idea than does either strict truth or possible utility.

The following quotes convey the thoughts of some people about connections between mathematics and the humanities.

The mathematician's best work is art, a high perfect art, as daring as the most secret dreams of imagination, clear and limpid. Mathematical genius and artistic genius touch one another.

—Gosta Mittag-Leffler

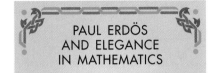

PAUL ERDÖS AND ELEGANCE IN MATHEMATICS

The mathematician best known in the mathematical community during the last half of the twentieth century died on September 20, 1996, while attending a mathematics conference in Warsaw. Paul Erdös was a Hungarian born, itinerant mathematician who packed most of his belongings in a suitcase and traveled the world posing mathematical problems, solving problems, and proving theorems. Those efforts led to the publication of well over 1000 papers, more than any other mathematician in history. He described a mathematician as a device for turning coffee into theorems.

Erdös loved elementary, elegant proofs of mathematical results. He "joked" that God has a book containing all the (infinite number of) theorems and the best proofs of each, and if God is well-intentioned, he will let us look at the book for a moment. Erdös added that you didn't have to believe in God, but you should believe that the book with elegant proofs exists.

Architecture is geometry made visible in the same sense that music is number made audible.

—Claude Bragdon

A man who is not somewhat of a poet can never be a mathematician.

—Karl Weierstrass

Mathematics possesses not only truth but supreme beauty—a beauty cold and austere, like that of a sculpture, without appeal to any part of our weaker nature, without the gorgeous trappings of painting or music, sublimely pure, and capable of a stern perfection such as only the greatest art can show.

—Bertrand Russell

The mathematician's patterns, like the painter's or the poet's, must be beautiful; the ideas, like the colors or the words, must fit together in a harmonious way. Beauty is the first test: there is no permanent place in this world for ugly mathematics.

—G. H. Hardy

Mathematics is an activity governed by the same rules imposed upon the symphonies of Beethoven, the paintings of da Vinci, and the poetry of homer. Just as scales, as the laws of perspective, as the rules of metre seem to lack fire, the formal rules of mathematics may appear to be without lustre. Yet ultimately, mathematics reaches pinnacles as high as those attained by the imagination in its most daring reconnoiters. And this conceals, perhaps, the ultimate paradox of science. For in their prosaic plodding both logic and mathematics often outstrip their advance guard and show that the world of pure reason is stranger than the world of pure fancy.

—Edward Kasner and James Newman, *Mathematics and the Imagination* (New York: Simon & Schuster, 1940), p. 362.

The basic affinity between mathematics and the arts is psychological and spiritual and not metrical and geometrical.

The first essential bond between mathematics and the arts is found in the fact that discovery in mathematics is not a matter of logic. It is rather the result of mysterious powers which no one understands, and in which unconscious recognition of beauty must play an important part. Out of an infinity of designs a mathematician chooses one pattern for beauty's sake, and pulls it down to earth, no one knows how. Afterwards the logic of words and of forms sets the pattern right. Only then can one tell someone else. The first pattern remains in the shadows of the mind. . . .

A second affinity between mathematicians and other artists lies in a psychological necessity under which both labor. Artists are distinguished from their fellows who are not artists by their overriding instinct of self-preservation as creators of art. . . .

The third type of evidence of the affinity of mathematics with the arts is found in the comparative history of the arts. The history of the arts is the history of recurring cycles and sharp antitheses. These antitheses set pure art against mixed art, restraint against lack of restraint, the transient against the permanent, the abstract against the nonabstract. These antitheses are found in all of the arts, including mathematics.

—Marston Morse, "Mathematics and the Arts," *Bulletin of the Atomic Scientist,* February 1959, p. 56.

We now look at relationships between mathematics and three of the humanities—art, music, and literature.

Mathematics and Art

In this subsection, by *art* we mean all forms of visual art. One description of the relationship of mathematics and art is given by Martha Boles and Rochelle Newman (1992, p. xii):

Nature is relationships in space.
Geometry defines relationships in space.
Art creates relationships in space.

How are mathematics and art alike? The noted American mathematician Paul Halmos writes (Halmos, 1968, p. 388):

Perhaps the closest analogy is between mathematics and painting. The origin of painting is physical reality, and so is the origin of mathematics—but the painter is not a camera and the mathematician is not an engineer. . . . How close to reality painting (and mathematics) should be is a delicate matter of judgment. Asking a painter to "tell a concrete story" is like asking a mathematician to "solve a real problem." Modern painting and modern mathematics are far out—too far in the judgment of some. Perhaps the ideal is to have a spice of reality always present, but not to crowd it the way descriptive geometry, say, does in mathematics, and medical illustration, say, does in painting.

Talk to a painter (I did) and talk to a mathematician, and you'll be amazed at how similarly they react. Almost every aspect of the life and of the art of a mathematician has its counterpart in painting, and vice versa. Every time a mathematician hears "I could never make my checkbook balance" a painter hears "I could never draw a straight line." . . . The invention of perspective gave the painter a useful technique, as did the invention of 0 to the mathematician. Old art is as good as new; old mathematics is as good as new. Tastes change, to be sure, in both subjects, but a twentieth century painter has sympathy for cave paintings and a twentieth century mathematician for the fraction juggling of the Babylonians. A painting must be painted and then looked at; a theorem must be printed and then read. The painter who thinks good pictures and the mathematician who dreams beautiful theorems are dilettantes; an unseen work of art is incomplete. In painting and mathematics there are some objective standards of good—the painter speaks of structure, line, shape, and texture, where the mathematician speaks of truth, validity, novelty, generality—but they are relatively the easiest to satisfy. Both painters and mathematicians debate among themselves whether these objective standards should even be told to the young—the beginner may misunderstand and overemphasize them and at the same time lose sight of the more important subjective standards of goodness. Painting and mathematics have a history, a tradition, a growth.

Halmos talks about similarities between mathematics and art. We often think of art as having great freedom and aesthetic value and think of mathematics as following strict rules and not being concerned with aesthetics. Beauty in mathematics is addressed in the introduction to this section in the quote by Hardy and the one by Kasner and Newman. Another statement pointing out that art has rules and that mathematics is involved with aesthetics is by F. Rudio (Moritz, 1914, p. 183):

It was not alone the striving for universal culture which attracted the great masters of the Renaissance, such as Brunellesco, Leonardo da Vinci, Raphael, Michelangelo and especially Albrecht Dürer, with irresistible power to the mathematical sciences. They were conscious that, with all the freedom of the individual phantasy, art is subject to necessary laws, and conversely, with all its rigor of logical structure, mathematics follows esthetic laws.

▶ FIGURE 13.9 Dürer's woodcut *Melencolia I*

16	3	2	13
5	10	11	8
9	6	7	12
4	15	14	1

▶ FIGURE 13.10 Magic square from *Melencolia I*

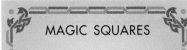
Rudio points out that many of the great art masters of the Renaissance were attracted to mathematics. One of those he mentions is Albrecht Dürer (1471–1528). Let's examine some of Dürer's use of mathematics.

One of Dürer's displays of mathematics is in his woodcut *Melencolia I,* shown in ▶ Figure 13.9. In the upper right-hand corner is an example of a magic square. A *magic square* is an $n \times n$ square arrangement of the (positive) integers 1 through n^2 that gives the same sum for each row, column, and diagonal. (Dürer's magic square is given in ▶ Figure 13.10.) Check that all these sums are the same. What is the sum?

One important way in which the art of the Renaissance differs from that of the Middle Ages is in the use of perspective during the Renaissance. One of the leader's in this was Dürer. Perspective is a way of representing three-dimensional objects in two dimensions. One form of it is shown in ▶ Figure 13.11. In it, what is painted is the intersection of the canvas with the line of sight from the object to the artist's eye. Another example of perspective is shown in Section 7.1, Figure 7.3.

Other artists involved in the development of perspective include Alberti, Brunelleschi, Francesca, and da Vinci. Their use of geometry in painting in the 1400s and 1500s helped lead in the 1800s to the development of a branch of geometry called projective geometry (see Section 7.1).

The Renaissance use of perspective was one of the artistic techniques that enable paintings to show the world realistically. This artistic viewpoint dominated styles of painting for about 500 years. One who changed that viewpoint was the French painter Paul Cézanne (1839–1906). He believed that a painting should reflect the

▶ FIGURE 13.11 Dürer's woodcut *Draftsman Drawing a Portrait*

artist's sensations or impressions. He introduced a new geometrization of forms and new spatial relationships that deviated from the realism of perspective. His work, together with concepts from African sculpture, influenced Pablo Picasso and Georges Braque in the early 1900s to found the painting movement known as cubism. Cubism utilized basic shapes of solids such as cubes, spheres, cylinders, and cones. The style was given its name by a French art critic who reviewed negatively Braque's style as "reduc[ing] everything to little cubes." An example of a cubist painting is shown in ▶ Figure 13.12.

▶ FIGURE 13.12 A cubist painting, *The Factory*, by Picasso

▶ FIGURE 13.13 Two drawings by M. C. Escher, *Day and Night* (left) and *Sky and Water I* (right)

Heading the list of well-known artists of the last century who have used signifi-cant amounts of mathematics in their work is M. C. Escher (1898–1972). He has been described as the most popular artist among college students. The two exam-ples of his drawings shown in ▶ Figure 13.13 illustrate what is called regular divi-sion of the plane. How did Escher fit the figures together? He comments (Schattschneider, 1990, p. ix):

At first I had no idea at all of the possibility of systematically building up my figures. I did not know any "ground rules" and tried, almost without knowing what I was doing, to fit together congruent shapes that I attempted to give the form of animals. Gradually, designing new motifs became easier as a result of my study of the litera-ture on the subject, as far as this was possible for someone untrained in mathemat-ics, and especially through the formulation of my own layman's theory, which forced me to think through the possibilities.

Escher was modest in his assessment of his basic geometric understanding. His use of this basic understanding is described by Doris Schattschneider, a geometer who has studied Escher for almost 20 years (Schattschneider, 1990, pp. x–xi):

It is an extraordinary story of an artist who was captivated by the rhythm and inter-play of repeating forms and who, frustrated with his naive attempts to create his own interlocking designs, sought and found in mathematics the fundamental keys to unlock the secrets of such patterns. His tenacious desire to understand the rules by which such patterns are governed was not satisfied, but only deepened by his study of the technical papers. The systematic analysis of such patterns by scientists addressed their concerns, not his, and so he embarked on fundamental mathematical research. He systematically explored the possibilities of regular division of the plane more fully than the mathematicians and crystallographers had done and developed his "layman's theory" in order to own the knowledge.

In his work, Escher not only used regular division of the plane, but he incorpo-rated ideas from non-Euclidean geometry (see Section 7.2) and ideas about the infi-nite. These mathematical ideas served as underpinnings for the marvelous impres-sions he created in his art.

Several current artists employ mathematics extensively in their work. One of these is Helaman Ferguson, who studied painting as an undergraduate and studied sculpture in graduate school. He also has a Ph.D. in mathematics. Ferguson creates sculptures that represent mathematical concepts. His sculptures in bronze, marble, onyx, and other materials show his love of mathematics and the beauty he sees in it. A photograph of one of his sculptures is shown in ▶ Figure 13.14. In the "Foreward" to a book about Ferguson's work, Richard Waller, director of the Marsh Art Gallery at the University of Richmond, says this:

> Helaman Ferguson is an artist who uses mathematics to create his work, or, also true he is a mathematician who uses art to create his work. . . . His art is rooted in the duality of mathematics as both art form and science.
>
> Art is often referred to as the universal language. Combined with the universality of mathematics, the art of Ferguson doubly echoes this thematic possibility of global communication.
>
> Mathematics as art, art as mathematics: the melding of these two very different disciplines is the essence of Ferguson's sculpture.

Ferguson and about 160 other people interested in art and mathematics gathered for a conference, Art & Mathematics 92, in June 1992. The sculpture and painting displayed there helped illustrate that both mathematics and art involve some constraint and some freedom. Mathematics provides a structure for artistic expression, while art can awaken and express mathematical intuition. Fractal geometry (see Section 7.3) has opened up new opportunities in art. In both art and mathematics, creativity is valued.

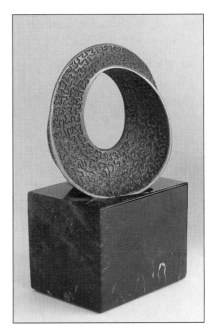

▶ FIGURE 13.14 Helaman Ferguson's sculpture *Umbilic Torus NC*

Mathematics and Music

Music has been called the most mathematical of the arts. Mathematics and music have an apparent relationship in that both involve counting and both employ symbols to tell how much. The music symbol might tell pitch by the placement of a symbol for a note on a music staff, or it might tell how long to play a note. In fact, both of these musical uses of symbols are alternate notations for numbers.

In music, the notes A, B, C, D, E, F, and G can be thought of as the numbers 1, 2, 3, 4, 5, 6, and 7. With the chromatic scale, the notes A, A-sharp, B, C, C-sharp, . . . , G, and G-sharp can be thought of as numbers 1, 2, . . . , 13.

The symbols for notes, shown in ▶ Figure 13.15, give an alternate way of expressing the number 1 and the fractions 1/2, 1/4, 1/8, and 1/16.

Significant developments in both mathematics and music can be traced back to Pythagoras on the Greek island of Samos around 550 B.C.E. In mathematics Pythagoras was the leader of a secret, communal group who called themselves Pythagoreans. Their ideas shaped mathematical thinking for hundreds of years. They played what many people would consider the critical role in the development of mathematics. Although we recognize Pythagoras' name from the Pythagorean theorem, this is only one example of an answer to a question posed by nature that the Pythagoreans explored. The Pythagoreans helped establish a way of thinking that questioned what they saw and tried to argue logically an explanation for these things. Although their mystical ways and their attempts to explain everything mathematically go beyond what most people today accept, they exerted a major influence on the development of Greek thinking generally and of mathematics in particular.

The writer Arthur Koestler described the orchestration by Pythagoras of the development of human ideas in the following musical image.

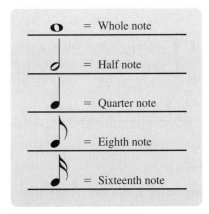

▶ FIGURE 13.15 Musical notes and their denominations

The sixth century [B.C.E.] scene evokes the image of an orchestra expectantly tuning up, each player absorbed in [the player's] own instrument only, deaf to the caterwaulings of the others. Then there is a dramatic silence, the conductor enters the stage, raps three times with his baton, and harmony emerges from the chaos. The maestro is Pythagoras of Samos, whose influence on the ideas, and thereby on the destiny, of the human race was probably greater than that of any single man before or after him.

 YOUR FORMULATION

1. Name two other people who in your estimation vie with Pythagoras for the title of "Person whose influence on the ideas, and thereby the destiny, of the human race was greater than any other person before or since."
2. How would each of your rivals fare against Pythagoras for the title? Support your answer.

Such a primordial position for Pythagoras is supported by Jamie James, an author of newspaper articles, magazine articles, and books. He views Pythagoras as standing at the source of both mathematics and music. (James, 1993, p. 20): "Music and science begin at the same point, where civilization itself begins, and standing at the source is the quasi-mythical figure of Pythagoras."

The Pythagoreans discovered that numbers describe the basis for harmony in music. When a string is plucked, it produces a sound. If the length of the string is cut in half, then a higher-pitched sound is produced. The original note and the higher-pitched note sound well together. They are what is called an *octave* apart, and we give the two notes the same name. For example, they might both be C's. (The Pythagoreans did not use our names A, B, C, D, E, F, and G for their notes, but it is convenient for us to do so.) The lengths of the two strings are in the ratio 2:1. The Pythagoreans found that two strings, when plucked, will produce harmonious sounds when the lengths form a simple ratio like 2:1 or 3:2 or 4:3, but they will produce discordant sounds with more complicated ratios like 9:8.

Different string lengths produce notes that we hear as a certain distance apart. That distance is called a musical *interval*. The interval in which the ratio of lengths is 2:1 is the octave. If a pair of notes are an octave (from the Latin *octavus*, meaning eighth) apart, in our scale one note is eight (full) steps above the other. This is illustrated on the piano keyboard shown in ▶ Figure 13.16.

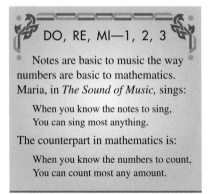

DO, RE, MI—1, 2, 3

Notes are basic to music the way numbers are basic to mathematics. Maria, in *The Sound of Music*, sings:

When you know the notes to sing,
You can sing most anything.

The counterpart in mathematics is:

When you know the numbers to count,
You can count most any amount.

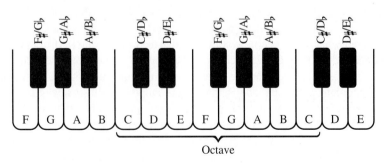

▶ FIGURE 13.16 A piano keyboard

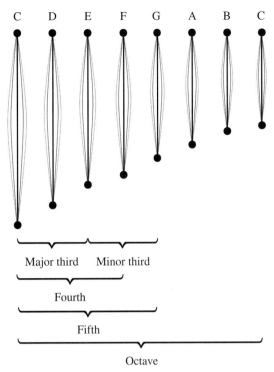

C D E F G A B C

Major third Minor third

Fourth

Fifth

Octave

▶ FIGURE 13.17 Relative lengths of strings for an octave

If the lengths are in the ratio 3:2, the interval is called a *fifth*. Notice that on the piano, if one note is played, the fifth is five (whole) steps higher, counting the first note as "one." For example, the interval from C to G is a fifth (count C-D-E-F-G as 1-2-3-4-5 to get a fifth).

If the lengths are in the ratio 4:3, the interval is called a *fourth*. On the piano, if one note is played, the fourth is four (whole) steps higher. For example, the interval from C to F is a fourth.

If the lengths are in the ratio 5:4, the interval is called a *major third;* and if the lengths are in the ratio 6:5, the interval is called a *minor third.* Notice that a major third crosses two black keys on the piano, while a minor third only crosses one. The relative lengths of strings for some notes are shown in ▶ Figure 13.17.

What we've been describing here is what is usually called the Pythagorean theory. There are strong indications that the same theory, attributed to Ling Lun, was known in China perhaps 2000 years before the time of Pythagoras.

Given the Pythagorean theory, we can use the length of a string for a given note to calculate the lengths of strings for other notes. This is illustrated in Example 13.10.

⌐•EXAMPLE 13.10

If a C is produced with a string of length 60 cm, what lengths of string are needed to produce the next-higher G, B, and C?

SOLUTION The G is a fifth higher, so the ratio of the C-length to the G-length is 3:2. Equivalently, the ratio of the G-length to the C-length is 2:3. Therefore, the length of the G string is

$$\left(\frac{2}{3}\right)(60 \text{ cm}) = 40 \text{ cm}$$

From this G to B is a major third (two black keys are passed), so the ratio of the B-length to the G-length is 4:5. Therefore the length of the B-string is

$$\left(\frac{4}{5}\right)(40 \text{ cm}) = 32 \text{ cm}$$

The higher C is an octave higher than the original C, so its string length is half that of the original, or 30 cm.

Notice in Example 13.10 that to go from the higher C to G and then to the lower (original) C, we add a fourth and a fifth. In terms of ratios, we multiply 4/3, the ratio for a fourth, by 3/2, the ratio for a fifth, to get

$$\frac{4}{3}\frac{3}{2} = \frac{4}{2} = \frac{2}{1}$$

which is the ratio for the octave. Similarly, if we add a major third and a minor third, we have a fifth. In terms of ratios, we multiply (5/4 by 6/5 to get 6/4, or 3/2), which is the ratio for a fifth. Thus, adding intervals corresponds to multiplying ratios of string lengths. This is exactly the property that is a key result for logarithms: We multiply numbers by adding their logarithms:

$$\log_b MN = \log_b M + \log_b N$$

The acoustically "pure" intervals that Pythagoras discovered create problems for tuning a stringed instrument. Let's illustrate by referring to Example 13.10. We started with a C-string of length 60 (we'll drop the unit of centimeters).

- Go up a fifth to a G with string length (2/3)60 = 40.
- Go up a fifth to a D with string length (2/3)40 = 80/3.
- Go down a fourth to an A with string length (4/3)(80/3) = 320/9.
- Go down a fourth to an E with string length (4/3)(320/9) = 1280/27.
- Go down a fourth to a B with string length (4/3)(1280/27) = 5120/81.

This B-string's length of 5120/81 = 63 + 17/81 should be double that of the B-string an octave higher. But we found in Example 13.10 that the octave-higher B-string had length 32, and $63\frac{17}{81}$ is not quite 32 doubled. This discrepancy and others like it cause problems for a tuner of stringed instruments.

Various tuning systems in which minor adjustments are made in the Pythagorean ratios have been developed to deal with this problem. Our discussion refers to Western music, with its 12 tones—the seven whole tones A, B, C, D, E, F, and G plus the five (black key) chromatic tones A♯, C♯, D♯, F♯, and G♯. Non-Western cultures use different tuning systems, ranging from some African tribal music with as few as two or three tones to some Asian music with more than 12 tones.

A system of equal intervals was first proposed in the 1500s. A form of it became standardized during the 1700s and early 1800s. We'll describe it by referring to the piano keyboard.

On the piano keyboard the distance from one key to the next, whether white or black, is called a *half step*. The half step above any white key is called its *sharp*, and the half step below any white key is called its *flat*. For example, in Figure 13.16, the first black key is called either F-sharp or G-flat. In fact, each of the black keys has two different names. The strings in the piano are made so that the lengths of the strings have the following ratios, called interval ratios.

interval ratios: $\dfrac{C}{C\sharp} = \dfrac{C\sharp}{D} = \dfrac{D}{D\sharp} = \dfrac{D\sharp}{E} = \dfrac{E}{F} = \dfrac{F}{F\sharp} = \dfrac{F\sharp}{G} = \dfrac{G}{G\sharp} = \dfrac{G\sharp}{A} = \dfrac{A}{A\sharp} = \dfrac{A\sharp}{B} = \dfrac{B}{C'}$

C′ denotes the octave above C. The 12 notes (black and white keys) referred to form the *chromatic scale*. Making all the ratios the same, as indicated in the interval ratios, produces what is called an *equal-tempered* or *even-tempered scale*.

Let r denote the common interval ratio. Then

$$\frac{C}{C\sharp} = r \qquad \text{so} \qquad C = rC\sharp$$

$$\frac{C\sharp}{D} = r \qquad \text{so} \qquad C\sharp = rD$$

Thus, from the preceding equations,

$$C = r(rD) = r^2D$$

We continue in this manner to get

$$C = r^{12}C'$$

However, since C′ is an octave above C,

$$\frac{C}{C'} = \frac{2}{1} \qquad \text{so} \qquad C = 2C'$$

Put this into the preceding equation for C:

$$2C' = r^{12}C'$$

Divide by C′ to get $\qquad\qquad r^{12} = 2$

This means that the common interval ratio r is the twelfth root of 2; i.e. $r = 2^{1/12}$. Enter this in your calculator to get $r \approx 1.059463$. To see that this is an approximation, calculate $(1.059463)^{12}$ and notice that you don't get exactly 2; you get a number a little smaller than 2. We could calculate $2^{1/12}$ to more decimal places, but we would always have an approximation. In fact, this equal-tempered scale causes the following "problems":

PROBLEMS WITH THE EQUAL-TEMPERED SCALE

1. The Pythagoreans wouldn't like it. The common interval ratio $2^{1/12}$ is an irrational number; i.e., it cannot be expressed as a ratio of two natural numbers (for details, see Exercise 6). This would have bothered the Pythagoreans, since they believed that everything could be explained in terms of ratios of natural numbers.
2. The equal-tempered scale differs from what is called the *natural diatonic* scale on a few notes by an amount that is harmonically objectionable. There are some inconsistencies with the ratios for a fifth, a fourth, a major third, and a minor third. The ratio for a fifth is 3:2, or 1.5. A fifth is the interval from C to G. Our earlier calculations show that $G = r^7C$. Equivalently,

$$G{:}C = r^7 = (2^{1/12})^7 = 2^{7/12} = 1.4983 \ldots$$

Note that this is off a little from a perfect fifth.

The equal-tempered scale we use was first proposed in the 1600s by the mathematician Marin Mersenne (Mersenne primes were mentioned in Section 12.5) and by the musician Andreas Werckmeister.

Some theorists have suggested a scale with 19 half steps instead of 12. That would create a need for a piano with 19 keys to an octave rather than 12. This would give the advantage of better harmony, plus it would have different keys, for example, for C-sharp and D-flat. Composers usually make this distinction. Violinists make the distinction but pianists cannot.

One of the present-day innovations in music is the computer. The computer can generate music. Complicated combinations of rhythms that would be difficult for players to perform can be programmed for the computer. Composers can have the computer check out or perform what they have written.

The contemporary Greek composer Iannis Xenakis uses not only the computer but also ideas from set theory, logic, and probability to create what he calls *stochastic* music. He describes music as follows (Xenakis, 1991, p. 4):

> Music . . . may be defined as an organization of these elementary operations [of intersection, union, and complementation of sets] and relations [of subset and equality] between sonic entities. We understand the first-rate position which is occupied by set theory, not only for the construction of new works, but also for analysis and better comprehension of the works of the past.

Diana S. Dabby, a musician and electrical engineer, has used the mathematical idea of chaos to compose music that changes from one performance to the next. A procedure based on the characteristics of a chaotic system generates the different versions.

In October 1994 a *Cathedral Dreams* music and light show in New York City featured a production in which intricate mathematical formulas, programmed into a specially designed computer by mathematican Ralph Abraham, were transformed into visual effects that were coordinated with music.

In addition to the use of mathematics in music and the use of music as a source of problems in mathematics and as a means of expression of mathematical ideas, mathematics and music, like mathematics and all the arts, share some common ways of looking at their subject matter. One commonality likens an equation to a musical score. A. Pringsheim says (Pringsheim, 1914, p. 364):

> Just as the musician is able to form an acoustic image of a composition which he [or she] has never heard played by merely looking at its score, so the equation of a curve, which [the mathematician] has never seen, furnishes the mathematician with a complete picture of its course. Yea, even more: as the score frequently reveals to the musician niceties which would escape his [or her] ear because of the complication and rapid change of the auditory impressions, so the insight which the mathematician gains from the equation of a curve is much deeper than that which is brought about by a mere inspection of the curve.

This commonality of a musical score and an equation can be expanded to the observation that both music and mathematics are languages. The concepts of mathematics are written as a sequence of logically argued statements. The concepts of a composer are written in a musical score as a sequence of sounds that flow from one to another. The understanding and appreciation of the composer's work come from listening to it played or sung. The understanding and appreciation of the mathematical ideas come from being able to internalize the mathematical ideas and use them as one's own. Some people have the talent to compose music; more people have only the ability to appreciate it when it is played or sung. Similarly, some people have the ability to create mathematics; more people have the ability to understand it and use it.

Mathematics and Literature

Just as there are similarities between mathematics and art and between mathematics and music, there are similarities between mathematics and literature. We list five of them.

SIMILARITIES BETWEEN MATHEMATICS AND LITERATURE

- *Both use metaphor.* For example, the animal farm in George Orwell's novel of the same name is a metaphor for a communistic society. A mathematical model, such as a model of a street network, is a metaphor for a real-world problem or situation.
- *Both use precision.* In poetry, some forms are defined very precisely. For example, haiku is a Japanese verse form in three lines of five, seven, and five syllables, respectively.

> Both use metaphor
> Do math and literature
> Both use precision

- *Both involve ambiguity.* In murder mysteries, the perpetrator of the crime is usually not clearly defined at the outset. Endings of some novels are ambiguous. Cynics may believe that poems are written with ambiguity so they can be interpreted in English classes. Ambiguity underlies concepts of probability. If the probability of passing a certain exam is 0.7, then it is not possible to say whether a randomly selected person will pass the exam.
- *Both have lasting value through the years.* What is regarded as good literature at one time and in one culture is generally regarded as good literature at all times and in most cultures. Mathematical theorems that are proved at one time and in one culture are true at all times and in all cultures. Compare this lasting value with what is regarded as true in the natural sciences or in the social sciences. There, what is true at one time might be established as incorrect at a later date. For example, the model of our planetary system with the Earth at the center served its purpose for many years, but was later shown to be incorrect. Ideas on proper treatment of prisoners change from one time to another and from one culture to another.
- *Both employ an economy of words.* Especially in poetry, literature values the minimal use of words. In mathematics, statements of theorems are worded efficiently, and a considerable amount of interpretation usually goes into understanding them.

Although there are similarities between mathematics and literature, there are at least two respects in which they have opposite goals.

DIFFERENCES BETWEEN MATHEMATICS AND LITERATURE

- *One to many vs. many to one.* It has been said that the job of the poet is to take a concept and express it in a variety of ways, while the job of the mathematician is to take a number of specific expressions of a concept and extract from them a commonality. For example, a mathematician looks at a number of objects that have a certain common property and applies to them the term *triangle*. Also, the mathematician looks at every right triangle with legs a and b and hypotenuse c and notes that $a^2 + b^2 = c^2$. The poet talks about love in a variety of ways or says, "How do I love thee? Let me count the ways."
- *Number of interpretations.* In literature, the author often writes so as to invite each reader to make his or her own interpretation. The mathematician writes mathematical results with the intent that all readers will have the same interpretation.

Developments in mathematics have influenced developments in literature. This was especially noticeable during the Age of Reason, which began in the 1600s and continued until the late 1700s. The time period is so named because Western society

ON A TANGENT

Mathematics in Literature

There are a number of works of literature in which reference is made to mathematics. We list some of them here, by author. These particular items were chosen because of their or their author's familiarity. [The listed works come from a larger list in Lew (1992).]

Lewis Carroll, *Alice's Adventures in Wonderland* and *Through the Looking Glass.* Lewis Carroll is the pen name of Charles Lutwidge Dodgson, who was a minister and mathematician. Reference is made to him in Section 4.6, Writing Assignment 1, and in a margin note in Section 11.5 entitled "C. L. Dodgson, aka Lewis Carroll."

Geoffrey Chaucer, *A Treatise on the Astrolabe.* Chaucer wrote to instruct his son in the use of the astrolabe, an instrument of ancient astronomers that required knowledge of geometry and ratios.

Fryodor Dostoevsky, *The Brothers Karamazov.* Reference to an axiomatic system is made when a character says, "Nor . . . am I going to analyze all the modern axioms laid down by the Russian boys on that subject." This comment is made while desribing why he believes in Euclidean, rather than non-Euclidean, geometry.

Benjamin Franklin, "On the Usefulness of Mathematics" in *Collection M.*

Arthur Koestler, *Darkness at Noon.* History finds the meaning of formulas in *x,* where *x* = the masses.

Vladimir Nabokov, *Pnin.* One chapter is a brief satire on mathematical linguistics.

Edgar Allan Poe, "Eureka," "The Purloined Letter," and "The Thousand-and-Second Tale of Scheherazade" in *Poetry and Tales.* These pieces mention algebra and geometry, algebra versus poetry, and birds and bees that "speak" mathematics.

William Shakespeare, *Henry V* and *The Winter's Tale.* Both contain reference to the place-value system.

George Bernard Shaw, *An Unsocial Socialist, Back to Methusaleh, Major Barbara, Mrs. Warren's Profession.* Mathematics requires postulates, but life doesn't present many. In the year 31920 the main human activity is mathematical contemplation. Professor Cusins, who is not a mathematician, seeks help to determine if 3/5 exceeds 1/2. Mrs. Warren's daughter is a mathematics student.

Alexander Solzhenitsyn, *The First Circle.* A mathematician is one of those kept at a Stalinist work camp for "unreliable" scientists.

Jonathan Swift, *Gulliver's Travels.* Mathematics of scale is involved, with the 6-inch people of Lilliput and the giants of Brobdingnag.

---

emphasized reason as the best method for discovering truth. It was during this time period that calculus was developed. It was observed that great discoveries were made in mathematics by following an axiomatic method that started with a limited number of axioms or assumptions and then logically deduced conclusions that had to follow. This axiomatic method encouraged and supported the scientific method in the sciences. Just as the axiomatic method went back to Euclid and other ancient Greeks, in literature the Greek and Roman classics gained new importance. Language came to be standardized. Two efforts in this direction were the publication, by the Academie Française, of the two-volume *Dictionary of the Academy* and, in England, of Samuel Johnson's *Dictionary.* The rules of form in poetry were observed to be like mathematical axioms, so they were followed strictly. A form known as the heroic couplet gained favor because it had the two additional mathematical characteristics of balance and symmetry. However, the focus on reason sub-

continued

Henry David Thoreau, "A Week on the Concord and Merrimac Rivers" in part in *Collection M.* This refers to the poetry of mathematics.

James Thurber, *Many Moons.* One of the King's three advisors is the Royal Mathematician.

Mark Twain, *Life on the Mississippi.* This novel contains a tongue-in-cheek estimation of the river's future length.

John Updike, *Roger's Version.* A computer scientist tutors a child in mathematics.

Jules Verne, *From the Earth to the Moon, Journey to the Center of the Earth,* and *The Mysterious Island.* These novels include geometric communication with extraterrestrials, mathematical plans for the trips, a method of coding, and a mathematical determination of location.

Voltaire, "Nature" in *Philosophical Dictionary.* Nature is not a mathematician but it obeys mathematical principles.

H. G. Wells, *Joan and Peter* and *The Undying Fire.* All students except the mathematically able are confused by the teaching of arithmetic. According to Satan, Job's descendants have increased geometrically to become the whole human race.

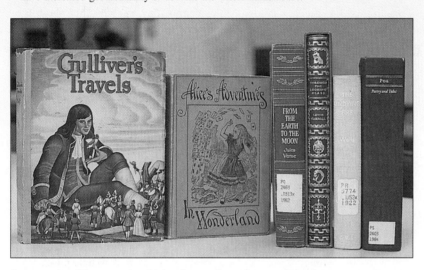

Some works of literature in which reference is made to mathematics

jugated the impressions of poetry to the more rational expositions of prose. The works of John Dryden, Jonathan Swift, Joseph Addison, and Alexander Pope reflect the emphasis on reason.

Sir Isaac Newton, one of the inventors of calculus in the late 1600s, came to symbolize the rational, quantitative style. Some writers who studied his works showed respect; others poked fun at him, as happens to people of great renown. Literature's respect for him, and by extension for all of mathematics, is seen in Pope's *Epitaph* for Sir Isaac Newton:

Nature and Nature's laws lay hid in night,
God said, *Let Newton be!* and all was light.

Other poets and writers have written on mathematical subjects. Here are two of their works, one by Vachel Lindsay and one by Edna St. Vincent Millay.

Euclid   by Vachel Lindsay

Old Euclid drew a circle
On a sand-beach long ago.
He bounded and enclosed it
With angles thus and so.
His set of solemn graybeards
Nodded and argued much
Of arc and of circumference,
Diameters and such.
A silent child stood by them
From morning until noon
Because they drew such charming
Round pictures of the moon.

—from *The Congo and Other Poems*

Though this next piece, by Edna St. Vincent Millay, is ostensibly about Euclid, its real emphasis is on the beauty of mathematics.

Euclid Alone Has Looked on Beauty Bare   by Edna St. Vincent Millay

Euclid alone has looked on Beauty bare.
Let all who prate of Beauty hold their peace,
And lay them prone upon the earth and cease
To ponder on themselves the while they stare
At nothing, intricately drawn nowhere
In shapes of shifting lineage; let geese
Gabble and hiss, but heroes seek release
From dustry bondage into luminous air.
O blinding hour, O holy, terrible day,
When first the shaft into his vision shone
Of light anatomized! Euclid alone
Has looked on Beauty bare. Fortunate they
Who, though once only and then but far away,
Have heard her massive sandal set on stone.

—from *Collected Poems*

In summary, mathematics and literature have both similarities and differences. In his letter to the editor of the *New York Times* written May 5, 1935, Albert Einstein said:

Pure mathematics is, in its way, the poetry of logical ideas. One seeks the most general ideas of operation which will bring together in simple, logical and unified form the largest possible circle of formal relationships. In this effort toward logical beauty spiritual formulae are discovered necessary for the deeper penetration into the laws of nature.

Although both mathematics and literature operate within constraints of form, both rely on the creativity of a person for their development.

1. Construct a 3×3 magic square.
2. A certain 5×5 magic square has the following rows:

   (4 6 13 20 22), (10 12 19 21 3), (11 18 25 2 9),
   (17 24 1 8 15), (23 5 7 14 16)

   Arrange them in the proper order to form a magic square.
3. In a 3×3 magic square made up of the numbers 1, 2, . . . , 9, the sum across any row, column, or diagonal is 15. In a 4×4 magic square using the number 1, 2, . . . , 16, the sum across any row, column, or diagonal is 34. In a 5×5 magic square with the numbers 1, 2, . . . , 25, what is the sum across any row, column, or diagonal?
4. In a 6×6 magic square made up of the numbers 1, 2, . . . , 36, what is the sum across any row, column, or diagonal? (See Exercise 3.)
5. In Example 13.10 find the lengths of the D, E, F, and A strings.
6. In Example 13.10, find the lengths of each of the strings an octave below the original C.

In Exercises 7–10, find the length of a string for the next-lower note below a 30-cm-long string whose interval is as stated. Calculate by multiplying ratios. Show your work.

7. An octave plus a fifth
8. An octave plus a major third
9. A fifth plus a minor third
10. A fourth plus a major third
11. Complete the following details to show that $2^{1/12}$ is not a rational number. Assume that $r = 2^{1/12}$ is a rational number, say, $r = a/b$, where a and b are natural numbers. Then $r^6 = a^6/b^6$ and both a^6 and b^6 are natural numbers, so r^6 is a rational number. Also (*Fill in the blank*):

$$r^6 = (2^{1/12})^6 = 2\underline{\qquad}$$

What contradiction does this cause? Because of this contradiction, $2^{1/12}$ is not a rational number.

Exercises 12 and 13 are in reference to "problem" 2 of the equal-tempered scale.

12. Calculate what the ratio of two notes a fourth apart should be and what it is in the equal-tempered scale.
13. Calculate what the ratio of two notes a major third apart should be and what it is in the equal-tempered scale.

1. The opening of this section quotes Gosta Mittag Leffler as saying that "the mathematician's best work is art." Give an example of a piece of mathematics that you regard as art, and write a paragraph explaining why.
2. Which two of the quotes in the opening of this section do you think come closest to expressing the same idea? Write a paragraph or two explaining why you believe this.
3. In the "Mathematics and Art" subsection is a long quote by Paul Halmos in which he says, "Talk to a painter . . . and talk to a mathematician, and you'll be amazed at how similarly they react." Do what he suggests, and write a two- to four-page paper on what you find. Ask your respondents some common questions concerning what their discipline tries to accomplish. Take some of Halmos' ideas as guides for your questions.
4. Halmos' quote in the "Mathematics and Art" subsection makes an analogy between mathematics and painting. Rewrite this quote by making an analogy between mathematics and composing (music). You may use as many of Halmos' words as you wish, as long as they are appropriate.
5. (*Library research*) Other artists besides Dürer were involved in the development of perspective, including Alberti, Brunelleschi, Francesca, and da Vinci. Write one to three pages on any one of these other artists' use of perspective and/or his contributions to the development of perspective drawing.
6. (*Library research*) The Pythagorean theory of music was apparently known in China considerably before the time of Pythagoras. The Chinese originator is usually acknowledged to be Ling Lun. Write one to three pages on Ling Lun's contributions to a theory of music. Include a bibliography.
7. The Age of Reason is so named because during that period Western society emphasized reason as the best method for discovering truth. Describe two other methods for discovering truth. Which do you prefer? Why?
8. The subsection "Mathematics and Literature" lists five similarities between mathematics and literature, one which is that both involve ambiguity. Another viewpoint on ambiguity is that authors create ambiguity whereas mathematicians try to bring some order out of ambiguity. Do you agree with this alternate viewpoint? If so, how do you reconcile it with the stated similarity that both mathematics and literature involve ambiguity? If you disagree with the alternate point of view, why?

13.3 MATHEMATICS AND THE HUMANITIES

9. Read one of the works listed in the On a Tangent box entitled "Mathematics in Literature," and write a one- to three-page description of the way that mathematics enters this piece of literature.

10. Several of the works of playwright Tom Stoppard contain mathematical references. For each of the following, provide more details on the mathematics mentioned and the setting for its mention.

a. In *Hapgood*, the double-agent physicist refers to the Königsberg bridge problem.

b. In *Every Good Boy Deserves Favour*, the boy has a geometry lesson. Later in the play the prisoner Ivanov expounds on his version of Euclid's axioms.

c. In *Rosencrantz* and *Guildenstern Are Dead*, the title characters toss a coin and muse on the probability of 85 heads appearing in a row.

d. In *Arcadia*, which was produced in London in 1993 and in New York in 1995, Thomasina works with iterative processes. There is also reference to Fermat's Last Theorem.

CHAPTER 13 REVIEW EXERCISES

1. a. When did general education (liberal arts education) begin?
   b. Name two of the original liberal arts subjects.

2. Describe similarities and differences between mathematical modeling and the scientific method.

3. Describe a way that mathematics is used in two of the following sciences: chemistry, geology, medical science, meteorology, physics.

4. Describe the difference between the establishment of truth in mathematics and the establishment of truth in the sciences.

5. The planet Jupiter is at an average distance of 480 million miles from the sun. It takes it almost 12 of our years to make one complete orbit of the sun.
   a. If the orbit of Jupiter were circular, what would be the length of its orbit?
   b. If the orbit were circular, what would Jupiter's average speed be, in miles per hour?

6. The constellation Leo is about 43 light-years from Earth. How many miles is that?

7. A certain optical microscope has a magnification factor (scaling factor) of 15. If a specimen is 12 mm wide under the microscope, what is its actual width?

8. A circle with radius r is scaled by a factor of k.
   a. What is the area of the original circle?
   b. What is the area of the scaled circle?
   c. (*Fill in the blank*) The area of the scaled circle is _____ times the area of the original circle.

9. A tree that is 20 feet tall produces about 400 square feet of shade at a certain time of day. When it grows to 25 feet tall, how much shade will it produce at the same time of the day?

10. The Earth's largest living thing on the surface or in the ocean is the General Sherman giant sequoia tree in Sequoia National Park in California, which stands 275 feet tall and weighs more than 1400 tons. If a smaller giant sequoia is 225 feet tall, estimate its weight.

11. The respiration rate of an animal is stated as an amount of oxygen per unit of body weight. It is proportional to the surface area of the animal divided by the animal's weight. Suppose an animal is scaled with a scaling factor k. Therefore, the respiration rate is proportional to _____.

12. A human's respiration rate is 0.20 milliliters of oxygen per gram of tissue weight per hour. If a giant were double the height of an average human, what would you expect the giant's respiration rate to be? (See Review Exercise 11.)

13. Write one or two paragraphs summarizing interrelationships between mathematics and science.

14. Two stores, K and W, are vying for more share of the market in a small city. They are both considering spending $5000 on advertising. A local economist predicts that if neither advertises, there will be no change in market share; if K advertises but not W, then K will gain 5% of the market and W will lose 5% of the market; if W advertises but not K, then W will gain 6% of the market and K will lose 6% of the market; and if both advertise, K will lose 2% of the market and W will gain 2% of the market.

a. Write out the payoff matrix for this game.

b. Is the game a pure-strategy game? If so, describe the best strategy for each player and the value of the game. If not, describe why not.

c. Is this game a zero-sum game? If so, give an example of a game that is *not* a zero-sum game. If not, give an example of a game that is a zero-sum game.

15. A game has the following payoff matrix:

$$\begin{bmatrix} 3 & 6 & 8 \\ 1 & 5 & 4 \\ 7 & 2 & 9 \end{bmatrix}$$

a. By means of dominant-strategy analysis, reduce the size of the payoff matrix as much as possible.

b. Use the reduced matrix to determine whether or not the game is a pure-strategy game. If so, describe the best strategy for each player and the value of the game. If not, describe why not.

16. a. Who was the developer of game theory?
    b. About when?
17. Describe some similarities between mathematics and art (painting, drawing, sculpture).
18. a. Name two (visual) artists who have employed significant amounts of mathematics.
    b. Describe how one of these two artists used mathematics.
19. Fill in the blanks to create a magic square.

4	_____	15	1
5	11	_____	8
9	7	6	_____
_____	2	3	13

20. Describe the use of mathematics in perspective in art.
21. Describe Pythagoras' contribution to music.
22. A 30-cm string is plucked and a note is sounded.

a. How long should the string be to sound a note an octave higher?
b. When two strings are in the ratio 3:2, the interval formed by the two notes is a fifth. How long should the second string be to produce a fifth?
c. When two strings are in the ratio 4:3, the interval is a fourth. If you start with the original string, then go up a fifth, and then go up a fourth, how long is the new string?
23. a. Describe what the equal-tempered scale is.
    b. Illustrate it with a fifth.
    c. Describe at least one problem with the equal-tempered scale.
24. Describe three ways in which mathematics and literature are similar.
25. a. Name three works of literature that involve mathematics.
    b. Describe the involvement of mathematics in two of the works you named.

 BIBLIOGRAPHY

Alexander, R. McNeill. *Size and Shape.* London: Edward Arnold, 1971.

Barrow, John D. *The World Within the World.* New York: Oxford University Press, 1988.

Boles, Martha, and Newman, Rochelle. *The Surface Plane.* Bradford, MA: Pythagorean Press, 1992.

Bosveld, Jane. "Cathedral Dreams: A Synthesis of Music, Mathematics, and Mysticism," *Omni,* October 1994, p. 8.

Boyer, Carl B. and Merzbach, Uta C. *A History of Mathematics,* 2nd ed. New York: Wiley, 1991.

Brams, Steven J. *Game Theory and Politics.* New York: Free Press, 1975.

Davis, Morton. *Game Theory,* rev. ed New York: Basic Books, 1983.

Devlin, Keith. *Mathematics, the Science of Patterns: The Search for Order in Life, Mind, and the Universe.* New York: Scientific American Library, 1994.

Dudley, Brian A. C. *Mathematical and Biological Interrelations.* New York: Wiley 1977.

Ferguson, Claire. Helaman Ferguson: *Mathematics in Stone and Bronze.* Erie, PA: Meridian Creative Group, 1994.

Ferguson, John. *Mathematics in Geology.* Boston: Allen & Unwin, 1988.

Grossman, Stanley I. and Turner, James E. *Mathematics for the Biological Sciences.* New York: Macmillan, 1974.

Halmos, Paul. "Mathematics as a Creative Art," *American Scientist 56* (Winter 1968): 388.

Haywood, Jr., O. G. "Military Decision and Game Theory," *Journal of the Operations Research Society of America 2* (1954): 365–385.

Hoppensteadt, F. C. and Peskin, C. S. *Mathematics in Medicine and the Life Sciences.* New York: Springer-Verlag, 1992.

James, Jamie. *The Music of the Spheres: Music, Science, and the Natural Order of the Universe.* New York: Grove Press, 1993.

Kasner, Edward, and Newman, James. *Mathematics and the Imagination.* New York: Simon & Schuster, 1940.

Koestler, Arthur. *The Sleep Walkers: A History of Man's Changing Vision of the Universe.* New York: Macmillan, 1959.

Laughlin, M. A., Hartoonian, H. M., and Sanders, N. M. *From Information to Decision Making: New Challenges for Effective Citizenship.* Bulletin 83. Washington, DC: National Council for the Social Studies, 1989.

Lew, John S. "Mathematical References to Literature," *Humanistic Mathematics Network Journal #7* (April 1992) pp. 26–47.

Lloyd, Llewelyn S., and Boyle, Hugh. *Intervals, Scales and Temperaments.* New York: St. Martin's Press, 1979.

McMahon, Thomas A. and Bonner, John Tyler. *On Size and Life.* New York: Scientific American Library, 1983.

Moritz, Robert Edouard. *Memorabilia Mathematica: The Philomath's Quotation Book.* Washington, DC: Mathematical Association of America, 1914.

Morse, Marston. "Mathematics and the Arts," *Bulletin of the Atomic Scientist,* February 1959, p. 56.

National Council of Teachers of Mathematics (NCTM). *Curriculum and Evaluation Standards for School Mathematics.* Reston, VA: NCTM, 1989.

Osserman, Robert. "Rational and Irrational: Music and Mathematics," in Alvin M. White, ed., *Essays in Humanistic Mathematics.* Washington, DC: Mathematical Association of America, 1993.

Peterson, Ivars. "Bach to Chaos: Charotic Variations on a Classical Theme," *Science News,* December 24, 31, 1994, pp. 428–429.

Pringsheim, A. *Jahresbericht der Deutschen Mathematiker Vereiningung.* Bd. 13, p. 364, quoted in Moritz, p. 191.

Project 2061, American Association for the Advancement of Science. *Benchmarks for Science Literacy.* New York: Oxford University Press, 1993.

Rose, Nicholas J., ed. *Mathematical Maxims and Minims.* Raleigh, NC: Rome Press, 1988.

Schattschneider, Doris. *Visions of Symmetry: Notebooks, Periodic Drawing, and Related Work of M. C. Escher.* New York: W. H. Freeman 1990.

Smith, John Maynard. *Mathematical Ideas in Biology.* Cambridge: Cambridge University Press, 1968, Chap. 1.

Steen, Lynn Arthur, ed. *Mathematics Today.* New York: Springer-Verlag, 1978.

Trefil, James S. *The Unexpected Vista: A Physicist's View of Nature.* New York: Scribner's, 1983, Chap. 10.

van der Waerden, B. L. *Science Awakening.* New York: Oxford University Press, 1963.

Walton, Karen Doyle. "Albrecht Dürer's Renaissance Connections Between Mathematics and Art." *The Mathematics Teacher 87* (April 1994), pp. 278–282.

Xenakis, Iannis. *Formalized Music: Thought and Mathematics in Composition,* rev. ed. Stuyvesant, NY: Pendragon Press, 1991.

ANSWERS TO SELECTED ODD-NUMBERED EXERCISES

 CHAPTER 1

Exercises 1.2

1. Make a table such as

Number of students	Number of handshakes
1	0
2	1
3	3
.	.
.	.
.	.

Continue adding students and recording results until you can explain what will happen for larger numbers.

3. The perimeter is a minimum when the rectangle is as close as it can be to a square.

5. One way:

Conclusions: Adding a square adds either 2, 0, or −2 to the perimeter. Adding a square that touches the existing figure on only one edge adds 2 to the perimeter, adding a square that fits in a corner adds 0 to the perimeter, and adding a square that fits in a slot adds −2 to the perimeter.

Here's the Table partially completed:

Number of added squares	One possible shape
3	Shown above
4	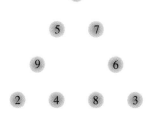

Number of added squares	One possible shape
14	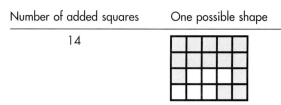

7. 8
9. 24
11. 301
13. One solution is

```
            1
        5       7
      9           6
    2     4     8     3
```

15. No. The sum $1 + 2 + \cdots + 10 = 55$ is an odd number, so it cannot be evenly divided into two parts.
17. 84
19. All amounts from 1 to 14 gallons are possible. To get 7 gallons, use the following sequence:

$5 \to 9$, $5 \to 9 = 1$ (in 5-gallon pail)
$1 \to 9 + 5 \to 9 = 6$ (in 9-gallon pail)
$6 \to 9$, $5 \to 9 = 2$ (in 5-gallon pail)
$2 \to 9 + 5 = 7$

In other words:

a. Fill the 5-gal pail, then empty all 5 gallons into the 9-gal pail.
b. Fill the 5-gal pail again, and empty 4 gallons of it by filling the already partly filled 9-gal pail to completely full. This leaves $5 - 4 = 1$ gallon in the 5-gal pail.
c. Empty the 9-gal pail, and move the 1 gallon just left in the 5-gal pail into the 9-gal pail.
d. Fill the 5-gal pail again, and add those 5 gallons to the 1 gallon already in the 9-gal pail, giving a total now of $5 + 1 = 6$ gallons in the 9-gallon pail.

e. Fill the 5-gal pail again, and empty it of 3 gallons by filling the already partly filled 9-gal pail to completely full, leaving 5 − 3 = 2 gallons in the 5-gal pail.

f. Empty the 9-gal pail once more, and move the 2 gallons just left in the 5-gal pail into the 9-gal pail.

g. Fill the 5-gal pail again, and empty it into the already partly filled 9-gal pail, yielding 2 + 5 = 7 gallons. Et voila!

21. $6 + 2\pi$ ft

23. Yes. Choose any line whose slope is *not* a rational number, for example, slope = $\sqrt{2}$.

Exercises 1.3

1. a. No
   b. If the pool does not have to be twice as long as it is wide, the dimensions are 18 ft. × 30 ft.

3. a. Yes
   b. The diagonal is approximately 12 yards, so a diagonal lap would require slightly less than 30 seconds.
   c. Assume that she swims at the same rate in both pools. In particular, assume that turns don't affect her rate.

5. a.

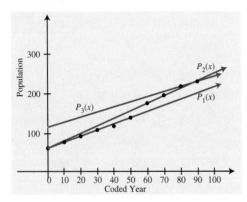

   b. $P_2(x)$
   c. $P_1(70) = 188$ million, $P_2(70) = 203$ million, $P_3(70) = 214$ million
   d. $P_1(100) = 236$ million, $P_2(100) = 258$ million, $P_3(100) = 253$ million
   e. $P_3(x)$. It uses the most recent data to make a prediction, so it reflects the decline of the birth rate.

Chapter 1 Review Exercises

1. Problem solving is the process of applying previously acquired knowledge to a new and unfamiliar situation in order to arrive at a solution to the situation.

3. a. George Polya.
   b. See "Historical Development" in Section 1.1.

5. a. A procedure or method for solving a problem
   b. Our usual procedure for adding by hand a one-digit and a two-digit number (such as 8 + 17)

7. **30:** 9 + 10 + 11, 6 + 7 + 8 + 9, 4 + 5 + 6 + 7 + 8
   300: 99 + 100 + 101, 58 + 59 + 60 + 61 + 62, 34 + 35 + 36 + 37 + 38 + 39 + 40 + 41, 13 + 14 + · · · + 27, 1 + 2 + · · · + 24
   1000: 198 + 199 + 200 + 201 + 202, 55 + 56 + · · · + 70, 28 + 29 + · · · + 52
   Let *n* denote the number (30, 300, 1000, etc.) and *t* denote the number of terms in the sum. If *t* is odd, then the factors of *t* must be factors of *n*. If *t* is even, then when *n* is divided by *t*, the decimal part is .5

9.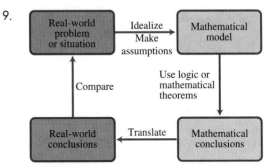

If necessary, refine the model

11. a. 1,073,741,824.
    b. All the grandparents are distinct; i.e., no two of them are the same.
    c. About 1100.
    d. Generations are about 30 years apart.

2 ▶ CHAPTER 2

Exercises 2.1

1.

3. 1025

5. LIV

7. 117

9. 1955

11. MXXXIV

15. **13:** ilinaito **14:** ilinaino

17. $(5 \times 10^4) + (1 \times 10^3) + (2 \times 10^2) + (0 \times 10^1) + (3 \times 10^0)$

Exercises 2.2

1. 476

3. 1130

5. 6

7. 36

9. 33,017

11. $(3041)_5$

13. $(10011)_2$

15. $(18T0)_{12}$
17. $(115C)_{16}$
19. $(2043)_5$
21. $(1010100011)_2$
23. $b = 11$
25. a. No. $4b + 2$ is always even.
    b. No. $5b + 2$ is always even.
27. 0 and 1
29. a. 4
    b. 7
    c. 10
31. Have the other player start first. On your subsequent turns, leave items according to the sequence 19, 16, 13, 10, 7, 4, 1.
33. a.
$$\begin{array}{r} 10 \\ +11 \\ \hline 21 \end{array}$$
    b.
$$\begin{array}{r} 12 \\ +11 \\ \hline 100 \end{array}$$
    c.
$$\begin{array}{r} 12 \\ +22 \\ \hline 111 \end{array}$$
    d.
$$\begin{array}{r} 210 \\ +122 \\ \hline 1102 \end{array}$$
    e.
$$\begin{array}{r} 222 \\ +222 \\ \hline 1221 \end{array}$$
    f.
$$\begin{array}{r} 1201 \\ +201 \\ \hline 2102 \end{array}$$
    g.
$$\begin{array}{r} 2102 \\ +1212 \\ \hline 11021 \end{array}$$

Section 2.3

1. Half a popsicle; three-quarters of a dollar
3. a. $\frac{1}{5}$ b. $\frac{1}{13}$ c. $\frac{1}{21}$ d. $\frac{1}{100}$
5. $\frac{1}{15}$
7. One method uses $42 \cdot 126 > 88 \cdot 60$; thus, $\frac{42}{60} > \frac{88}{126}$. Another method uses common denominators to show $\frac{441}{630} > \frac{440}{630}$.
9. a. $\frac{5}{16}$ b. $\frac{14}{32}$ c. $\frac{22}{133}$ d. $\frac{a+b}{b+1} > \frac{a}{b}$
11. a. bd
    b. $\frac{a}{b} = \frac{ad}{bd}$ and $\frac{c}{d} = \frac{bc}{bd}$
    c. $\frac{a}{b} > \frac{c}{d}$ in equivalent to $\frac{ad}{bd} > \frac{bc}{bd}$ which is equivalent to $ad > bc$.

Section 2.4

1. 6
3. 5
5. a. 1 b. 2 c. 3 d. −4 e. 0 f. 0 g. 0
   h. undefined
7. a. 1 b. 0 c. −1 d. 3 e. 0 f. −1
   g. −7 h. 3 i. 0 j. −2 k. 5 l. 8
   m. −8 n. −12 o. −14 p. −6 q. −4 r. 0
   s. 1 t. −14 u. −24 v. 20 w. 0
9. $\sqrt{2}$
11. a. $\frac{17}{12}$ b. $-\frac{1}{12}$ c. $\frac{1}{2}$ d. $\frac{8}{9}$

13. a. $\frac{-7}{20}$ b. $\frac{-17}{20}$ c. $\frac{-3}{20}$ d. $\frac{-12}{5}$
15. a. $\frac{10}{3}$ b. $\frac{-14}{3}$ c. $\frac{-8}{3}$ d. $\frac{-1}{6}$
17. Exponentiation (a^b) when neither a nor b is fixed.
19. 25
21. $82.81 < n < 84.64$
23. $\sqrt{-3}$
25. a. −3 b. −4 c. −125

Section 2.5

1. address, bank account number, PIN (personal identification number) for a bank card
3. 0
5. 3
7. Correct
9. 8
11. Check digit for 9901234567 is 1
13. Error
15. 7
17. 5
19. Any of 0, 1, 2, 3, 4, 5, 6, 7, 8, or 9
21. Error
23. 0
25. 3
27. No
29. c. Never
31. The sum of the digits is not affected by transposing two digits, so a transposition of two digits is never detected by the U.S. Postal Service scheme.
33. No
35. If they are an even number of digits apart and transposed, the UPC scheme will not detect the error.
37. 011110070548

Chapter 2 Review Exercises

1. The origins of counting predate written history. Counting probably started with the need to keep track of "how many"—how many children in our family, how many animals we have, etc. Pebbles were probably used to count: a pebble was matched to an object. Subsequently, tally marks probably were used to represent pebbles. Eventually, spoken and written names were given to numbers.
3. a. XXIV b. 1976
5. a. set of gestures
7. a. In a place-value system of numeration, the place a symbol occupies indicates its value, whereas in an additive system a symbol has the same value no matter what place it occupies. This allows most numbers to be written with fewer symbols in a place-value system. A place-value system requires a symbol for zero, whereas an additive system does not. Multiplication and division are easier to calculate in a place-value system.

b. Egyptian hieroglyphics, Roman numerals, and the Igbo system

c. Hindu-Arabic and Mayan

9. $(1 \times 10^3) + (9 \times 10^2) + (8 \times 10^1) + (5 \times 10^0)$

11. 4

13. a. There are only five symbols, 25 addition facts, and 25 multiplication facts to memorize in base 5, whereas in base 12 there are 12, 144, and 144, respectively.

b. It takes more symbols to record most numbers in base 5 than in base 12. For example, $(98)_{12} = (431)_5$.

15. a. $(100011)_2$ b. $(120)_5$ c. $(2E)_{12}$ d. $(23)_{16}$

17. a. one b. two

19. 1/3

21. Unit fractions are fractions with a numerator of 1. The ancient Egyptians used them. One speculation about why unit fractions were used is that they somehow seem natural: 1/2, 1/3, and 1/4 are in some sense more natural than 2/3, 2/4, or 3/4. Furthermore, notation for a unit fraction is simpler than for other fractions. In particular, if we dealt exclusively with unit fractions, we could write /2, /3, and /4 for 1/2, 1/3, and 1/4, respectively.

25. 189/456 < 51/123 < 330/789

27. a. No b. In England, among other countries, a comma is used in place of a decimal point. For example, what we write as 2.3 the English write as 2,3.

29. a. $i = \sqrt{-1}$ b. –2 c. No example d. $\sqrt{2}$
e. 1 f. 1/2 g. No example h. 1

31. The Mayans are the first known civilization to have both a symbol for zero and a place-value system. It dates back to at least the third or fourth century B.C.E. A zero appeared in Hindu-Arabic numerals sometime before 800. It was needed to make representation of numbers in a place-value system clear.

33. (1) temperatures below zero, (2) bank accounts that go in the hole, (3) a loss of yardage on a football play

35. If $3/0 = n$, then $3 = 0 \cdot n$, so $3 = 0$. Since this is impossible, there is no number n such that $3/0 = n$. In general, 3 can be replaced in this example by any nonzero number to show that division of a nonzero number by zero is impossible.

37. a. –27 b. 8/125

39. 27

43. (1) on U.S. Postal Service money orders, (2) on UPC symbols, (3) on ISBNs

45. a. No b. It detects no transposition errors whatsoever. The U.S.P.S. scheme can be described by saying the check digit is formed by taking the remainder when the sum of the digits of the number is divided by 9. A transposition of two digits of the number will not affect the sum of the digits. Hence, the check digit will still be correct when two digits of a number are transposed, so the error will go undetected.

47. a. no. $0 \cdot 10 + 4 \cdot 9 + 7 \cdot 8 + 1 \cdot 7 + 5 \cdot 6 + 4 \cdot 5 + 3 \cdot 4 + 9 \cdot 3 + 1 \cdot 2 + 7 = 197$, which is not divisible by 11.

b. 6

c. 4

CHAPTER 3

Exercises 3.1

1. $\{1, 2, 3, 4\}$

3. This will vary according to your state of residence.

5. $\{n \mid n \text{ is a positive integer and } n < 5\}$

7. This will vary according to your state of residence.

9. $\{9, 11, 13, 15\}$

11. $\{\text{Alaska, Texas}\}$

13. $\{x \mid x \text{ is an even integer and } 2 \leq x \leq 8\}$

15. $\{x \mid x \text{ is a vowel in the Roman alphabet}\}$

17. Not well defined; it is unclear what "heavy" means.

19. This is well defined; the set is $\{1, 2, 3, \ldots\}$.

21. Sets in Exercises 19 and 20.

23. True

25. True (as of the printing of this text)

27. False

29. True

Exercises 3.2

1. a. C

b. $7 \in B$ and $6 \in D$

c. $\varnothing$, $\{1\}$, $\{2, 3\}$

d. C

3. a. All sets S

b. $S = \varnothing$

5. $2^6 = 64$ subsets

7. $2^5 = 32$ pictures

9. $2^n - 1$

11.

Voter	1	2	3
Power	3	1	1

13.

Voter	1	2	3
Power	2	2	2

15.

Voter	1	2	3
Power	2	2	2

17.

Voter	1	2	3
Power	2	2	2

19.

Voter	1	2	3	4
Power	7	1	1	1

Exercises 3.3

1. $\{0, 1, 2, 4, 6, 8, 9\}$

3. O

5. $\{1, 9\}$

7. $\{1, 3, 5, 7, 9\}$

9. $\{2, 6, 8\}$

11. $\varnothing$

13. {2, 3, 5, 6, 7, 8}
15. ∅
17. {3, 5, 7}
19. {3, 5, 7}
21. {2, 6, 8}
23. {0, 1, 4, 9}
25. S
27. {0, 1, 2, 4, 6, 8, 9}
29. {4}
31. {0, 1, 3, 4, 5, 7, 9}
33. U
35. A
37. ∅
39. A
41. A'
43. ∅
45. A'
47. ∅
49. C
51. R ∩ M
53. M ∩ R'
55. S' ∩ M'

Exercises 3.4

1. a. {1, 2, 3, 4, 6, 7, 8, 9}
   b. {1, 3, 4}
   c. {0, 5, 6, 7, 9}
   d. {2, 8}
   e. {0, 1, 2, 3, 4, 5, 8}
   f. {0, 2, 5, 6, 7, 8, 9}
   g. v.

3.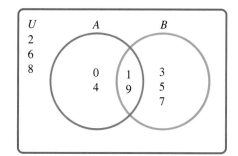

5. a. (A ∪ B)'
   b. B − A
   c. A'
   d. B'
   e. (A ∪ B) − (A ∩ B)

   e. (A ∩ B)
   f. (A ∪ B) − (A ∩ B)
   g. (A ∩ B)'
   h. (A ∪ B)

7. a. A − B
   b. (A ∩ B)
   c. (A ∪ B)'
   d. A
   e. (A ∪ B) − (A ∩ B)

   f. B
   g. (A ∩ B) ∪ (A ∪ B)'
   h. A ∪ B
   i. A ∪ (A ∪ B)'
   j. B ∪ (A ∪ B)'

9.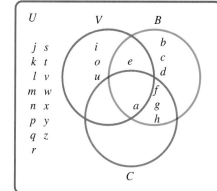

11. U = {1, 2, 3, . . . , 12}
    A = {2, 4, 6, 8, 10, 12}
    B = {3, 6, 9, 12}
    C = {4, 8, 12}
    a. LCM {2, 3} = 6
    b. LCM {2, 4} = 4
    c. LCM {3, 4} = 12
    d. LCM {2, 3, 4} = 12

13. a. A = {6, 3, 2, 1}
       B = {4, 2, 1}
       GCD {4, 6} = {2}
    b. A = {9, 3, 1}
       B = {4, 2, 1}
       GCD = {1}

15.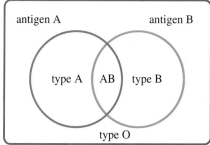

Exercises 3.5

1.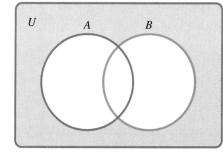

   (A∪B)'

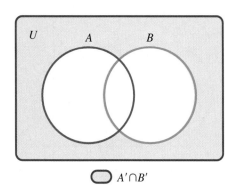

$A' \cap B'$

3.

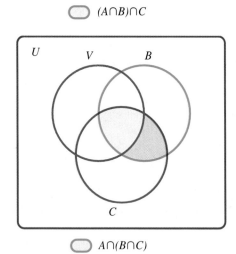

$(A \cap B) \cap C$

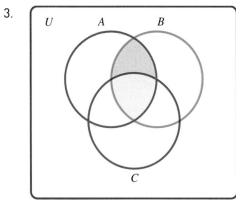

$A \cap (B \cap C)$

5.

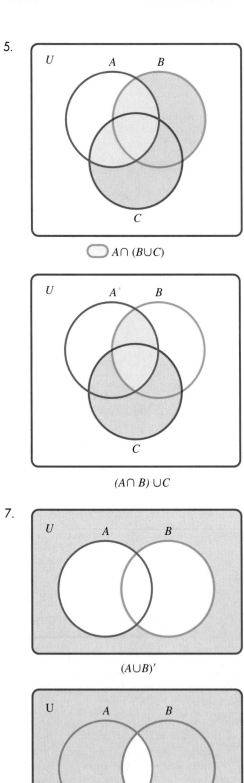

$A \cap (B \cup C)$

$(A \cap B) \cup C$

7.

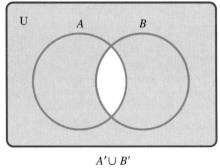

$(A \cup B)'$

$A' \cup B'$

9. They are not equal.

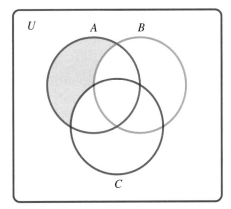

$A - (B \cup C)$

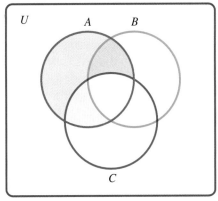

$(A - B) \cup (A - C)$

11. They are equal.

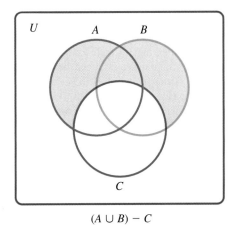

$(A \cup B) - C$

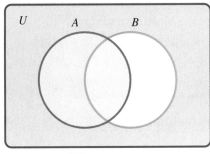

$(A - C) \cup (B - C)$

13. They are not equal.

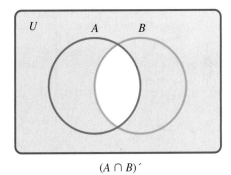

$A \cap B'$

$(A \cap B)'$

Exercises 3.6

1. a. 2
   b. 9
3. a. 4
   b. 5
   c. 3

5. Use the Venn diagram:

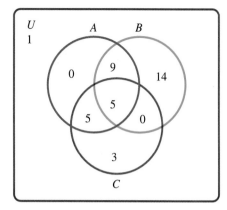

A: 300 or more career victories
B: 200 or more career losses
C: .600 or more winning percentage
a. 9
b. 5
c. 0
d. 5
e. They all had a winning percentage of .600 or more.
f. 14
g. They all had a winning percentage of less than .600.
h. 3
i. 1

7. Use the Venn diagram:

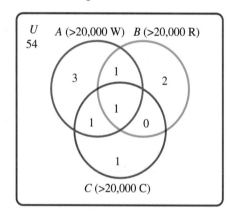

a. 3
b. 2
c. 1
d. 54

9. Use the Venn diagram:

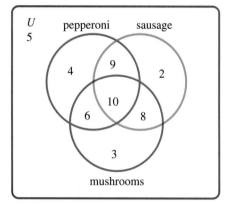

a. 5
b. 10
c. 10
d. 20
e. Order either pepperoni or sausage; 29 people prefer either one.

Chapter 3 Review Exercises

1. a. Georg Cantor   b. late 1800s
3. a. $\{x \mid x$ is an even positive integer$\}$
   b. $\{x \mid x$ is a state that starts with A$\}$
5. a. finite   b. infinite   c. finite
7. a. True   b. False   c. False   d. True
9. a. False   b. False   c. True   d. True   e. True
   f. True   g. False   h. False   i. False   j. False
   k. True   l. True
11. 32
13. The statement is incorrect. Although mathematics does deal with definite statements like "If $2x + 3 = 11$, then $x = 4$," it advances based on conjectures that are explored to see if they are true or false. For example, Cantor conjectured that there are different sizes of infinite sets. His exploration of this idea helped lead to the development of set theory. In this chapter, you were led to conjecture that if a set has n elements, then it has 2^n subsets. Truth in mathematics is established deductively. The definite statement "If $2x + 3 = 11$, then $x = 4$" can be established deductively. However, mathematics explores new territory by using conjectures that are often developed by means of inductive reasoning. Conjectures are essential in the development of mathematics.
15. The voting of stockholders in a corporation is done with weighted voting. Each stockholder has a number of votes equal to the number of shares of stock the person holds.
17. a. $\{C, E, F\}$ b. M c. $\{C\}$ d. T e. T f. T
    g. $\{A, B, D, F, G\}$ h. $\varnothing$ i. $\{A, B, D, E, G\}$
    j. $\{A, B, D, G\}$ k. $\{A, B, D, E, F, G\}$ l. $\{A, B, C, D, F, G\}$ m. $\{A, B, C, D, E, G\}$ n. $\{A, B, D, E, F, G\}$
19. a. $\{1, 2, 6, 8\}$ b. $\{1\}$ c. $\{2, 3, 4, 5, 6, 7, 9\}$
    d. $\{8\}$ e. $\{1, 3, 4, 5, 7, 8, 9\}$ f. $\{2, 3, 4, 5, 6, 7, 8, 9\}$

21. a. $A - B$ **b.** $A \cap B$ **c.** $B - A$ **d.** $(A \cup B)'$
e. B' **f.** $(A \cap B) \cup (A \cup B)'$ **g.** A' **h.** $(A \cap B)'$

23.

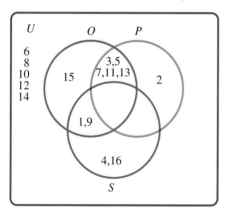

25. a. Not equal.

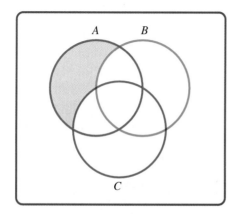

$(A \cap B)'$

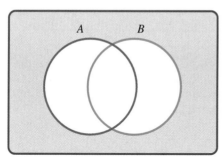

$A' \cap B'$

b. Equal.

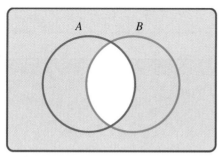

$A \cap (B \cup C)'$

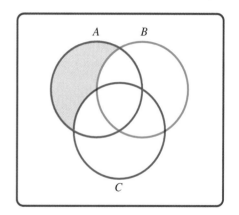

$A - (B \cup C)$

c. Not Equal.

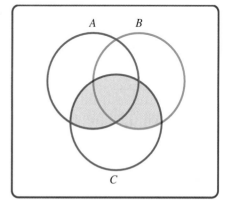

$(A \cup B) \cap C$

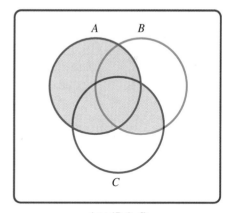

$A \cup (B \cap C)$

Exercises 4.2

1. A statement
3. Not a statement

5. Not a statement
7. Let the simple statements be
    p: Steve is going to Paul's house.
    q: Paul is coming to Steve's house.
    Then the sentence is $p \vee q$.
9. Let the statements be
    p: The party is on Friday evening.
    q: The party is on Staturday noon.
    r: We should bring candy.

    Then the sentence is $(p \vee q) \wedge r$.
11. Let the statements be

    p: The New York Giants played in the 1992 Super Bowl.
    q: The Los Angeles Rams played in the 1992 Super Bowl.

    Then the sentence is $\sim (p \wedge q)$.
13. Let the statements be

    p: Vote for Smith for Governor.
    q: Vote for Brown for Lieutenant Governor.
    r: Don't vote.

    Then the sentence is $(p \wedge q) \vee r$.
15. The United States is not in Europe.
17. Either Nigeria is in Africa or Suriname is in South America.
19. Suriname is in South America and either Nigeria is in Africa or the United States is in Europe.
21. Either Suriname is not in South America or Nigeria is not in Africa, and the United States is in Europe.

Exercises 4.3

1. No. It does not allow for p and q both to be true or both to be false.

3.

p	q	$p \wedge q$	$\sim (p \wedge q)$
T	T	T	F
T	F	F	T
F	T	F	T
F	F	F	T

5.

p	q	$p \wedge q$	$\sim q$	$\sim q \vee (p \wedge q)$
T	T	T	F	T
T	F	F	T	T
F	T	F	F	F
F	F	F	T	T

7.

p	q	$p \vee q$	$\sim q$	$(p \vee q) \vee \sim q$
T	T	T	F	T
T	F	T	T	T
F	T	T	F	T
F	F	F	T	T

9.

p	q	r	$\sim p$	$q \wedge r$	$\sim p \wedge (q \wedge r)$
T	T	T	F	T	F
T	T	F	F	F	F
T	F	T	F	F	F
T	F	F	F	F	F
F	T	T	T	T	T
F	T	F	T	F	F
F	F	T	T	F	F
F	F	F	T	F	F

11.

p	q	r	$p \vee q$	$(p \vee q) \wedge r$
T	T	T	T	T
T	T	F	T	F
T	F	T	T	T
T	F	F	T	F
F	T	T	T	T
F	T	F	T	F
F	F	T	F	F
F	F	F	F	F

13.

Statement	Value
$\sim r$	T
$p \wedge q$	T
$q \vee p$	T
$r \wedge \sim p$	F
$p \wedge (q \vee r)$	T
$\sim p \wedge (q \vee r)$	F
$(\sim p \vee \sim q) \wedge r$	F
$(p \vee q) \wedge (\sim r)$	T

Exercises 4.4

1.

p	q	$q \to p$
T	T	T
T	F	T
F	T	F
F	F	T

3.

p	$\sim p$	$p \to \sim p$
T	F	F
F	T	T

5.

p	q	$p \wedge q$	$(p \wedge q) \to p$
T	T	T	T
T	F	F	T
F	T	F	T
F	F	F	T

7. *p:* You can do anything with children.
   q: You only play with them.

$$q \to p$$

The sentence would be false if you did play with the children but they wouldn't do something for you.

9. *p:* You talk.
   q: You can improve the silence.

$$p \to q \qquad \text{(or, equivalently, } \sim q \to \sim p)$$

The sentence would be false if you did talk but did not improve the silence.

11. *p:* You know where you are going.
   q: You will end up somewhere else.

$$\sim p \to q$$

The sentence would be false if you don't know where you are going but end up where you want.

13. Answers will vary.

15.

p	q	$\sim p$	$p \to q$	$\sim p \vee q$
T	T	F	T	T
T	F	F	F	F
F	T	T	T	T
F	F	T	T	T

The truth tables are the same.

17.

p	q	r	$q \leftrightarrow r$	$p \to (q \leftrightarrow r)$
T	T	T	T	T
T	T	F	F	F
T	F	T	F	F
T	F	F	T	T
F	T	T	T	T
F	T	F	F	T
F	F	T	F	T
F	F	F	T	T

19.

p	q	$p \to q$	$q \to p$	$(p \to q) \leftrightarrow (q \to p)$
T	T	T	T	T
T	F	F	T	F
F	T	T	F	F
F	F	T	T	T

21. The converse (false) is: If x^2 is a nonzero real number, then x is a nonzero real number.
The inverse (false) is: If x is not a nonzero real number, then x^2 is not a nonzero real number.
The contrapositive (true) is: If x^2 is not a nonzero real number, then x is not a nonzero real number.

23. statement: $\sim p \to \sim q$; converse: $\sim q \to \sim p$; inverse: $p \to q$; contrapositive $q \to p$

25. statement: $p \to (p \vee q)$; converse $(p \vee q) \to p$; inverse: $\sim p \to \sim (p \vee q)$; contrapositive: $\sim (p \vee q) \to \sim p$

27. statement: $(p \wedge \sim p) \to q$; converse: $q \to (p \wedge \sim p)$; inverse: $\sim (p \wedge \sim p) \to \sim q$; contrapositive: $\sim q \to \sim (p \wedge \sim p)$

29. a.

p	q	p only if q
T	T	T
T	F	F
F	T	T
F	F	T

b. i. *p:* Enter this gate.
      q: You know geometry.
      p only if *q* which means $p \to q$
   ii. *p:* We will go.
      q: The probability of snow is less than 20%.
      p only is *q* which means $p \to q$

c. i. false
   ii. true
   iii. false
   iv. true
   v. false
   vi. false

d. *p:* It's 11 a.m.
   q: In two hours it will be 1 P.M.

31. a. true
   b. false
   c. true
   d. true

33. The punch line assumes $\sim p \to \sim q$ is true. However, the inverse is not logically equivalent to $p \to q$.

Exercises 4.5

1. b and e
3. It isn't springtime.

5. If the government is a lawbreaker, then the government breeds contempt for the law.

7. b

9. c

11. a

13. *p:* It is Friday
    q: I have to go to work.

$(p \vee \sim q) \wedge q \rightarrow \sim p$
This is not a valid argument.

p	*q*	*~q*	*~p*	*p* $\vee$ *~q*	*(p* $\vee$ *~q)* $\wedge$ *q*	*(p* $\vee$ *~q)* $\wedge$ *q* $\rightarrow$ *~p*
T	T	F	F	T	T	F
T	F	T	F	T	F	T
F	T	F	T	F	F	T
F	F	T	T	T	F	T

15. *p:* It is Friday.
    q: I have to go to work.

$[(p \vee \sim q) \wedge \sim q] \rightarrow p$

p	*q*	*~q*	*p* $\vee$ *~q*	*(p* $\vee$ *~q)* $\wedge$ *~q*	*[(p* $\vee$ *~q)* $\wedge$ *~q]* $\rightarrow$ *p*
T	T	F	T	F	T
T	F	T	T	T	T
F	T	F	F	F	T
F	F	T	T	T	F

This is not a valid argument

17. *p:* You have Devil Squares.
    q: You have Figaroos.
    r: You have a Little Debbie product.

$[((p \vee q) \rightarrow r) \wedge \sim r] \rightarrow \sim p$

p	*q*	*r*	*~ r*	*(p* $\vee$ *q)* $\rightarrow$ *r*	*((p* $\vee$ *q)* $\rightarrow$ *r)* $\wedge$ *~ r*	*[((p* $\vee$ *q)* $\rightarrow$ *r)* $\wedge$ *~ r]* $\rightarrow$ *~p*
T	T	T	F	T	F	T
T	T	F	T	F	F	T
T	F	T	F	T	F	T
T	F	F	T	F	F	T
F	T	T	F	T	F	T
F	T	F	T	F	F	T
F	F	T	F	T	F	T
F	F	F	T	T	T	T

This is a valid argument.

19. *p:* You are on the Pacific Ocean.
    q: You are on the Atlantic Ocean.
    r: You are on one of the world's two largest oceans.

$[((p \vee q) \rightarrow r) \wedge \sim r] \rightarrow \sim p$

p	q	r	$\sim r$	$(p \lor q)$	$(p \lor q) \to r$	$((p \lor q) \to r) \land \sim r$	$[((p \lor q) \to r) \land \sim r] \to \sim p$
T	T	T	F	T	T	F	T
T	T	F	T	T	F	F	T
T	F	T	F	T	T	F	T
T	F	F	T	T	F	F	T
F	T	T	F	T	T	F	T
F	T	F	T	T	F	F	T
F	F	T	F	F	T	F	T
F	F	F	T	F	T	T	T

This is a valid argument.

21. c
23. a
25. c

Exercises 4.6

1.

p	q	$\sim p$	$\sim q$	$p \lor q$	$\sim(p \lor q)$	$(\sim p) \land (\sim q)$
T	T	F	F	T	F	F
T	F	F	T	T	F	F
F	T	T	F	T	F	F
F	F	T	T	F	T	T

They are logically equivalent.

3.

p	q	$\sim q$	$p \to q$	$p \lor \sim q$
T	T	F	T	T
T	F	T	F	T
F	T	F	T	F
F	F	T	T	T

They are not logically equivalent.

5.

p	q	r	$q \lor r$	$p \land (q \lor r)$	$(p \land q)$	$(p \land q) \lor r$
T	T	T	T	T	T	T
T	T	F	T	T	T	T
T	F	T	T	T	F	T
T	F	F	F	F	F	F
F	T	T	T	F	F	T
F	T	F	T	F	F	F
F	F	T	T	F	F	T
F	F	F	F	F	F	F

They are not logically equivalent.

7. Either Amy is not single or she does claim some dependents.
9. Either Ben is 65 or older or Ben is not blind.
11. 1. $(A \cup B)' = A' \cap B'$
    4. A = B is equivalent to A ⊆ B and B ⊆ A
    6. $A \cup (B \cap C) = (A \cup B) \cap (A \cup C)$
13. Either you go to other men's funerals or they won't go to yours.
15. Either it is not the case that one man offers you democracy and another offers you a bag of grain or at some stage of starvation you will prefer the grain to the vote.
17. If the suspect was in Los Angeles at 9:00 P.M., then the suspect was not in San Francisco at 9:30 P.M.

Chapter 4 Review Exercises

3. a. $p \wedge q$, where p is "Chapter 4 is on logic" and q is "Chapter 3 is on sets."
   b. $p \wedge \sim q$, where p is "I love the game" and q is "I need that kind of turmoil."
   c. $p \to q$, where p is "Thomas Jefferson is alive today" and q is "I would appoint him secretary of state."
   d. $p \vee q$, where p is "This is night" and q is "We're having an eclipse."
   e. $(p \vee q) \wedge r$, where p is "She has a journalism major," q is "She has a journalism minor," and r is "She writes well."
   f. $\sim(p \wedge q)$, where p is "The Los Angeles Dodgers will win the 2001 World Series" and q is "The New York Mets will win the 2001 World Series."
   g. $p \to (q \vee r)$, where p is "This is the weekend," q is "Today is Saturday," and q is "Today is Sunday."
   h. $p \to q$, where p is "Speak" and q is "[You are] called on."
   i. $p \to q$, where p is "You get a C" and q is "You pass this course."

5. a.

p	q	$\sim q$	$p \wedge \sim q$	$\sim(p \wedge \sim q)$
T	T	F	F	T
T	F	T	T	F
F	T	F	F	T
F	F	T	F	T

b.

p	q	$\sim q$	$p \vee \sim q$
T	T	F	T
T	F	T	T
F	T	F	F
F	F	T	T

c.

p	q	$\sim p$	$p \wedge q$	$\sim p \vee (p \wedge q)$
T	T	F	T	T
T	F	F	F	F
F	T	T	F	T
F	F	T	F	T

d.

p	q	$\sim p$	$\sim p$ XOR q
T	T	F	T
T	F	F	F
F	T	T	F
F	F	T	T

e.

p	q	$\sim q$	$p \to \sim q$
T	T	F	F
T	F	T	T
F	T	F	T
F	F	T	T

f.

p	q	$p \wedge q$	$(p \wedge q) \to q$
T	T	T	T
T	F	F	T
F	T	F	T
F	F	F	T

g.

p	q	$p \vee q$	$q \to (p \vee q)$
T	T	T	T
T	F	T	T
F	T	T	T
F	F	F	T

h.

p	q	~p	~q	~p → q	~q → p	(~p → q) ↔ (~q → p)
T	T	F	F	T	T	T
T	F	F	T	T	T	T
F	T	T	F	T	T	T
F	F	T	T	F	F	T

i.

p	q	r	p ∨ q	(p ∨ q) ∧ r	q ∧ r	p ∨ (q ∧ r)	[(p ∨ q) ∧ r] ↔ [p ∨ (q ∧ r)]
T	T	T	T	T	T	T	T
T	T	F	T	F	F	T	F
T	F	T	T	T	F	T	T
T	F	F	T	F	F	T	F
F	T	T	T	T	T	T	T
F	T	F	T	F	F	F	T
F	F	T	F	F	F	F	T
F	F	F	F	F	F	F	T

7. a. $p \rightarrow q$, where p is "I had a hammer" and q is "I'll hammer in the morning."

b.

p	q	p → q
T	T	T
T	F	F
F	T	T
F	F	T

c. The sentence would be false when I did have a hammer but I didn't hammer in the morning.

9. You are either a sophomore or not a sophomore.

11. In state X only outlaws will have guns.

13. a.

$$p \rightarrow (q \wedge r)$$
$$\underline{\sim q \wedge r}$$
$$\sim p$$

where p is "A high school graduate meets the standards of Goals 2000," q is "A high school graduate knows about the functions of Amnesty International," and r is "A high school graduate knows about the functions of NATO."

p	q	r	q ∧ r	p → (q ∧ r)	~q	~q ∧ r	A [p → (q ∧ r)] ∧ (~q ∧ r)	~p	A → ~p
T	T	T	T	T	F	F	F	F	T
T	T	F	F	F	F	F	F	F	T
T	F	T	F	F	T	T	F	F	T
T	F	F	F	F	T	F	F	F	T
F	T	T	T	T	F	F	F	T	T
F	T	F	F	T	F	F	F	T	T
F	F	T	F	T	T	T	T	T	T
F	F	F	F	T	T	F	F	T	T

The argument is valid.

b.
$$(p \vee q) \rightarrow r$$
$$\dfrac{\sim p \wedge \sim q}{\sim r}$$

where p is "You do a crossword puzzle," q is "You lease an Acura," and r is "You demonstrate your intelligence."

p	q	r	$p \vee q$	$(p \vee q) \rightarrow r$	$\sim p$	$\sim q$	$\sim p \wedge \sim q$	B $[(p \vee q) \rightarrow r] \wedge (\sim p \wedge \sim q)$	$\sim r$	$B \rightarrow \sim r$
T	T	T	T	T	F	F	F	F	F	T
T	T	F	T	F	F	F	F	F	T	T
T	F	T	T	T	F	T	F	F	F	T
T	F	F	T	F	F	T	F	F	T	T
F	T	T	T	T	T	F	F	F	F	T
F	T	F	T	F	T	F	F	F	T	T
F	F	T	F	T	T	T	T	T	F	F
F	F	F	F	T	T	T	T	T	T	T

The argument is invalid.

15. a. If I could, I would.
 b. If this is Chapter 4, then it is on logic.
  $$\dfrac{\text{This is Chapter 4.}}{\text{This chapter is on logic.}}$$
 c. If this is Chapter 4, then it is on sets.
  $$\dfrac{\text{This is Chapter 4.}}{\text{This chapter is on sets.}}$$
17. a. She is either under 25 or less than 7 years a citizen of the United States.
 b. It's not the case that he was either young or wealthy.
19. $p \wedge (q \vee r)$ is logically equivalent to $(p \wedge q) \vee (p \wedge r)$.
21. a. Either the shoe doesn't fit or I wear it.
 b. If the steak is done, then I'm a poor judge of meat.

5 CHAPTER 5

Exercises 5.1

1. 21 and 51
3. 2
5. $\dfrac{8x^2 - 9x + 20}{(x + 3)(x - 4)}$
7. $x + 1$
9. $\dfrac{1}{(x + 1)(x + 3)}$
11. $\dfrac{x + 2}{x + 4}$
13. $x^2 + 3 = 4$
15. a. $28.43
 b. $39.80
 c. 33⅔%
17. 6
19. 20 ft × 40 ft
21. $12,342.00
23. a. 77°F
 b. 21.1°C
 c. 0°C
 d. 100°C
 e. −40°
25. $6.50
27. a. advances ≤ $80
 b. ≥$400
29. a. x
 b. $2x$
 c. $2x + 10$
 d. $\dfrac{2x + 10}{2}$
 e. $\dfrac{2x + 10}{2} - x$
 f. 5
31. $\dfrac{3}{10}$ hekat

Exercises 5.2

1. a, b, and c.
3. a. 3
 b. 6
 c. 4
 d. 9
5. $x = y^2$
7. $f(x) = 3x$
9. y is a function of x if and only if there is a rule $y = f(x)$ such that for each real x, there is a unique y.
11. a. 6
 b. 110
13. $f(x) = x + 1$
15. vertex: $(-1, 9)$; zeros: $x = 2, -4$

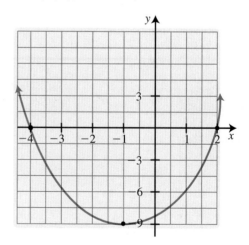

17. vertex: $(.1, -2.05)$; zeros: $x = \dfrac{1 \pm \sqrt{41}}{10}$

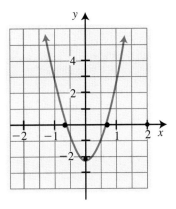

19. vertex: $(0, 0)$; zeros: $x = 0$

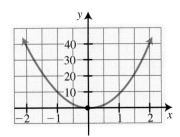

21. vertex: $\left(\dfrac{3}{8}, \dfrac{-23}{16}\right)$; no zeros

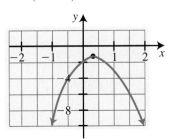

23. vertex: $\left(\dfrac{-3}{14}, \dfrac{9}{28}\right)$; zeros: $x = 0, \dfrac{-3}{7}$

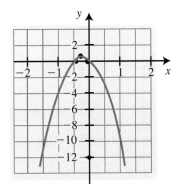

25. $y = x^3$

27. a. 214.8 ft
    b. 254.9 ft

29. 32.7 miles per hour

31. 7.5 ft $\times$ 15 ft

33. 0 or 1

Exercises 5.3

1. b

3.

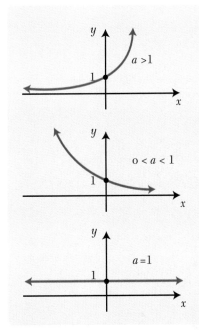

5. a. 4096
   b. 1296
   c. $\dfrac{1}{64}$
   d. 81
   e. 1.5874
   f. 1/81
   g. 81
   h. -81

7.

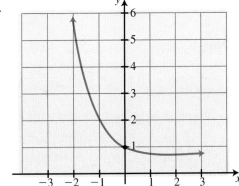

9. a. All values
   b. No values
   c. No values
11. The earlier example assumed the bank rounded $13.125 to $13.13.
13. $7051.10
15. $14,985.62
17. $12,202.51
19. $0.46
21. $84,415.87
23. 6,369,000,000
25. $1433.33
27. $6851.30
29. $14,544.67
31. $12,158.76
33. a. 1.36 g
   b. 0.82 g
   c. 0.000000836 g
   d. 1 hour
   e. 2.44 g

   f.

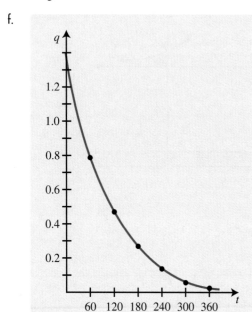

35. a. 15.8 times as powerful
   b. 63 times as much energy

Exercises 5.4

1. a. true
   b. true
   c. true
   d. true
   e. false
   f. false
   g. false

3. a. <
   b. >
   c. <
   d. >
   e. < for $c > 0$, > for $c < 0$, = for $c = 0$
   f. > for $c > 0$, < for $c < 0$, = for $c = 0$
   g. >
   h. >
5. $x \leq 10$
7. $x < -3$
9. $x > -12$
11. $-2 \leq x$
13. $5x + 4 > 2x + 10$

   $\quad 4 > -3x + 10 \qquad$ Subtract $5x$
   $\quad -6 > -3x \qquad\qquad$ Subtract 10
   $\quad 2 < x \qquad\qquad\quad$ Divide by -3

15. Yes, he needs an 87 or higher.
17. 1.5 gallons
19. 55
21. $y \leq 2x + 1$

Chapter 5 Review Exercises

1. Algebra is a system that generalizes arithmetic. In arithmetic we add, subtract, multiply, divide, raise numbers to powers, and take roots of numbers. In algebra we do the same operations but on either numbers or variables, which represent numbers. In this sense, algebra is generalized arithmetic. Algebra also deals with solving equations of any degree. In summary, algebra is characterized by a use of variables, it involves generalized arithmetic, and it is involved with solving equations.
3. 83
5. $\dfrac{11x^2 - 11x + 11}{5x^2 + 9x - 2}$
7. $\dfrac{x + 2}{x - 1}$
9. $45.95
11. i and ii
13. f given by $f(x) = 3x$
15. a. 3.24 b. 3.20
17. a. upward b. $y = -7x^2$ c. $(\dfrac{3}{7}, -\dfrac{16}{7})$
   d.

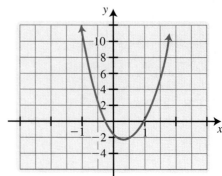

e. $-\frac{1}{7}$ and 1   f. i. 64   ii. There are two real zeros.

19. a. vi   b. vii

21.
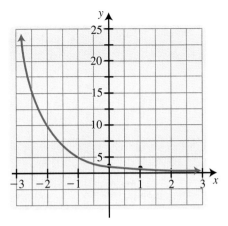

23. a. $2540.98   b. $2361.30   c. $2542.50
25. 2.95 billion people
27. The formula for the continuous compounding of interest uses e.
29. Q, t
33. $x \geq 2$
35. a. $<$ b. $>$ c. Depends on the value of c: If $c > 0$, then $<$ is appropriate; if $c = 0$, then $=$ is appropriate; if $c < 0$, then $>$ is appropriate.   d. Depends on the value of c: If $c > 0$, then $>$ is appropriate; if $c = 0$, then $=$ is appropriate; if $c < 0$, then $<$ is appropriate.
37. a. 2.385 or higher   b. 4.053 or higher, so if a course has a 4.00 maximum, he cannot get his overall GPA to 2.000 by the end of 60 credits.

6 ▸ CHAPTER 6

Exercises 6.2

1. No. You could have
3. $\overrightarrow{SR} \cup \overrightarrow{ST}$
5. Here is one of infinitely many solutions:

Exercises 6.3

1. Only iii
3. a. adjacent
   b. vertical
   c. adjacent
   d. adjacent

5. a.
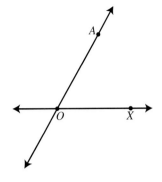

b.
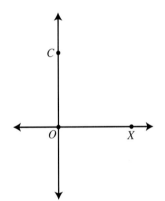

c.
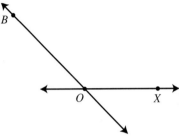

7. a.
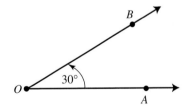

b.

c.

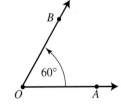

9. a. The complement is 80°; the supplement is 170°.
   b. The complement is 64°; the supplement is 154°.
   c. The complement is 1°; the supplement is 91°.

11.

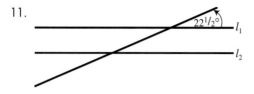

13. a. 60°
    b. 30°

Exercises 6.4

1. a. Figures a, b, and c
   b. None
   c. Figure a
   d. Figure b
   e. Figure d
3. a. Isosceles right triangle
   b. Isosceles obtuse triangle
   c. Isosceles acute triangle
   d. Scalene obtuse triangle
   e. Scalene right triangle
5. a. The hypotenuse is 10. This is a scalene right triangle.
   b. The third side is 12. This is a scalene right triangle.
7. $a = 4$, $b = 8$
9. 80 ft
11. a. square
    b. parallelogram
    c. trapezoid
    d. rectangle
13. a. A
    b. S
    c. A
    d. S
15. Pentagon

17. a.

b.

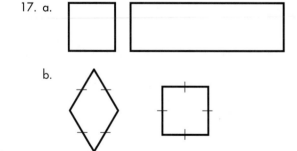

Exercises 6.5

1. a. 9
   b. 14
   c. $9\frac{1}{2}$
3. a. 360 ft
   b. 8100 sq ft
5. a. 6
   b. 15
   c. 20
7.

State	Perimeter	Area
Colorado	1340 mi	109,000 sq mi
Georgia	980 mi	59,000 sq mi
Kansas	1220 mi	82,000 sq mi
Nevada	1445 mi	105,000 sq mi

9. a. $C = 4\pi$ ft, $A = 4\pi$ ft^2
   b. $C = 8\pi$ cm, $A = 16\pi$ cm^2
   c. $C = 12\pi$ in., $A = 36\pi$ in.^{2}
11. The price per square inch of $0.0389 is better than that for two 10-in. pizzas or two 12-in. pizzas, but it's not as good as that for two 14-in. pizzas.
13. Between 840 and 841 revolutions
15. $A = 7.57$ ft^2, $P = 11.14$ ft

Chapter 6 Review Exercises

3. The ancient Egyptians approached geometry empirically. They saw what worked and recorded it. In contrast, the ancient Greeks used a deductive approach to geometry. They reasoned from a few basic principles what must be true.
5. a. Euclid   b. Next to the *Bible* it is the most widely published, translated, and studied single work. It set the shape of geometry teaching ever since. It is perhaps the most influential scientific work ever written.
7. $\overrightarrow{BA}$ and $\overrightarrow{BC}$
9. A line and a point that is not on the line, or two parallel lines

11.

13. a. $\angle CBA$ b. B c. $\overrightarrow{BA}$ and $\overrightarrow{BC}$ d. No   e. No
   f.   g.

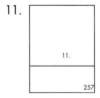

h. Yes   i. No   j. No   k. No   l. 30°

m. ∠DBA

15. a.

135°

b. No   c. 45°

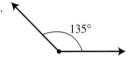
45°

17.

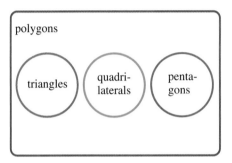
polygons

triangles   quadri-
laterals   penta-
gons

19.

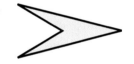

21. a.   b.

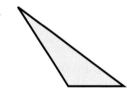

c. None exists   d.

23. 3,4,5
25. 90 ft
27. a. six   b. heptagon; seven
29. a. $36\frac{1}{2}$ in.   b. 82 in.^{2}
31. a. 644.0 mm b. Between 2498 and 2499

33.

35. 63 m and 66 cm

7 CHAPTER 7

Exercises 7.1

1.

Point	Parallel point
A	F
B	E
C	D
D	C
E	B
F	A

3. a.

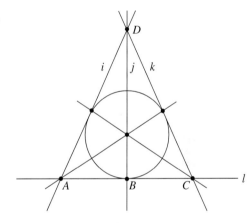

Exercises 7.2

1.

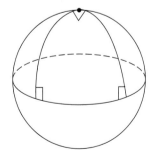

3. a. less than 360°
   b. 360°
   c. greater than 360°
5. more than one

Exercises 7.3

1. The next level (level 2) is:

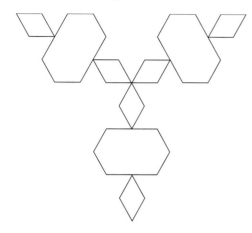

3. Level 2 is:

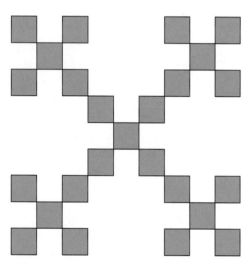

5. a. length = 4
   b. Level 3 produces a design with $8^3 = 512$ segments, each of which is $\frac{1}{64}$ of the original length.
   c. The total length is $512\left(\frac{1}{64}\right) = 8$.

Chapter 7 Review Exercises

1. An axiomatic system is a system for logically deducing some statements from other statements. It consists of undefined terms, definitions, axioms or postulates, and theorems. Undefined terms, terms for which there is an understood meaning, are used to define other terms. Axioms or postulates are statements that utilize the defined and undefined terms and that are assumed to be true. Theorems are proved on the basis of the defined and undefined terms and axioms. The following figure shows the relationship of the parts.

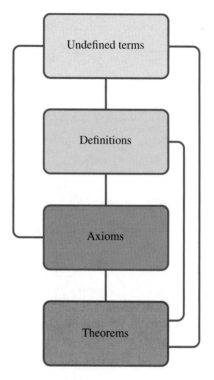

3. a. *ad, bc, bd,* and *cd*
   b. *ab, bc,* and *bd*

5.

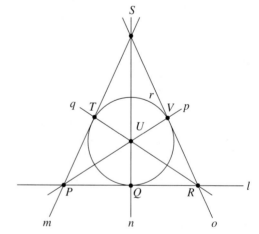

7. a. (a, c)
   b. $\sqrt{a^2 + c^2}$
   c. $\sqrt{a^2 + c^2}$
   d. From parts b and c, $AC = OB$.
9. In plane geometry, the sum of the angles of a triangle is 180°. In spherical geometry, a triangle can have an angle sum

greater than 180°. For example, the angle sum in the triangle shown is 270°.

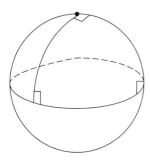

11. Euclid established a number of results (28 to be exact) before he used the fifth postulate.
13. Saccheri and Lambert, in the first seven decades of the eighteenth century, tried to prove Euclid's fifth postulate from Euclid's other axioms and postulates. Saccheri considered a quadrilateral with two opposite sides equal and their base angles with a third side both right angles. He showed (without the fifth postulate) that the remaining pair of angles are equal. He considered three possibilities: They are acute, they are right, and they are obtuse. He hoped to show that the acute-angle and the obtuse-angle hypotheses were incorrect, making the right-angle hypothesis true and thereby establishing Euclid's fifth postulate. Though he was unable to show they were incorrect, in exploring what would happen with these hypotheses he unknowingly laid the groundwork for non-Euclidean geometry. Lambert continued Saccheri's work.
15. a. Gauss, Bolyai, and Lobachevsky
    b. 1820s
17. *Elliptic geometry* is closely related to the geometry of a sphere; *hyperbolic geometry* is closely related to the geometry of a pseudosphere. In *elliptic geometry* the sum of the angles of a triangle is greater than 180°; in hyperbolic geometry the sum of the angles of a triangle is less than 180°.
19. Both are correct in the sense that they are logically consistent systems. Which fits the real world? On Earth, Euclidean geometry works very well. However, with figures that involve larger distances (going beyond the Earth) it is possible that non-Euclidean geometry might be the better model.
21. A fractal is a design that exhibits self-similarity. In other words, it is a design with the property that part of the design repeats itself over and over in ever-smaller forms.
23. i. Draw an H for which the sides are two-thirds the length of the line segment connecting the sides.
    ii. Let each of the original sides be a connecting line segment for a new H, for which the sides are two-thirds the length of the connecting segment.
    iii. Keep repeating Step ii.

CHAPTER 8

Exercises 8.2

1. {red, black}
3. {a, e, i, o, u}
5.

Outcome	A	B
Probability	0	1

7. Not possible because $P(B) < 0$.
9. Not possible because $P(A) + P(B) = 1.1$, so to make $P(A) + P(B) + P(C) = 1$, $P(C)$ must be -0.1, which is contrary to Rule 1.
11. Any answer where $P(B) + P(C) = .7$ and $P(B) \geq 0$ and $P(C) \geq 0$
13. a. yes
    b. no
    c. no
    d. no
15. a. $P(\text{black}) = .5$
    b. $P(\text{diamond}) = .25$
    c. $P(\text{queen}) = \dfrac{1}{13}$
    d. $P(\text{red jack}) = \dfrac{1}{26}$
    e. $P(\text{king of hearts}) = \dfrac{1}{52}$
17. Selecting a winner in an automobile race.
19. a. 0.172
    b. 0.916

Exercises 8.3

1. 17,576
3.

A, B, C, D	B, A, C, D	C, A, B, D	D, A, B, C
A, B, D, C	B, A, D, C	C, A, D, B	D, A, C, B
A, C, B, D	B, C, A, D	C, B, A, D	D, B, A, C
A, C, D, B	B, C, D, A	C, B, D, A	D, B, C, A
A, D, B, C	B, D, A, C	C, D, A, B	D, C, A, B
A, D, C, B	B, D, C, A	C, D, B, A	D, C, B, A

5. Answers will vary.
7. 90
9. 45
11. 1
13. 676
15. 35
17. 200

Section 8.4

1. a. $\dfrac{1}{2^{10}}$

   b. 0.24609375

3. a. $\dfrac{1}{390,625} \approx 0.000003$

   b. $\dfrac{70}{390,625} \approx 0.000179$

5. a. $\dfrac{1}{9,366,819} \approx 0.0000001068$

   b. $\dfrac{1}{1,533,939} \approx 0.0000006519$

7. a. $\dfrac{1}{5}$

   b. $\dfrac{2}{5}$

9. iii. $p > p'$

Exercises 8.5

1. 0.4
3. 0.61
5. 0.28
7. Let A and B be any events such that $P(A) = P(B) = 0$.
9. Let A be the event *you draw a red card from a deck of cards.* Let B be the event *you draw a black card from a deck of cards.*
11. a. 0.40
    b. 0.60
    c. 0.78
    d. 0.82
13. a. 0
    b. 0
15. a. 0.511
    b. 0.184
17. No
19. a. No
    b. Yes
    c. Yes
21. a. 0.85735
    b. 0.99451
    c. 0.00549
23. a. 1/5
    b. 2/15
    c. 2/15
25. a. 0.165
    b. 0.308
    c. 0.473
27. 0.008
29. 0.311

Exercises 8.6

1. a. 3/4
   b. 5/7
   c. 1/5
3. $\dfrac{1}{260,000,001}$
5. a. 391:99,609
   b. 99,609:391
   c. More than \$3.91
7. 5.8
9. \$50
11. a. 0.04
    b. 0.10

Chapter 8 Review Exercises

1. Galileo wrote in the late 1500s and early 1600s about some games of chance involving dice. Blaise Pascal corresponded with Pierre de Fermat in 1654 about theoretical reasoning on a game of chance, Jakob Bernoulli worked in probability theory during the last half of the seventeenth century. His *Ars Conjectandi* was the next major accomplishment in probability after the Pascal–Fermat correspondence. Jakob's younger brother, Johann, also made advances in probability.

3. Fermat's Last Theorem says that the equation $x^n + y^n = z^n$ has no solution among the natural numbers when n is a natural number greater than 2. Although Fermat claimed to have a proof of it, the first known proof was given by Andrew Wiles in 1994.

5. 9/10
7. a. i. 1/2
      ii. 1/6
   b. i. 4/31
      ii. 8/31
9. 125
11. 120
13. 504
15. a. 1330
    b. 1/70
17. a. 1/256
    b. 35/128
19. a. Roll a die and record what's on the top face. Let A be "get a 1" and B be "get a 2."
    b. Roll a die and record what's on the top face. Let A be "Get an odd number" and B be "Get a prime number."
21. "Get a freshman."
23. a. iii
    b. iii
25. 11/245
27. a. 2:3
    b. 3:2
29. 5.8

Exercises 9.1

1. Statistics is the science that deals with data—their collection, analysis, presentation, and inferences about the population based on information contained in a sample.

3. a. Between 40 and 44 years of age, inclusive.
   b. Between 65 and 69 years of age, inclusive.
   c. 24
   d. 32

5. a. Answers will vary according to your scale. We suggest the width of the land-area intervals be between 30,000 and 100,000 square miles.

   b.

Ten thousands digit	
0	4 1 6 9 7 1 9
1	
2	4
3	5 9 0 0 9
4	3 6 7 8 0 4 1
5	0 2 3 7 5 5 6 4
6	8 8 8 6
7	9 6 4 5
8	2 1 9 2
9	6 7
10	3 9
11	3
12	1
13	
14	5
15	5
over 15	0 1

7. *Population:* readers of that newspaper. *Sample:* 210 of those readers who responded.

9. *Population:* all physicians. *Sample:* 21,996 physicians.

11. Choosing the five students in the front row does not give every set of five students an equal chance of being chosen. This method of sampling might lead the teacher to believe that the class understood the topic better than they really did, since front-row students are likely to be paying better attention than other students.

Exercises 9.2

1. a. 1, 3, 5
   b. 1, 2, 7

3. mean $= \dfrac{23}{9} \approx 2.56$

   median $= 3$

5. Change either the 3 or the 5. For example: 2,2,4,5,8,8.

7. a. If $x < 2$, then the median is 2; if $2 \le x \le 3$, then the median is x; and if $3 < x$, then the median is 3. In any case, $2 \le$ median ≤ 3.
   b. mean $= 2 + x/5$.

9. $53.0 \le$ mean ≤ 57.0, or mean ≈ 55.0; $55 \le$ median ≤ 59.

11. 1, 1, 10

13. The median was 20 or below.

15. 2.5, 7

17. 10% of 11-year-old boys are taller than 60 inches.

Exercises 9.3

1. 8

3. 7.0%

5. $\dfrac{1}{4}, \dfrac{1}{2}, \dfrac{3}{4}$

7. 1, 2, 2, 2, 5

9. a. Yes
   b. Example: 1, 2, 2, 2, 2, 2, 2, 2, 2, 2, 2, 3

11.

13. Test 1 had better results.

	Test 1	Test 2
Median	75	70
Lower quartile	65	60
Upper quartile	84	84
Low score	40	40
High score	100	100

15. $\sigma^2 = 6.5$, $\sigma \approx 2.55$

17. i. $\sigma_s < \sigma_f$

19. $16^2 = 256$

21. Range approximation $= \dfrac{6}{4} = 1.5$; $\sigma \approx 2.55$

23. 0.1

25. No. The standard deviation is the square root of the variance. The square root of a real number cannot be negative.

Exercises 9.4

1. Discrete
3. Discrete
5. a. 0.054
   b. 0.042
7. .4977
9. .2912
11. .5478
13. .0361
15. .9547

17. a. .0336
    b. .9664
    c. .7492
    d. .2501
19. a. .0038
    b. .0228
21. a. A: 6.7%
       B: 24.2%
       C: 38.3%
       D: 24.2%
       E: 6.7%
    b. A: $100 \geq x \geq 85$
       B: $85 > x \geq 77$
       C: $77 > x \geq 68$
       D: $68 > x \geq 60$
       E: $60 > x$
23. i

Exercises 9.5

1. $\mu_{\bar{x}} = 14.93\left(\frac{1}{10}\right) + 15.00\left(\frac{2}{10}\right) + 15.07\left(\frac{3}{10}\right) + 15.13\left(\frac{2}{10}\right) +$

   $15.20\left(\frac{2}{10}\right) = 15.08 = \mu$

3. .0174
5. .0384
7. 49 or more
9. .0344
11.

If $\hat{p}$ is:	Then $\sigma_{\hat{p}} = \sqrt{\hat{p}\hat{q}/50}$ is:
.61	.0690
.62	.0686
.63	.0683
.64	.0679
.59	.0696
.58	.0698
.57	.0700
.56	.0702

13. (21.5%, 28.5%)
15. iii
17. a. (43%, 49%)
    b. We are 95% sure that the percentage of all adult Americans who would have answered "better" at that time is within 3 percentage points of the percentage found in the survey; i.e., we are 95% sure it is between 43% and 49%.

Chapter 9 Review Exercises

1. Descriptive statistics deals with describing a set of data; inferential statistics deals with making inferences about a population based on information contained in a sample.

3.

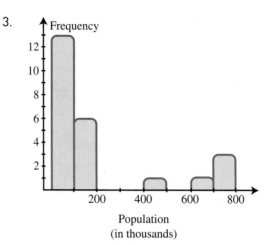

5. a. 312 ft
   b. 355 ft
   c. 28
   d. 19
   e. 331.7 ft
   f. 330 ft
   g. 43 ft
   h. iv
7. The population is the set of all students at the school; the sample is the set of five students interviewed.
9. a. 7
   b. 8.5
   c. 6
   d. 2.5
   e. i. It will get smaller by (3 − new number)/6
      ii. It will stay the same.
      iii. It will get larger by (3 − new number)
      iv. It will get larger.
   f. Skewed to the left
11. 32,251
13. On the SAT exam, 63% of the students scored higher than this student and 37% scored lower.
15. 739,185

17.

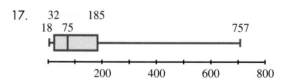

19. a. 7.6
    b. 2.8
21. a. 16
    b. 2
23. a. Discrete
    b. The length of a string wrapped around people's heads
25. iii
27. a. .1915
    b. .9554
    c. .0885
    d. Between .0000 and .0010

29. A sampling distribution is a probability distribution of a sample statistic such as $\bar{x}$. For example, suppose the population variable x has three possible values, 1, 3, and 7, each equally likely. Then $\bar{x}$ can take on the values 1, 2, 3, 4, 5, and 7 with probabilities 1/9, 2/9, 1/9, 2/9, 2/9, and 1/9, respectively.

31. a. .0104
    b. .8849
    c. .0217

33. a. (15.4%, 22.6%)
    b. There is a 95% chance that the proportion of all adult Americans who believed (on the dates the survey was taken) that you can trust the government to do what's right always or most of the time was between 15.5% and 22.5%.

35. iii

 CHAPTER 10

Exercises 10.2

1. The 40-lb bag is $0.10725 per lb; the 50-lb bag is $0.107 per lb. So the 50-lb bag is the better buy.
3. The eight-pack costs $0.015625 per oz; the 2-liter bottle costs $0.014645 per oz. Therefore, the 2-liter bottle is the better buy.
5. 98 CCF
7. $0.075472
9. $37,400
11. $28.00
13. 52%
15. 12.5%
17. a. one 48-oz
    b. one 22-oz

Exercises 10.3

1. a. $7.70
   b. $387.70
3. a. 0.0175%
   b. 1.000175X amount borrowed
5. $85.33
7. $210
9. a. 0.003333%
   b. 121.67%
11. 19.80%
13. a. 1.367%
    b. 0.00045
15. $3.83
17. 365

Exercises 10.4

1. $1085.76
3. $767.20

5. $1741.10
7. $90.51
9. 4.688%
11. $6.08
13. $16,337
15. a. $787,273,000
    b. $5,540,340,000
17. 9.05%
19. The average annual inflation rate for the decade was greater than 7.1773%.

Exercises 10.5

1. $17,147.60
3. $28,594
5. $28,221
7. $65,097
9. $537
11. $19,667

Exercises 10.6

1. a. $44.86
   b. $1076.72
   c. $76.72
3. a. $760.70
   b. $136,926
   c. $378.70
   d. $126,486
   e. $10,439.70
5. 2, $178.82, $32.40, $5622.13
7.

Payment Number	Amount of Payment	Interest per Period	Portion to Principal	Principal Owed at End of Period
0	—	—	—	$60,000.00
1	$534.74	$462.50	$72.24	59,927.76
2	534.74	461.94	72.80	59,854.96
3	534.74	461.38	73.36	59,781.60

9. a. $468.14
   b. $366.51
   c. $305.86

Chapter 10 Review Exercises

1. One area of consumer mathematics deals with taxes calculated from millage rates. Another is using unit costs to compare prices. Another is calculating prices and savings with sales (percentage off). Another is comparing prices with coupons and double coupons. Calculations with interest, including annuities and loans, is another area of consumer mathematics.

3. Best buy—2-liter bottle; second best—eight-pack of 20-oz bottles; poorest buy—12-pack of 12-oz cans
5. $5.00
7. $2320
9. 41 mills
11. 27.49%
13. a. 15 oz
    b. 10 oz
15. a. $161.99
    b. 0.015238%
17. a. 1.596%
    b. 0.05247%
19. Simple interest is calculated on the amount invested or borrowed; compound interest is calculated on both the amount invested or borrowed and the intermediate interest earned or owed. For example, if $100 is invested for 1 year at 5% simple interest, $5.00 interest is earned during the year. If interest is earned at 5% compounded semiannually, then at the end of half a year the interest earned is $100 \times 0.05 \times 1/2 = $2.50. For the second half-year, interest is earned on $102.50, so the interest earned for the second half-year is $102.50 \times 0.05 \times 1/2 = $2.56. The total interest earned for the year with compound interest is $5.06, whereas the total interest earned for the year with simple interest is $5.00.
21. $2377.14
23. $1888.50
25. $158.29
27. About 14%
29. An ordinary annuity has payments made at the end of each time period, so the last payment is made at the time the annuity pays out. An annuity due has payments made at the start of each time period.
31. $3940.82
33. a. $242.96
    b. $11,662.08
    c. $1662.08

11 CHAPTER 11

Exercises 11.2

1.

Vertex	Degree
A	3
B	2
C	2
D	3

3.

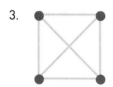

5. a. Yes
   b. *BD, DC*
7. Yes. *C, B, A, E, D, C, A, D, B, E, C*
9. No. If the road intersections serve as vertices of a graph, then there are several of odd degree.
11. Yes. Label the star's vertices *A, B, C, D, E*, clockwise. One circuit is *A, D, B, E, C, A*.
13. No. Construct a graph where the doorways are represented by edges and the vertices are represented by the rooms and the outside. Two vertices have degree 3.
15. a. Yes. Remove *AD*.
    b. No. Either the graph will not be connected or it will have a vertex of odd degree.
17. a. No. To have an Euler circuit, every vertex must have even degree. Both *C* and *D* have degree 3. The only way to remove exactly one edge so that *C* and *D* both have even degree is to remove edge *CD*. However, this would disconnect the graph, so it would have no Euler circuit.
    b. No. To have a graph with an Euler circuit, every vertex must have even degree. Both *C* and *D* have degree 3. Each must have its degree reduced to 2. This means that either *BC* or *AC* would need to be removed. Removing *BC* leaves *B* with degree 1 and removing *AC* leaves *A* with degree 1. Neither possibility gives an Euler circuit.

Exercises 11.3

1. Yes. 1, 2, 3, 1
3. No
5. Yes. 8, 3, 2, 1, 6, 5, 4, 7, 8
7. Yes. Chicago, Champaign, Effingham, Mt. Vernon, E. St. Louis, Bloomington, Joliet, LaSalle, Moline, Rochelle, Rockford, Chicago
9. $\frac{13!}{2} = 3{,}113{,}510{,}400$
11. No
13. a. 2080 miles (Birmingham → St. Louis → Denver → Houston → Birmingham)
    b. 2080 miles
    c. 2080 miles
15. 7968 miles
17. Answers will vary.

19. a.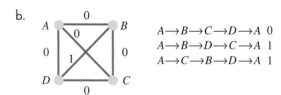

    b.

c. Yes

d. The original graph has a Hamiltonian cycle
   (A → B → C → D → A)

Exercises 11.4

1. a. C and Y
   b. A and X
   c. A and X

3.

Set of marriages	Stable/unstable	Unstable man/woman
AX, BY, CZ	stable	
AX, BZ, CY	unstable	C and Z
AY, BX, CZ	unstable	A and X
AY, BZ, CX	unstable	A and X
AY, BX, CY	unstable	A and X
AZ, BY, CZ	unstable	A and X

5. At the conclusion of each step, at least one woman will have a proposal that will ultimately be accepted. Thus, the process ends in at most n steps.

7. The process ends after one step with marriages XA, YB, ZC.

9.
Proposals by: A B C D	W X Y Z holds:	W X Y Z rejects:
X X W X	C B	A,D
Y Z	A D	

The stable set of marriages is AY, BX, CW, DZ.

11.
Proposals by: A B C D E	X Y Z holds:	X Y Z rejects:
1. X X Y X X	E C	A, B, D
2. Y Z Z	D	A B

A: ⊗⊗Z
B: ⊗⊗Y
C: ⓎX X
D: ⊗⊗Y
E: ⊗Y Z

The stable set of marriages is XE, YC, and ZD.

13. a. 360

b.
Proposals by: A B C D	U V W X Y Z holds:	U V W X Y Z rejects:
1. U U V Y	A C D	B
2. Y	B	D
3. W	D	

A: Ⓤ V W X Y Z
B: ⊗Ⓨ V W Z X
C: Ⓥ X U Z Y W
D: ⊗Ⓦ Z U V X

The stable set is UA, VC, WD, and YB.

c.
Proposals by: U V W X Y Z	A B C D holds:	A B C D rejects:
1. A A B C A B	U W X	V, Y Z
2. C B A	Y V	Z W X
3. A A D	Z	W, X

U: Ⓐ B C D
V: ⊗Ⓒ B D
W: ⊗⊗C D
X: ⊗⊗B D
Y: ⊗Ⓑ D C
Z: ⊗⊗Ⓓ C

The stable set is AU, BY, CV, and DZ.

Exercises 11.5

1. 3
3. 4! = 24
5. a. A
   b. A
   c. A
   d. A
   e. A
   f. A (ACB)
7. a. No winner
   b. A
   c. E
   d. C
   e. B
   f. D
9. a. 75
   b. 105
   c. 5500
11. The Borda method produces the winner B, but the withdrawal of alternative C produces the Borda winner A.
13. T
15. T
17. F
19. F
21. F
23. F
25. F
27. T
29. F

Chapter 11 Review Exercises

1. Discrete mathematics deals with sets of objects that can be counted or processes that consist of a sequence of individual steps, for example, check digits, logic, counting arguments, Euler circuits, Hamiltonian circuits, the traveling salesperson problem, the marriage problem, and the study of voting methods all fall in the province of discrete mathematics. Discrete

mathematics differs from continuous mathematics, which deals with infinite sets containing an interval and infinite processes other than those composed of a sequence of individual steps.

3. a. No
   b. If you go down the top aisle first, then halfway down the back aisle (call this point *A*), you must choose between going all the way down the back aisle or going down the middle aisle. The first option leaves no way to get down the middle aisle without traveling in some aisle more than once. The second option forces you to turn left or right when you come to the left side of the store (call this point *B*). A left turn leads you to *A* with no untraveled aisles left; a right turn leads you back to the door, with the bottom aisle untraveled. If you start out by going down the left aisle, at *B* you must decide whether to go all the way down the left aisle or to go down the middle aisle. If you go all the way down the left aisle, your next decision point is at *A*. If you go left (down the middle aisle), then you're led back to *B*, with only untraveled aisles facing you. If you go straight, you're led back to the door without traveling the middle aisle. If initially at *B* you go down the middle aisle, then at *A* you must go left or right. If you go left, you're led back to the door with the bottom aisle untraveled. If you go right, then you're led back to *B*, with only untraveled aisles ahead. Thus, in no case is the desired route possible.

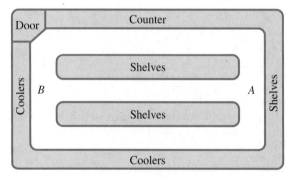

The blank areas denote aisles.

5. See the proof of Theorem 11.1, in Section 11.2.
7. a. No
   b. What is sought is an Euler circuit. Since some vertices have odd degree (for example, *E* has degree 3), the graph has no Euler circuit.
9. 360
11. An exact method gives an answer that solves the problem. A heuristic method arrives at an answer that may or may not solve the problem. A heuristic method seems reasonable and will often give an answer that solves the problem. It is usually a faster method of solving the problem than an exact method.
13. a. Stable
    b. Unstable—A and X, B and X, or C and Y

c. Suppose women propose.

Proposals by: A B C	X Y Z holds:	X Y Z rejects:
X X Y	B C	A
Y	C	A
Z	A	

The set of marriages is AZ, BX, CY. If men propose, the set of marriages is the same.

15. a. Every man has his most preferred woman, so there is no male–female pair in which the man would rather be with another woman. Therefore, the set of marriages is stable.
    b. iii and iv
17. No, a plurality winner receives more votes than any other alternative but does not necessarily receive a majority (over half) of the votes cast. For example, if there are five voters and four alternatives, A, B, C, and D, and two voters rank the alternatives (ABCD), one ranks them (BACD), one ranks them (CDAB), and one ranks them (DCBA), then A is a plurality winner but not a majority winner.
19. a. Jordan
    b. Jordan
    c. In any format, Jordan wins.
    d. Jordan
    e. Jordan, Olajuwon, Malone, Barkley, Price
21. The paradox of voting says that even if each individual voter's preference is transitive (if A is preferred to B and B is preferred to C, then A is preferred to C), the group preference need not be transitive.
23. The Marquis de Condorcet (eighteenth century) was a leader in a widespread attempt to apply the scientific method to all aspects of human and social affairs. The paradox of voting is also called Condorcet's paradox, and the voting method of picking an alternative that beats all other alternatives in pairwise voting is called the Condorcet method. Jean-Charles de Borda (eighteenth century) pointed out that the plurality method might lead to a choice that was not the real preference of the group. His method of "marks" is now known as the Borda method. He was perhaps the first person to start a discussion of the mathematical analysis of voting. C. L. Dodgson, who wrote under the pseudonym Lewis Carroll (nineteenth century), wrote at least nine publications on the subject of elections. Kenneth Arrow (twentieth century) proved a famous result known as the Arrow Impossibility Theorem that states that no voting method satisfies four rather basic considerations.
25. The voting method used by a group can determine the winner of an election. For example, if there are five voters with preferences (ABCD), (ACBD), (BDCA), (CBDA), and (DCBA), then A is a plurality winner, and C is a Condorcet winner, and under the Borda method B and C tie. It is possible to have an election with five alternatives in which the voters' preferences can stay fixed and any of the five alternatives can win, depending on which voting method is used.

Exercises 12.1

1. 1.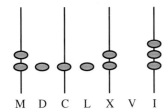

M D C L X V I

2.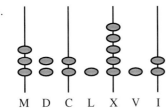

M D C L X V I

3.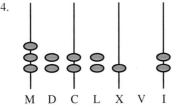

M D C L X V I

4.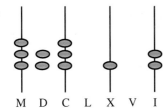

M D C L X V I

5.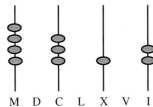

M D C L X V I

6.

M D C L X V I

The sum is MMMMCCCXII.

3. a. 4.0000434
   b. 4.0013875
   c. 4.0030203

5. log 10,040 + log 10,400 = 4.0017337 + 4.0170333 = 8.0187670

   The number whose log is 8.0187670 is 104,320,000.

7. 0.113943

9. 2.11394

11. Take the distance for 3 (.48) and add it to the distance for 3 (.48) to get .96. Look above .96 to get the answer of 9.

13.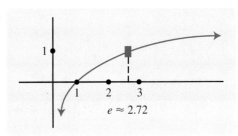

$e \approx 2.72$

Exercises 12.2

1. 618

3. $83 - 76 = 83 + (23 - 100 + 1)$
   $= 106 - 100 + 1$
   $= 107 - 100$
   $= 7$

5. $4032 - 2164 = 4032 + (7835 - 10,000 + 1)$
   $= 11,867 - 10,000 + 1$
   $= 11,868 - 10,000$
   $= 1868$

7.

x	$f(x) = x^2$	First difference	Second difference
0	0		
		1	
1	1		2
		3	
2	4		2
		5	
3	9		2
		7	
4	16		2
		9	
5	25		

9.

x	$f(x) = x^4$	First difference	Second difference	Third difference	Fourth difference
0	0				
		1			
1	1		14		
		15		36	
2	16		50		24
		65		60	
3	81		110		24
		175		84	
4	256		194		
		369			
5	625				

11.

x	$f(x) = 2x^3 - 4x^2 + 5x - 6$	First difference	Second difference	Third difference
0	−6			
		3		
1	−3		4	
		7		12
2	4		16	
		23		12
3	27		28	
		51		12
4	78		40	
		91		
5	169			

13. $n!$

15. a. No

b. Push the red slide (Bs) six units across the window and the gray slide (Cs) three units across the window from the opposite side. They will not overlap.

c.

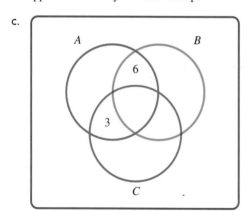

17.

0	0	1	1
+0	+1	+0	+1
0	1	1	10

Exercises 12.3

1. 34
3. a. 1
   b. 2
   c. 2
5. a. nonteenager
   b. teenager
   c. nonteenager
   d. nonteenager
7. a. 4
   b. 3
   c. 2
   d. 3
   e. 3
9. a. 6
   b. −6
   c. 0
   d. 0
   e. 6
   f. 0
   g. 8
   h. It computes the product ab.
11.

i	s
1	1
2	4
3	9
4	16
5	25

13. a.

i	c	a	b
		1	1
3	2	1	2
4	3	2	3
5	5	3	5
6	8	5	8
7	13	8	13

b. This produces the Fibonacci sequence, where each term is the sum of the previous two terms.

Exercises 12.4

1. a. 36,000
   b. 600
3. The algorithm is:

   1. $N = 4$.
   2. $N \neq 1$, so go to step 3.
   3. $N \neq 2$, so go to step 4.
   4. The trial divisor is 2.
   5. First time.

6. 2 divides 4, so 4 is not prime.

5. The algorithm is:

  1. $N = 7$.
  2. $N \neq 1$, so go to step 3.
  3. $N \neq 2$, so go to step 4.
  4. The trial divisor is 2.
  5. First time.
    6. 2 is not a divisor of 7. Go to step 7.
    7. Increase the trial divisor to 3.
    8. Since 3 does not equal 7, go back to step 6.
  Second time.
    6. 3 is not a divisor of 7. Go to step 7.
    7. Increase the trial divisor to 4.
    8. Since 4 does not equal 7, go back to step 6.
  Third time.
    6. 4 is not a divisor of 7. Go to step 7.
    7. Increase the trial divisor to 5.
    8. Since 5 does not equal 7, go back to step 6.
  Fourth time.
    6. 5 is not a divisor of 7. Go to step 7.
    7. Increase the trial divisor to 6.
    8. Since 6 does not equal 7, go back to step 6.
  Fifth time.
    6. 6 is not a divisor of 7. Go to step 7.
    7. Increase the trial divisor to 7.
    8. Since the trial divisor equals 7, 7 is prime.

7. 96

9. The algorithm is:

  1. $N = 7$.
  2. $N \neq 1$, so go to step 3.
  3. $N \neq 2$ and $N \neq 3$, so go to step 4.
  4. 2 is not a divisor. Go to step 5.
  5. The trial divisor is 3.
  6. First time.
    7. 3 is not a divisor of 7. Go to step 8.
    8. Increase trial divisor to 5.
    9. $5 > \sqrt{7}$, so 7 is prime.

11. The algorithm is:

  1. $N = 103$.
  2. $N \neq 1$, so go to step 3.
  3. $N \neq 2$ and $N \neq 3$, so go to step 4.
  4. 2 is not a divisor. Go to step 5.
  5. The trial divisor is 3.
  6. First time.
    7. 3 is not a divisor of 103. Go to step 8.
    8. Increase trial divisor to 5.
    9. $5 < \sqrt{103}$. Go to step 7.
  Second time.
    7. 5 is not a divisor of 103. Go to step 8.
    8. Increase trial divisor to 7.
    9. $7 < \sqrt{103}$. Go to step 7.
  Third time.
    7. 7 is not a divisor of 103. Go to step 8.

8. Increase trial divisor to 9.
9. $9 < \sqrt{103}$. Go to step 7.
Fourth time.
  7. 9 is not a divisor of 103. Go to step 8.
  8. Increase trial divisor to 11.
  9. $11 > \sqrt{103}$, so 103 is prime.

13. a. At most 2^{4096} trial divisors would be used (otherwise, $2^{8191} - 1$ is prime).
   b. Over 33×10^{1215} years

15. First pass:

  1. $N = 9$.
  2. $N \neq 1$, so go to step 3.
  3. $N \neq 2$ and $N \neq 3$, so go to step 4.
  4. 2 is not a divisor. Go to step 5.
  5. Trial divisor is 3.
  6. First time.
    7. 3 is a divisor of 9. Print this and go to step 10.
  10. Let $N = 9/3 = 3$. Go to step 1.
  Second pass:

  1. $N = 3$.
  2. $N \neq 1$, so go to step 3.
  3. $N = 3$. Print this.

17. First pass:

  1. $N = 15$.
  2. $N \neq 1$, so go to step 3.
  3. $N \neq 2$ and $N \neq 3$, so go to step 4.
  4. 2 is not a divisor. Go to step 5.
  5. Trial divisor is 3.
  6. First time.
    7. 3 is a divisor of 15. Print this and go to step 10.
  10. Let $N = 15/3 = 5$. Go to step 1.
  Second pass:

  1. $N = 5$.
  2. $N \neq 1$, so go to step 3.
  3. $N \neq 2$ and $N \neq 3$, so go to step 4.
  4. 2 is not a divisor. Go to step 5.
  5. Trial divisor is 3.
  6. First time.
    7. 3 is not a divisor. Go to step 8.
    8. Trial divisor is 5. Go to step 9.
    9. $5 > \sqrt{5}$, so 5 is prime. Print this factor and quit.

19. First pass:

  1. $N = 63$.
  2. $N \neq 1$, so go to step 3.
  3. $N \neq 2$ and $N \neq 3$, so go to step 4.
  4. 2 is not a divisor. Go to step 5.
  5. Trial divisor is 3.
  6. First time.
    7. 3 is a divisor. Print this and go to step 10.
  10. Let $N = 63/3 = 21$. Go to step 1.
  Second pass:

  1. $N = 21$.

2. $N \neq 1$, so go to step 3.
3. $N \neq 2$ and $N \neq 3$, so go to step 4.
4. 2 is not a divisor. Go to step 5.
5. Trial divisor is 3.
6. First time.
 7. 3 is a divisor. Print this and go to step 10.
10. Let $N = 21/3 = 7$. Go to step 1.

Third pass:

1. $N = 7$.
2. $N \neq 1$, so go to step 3.
3. $N \neq 2$ and $N \neq 3$, so go to step 4.
4. 2 is not a divisor. Go to step 5.
5. Trial divisor is 3.
6. First time.
 7. 3 is not a divisor. Go to step 8.
 8. Trial divisor is 5. Go to step 9.
 9. $5 > \sqrt{7}$, so 7 is prime. Print this factor and quit.

21. First pass:

1. $N = 121$.
2. $N \neq 1$, so go to step 3.
3. $N \neq 2$ and $N \neq 3$, so go to step 4.
4. 2 is not a divisor. Go to step 5.
5. Trial divisor is 3.
6. First time.
 7. 3 is not a divisor. Go to step 8.
 8. Trial divisor is 5. Go to step 9.
 9. $5 < \sqrt{121}$. Go to step 7.
Second time.
 7. 5 is not a divisor. Go to step 8.
 8. Trial divisor is 7. Go to step 9.
 9. $7 < \sqrt{121}$. Go to step 7.
Third time.
 7. 7 is not a divisor. Go to step 8.
 8. Trial divisor is 9. Go to step 9.
 9. $9 < \sqrt{121}$. Go to step 7.
Fourth time.
 7. 9 is not a divisor. Go to step 8.
 8. Trial divisor is 11. Go to step 9.
 9. $11 \leq \sqrt{121}$. Go to step 7.
Fifth time.
 7. 11 is a divisor. Print this and go to step 10.
10. Let $N = 121/11 = 11$ and go to step 1.
Second pass:

1. $N = 11$.
2. $N \neq 1$, so go to step 3.
3. $N \neq 2$ and $N \neq 3$, so go to step 4.
4. 2 is not a divisor. Go to step 5.
5. Trial divisor is 3.
6. First time.
 7. 3 is not a divisor. Go to step 8.
 8. Trial divisor is 5. Go to step 9.
 9. $5 > \sqrt{11}$, so 11 is prime. Print this and quit.
23. 10 gets printed.

Exercises 12.5

1. 8
3. No
5. $2^{11} - 1 = 23 \cdot 89$
7. $2^{12} - 1 = (2^4 - 1)(2^{2 \cdot 4} + 2^4 + 1)$
9. iv and v

Chapter 12 Review Exercises

1. The invention of a medium on which to write (for us, paper) facilitated algorithmic calculation by enabling calculations to be preserved (rather than wiped out, as when written in sand or done with sticks).
5. c
7. $\log(10{,}077 \times 10{,}382) = \log 10{,}077 + \log 10{,}382 = 4.0037620 + 4.0162810 = 8.020043$. So $10{,}077 \times 10{,}382 = 104{,}720{,}000$.
9. $\log(4 \times 2) = \log 4 + \log 2$. Add $\log 4 + \log 2$ as shown.

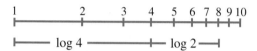

$\log 4 + \log 2 = \log 8$. So $4 \times 2 = 8$.
11. A slide rule is a device for doing mathematical calculations, including multiplication and division. It consists of a piece about the size of a ruler containing a sliding inner piece. Both the outer and inner pieces have a logarithmic scale marked on them.
13. d, g, h, a, b, c, f, e
15. a. Since the dials on early calculating machines turned only one way, subtraction was done by means of addition.
 b. $423 + 721 = 1144$. Drop the left-most 1 to get 144, then add 1 to get 145. So $423 - 278 = 145$.
17.

x	$f(x) = 2x^3 - 4x$	First difference	Second difference	Third difference
0	0			
		−2		
1	−2		12	
		10		12
2	8		24	
		34		12
3	42		36	
		70		12
4	112		48	
		118		
5	230			

19. There is about a hundredfold increase in speed every 10 years.

21. 19

23. a. 0
    b. 1
    c. 1
    d. 1
    e. 1
    f. 0

25. a. 5
    b. 32
    c. i := 0, p := 1

    while (i < 20)

        i := i + 1

        p := p * 3

27. a.

	Iteration Number			
	0	1	2	3
n	1	2	3	4
p	90	81	72.9	65.61

b. p := 150
   for n := 1 to 10
   p := 0.9*p
   next n

29. 1. $N = 8$
    2. $N \neq 1$, so go to step 3.
    3. $N \neq 2$, so go to step 4.
    4. The trial divisor is 2.
    5. First time
       6. The trial divisor is a divisor of N. Therefore, 8 is not prime.

31. 1. $N = 35$
    2. $N \neq 1$, so go to step 3.
    3. $N \neq 2$ and $N \neq 3$, so go to step 4.
    4. 2 is not a divisor, so go to step 5.
    5. Initialize the trial divisor to 3.
    6. First time
       7. The trial divisor is not a divisor of 35, so go to step 8.
       8. Change the trial divisor to 5.
       9. 5 does not exceed $\sqrt{35}$, so go back to step 7.
       Second time
       7. 5 is a divisor of 35, so 35 is not prime.

33. Improvements in computer hardware have made more computer power and speed available to the user. Another way to increase a computer's ability is to write a clever program that takes less time than other programs to achieve a result. Thus, a computer's ability is based on both its hardware and its software.

35. a. 9
    b. 10 A = 1

       20 C = 1

       25 D = 1

       30 B = A + 2

       40 C = C + B

       50 A = B

       55 D = D + 1

       65 if D < 100 go to 30

       70 print C

       80 end

37. Is there an infinite number of pairs of twin primes?

39. a. A Mersenne prime is a prime of the form $2^p - 1$.
    b. Let $p = 2$ to get 3; 3 is a Mersenne prime.

41. The four-color theorem has been proved via a computer-assisted proof. It states that any map can be colored with no more than four colors so that countries that share a border in more than a point have different colors.

13 CHAPTER 13

Exercises 13.1

1. a. 420,000,000 miles
   b. 21.7 miles/second

3. 13,000 miles

5. Example 13.2 uses all integers for the calculation. The conversion key uses decimals for better accuracy. For example, 1 year is slightly longer than 365 days.

7. .48

9. $7\frac{3}{16}$ inches

11. 7.2 feet

13. a. $\frac{1}{2}bh$

    b. $\left(\frac{1}{2}bh\right)k^2$

    c. k^2

15. a. About 1,000 sq ft
    b. The 30-ft tree is similar to the 20-ft tree.

17. a. 4.6 ft
    b. A female gorilla is a scaled-down version of a male gorilla.

19. a. 92 lb
    b. 177 lb

21. Crayfish are cold-blooded. If they were warm-blooded, their respiration rate would be about 2.4 milliliters of oxygen per gram of tissue weight per hour.

Exercises 13.2

1. a. Yes
   b.

		Officer	
		Checks	Doesn't check
You	Yes	−0.25	−0.25
	No	−5.00	0

   c. It is a pure-strategy game. You will put the money in the meter, and they will check the meter.

3. a. Yes
   b.

		Defense expects:	
		Run	Pass
Offense calls:	Run	+2	+6
	Pass	+12	−7

   c. This is not a pure-strategy game. If the offense runs, a run defense is best. If the offense passes, a pass defense is best. Therefore, if the defense knows what the offense is going to do, it can adjust its strategy and come out ahead.

5. a. No
   b. Yes. Column 1 dominates $\begin{bmatrix} 2 \\ 3 \end{bmatrix}$.
   c. The game in a zero-sum game, and it is a pure-strategy game. The best strategy for Kenney is to go on the northern route, the best strategy for the Japanese is to go on the northern route, and the value of the game is 2.

7. a. No row dominates.
   b. No column dominates.

Exercises 13.3

1. 4 9 2
   3 5 7
   8 1 6

3. 65

5. D: 54 cm; E: 48 cm; F: 45 cm; A: 36 cm

7. $\dfrac{2}{1} \times \dfrac{3}{2} = \dfrac{3}{1}$ $\dfrac{3}{1} \times 30$ cm = 90 cm

9. $\dfrac{3}{2} \times \dfrac{6}{5} = \dfrac{9}{5}$ $\dfrac{9}{5} \times 30$ cm = 54 cm

11. $r^6 = 2^{1/2} = \sqrt{2}$, which is irrational. This contradicts the fact that r^6 is rational.

13. Two notes a major third apart should have a ratio of 5:4 or 1.2500. In the equal-tempered scale, they have a ratio of $2^{4/12} \approx 1.2599$.

Chapter 13 Review Exercises

1. a. General education began at least as far back as the time of Plato.
   b. Arithmetic, geometry, astronomy, music

5. a. 1.5 billion miles

b. 14,000 miles per hour

7. 0.8 mm

9. 625 square feet

11. $1/k$

13. Mathematics is used in science to make statements precise. In mathematical form the statements can be tested by experiment and in theories. Attempts to explain phenomena are sometimes made using mathematical models. This often requires the creation of new mathematics. In turn, mathematics created for other purposes often can be applied to describe scientific phenomena.

15. a. $\begin{bmatrix} 3 & 6 \\ 7 & 2 \end{bmatrix}$
    b. The game is not a pure-strategy game. The row player's strategy would be to play row 1; the column player's strategy would be to play column 2. However, if the column player knew that the row player were going to play row 1, then the column player could switch to column 1 and come out better (lose 3 instead of losing 6).

17. Both mathematics and art encourage creativity in the production of new works, yet both are governed by rules. Both study patterns. Both value elegance. Both originate out of physical reality.

19. From top to bottom: 14, 10, 12, 16

21. Pythagoras (or at least the Pythagoreans) discovered that numbers describe the basis for harmony in music. When a string is plucked, it produces a sound. If a second string whose length makes an appropriate ratio with the length of the original string is plucked, then a pleasing sound occurs. If the ratio is not of a certain type, then a discordant sound occurs.

23. a. An equal-tempered scale is a musical scale in which string lengths are such that the ratio of any two consecutive string lengths is a constant. That is,

$$\frac{C}{C\#} = \frac{C\#}{D} = \frac{D}{D\#} = \cdots = \frac{B}{C'}$$

   where C' denotes the octave above C.
   b. The interval from C to G is a fifth. Let the common ratio in part a be denoted by r. Then C/C# = r, so C = rC#. C#/D = r, so C# = rD. Therefore, C = rC# = $r(r$D$) = r^2$D. Continuing this argument gives C = r^7G.
   c. The equal-tempered scale makes some intervals off a bit from their natural ratios. For example, for a fifth the notes have a natural ratio of 3:2, or 1.5. With the equal-tempered scale, an argument like that in part b shows that C = r^{12}C'. Since the length of C is twice the length of C', we have 2 = $r^{12} \cdot 1$, so $r = 2^{1/12}$. In an equal-tempered scale, the notes in a fifth have a ratio of $r^7 = (2^{1/12})^7 = 2^{7/12} \approx 1.4983$. Notice that this is off a bit from the natural ratio of 1.5, which creates a problem.

INDEX

Page numbers in bold indicate definitions. Page numbers in italic indicate illustrations.

A

continuous probability distributions, 362–363

continuous random variables, 362

probability distribution, *368*

Normal random variable, 363, *369*

Noyce, Robert, *509*

Null set, 70

Number

binary representation of, 33–35

concept of, 20

Number line, set of all real numbers on, 423

Number systems, 43–44

evolution of, *52*

Pythagoreans' beliefs about, 48–50

relationships among, *51*

Number theory, 298

Numbers

complex, 50–53

composite, 156

integers, 44–46

natural, 44

prime, 156

rational, 46–47

real, 47–48

Numbers Through the Ages, *31*

Numeral, 20, 21

Numeration systems

additive, 23, 26

African, 25–26

of ancient civilizations, 20–21

Babylonian, 27

base in, 29–33

comparison of, *60*

computer, 33–35

Egyptian hieroglyphics, 22–23, *22*, *24*, 27, 38, 39

geographical distribution of, *31*

Hindu-Arabic, 25, 26–28, *27*, 29, 40

Igbo, 26, 27, 29

Mayan, 19–20, *20*, 27–28, 30

modified additive, 23

place-value, 27

Roman numerals, 23–25, *23*, *24*, 27

Numerator, 38

Numerical syllogism, 491, *491*

O

Oberhofer, E. S., 194

Objects, similar, 537–538, *538*

Observation, in scientific method, 534

Obtuse angle, 227, *227*

hypothesis of, 273, 275

Octagon, *240*

Octave, 560, *560*

Odds, 330–331, *330*

Odhner, Willgodt, 488

Oliver, Dick, 290

"On Computable Numbers," 493

On the Heavens, 531

Open line segment, *220*

Operations research, 491

Ordinary annuity, 410–411

present value of, 413–414

Oughtred, William, 484, *484*

Outcomes (probability), 301–302

P

$\hat{p}$, sampling distribution for, 379–384

Padberg, Manfred, *444*

Painting, geometry in, 556

Paper, invention and development of, 475, 478

Papyrus, 475

Parabolas, 169–170

Paradox of voting, 460

Parallel lines, angles and, 228–229, *229*

Parallel planes, 223

Parallel points, 267

Parallel postulate, 272–274, *272*

Parallelogram, *239*

area of, *246*, 247–248, **247**, *248*

Pareto principle, 466

Pascal, Blaise, 297, *297*, 334, 487, 499, *513*

Pascal computer language, 499, *513*

Pathological curve, 285

Payoff matrix, 548, *549*, 550

Peano, Guiseppe, 109, *110*

Pearson, Karl, 305

Pedny, Joseph, 444

Pentagon, 232, *240*

Pentium processor, *517*

Pepys, Samuel, *35*, 488, *488*

Perimeter, 243–245

Permutations, 310–312

Perspective, 267–268, *267*, *268*

development of, 556–557

Pi (π), 249–250, **249**, *250–251*

Picasso, Pablo, 557, *557*

Pitman, Isaac, *35*

PL/1 computer language, 499

Place-value system of numeration, 27

Planes, 222–223, **222**

half-planes, *223*

parallel, 223

Plato, 3, *217*, 218, 545, *545*

Playfair, John, 274

Playfair's axiom, 553

Plurality method of election, 458, 465

Poincaré, Henri, 274, 280, *364*, *368*

Points, **219**, *220*

noncollinear, 222

parallel, 267

Political Arithmetic, *338*

Polya, George, 2–3, *3*

Polygons, *232*

areas of, 245–249, *246*, *247*, *248*, 254–255

convex, 231

general, 241–242

congruent, 241

similar, 241

sum of interior angles for, 241

CREDITS

TEXT CREDITS

Pages 3, 21, 67, 107, 155, 217, 263, 297, 339, 391, 425, 475, and 533 (Etymologies): Reprinted from "The Words of Mathematics" by Steven Schwartzman with the permission of The Mathematical Association of America, Washington, DC, 1994.

Page 31: Flegg graph of bases of systems of numeration: Reproduced with permission, from Graham Flegg (ed.) *Numbers Through the Ages* (Macmillan Press, Ltd. Basingstoke).

Pages 46, 70, and 80: Excerpts by permission from Nicholas J. Rose, *Mathematical Maxims and Minims* (Raleigh, NC: Rome Press, 1988).

Page 49: Steven Cushing poem "Mathematical Clerihew" from *Mathematics Magazine,* Feb. 1988, Mathematical Association of America, Washington, DC 20036.

Page 77: Ann Landers column excerpt: Permission granted by Ann Landers and Creators Syndicate.

Page 111: John L. Casti excerpt: From *Searching for Certainty* by John L. Casti, copyright 1990, by permission of Wm Morrow and Company, Inc.

Page 216: "Euclid Alone Has Looked on Beauty Bare" by Edna St. Vincent Millay. From COLLECTED POEMS, HarperCollins. Copyright 1923, 1951 by Edna St. Vincent Millay and Norma Millay Ellis. All rights reserved. Reprinted by permission of Elizabeth Barnett, literary executor.

Page 555: Paul Halmos excerpt: From "Centennial Lecture" at University of Illinois in 1967, *American Scientist,* Winter 1968, p. 388 as "Mathematics as a Creative Art."

Page 568: Vachel Lindsay poem "Euclid": Reprinted with the permission of Simon & Schuster from THE COLLECTED POEMS OF VACHEL LINDSAY (New York: Macmillan, 1925).

Page 568: "Euclid Alone Has Looked on Beauty Bare" by Edna St. Vincent Millay. From COLLECTED POEMS, HarperCollins. Copyright 1923, 1951 by Edna St. Vincent Millay and Norma Millay Ellis. All rights reserved. Reprinted by permission of Elizabeth Barnett, literary executor.

PHOTO CREDITS

Page 1: © David Madison 1996.
Page 2: Photo Researchers, Inc./Van Boucher.
Page 3: Stanford News Service.
Page 12 (top): Mehau Kulyk/Science Photo Library.
Page 12 (bottom): René Magritte, *The Blank Signature,* Collection of Mr. and Mrs. Paul Mellon, © 1996 Board of Trustees, National Gallery of Art, Washington.

Page 168: © of the Trustees of the British Museum.
Page 22: Egyptian Expedition of The Metropolitan Museum of Art, Rogers Fund, 1930. (30.4.79).
Page 25 (top): Woodfin Camp & Associates, Inc./Michal Heron.
Page 25 (bottom): © Jason Lauré.
Page 39: e.t.archive/British Museum.
Page 65: Photo Researchers, Inc./Daniel Boiteau/Explorer.
Page 66: The Granger Collection, New York.
Page 74: PhotoEdit/Robert Brenner.
Page 86: The Granger Collection, New York.
Page 105: Todd Siler, *States of Mind,* Solomon R. Guggenheim Museum, New York, Gift, Stewart and Judy Colton, 1987. Photo by Sally Ritts, © The Solomon R. Guggenheim Foundation, New York.
Page 108: The Granger Collection, New York.
Page 109: Science Photo Library.
Page 110: Corbis-Bettmann.
Page 111: UPI/Corbis-Bettmann.
Page 127: SuperStock, Inc./Elizabeth Strenk.
Page 130: Visuals Unlimited/Bill Banaszewski.
Page 153: Miriam Schapiro, *Big Ox No. 2,* 1968, acrylic on canvas, 90″ × 108″, Collection Museum of Contemporary Art, San Diego, Gift of Harry Kahn.
Page 166: Photo Researchers, Inc./Gerard Vandystadt.
Page 167: West.
Page 170: SuperStock, Inc./Steve Vidler.
Page 193: U.S. Geological Survey, U.S. Department of the Interior (DOI).
Page 201: Visuals Unlimited/Mike Eichelberger.
Page 204: The Granger Collection, New York.
Page 209: The Granger Collection, New York.
Page 215: Visuals Unlimited/Richard C. Walters.
Page 217: Photo Researchers, Inc./George Holton.
Page 223: The Granger Collection, New York.
Page 225: West.
Page 233: Courtesy of John Robinson (http://www.bangor.ac.uk/~mas007).
Page 240: Photri, Inc.
Page 261: © 1996 M.C. Escher/Cordon Art-Baarn-Holland. All rights reserved.
Page 267: "Last Supper" by Leonardo DaVinci Santa Maria Delle Grazie, Milan/L.K.G., Berlin/SuperStock, Inc.
Page 268: Photo Researchers, Inc./Jim Corwin.
Page 269: e.t.archive/Louvre Paris.
Page 274 (top): The Granger Collection, New York.
Page 274 (bottom): The Bettmann Archive.
Page 276: Visuals Unlimited/Art Matrix.
Page 280: West.
Page 287 (Figs. 7.23 and 7.24): Visuals Unlimited/Art Matrix.

MARGIN NOTES